Forensic DNA /

Forensic DNA Applications: An Interdisciplinary Perspective, Second ⸺ ⸺., is fully updated to outline the latest advances in forensic DNA testing techniques and applications. It continues to fill the need for a reference book for people working in the field of forensic molecular biology testing and research as well as individuals investigating and adjudicating cases involving DNA evidence, whether they be civil or criminal cases.

DNA techniques have greatly impacted obvious traditional forensic areas, but such advances have also positively affected myriad new areas of research and inquiry. It is possible today to think about solving forensic problems that were simply unheard of even a few years ago. As such, the book pulls all relevant research and applied science together into a detailed and comprehensive collection.

Part I begins with the history and development of DNA typing and profiling for criminal and civil purposes. It discusses the statistical interpretation of results with case examples, mitochondrial DNA testing, Y single-nucleotide polymorphisms (SNPs) and short tandem repeats (STRs), and X SNP and STR testing. It also explores low copy number DNA typing, mixtures, and quality assurance and control. Part II moves on to cover the various uses and applications of analyzing collected physical evidence, victim identification in mass disasters, analyzing animal DNA, forensic botany, and other unique applications. Part III is dedicated to the latest advances and developments in human molecular biology, and Part IV looks at policies and laws and ethics governing DNA evidence, and its utilization in various cases and the courts.

Forensic DNA Applications covers cutting-edge research and advancements in the field and is the most up-to-date reference available. Edited and contributed to by the world's foremost leaders in the field, it is a must-have reference for established professionals; an essential resource to legal professionals—lawyers and judges dealing with civil and criminal cases involving DNA technology; as well as students entering the fields of genetics and forensic DNA analysis.

Dragan Primorac, MD, PhD, is a pediatrician, forensic expert, and geneticist. Currently, he serves as an adjunct professor at Eberly College of Science, Pennsylvania State University; and the Henry C. Lee College of Criminal Justice and Forensic Sciences, University of New Haven in the United States,; and at medical schools at universities in Split, Osijek, and Rijeka in Croatia. He is also a professor at the Faculty of Dental Medicine and Health, Osijek, Croatia; Medical School REGIOMED, Coburg, Germany; and professor emeritus at the National Forensic Sciences, University Gandhinagar, Gujarat, India.

Moses S. Schanfield, PhD, holds undergraduate and master's degrees in anthropology and a PhD in human genetics. A professor of forensic science and anthropology at George Washington University, Schanfield made widespread contributions to the field and, sadly, passed away in January 2021.

Forensic DNA Applications
An Interdisciplinary Perspective
Second Edition

Edited by
Dragan Primorac and Moses S. Schanfield

Editorial assistants:
Petar Projić and Vedrana Škaro

CRC Press
Taylor & Francis Group
Boca Raton London New York

CRC Press is an imprint of the
Taylor & Francis Group, an **informa** business

Designed cover image: Shutterstock image

Second edition published 2023
by CRC Press
6000 Broken Sound Parkway NW, Suite 300, Boca Raton, FL 33487-2742

and by CRC Press
4 Park Square, Milton Park, Abingdon, Oxon, OX14 4RN
CRC Press is an imprint of Taylor & Francis Group, LLC

© 2023 selection and editorial matter, Dragan Primorac and Moses Schanfield; individual chapters, the contributors

First edition published by CRC Press 2014

Library of Congress Cataloging-in-Publication Data

Names: Primorac, Dragan, editor. | Schanfield, Moses S., editor.
Title: Forensic DNA applications : an interdisciplinary perspective /
edited by Dragan Primorac and Moses Schanfield.
Identifiers: LCCN 2022031052 (print) | LCCN 2022031053 (ebook) | ISBN
9780367030261 (hardback) | ISBN 9781032392028 (paperback) | ISBN
9780429019944 (ebook)
Subjects: LCSH: Forensic genetics. | DNA fingerprinting. | DNA--Analysis.
Classification: LCC RA1057.5 .F658 2023 (print) | LCC RA1057.5 (ebook) |
DDC 614/.1--dc23/eng/20221013
LC record available at https://lccn.loc.gov/2022031052
LC ebook record available at https://lccn.loc.gov/2022031053

ISBN: 9780367030261 (hbk)
ISBN: 9781032392028 (pbk)
ISBN: 9780429019944 (ebk)

DOI: 10.4324/9780429019944

Typeset in Minion
by Deanta Global Publishing Services, Chennai, India

A meaningful, passionate life

Colleagues, friends, and family celebrate the life of
Moses Schanfield (1944–2021)

I dedicate my contribution to a friend, colleague, and co-editor of the book, Professor Moses Schanfield, who passed away before the second edition of the book had been published. It was a massive loss to the forensic science community, but his legacy will continue for generations, including those reading this book.

Also, I bring this volume to the forensic community in honor and memory of the innocent who fell victim to violence and brutality around the world, and for their loved ones who suffered the pain of loss. I dedicate this volume to all forensic scientists and practitioners whose work is essential to uphold justice in the hope that the world will become a better and less violent place.

Finally, I dedicate my contribution to this volume to the memory of my father Marinko, and to my mother Dragica, wife Jadranka, daughters Lara and Matea, brother Damir and his family, and to all my friends and colleagues who have been a source of love, inspiration, and strength.

—Dragan Primorac

Contents

Part I

GENERAL BACKGROUND AND METHODOLOGICAL CONCEPTS

Foreword

I appreciate this second opportunity to introduce the new second edition of *Forensic DNA Applications*, the most comprehensive volume on forensic and ancestral DNA analysis yet published. Drs. Schanfield and Primorac deserve considerable credit for having gathered this wide-ranging material comprehensively under one cover. The sense of accomplishment the editors and authors can take from the publication of this edition is somewhat overshadowed by the untimely death of Dr. Schanfield prior to its completion. Many of the contributors to this book and many readers will have known him and worked with him. I knew him personally going back to his Denver days, and we had some research collaborations over the years. The passing of long-time friends, colleagues, and collaborators continues to be one of the less attractive aspects of being in my age cohort.

The coverage is comprehensive both in terms of the topics and of its international scope. Four logical sections comprise the book: General Background and Methodological Concepts; Uses and Applications; Recent Developments and Future Directions in Human Forensic Molecular Biology; and Law, Ethics, and Policy.

The first section reviews the history and development of DNA typing and profiling for criminal and civil purposes. It includes chapters on the statistical interpretation of results with case examples, mitochondrial DNA, X and Y chromosome STRs and SNPs, profiling of challenging forensic specimens (e.g., low copy number, degraded samples, presence of PCR inhibitors, etc.), DNA typing of mixtures and probabilistic genotyping, and rapid DNA profiling, particularly for potential use in the field.

The second section, Uses and Applications, has chapters on collection and preservation of biological evidence under a variety of different circumstances, identification of human remains in mass disaster situations, forensic microbiology and its applications to bioterrorism investigations, forensic animal DNA testing in criminal cases, pedigree questions and wildlife forensic problems, and applications in both forensic entomology and forensic botany.

The third section on recent developments and future directions contains chapters on developing technologies, including the rigorous identification of body fluid and tissue of origin (e.g., saliva, vaginal, skin, etc.), evolving technologies, especially next-generation sequencing (massively parallel sequencing) applied to both genomic and mtDNA, use of SNPs to ascertain phenotypic characteristics such as eye color, skin color, etc., and finally the "molecular autopsy" that looks at aspects of toxicogenetics and pharmacogenetics. A final chapter covers the new practice of genetic genealogy.

The fourth and final section on law, ethics, and policy, has chapters on the use of DNA evidence in the criminal justice system and the courts of both the US and Europe, and ethical issues in forensic genetics. In addition, there are chapters on DNA applications in immigration and human trafficking cases, and international perspectives on DNA databases.

The beauty and value of this book is that it treats and tries to connect all the disparate dots in present-day forensic molecular biology, taken in its broadest possible sense. This feature makes it both a valuable reference volume for practitioners and a potential textbook for upper division and graduate courses in forensic biology.

Human identification by an agreed-upon set of anonymous STRs is now a robust and reliable technology, but there are still issues surrounding mixture interpretation and obtaining reliable types with less than optimal quantities of target DNA (at least using CE-laser dye techniques). The level of sensitivity of DNA typing is a two-sided coin, at times helpful to sorting out a case, and at other times confusing as to what is signal and what is background noise. The introduction of Y STRs is a promising approach to sexual assault evidence analysis, because the commonly used differential extraction procedure is not always completely successful. And, although these markers are not sufficiently polymorphic in most populations to be very useful for individualization, they could clarify a case involving a small universe of potential donors. Coverage of DNA "statistics" is welcome. The probability of chance match is key to evaluating DNA results, and there are still some disagreements about different approaches, particularly Bayesian ones.

Applications to human identification in criminal cases and to resolving disputed parentage cases coalesce in the identification of mass disaster victims. Similarly, animal DNA techniques have helped in cases involving pet hair, and are obviously the sine qua non of game regulation, poaching, and endangered species law enforcement. Plant DNA technologies can help both controlled substance and some criminal investigations. DNA techniques applied to insect classes that are forensically important can assist forensic entomologists in estimating time since death. Microbial DNA analysis is perhaps the only fast method for diagnosing bioterrorist toxins or organisms. In a large-scale attack or epidemic, these techniques provide perhaps the only hope of containing the problem.

Forensic body fluid and tissue of origin identification has been a long-standing problem in forensic biology. Methods based on RNA and epigenetic differences may finally provide the unique tissue identification markers needed to do a thorough criminalistics workup in criminal cases. If robust, reliable techniques for common descriptive phenotypic traits (eye color, hair color, etc.) are developed, they would be very helpful in criminal investigations and considerably more reliable than eyewitness accounts. As molecular biological techniques are applied to postmortem examinations, pharmacogenetics and molecular toxicology will become incorporated into autopsy protocols. The recent use of "genetic genealogy," searching databases for persons potentially related to a suspect where a suspect's profile is not in the database, has established its worth in many cases. In the USA, apprehension of Joseph DeAngelo, the "Golden State killer," is a recent example.

It is widely recognized that the law can be slow to catch up with new, fast-moving science and technology. DNA technologies have proven no exception. Thus, coverage of this important subject is key to societal acceptance and use of these technologies in democratic countries. The public must accept that the technologies serve a compelling societal good, or their use will be curtailed or prohibited.

DNA techniques and resulting technologies have spread their wings into the obvious traditional forensic areas, but also into many new and unusual ones. It is possible today to think about solving forensic problems that were simply intractable with the techniques available even 30 years ago. This book is the only one I know about that pulls all the material together in a detailed, organized, and comprehensive manner. For this, the editors and contributing authors deserve our gratitude and appreciation.

R. E. Gaensslen
Professor Emeritus, Forensic Science
University of Illinois, Chicago

Preface

Forensic DNA Applications: An Interdisciplinary Perspective was created to fill a void we perceived in the literature for a book that could be a textbook for forensic molecular biology students, and a reference book for practitioners of forensic molecular biology, as well as lawyers and judges dealing with civil and criminal cases involving DNA technology. The book provides up-to-date coverage of the ever-broadening field of forensic DNA testing. The individual chapters written by multiple authors, all of whom are experts in their particular areas, provide a compact review of the state of the art for that particular topic. Every effort has been made to have each topic current at the time of submission. Unfortunately, a dear colleague and friend of mine who is also the book's editor, Prof. Moses Schanfield, passed away before the second edition of the book had been published. It was a massive loss to the forensic science community, but his legacy will continue for generations, including those reading this book.

The book was developed as an outgrowth of the biannual educational conference organized by the International Society for Applied Biological Sciences (ISABS) and held in Croatia, whereupon the organizers and speakers at the meeting felt that there was a need for a book that could be used as a textbook but also as a reference book for people working in the field of forensic molecular biology as well as individuals investigating and adjudicating cases involving DNA evidence, whether they be civil or criminal cases. The approach is international, so some things may not be relevant to analysts working only in the United States or in Europe, but as the title states, the approach is "interdisciplinary."

There are many texts on forensic molecular biology—some authored by a single author, others edited volumes. This particular volume is unique in the sense that the more than 50 authors who provided the 23 chapters are experts in their specific areas. The authors of this book span those who worked during the period before DNA testing was done through to the present, and those now working in developing some of the newer technologies mentioned and who therefore can provide a unique perspective on the history and practice of forensic DNA testing. The authors come from Australia, Austria, Bosnia and Herzegovina, Croatia, Finland, the Netherlands, and the United States of America.

This edition has four sections and 23 chapters, representing all aspects of forensic DNA methodology, ethics, law, and policy.

Part I, General Background and Methodological Concepts, consists of eight chapters reviewing the overall background of the field.

Chapter 1 "Basic Genetics and Human Genetic Variation" is a review of the history of forensic DNA testing, including screening tests for biological fluids, a review of DNA methods from the introduction of restriction fragment length polymorphism (RFLP) testing; to the beginning of the use of the polymerase chain reaction (PCR); to the use of modern PCR for both small tandem repeat (STR) testing, single-nucleotide polymorphism (SNP) testing, DNA sequencing procedures, DNA phenotyping, forensic analysis of plant DNA, and forensic analysis of animal DNA.

Chapter 2 "Forensic DNA Analysis and Statistics" is a review of forensic DNA analysis and statistical analysis of the data generated, including parentage testing, identification of remains, forensic identification using nuclear markers, as well as X and Y markers and mtDNA. Forensic genetics truly represent an excellent example of the synthesis of scientific knowledge with everyday life applications. This section of the book represents a few basic postulates that every scientist working in this area, but also every person that uses the results of DNA analysis in their daily work, should know.

Chapter 3 "Forensic Aspects of mtDNA Analysis" provides an overview of the forensically relevant characteristics of the mtGenome; how an mtDNA sequence profile (mitohaplotype) is generated and interpreted, including the emergence of massively parallel sequencing (MPS); and how mtDNA sequence analysis can be used to answer questions raised in forensic investigations.

Chapter 4 "Y Chromosome in Forensic Science" focuses on different forensic applications of Y-chromosome DNA. The underlining features that make the human Y chromosome attractive for forensic applications are the male specificity (hitherto absence in females), the (mostly) nonrecombining nature, and the rich content of different types of polymorphic markers with different underlying mutation rates. As described in this chapter, the human Y chromosome provides an increasing amount of forensically useful information, which is highly relevant for law enforcement given that the vast majority of criminal offenders are males.

Chapter 5 "Forensic Application of X Chromosome STRs" discuss some of the unique properties of XSTR testing, and provide some areas of future applications and research. Short tandem repeat (STR) markers on the X chromosome (XSTRs) have found an important niche in forensic DNA testing, principally in kinship and paternity testing where complex relationships using autosomal STR markers provide limited statistical support (e.g., paternal half-sisters). XSTR multiplexes have been developed and are commercially available. The interpretation of XSTR results and the determination of the strength of evidence is different than autosomal STRs as the XSTR markers are clustered into linkage groups.

Chapter 6 "Increasing the Efficiency of Typing Challenged Forensic Biological Samples" deals with challenged forensic biological samples. DNA extracts from challenged samples, such as environmentally insulted stains, bones, and other anthropological materials that are characterized by containing a low copy number of severely degraded DNA and/or the presence of PCR inhibitors, may pose challenges to downstream DNA analyses. PCR amplification, a critical step in the DNA typing process, generates millions to billions of copies of targeted DNA sequences. These amplified products will be subjected to further analyses, e.g., sequencing or capillary electrophoresis (CE), to determine genotypes or haplotypes of the samples. Therefore, enhancing recovery of DNA, purifying extracts to remove inhibitors, and making PCR robust are crucial in successful DNA data generation. Several methods for enhancement of the efficacy of PCR with challenged DNA samples have been proposed. Some techniques target improvement of amplification of degraded DNA, while others enrich amplification when PCR inhibitors are co-extracted. Herein, methods that can improve the recovery of genetic information from challenged samples will be discussed. The examples primarily will focus on bones; however, the approaches apply equally well to biological stain evidence recovered from crime scenes. Each strategy should be applied with an understanding of the particular traits of challenged samples and the techniques, and each requires validation.

Chapter 7 "Mixtures and Probabilistic Genotyping" discusses the basics of mixture interpretation using manual methods through the introduction of probabilistic genotyping. Each of the steps of interpretation are outlined along factors that contribute to the complexity of the process making it more challenging. The interpretation of a mixture is influenced by the statistical approach employed by the laboratory. The combined probability of inclusion (CPI), random match probability (RMP), likelihood ratio (LR), and different probabilistic genotyping (PG) approaches are described and compared.

The final chapter in this part, Chapter 8 "Rapid DNA", discusses an innovative technology that can generate scientifically valid DNA profiles in a field location in 90 minutes, and has the potential to significantly impact the availability of DNA for law enforcement. This technology removes the barriers of traditional DNA processing and allows end users to obtain scientific support for immediate intelligence/investigation and law enforcement decisions. As described in this chapter, Rapid DNA instruments add additional features such as mixture interpretation software, and law enforcement agencies can increase their reliance on DNA as a powerful investigative tool.

Part II, Uses and Applications, focuses on a multitude of applications of the DNA technologies described in the previous chapters, ranging from the collection and preservation of biological evidence, identification of missing persons in single and mass disasters, bioterrorism, animal DNA analysis, forensic entomology, and forensic botany.

Chapter 9 "Collection and Preservation of Physical Evidence" is an international approach to the collection and preservation of physical evidence, but provides basic guidelines on the preservation of biological evidence, although the title says physical evidence. The chapter also goes into the various forms of crime scene reconstruction that aid in the interpretation of the aggregate physical evidence collected at that scene.

Chapter 10 "Mass Disaster Victim Identification by DNA" deals with the seemingly unending job of identifying remains, either of a single missing person or of a mass disaster. Mass fatality incident (MFI) response is a multidisciplinary effort providing support for survivors, organizing the recovery of deceased individuals, and collecting forensic evidence needed for investigating criminal or negligent actions. Mass disasters are complex events; a preparedness plan and regular training exercises are parts of an effective response. The goal of postmortem operations is to determine cause and manner of death, as well as the identity of the victims. DNA testing, fingerprints, and dental chart comparisons are primary identification modalities, but which method is most applicable depends on the availability of antemortem data. A family assistance center (FAC) must be established to meet with families and collect antemortem information and family reference DNA. Documentation, chain of custody, and data handling are critical components. As is explained using the example of victim identification after the 2001 World Trade Center attack, DNA becomes the only identification method for fragmented remains.

Chapter 11 "Bioterrorism and Forensic Microbiology" provides a history of bioterrorism, which is older than one would think, and includes a classification of these "select agents." Forensic microbiology encompasses the tools used to identify these agents in a timely fashion so that an appropriate response can be initiated. The areas of biosafety and biosecurity are also included.

Chapter 12 "Forensic Animal DNA Analysis" describes the development of DNA identity testing systems in cats, dogs, cattle, and wildlife, with the inclusion of groundbreaking case studies. There are numerous examples of forensic applications of animal DNA testing. Animals can be the victims of crimes such as abuse, the perpetrators of violent acts against people or other animals, and implicate their owner in a crime through transfer of trace evidence such as fur. Animal forensics requires the development of species-specific genetic markers, and to be accepted in court it will be required to meet the same rigorous standards as for human identity testing including, for sample collection, maintenance of chain of custody, marker validation, database development for match probability calculations, and laboratory proficiency testing.

Chapter 13 "Application of DNA-Based Methods in Forensic Entomology" will show how DNA-based methods are used as an auxiliary tool in human forensics, not only to identify forensically important species in any life stage, but also to possibly identify human DNA that may be present in the insects' guts. Forensic entomology uses insects to help law enforcement determine the cause, location, and time of death of a human corpse as well as to detect poisons, drugs, and physical neglect and abuse. Considered as physical evidence at the crime scene, insect specimens should be processed as any other biological material, following the recommended procedures for quality assurance for collection, preservation, packaging, and transport in order to prevent contamination or destruction of evidence and to guarantee the chain of custody. The study of present necrophage species and developmental stages of insects that feed on the corpse can provide information such as postmortem interval (PMI) and time of colonization. Taxonomic keys, which rely on morphological characteristics, are commonly used for identification of forensically important insects.

Chapter 14 "Forensic Botany: Plants as Evidence in Criminal Cases and as Agents of Bioterrorism" concentrates on the subject of forensic botany. Bioterrorism is defined as terrorism involving the release of toxic biological agents. This is a potentially large group of items including plants and plant-derived products that can be whole, fragmented, or in powder form. There are approximately 391,000 vascular plant species with approximately 94% angiosperms or flowering plants. Basic classification schemes include angiosperms (i.e., flowering plants), gymnosperms (e.g., conifers, cycads), and algae (e.g., diatoms). This chapter does not discuss specifically all the myriad types but provides an overview of plant associative evidence and the basic principles for evidence collection, storage, and analysis; and describes a few examples of common species that could be weaponized.

Part III, Recent Developments and Future Directions in Human Forensic Molecular Biology, looks at the evolving technologies in forensic DNA testing that may not be widely used at present but are being developed as future technologies, including improved identification of specific tissues; new applications of mtDNA testing; the future of next-generation sequencing in forensic analysis; the use of SNPs to predict physical characteristics such as eye, hair, and skin color; and finally the area of molecular autopsy.

Chapter 15 "Forensic Body Fluid and Tissue Identification" provides the foundations of forensic identification of body fluids and tissues. It first introduces well established, widely used "classical tests" for body fluid identification, and then discusses in detail the advantages, principles, and methodologies of the more progressive, molecular approaches of RNA analysis and DNA methylation analysis. Subsequently, various other approaches to body fluid and tissue identification are presented, including fluorescence and mass spectrometry and microbiomics. The chapter closes with an outlook on future developments and new technologies.

Chapter 16 "Evolving Technologies in Forensic DNA Analysis" discusses some of the more recent developments in forensic DNA analysis such as next-generation sequencing, also known as massively parallel sequencing. The clonal nature of this technology allows the deconvolution of mixtures, a common and challenging category of crime scene evidence. The strategy discussed here is the construction of a shotgun library, targeted capture by hybridization with a probe panel, and sequencing on the IIlumina MiSeq platform, a protocol suited to the small DNA fragments found in many forensic samples. Sequencing the entire mitochondrial DNA genome, a haploid lineage marker allows deconvolution of mixtures by counting individual clonal sequence reads and assigning them to one of the contributors, aided by a phylogenetic-based software, Mixemt. The chapter also discusses the use of massively parallel sequencing using a probe capture strategy for the analysis of short tandem repeats (STRs) using a web-based software, toaSTR. In addition, this chapter reviews the implementation of Rapid DNA technology, instruments capable of automating the DNA extraction, amplification of the short tandem repeat markers, and analysis by capillary electrophoresis.

Chapter 17 "Prediction of Physical Characteristics, such as Eye, Hair, and Skin Color, Based Solely on DNA" discusses the use of SNPs, some of the complexities of the inheritance of these traits, and the future directions of the use of these assays.

The next chapter in this part, Chapter 18 "Molecular Autopsy", deals with principles and standards of molecular genetics and their application in the cases of sudden, unexpected deaths. Multidisciplinary investigation of death involves forensic pathology, forensic toxicology, and forensic genetics, which are crucial for public safety and legal medicine. Molecular autopsy is emerging from molecular pathology, a study of biochemical and biophysical cellular mechanisms as the basic factors in diseases. Although molecular genetic testing is not yet commonly performed at autopsy, it is expected to become an integrated component in forensic investigations in the near future.

Chapter 19 "Genetic Genealogy in the Genomic Era" introduces different testing options available which focus on different parts of the human genome and discuss the relative merits of each type. The underlying inheritance patterns for Y, mitochondrial, and autosomal DNA are then described, and how these patterns determine what can be inferred about historical origins from each test type. In general, haploid tests provide a deeper look into the past, but with results that are only relevant to one family lineage. On the other hand, while autosomal tests trace all genetic lineages, the semi-conservative nature of autosomal inheritance implies a much shallower look into the past. In addition, this chapter discusses what these tests can tell us about living relatives by looking for long contiguous blocks of DNA shared between two or more test-takers due to a shared ancestor. These are said to be "identical by descent" or IBD blocks. The explosive growth of the human population in recent generations implies tremendous amounts of IBD sharing, and this network of sharing, as revealed in vast direct-to-consumer databases, is quickly becoming an appealing and powerful forensic tool.

Part IV, Law, Ethics, and Policy, helps to set this book aside from others as it deals with the legal, ethical, and policy issues on the use of DNA evidence. This part is a good refresher for lawyers and judges as well as practitioners who are old and new to the field.

Chapter 20 "DNA as Evidence in the Courtroom" addresses contemporary issues involving the use of DNA evidence in the courtroom, utilizing perspectives from the American, European, and Croatian national experiences. Topics include the legal standards by which DNA may be admitted in court; complex challenges faced by practitioners ranging from partial DNA profiles and "cold hit" cases to database searches and the accuracy of random match probability figures; the statutory elements genuinely required to obtain post-conviction DNA testing; important privacy considerations concerning arrestees, defendants, and exonerees; the collection, storage, retention, and destruction of DNA samples, which should adhere to the principle of proportionality; how various European instruments and human rights principles affect national legislation; the importance of fostering cross-border cooperation without compromising data protection; the procedural steps necessary for countries to qualify for the exchange of DNA evidence; and the use of expert reports in domestic practice.

Chapter 21 "Some Ethical Issues in Forensic Genetics" deals with ethical issues in forensic genetics. Identifying individuals, living or dead, from their genotypes can help to convict offenders or exonerate innocent suspects. It can confirm one's presence at a crime scene or one's place in a family tree. However, the extensive and sensitive nature of the genetic information locked in the coils of the DNA molecule also gives rise to ethical dilemmas. This chapter surveys such issues from the perspectives of justice, privacy and confidentiality, autonomy, and informed consent. It explores topics including acquiring DNA samples, DNA databanks and biobanks, inferences of phenotypes from genotypes, identifying human remains, and professional and ethical standards for reporting forensic laboratory results and testifying about them.

Chapter 22 "DNA in Immigration and Human Trafficking" delves into the area of DNA testing in immigration cases and human trafficking. One of the very first applications of genetic identity markers was to establish biological connections for an immigration case. Several decades later, the use of forensic DNA to establish identity and to verify claimed relationships has expanded around the globe, to improve border security measures, to detect fraud, and to investigate cases of human trafficking. But it also raises substantial questions around ethical applications of the technology and the potential for abuse of the technology, leading to stigmatization or discrimination. This chapter reviews the myriad applications of DNA in immigration contexts in a rapidly expanding political and migratory population.

The final chapter in this part, Chapter 23 "DNA Databases", covers international perspectives on forensic DNA databases. The development of DNA technology and the establishment of a corresponding DNA database on national and transnational levels is one of the most efficient ways to detect and prevent crime. All information contained in a DNA database should be in compliance with the national legislation. With the proper establishment of DNA databases and their maintenance as defined by law, ethical issues related to threats to a number of civil rights are minimized. As legislation in countries worldwide is different, the amount of DNA data in national databases greatly varies. The rapid growth in the volume of data in databases worldwide requires additional infrastructure support and close cooperation at the international level for improvement of existing and the inclusion of new categories of data. Cross-border exchange of DNA profiles has been of great assistance in improving the resolution of criminal offenses and international crime.

Each of the chapters includes an extensive list of references for the reader.

Acknowledgments

We acknowledge the International Society of Applied Biological Sciences for support essential in bringing this volume to the light of day.

Editors

Dragan Primorac, MD, PhD, is a pediatrician, forensic expert, and geneticist. Primorac is the first recipient of the title "Global Penn State University Ambassador" since the university was established in 1855. Currently he serves as an adjunct professor at Eberly College of Science, Pennsylvania State University; and the Henry C. Lee College of Criminal Justice and Forensic Sciences, University of New Haven in the United States; and at medical schools at universities in Split, Osijek, and Rijeka in Croatia. He is also professor at Faculty of Dental Medicine and Health, Osijek, Croatia, Medical School REGIOMED, Coburg, Germany; and National Forensic Sciences, University Gandhinagar, Gujarat, India. He is founder of St. Catherine Specialty Hospital, Zagreb, Croatia.

Primorac is a pioneer in the application of DNA analysis for identification of bodies in mass graves. He has authored close to 300 scientific papers, book chapters, and abstracts in areas of forensic science, clinical medicine, molecular genetics, population genetics, genetic legacy of *Homo sapiens sapiens,* education, science, and technology policy. His papers have been cited more than 7500 times (Google Scholar). He is among in the top 2% of the most influential scientists for the years 2020–2022, according to a Stanford University research group.

Primorac has received 21 domestic and international awards including the Young Investigator Award of the American Society for Bone and Mineral Research in 1992, the Michael Geisman Fellowship Award of the Osteogenesis Imperfecta Foundation in 1993, the Life Time Achievement Award by the Henry C. Lee's Institute of Forensic Science in 2002, the Award of the Italian Region Veneto for Special Achievements in Promoting Science in the EU in 2007, the University of New Haven's International Award for Excellence in 2010, Mary E. Cowan Outstanding Service Award by the American Academy for Forensic Sciences, as well as two decorations by the President of the Republic of Croatia: "The Order of Croatian Star with the Effigy of Ruđer Bošković" and "The Order of Ante Starčević," both for his extraordinary achievements in science, education, politics.

Primorac is the cofounder of the International Society of Applied Biological Sciences (ISABS). He is also the founder of the "Nobel Spirit" session held during the ISABS conferences when Nobel prize laureates stimulate public discussion on science's role in solving global health issues, acute regional problems such as brain drain, demographic decline, as well as cultural and social change. He has been an invited speaker to more than 150 national and international scientific meetings. In 2011, he established St. Catherine Specialty Hospital, the European center of excellence. Several renowned media outlets, both electronic and print, have reported on his work, including the *New York Times, JAMA, Science, Profiles in DNA, Die Presse, Haaretz, Kleine Zeitung,* and many others. Furthermore, Connecticut TV Station Channel 8 filmed a television serial on forensic work in Bosnia and Herzegovina and the Republic of Croatia featuring Primorac and his colleagues.

From 2003 to 2009, he served as the Minister of Science, Education, and Sports of the Republic of Croatia. According to the International Republican Institute survey of October

1, 2007, he was rated as the most successful minister in the Croatian government with 31% approval rate. In recognition of the numerous efforts made by Primorac and his team, the Croatian educational system was rated 22nd in the world, ahead of 12 countries from the G20 group (*Newsweek*, August 16, 2010).

Moses S. Schanfield, PhD, holds undergraduate and master's degrees in anthropology and a PhD degree in human genetics. He has been involved full time in forensic testing for over 25 years. His career in forensic biology began before DNA testing was done. He was involved in some of the earliest forensic DNA cases and in some famous forensic cases, including the O.J. Simpson case and the JonBenet Ramsey case. Since 1995, Schanfield has been involved in repatriation of remains in the Balkans and the identification of remains in mass casualty events. This led to the creation of the first intensive course in forensic genetics, which has now been replaced by the biannual education conference of the International Society of Applied Biological Sciences (ISABS), in which Schanfield is one of the permanent organizers. Schanfield has been to Croatia many times as part of his ISABS and other duties.

He was involved in four of the first eight polymerase chain reaction (PCR) cases admitted in court and reviewed at the appellate level in the United States and testified in many of the early DNA admissibility hearings. Schanfield has been involved in the development of some of the critical functions of modern forensic DNA testing including the in-lane size ladder, which is used in all human identification testing. He has published extensively on forensic DNA testing and has testified in state (39 states), federal, and military courts in the United States and Canada, Puerto Rico, and Barbados on forensic cases over 100 times.

Schanfield has lectured all over the world on blood transfusion, forensic science, and anthropological issues. He was a professor of forensic science and anthropology at George Washington University in Washington, DC. Schanfield has published over 190 articles and books in the areas of anthropology, forensic science, immunology, and other scientific areas.

He was awarded the 2021 Paul Kirk award, the highest award by the Criminalistics Section of the AAFS. He passed away a month before he could receive it in person. Schanfield passed away in January 2021.

Contributors

Angie Ambers
University of New Haven, USA
and
Henry C. Lee Institute, USA

Šimun Anđelinović,
University of Split, Croatia

Kaye N. Ballantyne
Victoria Police Forensic Services Department, Australia

Gunmeet Kaur Bali
UCSF Benioff Children's Hospital, USA

Steven W. Becker
Law Office of Steven W. Becker LLC, USA

Frederick R. Bieber
Harvard Medical School, USA
and
Brigham and Women's Hospital, USA

Todd Bille
National Laboratory Center, USA

Zoran M. Budimlija
NYU Langone Health, USA

Bruce Budowle
University of North Texas, USA

Magdalena M. Buś
University of North Texas, USA

Jake K. Byrnes
Ancestry.com DNA, USA

Cassandra D. Calloway
UCSF Children's Hospital Oakland Research Institute, USA

Michael Coble
Health Science Center, University of North Texas, USA

Cornelius Courts
University Hospital of Cologne, Institute of Legal Medicine, Germany

Josip Crnjac
University of Split, Croatia

Lidija Cvetko Krajinović
University Hospital for Infectious Diseases, Croatia

Victor A. David
National Cancer Institute, USA

Davor Derenčinović
University of Zagreb Law School, Croatia

Henry Erlich
UCSF Children's Hospital Oakland Research Institute, USA

Sheila Estacio Dennis
John Jay College of Criminal Justice, USA

Mitchell M. Holland
Pennsylvania State University, USA

Sree Kanthaswamy
University of California–Davis, USA
and
Arizona State University, USA

Sara Huston Katsanis
Ann & Robert H. Lurie Children's Hospital of Chicago, USA
and
Northwestern University, USA

Manfred Kayser
Erasmus MC University Medical Center Rotterdam, Netherlands

Ivan-Christian Kurolt
University Hospital for Infectious Diseases, Croatia

Gordan Lauc
Genos, DNA Laboratory, Croatia

James W. Le Duc
Galveston National Laboratory, University of Texas
    Medical Branch, USA

Andrea Ledić
Ivan Vučetić Forensic Science Centre, Croatia

Henry C. Lee
Henry C. Lee College of Criminal Justice and
    Forensic Sciences, University of New Haven, USA
and
Henry C. Lee Institute of Forensic Science, USA

Adrian Linacre
Flinders University, Australia

Adela Makar
Ivan Vučetić Forensic Science Centre, Croatia

Damir Marjanović
International Burch University, Bosnia and
    Herzegovina
and
Institute for Anthropological Research, Croatia

Alemka Markotić
University Hospital for Infectious Diseases, Croatia

Marilyn A. Menotti-Raymond
National Cancer Institute (retired), USA

Christopher Miles
Department of Homeland Security Science and
    Technology Directorate, USA

Heather Miller Coyle
Henry C. Lee College of Criminal Justice & Forensic
    Sciences, University of New Haven, USA

Igor Oblešcuk
Ivan Vučetić Forensic Science Centre, Croatia

Timothy M. Palmbach
Henry C. Lee College of Criminal Justice &
    Forensic Sciences, University of
    New Haven, USA

Walther Parson
The Institute of Legal Medicine, Medical University
    of Innsbruck, Austria

Daniele Podini
George Washington University, USA

Mechthild Prinz
John Jay College of Criminal Justice, USA

Petar Projić
Institute for Anthropological Research, Croatia
and
Genos, DNA Laboratory, Croatia

Antti Sajantila
University of Helsinki, Finland

Vedrana Škaro
Greyledge Europe Ltd., Croatia
and
Genos, DNA Laboratory, Croatia

Peter Underhill
Stanford University School of Medicine, USA

Susan Walsh
Indiana University–Purdue University Indianapolis,
    USA

Jeffrey D. Wells
Florida International University, USA

Erin D. Williams
Center for Transforming Health, The MITRE
    Corporation, USA

General Background and Methodological Concepts

I

Basic Genetics and Human Genetic Variation

1

MOSES S. SCHANFIELD, DRAGAN
PRIMORAC, AND DAMIR MARJANOVIC

Contents

DOI: 10.4324/9780429019944-2

3

1.1 Introduction

Nowadays, deoxyribonucleic acid (DNA) undoubtedly has an inimitable role in forensic science. Since 1985, when Alec Jeffreys and colleagues first applied DNA analysis to solve forensic problems, numerous medicolegal cases have been won based on this method (Jeffreys et al. 1985a, 1985b). It is beyond a doubt that DNA analysis has become "a new form of scientific evidence" that is being constantly evaluated by both the public and professionals. More and more courts around the world accept the results of DNA analysis, and nowadays this technology is almost universally accepted in most legal systems. The foremost applications of DNA analysis in forensic medicine include criminal investigation, personal identification, and paternity testing. According to some sources, more than 300,000 DNA analyses in different areas of expertise are performed annually in the United States. The fact that more than 30% of men who were identified as possible fathers are excluded using DNA technology articulates the importance of DNA analysis. Through the "DNA Innocence Project," launched in the United States to acquit wrongfully convicted people, more than 300 persons have been exonerated by DNA testing (February 2013), including several individuals who were sentenced to death. DNA played an extremely important role in projects involving identification of war victims in Croatia and surrounding countries (Primorac et al. 1996; Primorac 2004; Džijan et al. 2009; Gornik et al. 2002; Alonso et al. 2001). Using this powerful "molecular" tool, identities were determined for thousands of skeletal remains (some of them even from the Second World War), and their families had the opportunity to bury their loved ones with dignity (Anđelinović et al. 2005; Marjanović et al. 2007; Marjanović et al. 2015). Although the development of DNA typing in forensic science was extremely fast, the process is still not finished, and today we are witnessing a new era in the development of DNA technology that involves the introduction of automation and "chip" technology. In this chapter, we will explain the structure of DNA, principles of inheritance of genetic information, technological advances in DNA analysis, as well as the common application of mathematical methods in forensic practice.

1.2 Historical Overview of DNA Research

In 1944, O. Avery and collaborators conclusively showed that DNA was the genetic material, whereas the structure of DNA was first presented by Nobel Prize winners James Watson and Francis Crick in 1953, the groundwork for development of modern molecular genetics (Tamarin 2002). In 1980, David Botstein and his colleagues demonstrated that there are small variations in genetic material, which differ from person to person, and Alec Jeffrey in 1985 showed that certain regions of DNA contain repetitive sequences that are variable in different individuals. It is this fact that was critical in resolving the first forensic case using DNA analysis (Butler 2010). After the murder of two girls, Lynda Mann and Dawn Ashworth, in 1983 and 1986, respectively, the police organized the testing of more than 4000 men and eventually found the killer (Wambaugh 1989). However, the discovery of the polymerase chain reaction (PCR) method in 1983 certainly determined the future of DNA analysis in both clinical and forensic medicine. In particular, with the discovery of this technology, it has become possible to analyze biological samples containing minuscule amounts of DNA. In 2001, the world witnessed the revolutionary discovery of the human genome structure, which was made by two separate groups of researchers: The International Human Genome Sequencing Consortium, led by Eric Lander; and Celera Genomics, which was spearheaded by Craig Venter (International Human Genome Sequencing Consortium 2001; Venter et al. 2001). The aim of both groups was to determine the complete nucleotide sequence of the human haploid genome, which contains about 3.0×10^9 base pairs (3.0 Gbp). However, they estimated that the size of the human diploid male genome is 6.294×10^9 bp, whereas for the human diploid female genome the size is 6.406×10^9 bp. The real surprise came with the finding that the human genome has about 20,000 protein-coding genes (Willyard 2018) and not, as previously predicted, 100,000 genes. Particularly surprising was the fact that just more than 1% of the DNA contains protein-coding sequences (Cooper and Hausman 2004). On the other hand, non-protein-coding regions (ncDNA) are still poorly understood, and at least four hypotheses explain the role of these nonfunctional regions of the genome (Wagner 2013). The *selectionist hypothesis* suggests that ncDNA regulate gene expression, whereas the *neutralist hypothesis* (junk DNA) implies that these regions are without function but are transmitted passively as relics of evolutionary processes. The *intragenomic selection hypothesis* (selfish DNA) posits that ncDNA stimulate their own transmission and accumulate because of their prominent reproduction rate compared to protein-coding regions. Finally, the *nucleotypic hypothesis* posits that ncDNA regions act to preserve the organizational/structural integrity of genome (Graur and Li 2000). Table 1.1 shows a summary of non-protein-coding genomic elements (Wagner 2013).

Doležel et al. (2003) proposed that 1 pg of DNA would represent 0.978×10^9 bp, and therefore diploid human female and male nuclei in the G_1 phase of the cell cycle should contain 6.550 and 6.436 pg of DNA, respectively. In the future, a great task lies before scientists: to discover the functions and interactions of many genes, as well as to exploit the possibility of gene and cell therapy. All advances in the field of molecular genetics will have an important role in the development of new technologies and methods in forensic DNA analysis (Primorac 2009).

1.2.1 Introduction to Human Genetics

All living creatures, including humans, are built of cells, which represent the smallest structural and functional units of our body.

1.2.2 Genome Structure

The basic genetic information in eukaryotes is located in the nucleus, whereas the cytoplasm contains numerous cellular structures that keep the cell alive. Figure 1.1 shows the cellular organization of eukaryotic cell as well as the structure and location of the nuclear and mitochondrial DNA (mtDNA). One of these cellular structures is the mitochondrion;

Table 1.1   Summary of Non-Protein-Coding Genomic Elements

Non-Protein-Coding Genomic Element			Brief Description
Transcription regulatory elements			Molecular elements considered typical of gene structure, such as promoters, enhancers, and intronic splicing signals (Slack 2006)
Introns			Segments of DNA located within genes that interrupt or separate exons from one another
5′ and 3′ untranslated regions		UTRs	Transcribed DNA sequences preceding (5′ UTR) and following (3′ UTR) coding sequences containing regulatory elements, such as binding sites for microRNAs (miRNAs) and polyadenylation signals (Maroni 2001)
RNA-specifying genes	MicroRNAs	miRNAs	Interact with specific mRNAs through complementary base-pairing to influence the translation or stability of the target mRNA molecule (Dexheimer and Cochella 2020)
	Transfer RNAs	tRNAs	Facilitate translation by transporting specific amino acids to the ribosome; ca. 80 nucleotides in length (Neilson and Sandberg 2010)
	Ribosomal RNAs	rRNAs	Facilitate the movement of tRNAs along the mRNA during translation; four types (18S, 28S, 5.8S, and 5S) (Neilson and Sandberg 2010)
	Spliceosomal RNAs	snRNAs	Facilitate the processing of pre-mRNAs (i.e., help splice introns that are not self-splicing); five types (Ul, U2, U4, U5, and U6) (Neilson and Sandberg 2010)
	Small nucleolar RNAs	snoRNAs	Facilitate post-transcription modifications of rRNAs, tRNAs, and snRNAs; two types (H/ACA box and C/D box) (Neilson and Sandberg 2010)
	Piwi-interacting RNAs	piRNAs	Protect the integrity of the genome in germline cells during spermatogenesis; 25–33 nucleotides in length (Neilson and Sandberg 2010)
	RNAse P/MRP genes		Process tRNA and rRNA precursors (Neilson and Sandberg 2010)
	Long noncoding RNAs	lncRNAs	About 200-plus nucleotides in length, such as XIST, which silences an X chromosome during X inactivation (Neilson and Sandberg 2010)
Repeat elements	Satellite DNA		DNA sequences often near centromeres and telomeres α-Satellite or alphoid DNA, a 171-bp sequence that is repeated in tandem and clustered at the centromeres of all chromosomes
			Repeat size of satellite DNA may be between 2 and 2000 bp and the size of the repeat array may be greater than 1000 bp (10,21)
	Minisatellites or variable number tandem repeats	VNTRs	Repeat units of 10–200 bp clustered into repeat arrays of 10–100 units
			Found near the telomeres (the terminal ends of chromosomes), but are also distributed across the chromosomes (Graur and Li 2000; Slack 2006)
	Microsatellites or short tandem repeats	STRs	Repeat units of 2–5 bp arranged in arrays of 10–100 units (Graur and Li 2000; Slack 2006)

(Continued)

Table 1.1 (Continued)   Summary of Non-Protein-Coding Genomic Elements

Non-Protein-Coding Genomic Element			Brief Description
	Short interspersed nucleotide elements	SINEs	About 1,500,000 copies of SINEs present in the genome account for more than 10% of the genome (Graur and Li 2000; Slack 2006)
	Long interspersed nucleotide elements	LINEs	About 850,000 copies of LINEs present in the genome, account for roughly one-fifth of the genome (Graur and Li 2000; Slack 2006)
	Retrovirus-like elements		About 450,000 copies present in the genome (Slack 2006)
	Transposons		About 300,000 copies present in the genome (Slack 2006)
Pseudogenes			Exhibit similarity to genes but lack introns and promoters and contain poly-A tails
			Most pseudogenes have lost the ability to be transcribed (Graur and Li 2000; Slack 2006; Jobling et al. 2004)

Source: Wagner, J.K. 2013. *J Forensic Sci* 58:292–294.

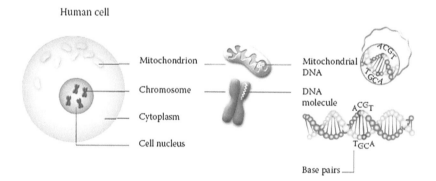

Figure 1.1 Sources of DNA in nucleated human cells include the nucleus (2n) and mitochondrion (100–1000 copies of mitochondrial DNA).

mitochondria act as a "cellular power plant" because they generate most of the cell's supply of adenosine triphosphate (ATP), which transports chemical energy within cells for metabolism. From the forensic perspective, mitochondria are important because they contain mtDNA, which is inherited directly through the maternal line (mtDNA will be discussed in greater detail in Chapter 3). Briefly, mtDNA does not contain introns, is not subject to classical recombination, and has a higher mutation rate than nuclear DNA. A human cell has between 100 and 1000 mitochondria, each of them containing multiple copies of circular mtDNA, which is 16,569 bp long. Mitochondria comprise one of the basic lines of evidence for the symbiogenetic theory of evolution, because mitochondria and mtDNA have a number of features that are characteristic to prokaryotic organisms. This suggests that this organelle derived from an ancient, engulfed (endosymbiotic) prokaryotic organism, which was one of the most important moments for the development of the first unicellular and later multicellular eukaryotic organisms. Although the average cell size corresponds to about one-tenth of the diameter of a hair, finding biological evidence that contains a sufficient number of cells can undoubtedly be crucial in solving forensic cases (Primorac and Marjanović 2008).

1.2.3 Chromosomes and Genes

Given the fact that every person inherits half of the genetic material from their father and half from their mother (excluding mtDNA, which is normally inherited exclusively from the mother, and the Y chromosome, which is only inherited paternally), DNA testing can be used to determine the genetic relationship between individuals. Another important feature is that DNA is an extremely stable molecule, and that over time, if properly stored in vitro, the order of its constituent units does not change, so a suitably deposited sample can be used to compare its DNA profile with the profile of another sample taken years later. As noted earlier, a small portion of nuclear DNA carries the genetic message, encoded in the *genes*. In other words, genes are active segments located at specific places (*loci*) in DNA strands. Because of the normal functioning of genes, each individual has certain parameters of growth and development (e.g., it is genetically specified that a person has two hands and two feet), but also some other traits, such as eye color and height. Thus, it is obvious that DNA represents the central molecule of life, which—as the primary (main) carrier of hereditary information—controls the growth and development of every living being. Total nuclear DNA is almost 2 m long and is located in structures called *chromosomes*. The word *chromosome* comes from two Greek words: *chromos* meaning "color" and *soma* meaning "body." From a total of 46 chromosomes found in the somatic cells of each individual, 23 is inherited from the mother and 23 from the father. In particular, during the formation of a zygote, 23 chromosomes come from the egg, and 23 from the sperm cell. Chromosomes are found in the nucleus in pairs, and humans have a total of 23 pairs of chromosomes. One of these pairs is the sex chromosomes (X, Y), which are important in determining the gender of each individual and how they differ from one another. The remaining 22 pairs are called *autosomes,* and they are homologous in both sexes. Excluding possible chromosomal abnormalities, in general, the rule is that two X chromosomes (XX) determine the female, and the combination of X and Y chromosomes (XY) determines the male sex. It is important to note that all somatic cells (e.g., bone cells, skin cells, white blood cells) contain 46 chromosomes (23 pairs), whereas gametes (sperm and ovum) contain half as much, or more precisely, 23 chromosomes. A large number of scientific studies have shown that there are certain genes located in mtDNA, so the study of their functions has become very intense in recent years (Report of the Committee on the Human Mitochondrial Genome 2009). The physical location of a gene on a chromosome is called a *locus.* Two autosomes that make a pair match each other both by structure and function. This is why two such chromosomes, which are similar in structure and carry the same genes, are called *homologous chromosomes.* Furthermore, variants of genes that occupy the same position or locus on homologous chromosomes and determine different forms of the same genetic trait are called *alleles.* Alleles are, in fact, alternative forms of the same gene or genetic loci with differences in sequence or length (Primorac and Marjanović 2008; Primorac and Paić 2004). The following terms, often used in genetics, including forensic medicine, are *homozygote* and *heterozygote.* Homozygote indicates the genotype (genetic version) that has identical alleles or variants of the same gene at a particular locus on a pair of homologous chromosomes. On the contrary, the heterozygote genotype indicates the person or genotype that has two different alleles at a particular locus on a pair of homologous chromosomes (Figure 1.2).

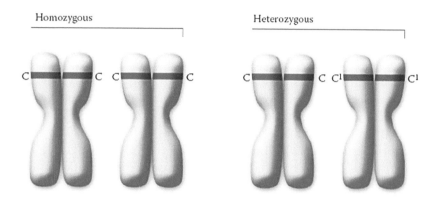

Figure 1.2 Example of homozygous and heterozygous loci on chromosomes. Note: Chromosomes are represented as sister chromatid before cell division, so each chromosome is represented twice, and as such, each allele is represented twice. Once division has occurred there will be only two copies of each nuclear gene.

1.2.4 Deoxyribonucleic Acid

DNA molecule is constructed in the form of double-stranded helix, and it is constituent of all 46 chromosomes. DNA consists of units called *nucleotides,* and the human genome (haploid) contains approximately 3.2×10^9 of such units. The nucleotide itself is composed of three subunits: a five-carbon sugar; a phosphate group; and the nitrogen-containing nucleotide bases *adenine* (A), *guanine* (G), *cytosine* (C), and *thymine* (T). It is good to remember that in a double helix, adenine always pairs with thymine (A–T), and cytosine with guanine (C–G). Under normal circumstances, base mating under another scheme is not possible. This occurs because adenine and guanine are a chemical structure called a *purine*, whereas thymine and cytosine are a chemical structure called a *pyrimidine*. The purine and pyrimidine can form complementary structure held together by hydrogen bonds. There are two hydrogen bonds between adenine and thymine, and three between guanine and cytosine holding the two strands of nucleotides in DNA together (Figure 1.3) and providing exceptional stability to this molecule. Every single contact between these units is called a *base pair* (bp), and the entire human genome (haploid) has about 3.2 billion bp. Changes in these bases or variations in the number of base pair repetitions are the basis for personal identification (Marjanović and Primorac 2013).

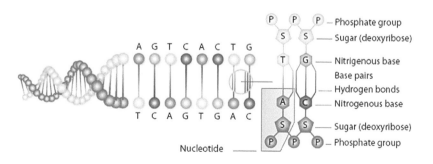

Figure 1.3 Structure of DNA molecule from double helix to base pairs.

1.2.5 Genetic Diversity

Genes (protein-coding sequence) make up only about 2% of the human genome. However, the rest (noncoding DNA sequence) do have important biological functions including the transcriptional and translational regulation of protein-coding sequences. Just as genes can contain genetic variation, noncoding DNA contains many types of genetic variation. These include differences in a single nucleotide referred to as *single nucleotide polymorphisms* (SNPs), repeated sequences of DNA that can range in size from 1 to 70 bp called *variable number of tandem repeats* (VNTRs; Inman and Rudin 1997; Budowle et al. 1991, 1992; FBI 1990) and *short tandem repeats* (STRs), as well as insertion and deletion polymorphisms (InDels) that can vary between 1 and thousands of bp.

1.2.6 Variability of DNA

Analysis of DNA in forensic science is based on the fact that only 0.5% of the DNA varies in every person. Nevertheless, that small part of DNA contains a great number of so-called *polymorphisms* (Greek: *poly* for "many," *morfoma* for "form"), or differences in DNA sequence among individuals, and therefore we can almost certainly claim that each of us has a unique genetic material, with the exception of identical twins. The fact that we are genetically different is the basis of analysis of evidence found at a crime scene, identification of victims, identification of rape offenders, or routine determination of kinship. The majority of loci that encode proteins have only one form of the gene. This is because most genes are not tolerant of mutations. Those genes that tolerate mutations have more than one form, that is, have their alleles. Loci containing alleles with relatively high frequency are called *polymorphic loci*. Genetic variation in blood groups, serum proteins, and transplantation antigens on the protein level is a reflection of that particular polymorphism, namely, the variability at the DNA level. Advances in DNA technology enabled the detection of variability (polymorphisms) in specific DNA sequences. Figure 1.4 shows an example of gene polymorphisms (i.e., sequential polymorphism) and displays the existence of two alleles for one gene where the change in a base pair is framed.

Other forms of polymorphisms are changes in the length of DNA between two homologous DNA segments (Figure 1.5).

During laboratory testing, the differences in alleles within a specific segment of DNA must be shown with a plain and understandable method that will clearly demonstrate the difference in the length of DNA between the two microsatellite alleles.

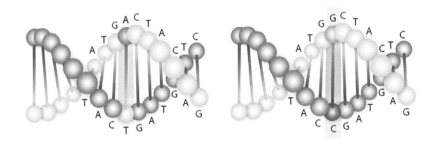

Figure 1.4 Example of a single nucleotide polymorphism (SNP) on DNA double helix.

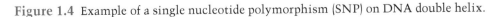

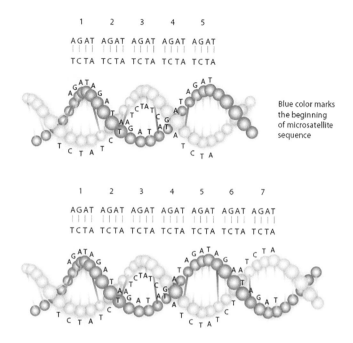

Blue color marks the beginning of microsatellite sequence

Figure 1.5 Example of a heterozygote for short tandem repeat (STR) and variable number tandem repeat (vNTR) with five and seven repeats, respectively, on DNA double helixes.

1.2.7 Structure and Nomenclature of STR Markers

The development of the Human Genome Project, which marked the beginning of the comprehensive studying and mapping of human genes, has imposed a need for developing guidelines on the identification and designation of new genes, and the inclusion of old, existing systems. The very beginning was the establishment of the International System of Gene Nomenclature in 1987 (Shows et al. 1987; Primorac et al. 2000). Meanwhile, the DNA Commission of the International Society for Forensic Human Genetics issued specific recommendations regarding the use of *restriction fragment length polymorphism* (RFLP) and PCR (1991 Report of the DNA Commission 1992; Mayr 1993).

There are thousands of STRs that can be used for forensic DNA analysis. STR markers consist of sequences of length 2–7 (according to some sources, 1–10) bp that are repeated numerous times at the locus. However, the majority of loci used in forensic genetics are tetranucleotide repeats (Figure 1.6), which include a 4 bp repeat motif (Goodwin et al. 2011). Figure 1.6 shows the homozygote genotype (a) for TH01 locus with two identical alleles with nine repeats, whereas the heterozygote genotype (b) indicates that the person has two different alleles, one with six and another with nine repeats. The number of repeats varies from person to person in a pseudo-normal distribution with usually one common allele, and increasing and decreasing numbers of repeats moving away from the most common allele. Because of this, it is not uncommon for two individuals to share the same allele at the observed STR locus or even two alleles at a locus, but the least theoretical probability that there are two unrelated people

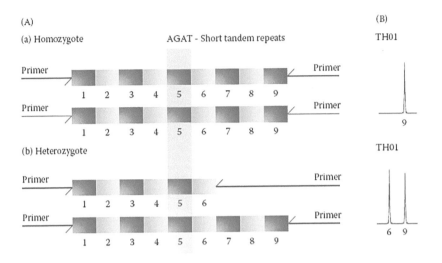

Figure 1.6 (A) Amplification and (B) electropherograms of the STR THO1 in (a) homozygous and (b) heterozygous samples.

with an identical profile at, say, 15 STR loci represented in the PowerPlex 16 system for the Caucasian population is $1/1.83 \times 10^{17}$ (Sprecher et al. 2000).

However, the real value of the application of these markers lies in the simplicity and rapidity of the process and the possibility of simultaneously testing of a large number of STR markers in multiplex STR systems, enabling an extremely high degree of individualization in identifying biological evidence. In addition to its wide application in forensic DNA analysis, STRs have become very attractive as a subject of genetic research from a medical point of view, because it was shown that the trinucleotide STR loci, in the case of their hyperexpansion, are associated with certain genetic disorders (Read and Donnai 2011).

Tetranucleotide STR loci, that is, those STR markers that have four bases in the repeat sequence are the best studied and most frequently used in the analysis of individual and population diversity. Also, several trinucleotide and pentanucleotide systems have found practical applications. The tetra and penta loci are often included in commercial multiplex analysis systems. Such systems provide the results with high index of exclusion.

There are several types of STRs (Urquhart et al. 1994):

1. Simple consisting—one repeating sequence (*D5S818, D13S317, D16S539, TPOX* [Figure 1.7a], *CSF1PO*, etc.)
2. Simple with nonconsensus alleles—one type of repeating sequence (*TH01* [Figure 1.7b], *D18S51, D7S820*, etc.)
3. Compound consisting—two or more different repeat sequences (*GABRB15*)
4. Compound with nonconsensus alleles—two or more different repeat sequences (*D3S1358, D8S1179, FGA, vWA* [Figure 1.7c])
5. Complex repeats—more repetitive sequences with the presence of DNA insertions (*D21S11*, etc.) (Figure 1.7d)
6. Hypervariable repeats (*SE33*)

(a) TPOX
      6      $[AATG]_6$
      7      $[AATG]_7$
      8      $[AATG]_8$
      9      $[AATG]_9$

(b) THO1
      8      $[AATG]_8$
      8.3 $[AATG]_5 ATG[AATG]_3$
      9      $[AATG]_9$
      9.3 $[AATG]_6 ATG[AATG]_3$
      10    $[AATG]_{10}$
      10.3 $[AATG]_6 ATG[AATG]_4$

(c) VWA03
      10   TCTA TCTG TCTA $[TCTG]_4[TCTA]_3$
      11   TCTA$[TCTG]_3[TCTA]_7$
      12   TCTA$[TCTG]_4[TCTA]_7$
      13    $[TCTA]_2[TCTG]_4[TCTA]_3 TCCA[TCTA]_3$

(d) D21S11
      29    $[TCTA]_4 [TCTG]_6 [TCTA]_3$ TA $[TCTA]_3$ TCA $[TCTA]_2$ TCCA TA $[TCTA]_{11}$
      29    $[TCTA]_6 [TCTG]_5 [TCTA]_3$ TA $[TCTA]_3$ TCA $[TCTA]_2$ TCCA TA $[TCTA]_{10}$
      29.2 $[TCTA]_5 [TCTG]_5 [TCTA]_3$ TA $[TCTA]_3$ TCA $[TCTA]_2$ TCCA TA $[TCTA]_{10}$ TA TCTA

Figure 1.7 (a) Simple repeat sequences in loci such as TPOX, CSF1PA, D5S818, D13S317, D16S539, Penta D, and Penta E. (b) Simple repeat sequences with nonconsensus repeats in loci such as THO1, D18S51, and D7S820. (c) Compound consisting of two or more different repeat sequences with nonconsensus repeats in loci such as VWA03, D3S1358, D8S1179, and FGA. (d) Complex repeats consisting of two or more different repeat sequences with nonconsensus repeats and DNA insertions found in loci such as D21S11. (From Van Kirk, M.E., McCary, A., and Podini, D., 2009, *Proceedings of the American Academy of Forensic Sciences, Annual Scientific Meeting*, February 16–21, Denver, CO, Abstract A134, p. 118. With permission.)

1.2.8 Analysis of Sex Chromosomes

Analysis of the sex chromosomes is important in the determination of gender and instantly excludes 50% of the population. The human gene usually analyzed to determine gender is the Amelogenin (AMEL) locus. After its amplification using PCR, DNA fragments of different lengths can be generated. The sequence on the *X chromosome* is shorter by 6 bp compared to the allele on the *Y chromosome* (male sex). The results of gender determination using Amelogenin are presented in Figure 1.8 (Primorac and Marjanović 2008).

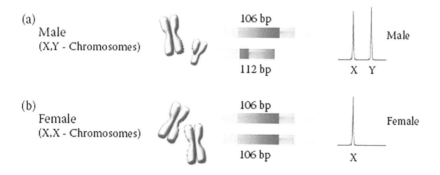

Figure 1.8 Schematic drawing of (a) male and (b) female Amelogenin locus at the chromosome, amplicon, and electropherogram levels.

1.2.8.1 Y Chromosome DNA Testing

According to the criteria of Denver convention, the human *Y chromosome* belongs to group G chromosomes, that is, the category of the shortest chromosomes in human genome, in which the 21st and 22nd chromosomes also belong. The Y chromosome contains about 50 million bp, representing about 1.8% of the entire human genome. Y chromosome can provide important information when it comes to determining the lineages of a specific man. This is possible because the Y chromosome contains highly polymorphic regions (Jobling et al. 1999, 2004). The human Y chromosome is present in normal males in a single copy and is passed from father to son; 95% of the chromosome is not subject to recombination and is referred to as NRY. Only 5% of the chromosome can potentially recombine with the X chromosome, and this region is called the pseudo-autosomal region of the X and Y chromosomes. Forensic analysis of the Y chromosome can play an important role in rape cases, particularly those involving more than one male (Kayser and Sajantila 2000; Prinz and Sansone 2001). These markers can also be useful in cases of paternity testing of male children and in the process of identification, when relatives only from the father's side are present. Additionally, the Y chromosome is increasingly used in determining the migration routes of some people in the past, because the Y chromosome does not undergo recombination during the generational transfer of genetic material (Semino et al. 2000). In recent years, a considerable number of population studies have been conducted using the NRY of the Y chromosome (Y Chromosome Consortium 2002), both worldwide and in Europe. Besides offering very interesting models and scenarios of human history in this region (Barać et al. 2003; Marjanović et al. 2005; Primorac et al. 2011), these studies promoted the utilization of *SNP markers,* which are based on substitution, at only 1 bp. SNPs continue to be explored as potential supplements to STR markers already in use but will probably not replace STRs in the near future. These markers will be discussed further in subsequent sections of this chapter. Furthermore, analysis of Y-STR loci plays an important role in rare but important cases where there is a lack of Amelogenin gene in man (Prinz and Sansone 2001).

From the standpoint of jurisprudence, it is important to emphasize that the identification and possible matching of Y-STR DNA profiles from two evidence samples, regardless of how many molecular markers were analyzed, does not mean complete individualization. That is, the Y chromosome is inherited by the male line—from father to son—so every male related through "father parental line" will share the same profile. Markers found on the Y chromosome exhibit less diversity than those on other chromosomes because 95% of this chromosome is not subject to recombination. Consequently, STR diversity on Y chromosomes results solely from mutation. Therefore, the most common interpretation of the results in court, in a case when the Y-STR profile of two samples match, is that this person cannot be excluded as a potential biological source of the analyzed evidence material. Usually, the DNA profile report contains the relative frequency of occurrence of the observed Y-STR profile (i.e., haplotype) in a specific, regional, or global population. Statistical interpretations of Y chromosome markers are discussed in Chapter 2.

One indicator of seriousness and importance of the Y chromosome analysis in forensics is a presence of several commercial multiplex systems that allow simultaneous investigation of more than 10 STR loci, such as PowerPlex® Y with 12 and the recently released Prototype PowerPlex® Y23 System with 23 loci (Davis et al. 2013), both manufactured by Promega Corporation. The PowerPlex® Y23 System provides all materials necessary to amplify

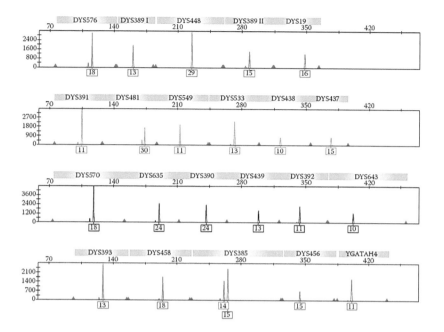

Figure 1.9 Electropherogram of PowerPlex® Y23 (23 Y chromosome SNPs) PCR amplification kit (Promega Corporation). Note: Five-color detection system (four colors for loci and fifth color for sizing ladder, not shown).

Y-STR regions of human genomic DNA. Furthermore, the PowerPlex® Y23 System allows coamplification and four-color fluorescent detection of 23 loci (Figure 1.9). The AmpFLSTR®Yfiler PCR Amplification Kit, manufactured by Life Technologies (Applied Biosystems), allows typing of 17 Y-STR loci, and the Investigator Argus Y-12 QS Kit (made by Qiagen) allows typing of 12 Y-STR loci (only available in Europe).

1.2.8.2 X Chromosome DNA Testing

In recent years, more importance has been given to the application of the human X chromosome, both in forensic and population genetics (Szibor et al. 2003). Unlike the Y chromosome, such an approach can be effective in cases of paternity testing of female offspring in conditions where the potential father is not present and results can be obtained by testing his relatives. In addition, the results can be achieved in the analysis of motherhood and most effectively in the analysis of mother–son relationships, which really is rare, but is still carried out for different purposes.

Application of this marker is still relatively limited, and it is mainly based on the combined use of X-related markers with common autosomal loci (Pereira et al. 2007). One reason for this limitation was the insufficient number of molecular markers examined on the X chromosome, as well as limited population data and calculation of forensic statistics. An increasing number of research groups are looking into this problem, so it is logical to predict that the application of this intriguing chromosome will soon become much more intense. However, it has been shown that X STRs can distinguish pedigrees that are otherwise indistinguishable using only unlinked autosomal markers (Pinto et al. 2011). (See Chapter 2 for comments on forensic calculations using the X chromosome markers and Chapter 5 for a review of X chromosome markers.)

1.2.9 Mitochondrial DNA

mtDNA—or, as some call it, "cytoplasmic chromosome"—is also a form of DNA that is analyzed and used in forensics, and makes up approximately 1% of the total cellular DNA. The mtDNA will be discussed in further detail in the next section, but here we will briefly point to some of its most important characteristics. An mtDNA molecule is 16,569 bp long, circular in shape, and does not contain intron sequences. Two highly polymorphic noncoding regions are especially important for forensic analysis: HV1 and HV2. Unlike nuclear DNA, mitochondria and their DNA come from the cytoplasm of an oocyte that contributed to the development of zygote and therefore are of maternal origin. Therefore, mtDNA is inherited from the mother and indicates the female ancestors of an individual. In contrast to nuclear DNA, mtDNA may be present in a few thousand copies depending on the energy requirements of the tissue. Because mtDNA molecules replicate independently of each other, unlike nuclear chromosomes in which the replication of individual chromosomes occurs first, while pairing and recombination of genes happen in meiosis 1, there is no mechanism for recombination of mtDNA. Mutations that can lead to changes in the sequence of base pairs are the only source of variability in mtDNA. Because mtDNA is inherited from the mother, a person cannot be a heterozygote, and this fact is useful in determining maternal lineages within families and populations. mtDNA is mainly used in forensic cases where it is necessary to analyze the evidence that does not contain enough nuclear DNA (Holland et al. 1993; Coble et al. 2004). mtDNA is also found in tissues that do not contain nuclei, such as a strand of hair. However, one of the biggest problems in working with mtDNA from hair is the presence of two or more subpopulations of mtDNA in one individual (heteroplasmy). Point heteroplasmy (PHP) is observed in approximately 6% of blood and saliva samples and more frequently in hair and metabolically active tissues such as muscle (Irwin et al. 2009). Heteroplasmy is most likely the result of frequent errors or mutations in mtDNA replication, which are rarer in nuclear DNA. The reason may be that the mtDNA molecules replicate independently of each other and are not strictly related to meiotic or mitotic cell division, which has proofreading functions. Since each cell contains a population of mtDNA molecules, a single cell may contain some molecules that have a particular mtDNA mutation that others do not have. It is possible that this phenomenon is responsible for the expression of different diseases that are inherited through the mitochondria. However, it is also possible that the number of mutated mtDNA molecules changes during the segregation in cell division, while the cells are multiplying and the number of mitochondria increases. Heteroplasmy is important for forensic purposes because it can help in the forensic examination of identity, but it can also make the analysis very complex. Still, it is obvious that heteroplasmy represents a new level of variability, which in most cases can increase the accuracy of mtDNA testing. Additional information on this issue may be found in an excellent paper by Holland and Parsons (1999). mtDNA has an important role in identifying human remains, especially skeletal, as well as in cases of decomposed bodies. One of the most famous cases solved by using this technology was the identification of Czar Nicholas II, when it was confirmed that he had the same heteroplasmy as the remains of his brother, Georgij Romanov, grand duke of Russia (Ivanov et al. 1996). Analysis of mtDNA is nowadays performed using the common method of sequencing, and the obtained results are compared with the so-called revised Anderson sequence (International Society for Forensic Haemogenetics; Andrews et al. 1999). In recent years, *sequence-specific oligonucleotide probe* (SSOP) analysis, in which the amplified DNA hybridizes with the existing probes previously bound to the nylon or another membrane, has been applied to mtDNA testing. Previously, it had been used for DQA1 and polymarker typing (see later). Using this analysis method, the time needed to examine a large number of specimens is significantly shortened (Gabriel et al. 2001); however, the level of individualization is lower than that generated by DNA sequencing, because a limited number of polymorphic sites are surveyed (Škaro et al. 2011). Given the large number of copies of mtDNA in the cell, the sample can easily be contaminated with foreign mtDNA if handled recklessly.

1.2.10 RNA Profiling

RNA differs from DNA in several respects: it has the sugar ribose in place of deoxyribose, it has the base uracil (U) instead of thymine (T), and it usually occurs in a single-stranded form (Tamarin 2002). Messenger RNA (mRNA) is a large family of RNA molecules that convey genetic information from DNA to ribosome. During transcription, the DNA serves as a template and an enzyme called RNA polymerase II catalyzes the formation of a pre-mRNA molecule, which is then by a process called splicing processed to form mature mRNA. Ribosomes link amino acids together in the order specified by mRNA and generate new proteins. Eukaryotes have segments of DNA within genes (introns) that are transcribed into mRNA but never translated into protein. The segments of the gene between introns that are transcribed and translated and hence exported to the cytoplasm and expressed to proteins are called exons. The normal mechanism by which introns are excised from unprocessed RNA and exons spliced together to form a mature mRNA is dependent on particular nucleotide sequences located at the intron–exon (acceptor site) and exon–intron (donor site). It has been known for years that either improper splicing or inadequate mRNA transport can cause cancer or numerous diseases (Kaida et al. 2012; Stover et al. 1994; Primorac et al. 1994, 1999; Johnson et al. 2000). However, alternative splicing can play a role even before life and after death (Kelemen et al. 2013). Identification of the tissue of origin for biological stains continues to be pursued using RNA profiling with reverse transcription (RT) and point PCR and real-time PCR (Brettell et al. 2011). Recently, the European DNA Profiling Group (EDNAP) performed a collaborative exercise involving RNA/DNA coextraction and showed the potential use of an mRNA-based system for the identification of saliva and semen in forensic casework that is compatible with current DNA analysis methodologies (Haas et al. 2013).

1.2.11 Application of New Molecular Markers

Forensic genetics is an extremely dynamic scientific discipline, and one of its basic features is certainly the almost daily evolution in terms of discoveries of new procedures, molecular markers, or improvement of existing systems so as to enhance their utility value. One of the newest approaches that is slowly finding its application in forensic DNA analysis is the use of SNP *molecular markers* (Figure 1.10). These are the most abundant forms of DNA polymorphisms that occur within the human genome. In recent years, especially with the discovery and use of new technologies, these features of nucleic acids have achieved full recognition not only in population-genetic research, but also in medical diagnostics and testing of identity. The mainstream of SNP usage in forensic purposes could be recognized in the analysis of highly degraded DNA and novel phenotyping approach.

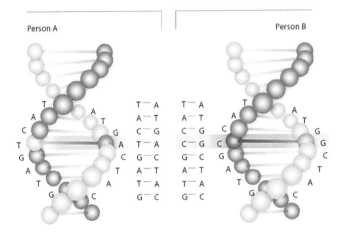

Figure 1.10 Example of DNA sequence variation between two different individuals A and B.

As already noted, this polymorphism is based primarily on the substitution of only one base in a standard order of observed sequences. Although less polymorphic compared to the now widely applied STR markers, their presence in almost every 1000 bp makes them extremely informative. One of the primary directions for development of testing procedures of these markers is directed to an additional analysis of mtDNA (Parsons and Coble 2001). In fact, as already mentioned in previous sections, mtDNA has three main features that give it the status of an extremely favorable molecule for use in DNA identification: (1) successful isolation from very degraded samples, (2) significant degree of informativeness when the reference samples are only those from the maternal side, and (3) application, when the possibility of using nuclear molecular markers does not exist. However, recently, the Spanish and Portuguese Speaking Working Group of the International Society for Forensic Genetics (GHEP-ISFG) during collaborative exercise in order to monitor the current practice of mtDNA reporting, noticed a tenfold range of reported likelihood ratio (LR) values, mainly due to the selection of different reference datasets in EDNAP Forensic mtDNA Population Database (EMPOP) but also due to different applied formulae. Prieto et al. (2013) suggested that more standardization and harmonization of mtDNA reporting is needed. On the other hand, its application can be limited because of the extremely low discriminatory power. The analysis of a large number of SNP markers positioned not only on the hypervariable region but also in the constant region can increase the discriminatory power, and therefore, may give a whole new dimension to the application of mtDNA in forensic genetics.

Also, one of the new interesting approaches in forensic genetics, which is still in its preliminary stages, is using InDels. These genetic markers are mostly considered short amplicons, and their usage could be suited for degraded DNA. Another relatively new approach in forensic DNA testing already provides extremely useful and promising results. The focus is not on the application of some new molecular markers, but rather on a different approach to the analysis of already existing ones. The so-called miniSTR molecular markers represent the modified version of existing standard STR DNA sequences, which are primarily based on moving the forward and reverse PCR primers closer to the STR polymorphic region. In this way the total molecular mass, that is, the size of these markers, is reduced so there is a higher likelihood for the successful amplification of samples in cases when analyzing highly degraded DNA. These markers have already been used in the analysis and identification of victims of terrorist attacks on the World Trade Center in New York. The current research objective is to develop and optimize a miniSTR multiplex system that will enable the simultaneous analysis of a large number of these markers, which has yielded the MiniFiler kit (Life Technologies) (Coble and Butler 2005).

A final area of research is the identification of SNPs that can be used as *ancestry informative markers* (AIMs) and *phenotype informative markers* (PIMs). AIMs provide information about the geographic population of origin, such as Europeans, Africans, and East Asians, of a DNA profile that does not match any samples in the DNA databases. This is of less use in ethnically homogeneous populations but is very useful in countries with highly diverse populations. Similarly, phenotypic informative markers provide information about skin, eye, and hair color, as well as some other traits. Our recent study on over 5000 individuals from Europe and China revealed very strong association of some immunoglobulin G (IgG) glycans with age, and currently, we are validating IgG glycosylation analysis for possible forensic applications (Primorac et al. 2013). At present, these markers and therefore multiplexes are still the subject of research. (See Chapter 16 for information on these markers.)

1.3 Potential Biological Sources of DNA

Forensic laboratories receive different types of biological evidence for testing. Evidence that can be tested using some of the methods of DNA analysis (with the exception of the mtDNA) is limited to those containing cells with a nucleus. In this sense, it is possible to successfully isolate and analyze DNA from the following biological sources:

1. Whole blood and blood cells
2. Semen and sperm cells
3. Tissues and organs
4. Bones and teeth
5. Hair roots and dandruff
6. Saliva, urine, feces, and other bodily secretions
7. Epithelial cells found on clothes

Biological evidence that lacks nuclei such as sweat (if there are no epithelial cells), tears, or hair shafts cannot be tested using standard DNA analysis. DNA can be extracted even from materials such as gastric juice and feces. However, it is sometimes difficult to obtain a sufficient amount of DNA from these sources. It should be also noted that, although good results of DNA analysis can be obtained from the listed evidence, in many cases the quality and/or quantity of the sample turn out to be unsuitable for DNA analysis. The success of the detection of a DNA profile from a certain item of evidence depends primarily on three basic conditions:

1. Amount of sample. Methods of DNA analysis, especially PCR are very sensitive, but still have limitations.

2. The level of degradation of DNA. If an item of evidence, even a large blood stain, is exposed to harsh outdoor conditions for a prolonged time there can be enough DNA degradation due to environmental insult or bacterial contamination that makes the sample unsuitable for further analysis.

3. Purity of the sample. Sometimes, dirt, grease, some fabric colors, and similar inhibitors of certain phases of DNA analysis (mainly amplification) can seriously affect the performance of DNA analysis.

1.3.1 Basic Models and Steps of Forensic DNA Analysis

Forensic DNA analysis takes place in several successive, interrelated steps, which are defined by clearly specified procedures.

1.3.2 Collecting and Storing Samples

The development of modern DNA methods made it possible to analyze biological evidence with a small quantity of DNA or highly degraded DNA; however, extraordinary precautions must be taken while handling the evidence because of the possibility of contamination. The collection, storage, and transfer of such biological evidence are the initial—but also the most critical—phases of a successful implementation of DNA testing. If the evidence is not properly documented, collected, and stored, it probably will not be accepted in court. The documentation of evidence must be detailed and must follow all guidelines for the use of these materials in such purposes. All collected samples must have a label with the number, date, time, place, and name of the person who collected the sample. If applicable, the case identification number should also be recorded on the evidence. A detailed description of how to collect and preserve evidence for DNA testing can be found in the work of Lee and colleagues (1998). Thus, we will now mention only a few basic rules that must be followed when collecting biological evidence intended for DNA analysis (Primorac and Marjanović 2008):

- Any biological evidence found in liquid or moist condition must first be dried, then packed, and transported to the place of its analysis.
- Biological evidence should never be permanently packed in plastic (PVC) packaging. This form of packaging may be used only for short-term transport to the place where it will be dried. Such transport must be quick to avoid accelerating the process of (irreversible) degradation (progressive destruction) of the evidence.
- In the process of collecting biological evidence, it is of the utmost importance to use sterile (latex, powder-free) gloves. Gloves must be changed when collecting more than one item of evidence, following the principle: one pair of gloves—one biological evidence! This way, the possible contamination of collected biological material will be avoided.
- During the collection of biological evidence, it is recommended (implied as required) to wear face masks covering the mouth and nose. In some cases, it is necessary to wear a jumpsuit with a hood to prevent any potential contamination

of biological evidence (e.g., by the hair), but also for the personal safety of crime scene technicians and investigators from a possible source of infection.

- If one does not have a mask (for objective or subjective reasons), possible contamination of the collected evidence can be well avoided by eliminating the talking and coughing, and preventing other nonessential personnel from entering the scene during the sample collection.
- When collecting a larger number of biological evidence samples from a single location, care must be taken to process each item of evidence individually, that is, each of them should be collected using new, sterile material and then packaged separately.
- When sending biological evidence to the laboratory, it is recommended to also provide, if possible, a sample of the adjacent material (or part of the material) from which the evidence was collected (clothes, smaller objects, etc.) to provide substrate controls. Substrate controls aid in the interpretation of results as they can verify that the DNA profile came from the stain and not the substrate.
- Packaging in which collected evidence is placed must be clearly and legibly marked with understandable labels that are more extensively described and recorded in the accompanying transmittal forms. Labels on the packaging and labels in the accompanying communication must match!
- When collecting samples it is obligatory to use a sample collection kit.
- The collection, packaging, and storage of these and other biological evidence will be further discussed in Chapter 9.

1.3.3 Determination of Biological Evidence

Forensic biological evidences may appear in different forms and states. Nearly every evidence sample has to be seen as an individual phenomenon and carefully examined before it is submitted for DNA analysis. There are a number of different methods for the isolation of DNA from a wide range of biological evidences. Most of these procedures are adapted and optimized for a particular type of evidence, for example, blood stain, semen stain, saliva from a cigarette butt or chewing gum, and epithelial cells collected from the clothing.

In order to select an appropriate procedure for DNA isolation, it is necessary to assess the type of biological evidence in question.

1.3.3.1 Blood

Blood is a liquid tissue composed of watery plasma and cells immersed in it: red blood cells (erythrocytes), white blood cells (leukocytes), and platelets (thrombocytes). From the standpoint of DNA analysis, red blood cells have no value because they do not contain DNA. However, they are extremely important in testing for the presence of hemoglobin, which is the basis of most tests for the detection of blood as biological evidence.

The fastest method—but not the most reliable one considering the high possibility of false-positive tests—is the detection of blood with specially designed strips that detect the presence of hemoglobin, that is, iron in hemoglobin. This method is commonly used only as a preliminary test because of its nonspecificity (Fisher and Fisher 2012).

One of the most widespread and—thanks to a great number of *CSI*-type movies and TV series—the most popular method is the luminol (3-aminophthalhydrazide) test. This test allows the rapid investigation of large areas and the possibility of detection of blood that was diluted up to 10 million times (Saferstein 2001).

It must be borne in mind that in certain cases a false-positive result can occur. Specifically, hydrogen peroxide can decompose in reaction with other chemicals that can be present, for example, in potato. Hence, modern-day tests are increasingly based on the specific reactions that not only confirm the presence of blood, but also successfully differentiate its human origin. These tests can be generally divided into two basic categories: diffusion reactions and electrophoretic methods. It is important to remember that the preceding chemical reagents (Leucocrystal violet and Leucomalachite green) do not allow for DNA typing when blood is found in small amounts (Geberth 2006).

1.3.3.2 Semen

Semen is a complex gelatinous mixture produced in male sexual organs and is ejaculated as a result of sexual stimulation. At least four male urogenital glands produce seminal fluid: seminal vesicle glands, the prostate, the epididymis, and bulbourethral glands. Semen consists of male sex cells (spermatozoa), amino acids, sugars, salts, ions, and other components produced by sexually mature males. The volume of ejaculate varies from 2 mL to 6 mL and typically contains between 100 million and 150 million spermatozoa/mL (Houck and Seigel 2006). Sperm cells are an interesting biological structure, approximately 55 pm in length, with a head containing DNA and a mobile tail that enables movement.

Semen contains an acid phosphatase, a common enzyme in nature that occurs at a very high level in semen. This feature is a basis of a number of commercial kits for the detection of semen in biological evidence. However, false-positive reactions are possible since acid phosphatase is not exclusive for semen, but can be found in some amount even in vaginal secretions.

Semen can be visualized by special lamps (wavelength of 450 nm) regardless of whether the stain is on a light or dark surface. This method is performed in the dark, and its major advantage is that it allows quick examination of a relatively large area.

Confirmatory tests for semen are based either on microscopic visualization of sperm or on the detection of prostate-specific antigen (PSA) or p30 (Gunn 2006).

From the standpoint of DNA analysis, the most conclusive and the only usable component of semen for generating DNA profile are certainly spermatozoa. Previous experience has shown that the presence of semen does not always mean that the DNA profile will be obtained from the examined evidence. In particular, the lack of spermatozoa (azoospermia) or their low concentration (oligospermia) can cause the absence or insufficient presence of DNA in the analyzed semen sample.

The most common test for sperm identification is the Christmas Tree Stain (CTS) method that stains the tip of the sperm's head pink, the bottom of the head dark red, the middle portion blue, and the tail yellowish-green.

1.3.3.3 Vaginal Body Fluid

Vaginal body fluid can be detected and separated from various other body fluids by microbial (bacteria of the female genital tract) signature detection using a multiplex real-time PCR assay (Giampaoli et al. 2012).

1.3.3.4 Saliva

Saliva is commonly found biological evidence. It can be recovered from cigarette butts, chewing gums, stamps, envelopes, bottles, drinking glasses, etc. (Abaz et al. 2002). Detection of saliva is based on the presence of the enzyme amylase. The problem is that amylase can occur in many other body fluids, so the test is not as nearly specific as for blood and semen. Saliva contains large numbers of epithelial cells from the buccal mucosa and therefore is easy to type for DNA analysis. Today, the most widely used commercial test for detection of saliva is the Phadebas amylase test.

1.3.3.5 Urine

Urine is the waste fluid produced by the kidneys through excretion. It is presumptively tested based on the presence of urea (using the enzyme urease) or creatinine (using picric acid). Urine has few epithelial cells, so DNA analysis has to be optimized for potentially minute amounts of sample evidence.

1.3.3.6 Feces

Feces is a waste product from the digestive tract expelled through the anus or cloaca during defecation. However, identification of feces is very important in a variety of crime investigations. Recently, a novel fecal identification method by detection of the gene sequences specific to fecal bacteria (*Bacteroides uniformis*, *Bacteroides vulgatus*, and *Bacteroides thetaiotaomicron*) in various body (feces, blood, saliva, semen, urine, vaginal fluids, and skin surfaces) and forensic (anal adhesions) specimens have been developed (Nakanishi et al. 2013).

1.4 DNA Isolation

In cells, DNA is not in a pure form, but is associated with many other molecules such as proteins, lipids, and many other contaminating substances. The first step in the isolation of DNA requires that the cell membranes are broken down and the cellular content with molecules of carbohydrates, proteins, lipids, etc., is released. Given that these molecules can interfere with PCR, they must be removed using special techniques. After the cell lysis, the DNA is initially released from the nucleus, and in the next step, large quantities of protein are removed. Today, several methods of DNA extraction are routinely implemented, but before giving a brief listing of these methods, it is essential to point out several important facts.

A large number of cases coming to the laboratory for analysis contain minimal amounts of DNA, and any unnecessary manipulation can have a direct impact on the quality and quantity of DNA. All persons involved in the process of DNA analysis must be made aware of the possibility of contamination of the existing DNA with some other DNA. Some of the substances used in the isolation of DNA are potentially harmful to the DNA if not removed completely and as soon as possible. Figure 1.11 shows the schematic representation of some DNA extraction methods applied in forensic genetics.

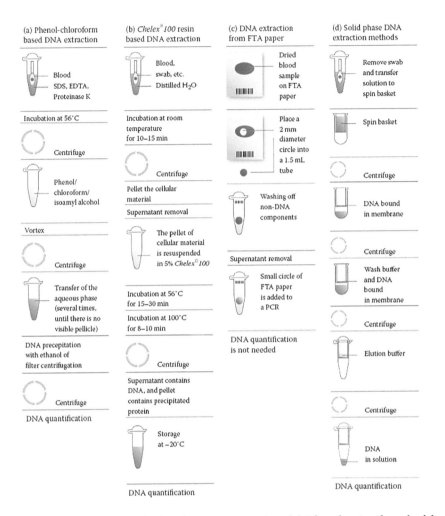

(a) Phenol-chloroform based DNA extraction

Blood
SDS, EDTA, Proteinase K

Incubation at 56°C

Centrifuge

Phenol/chloroform/isoamyl alcohol

Vortex

Centrifuge

Transfer of the aqueous phase (several times, until there is no visible pellicle)

DNA precepitation with ethanol of filter centrifugation

Centrifuge

DNA quantification

(b) Chelex®100 resin based DNA extraction

Blood, swab, etc.
Distilled H₂O

Incubation at room temperature for 10–15 min

Centrifuge

Pellet the cellular material

Supernatant removal

The pellet of cellular material is resuspended in 5% Chelex®100

Incubation at 56°C for 15–30 min

Incubation at 100°C for 8–10 min

Centrifuge

Supernatant contains DNA, and pellet contains precipitated protein

Storage at –20°C

DNA quantification

(c) DNA extraction from FTA paper

Dried blood sample on FTA paper

Place a 2 mm diameter circle into a 1.5 mL tube

Washing off non-DNA components

Supernatant removal

Small circle of FTA paper is added to a PCR

DNA quantification is not needed

(d) Solid phase DNA extraction methods

Remove swab and transfer solution to spin basket

Spin basket

Centrifuge

DNA bound in membrane

Centrifuge

Wash buffer and DNA bound in membrane

Centrifuge

Elution buffer

Centrifuge

DNA in solution

DNA quantification

Figure 1.11 Four different methods of DNA extraction: (a) The classic phenol–chloroform method; (b) Chelex®100 resin-based extraction; (c) DNA extraction from FTA® card; (d) solid phase DNA extraction methods.

1.5 DNA Quantification

DNA quantification is a required step in forensic PCR-based testing in the Federal Bureau of Investigation (FBI) DNA standards. The outcome of DNA analysis depends on its quality and quantity. PCR is the most common method used for the analysis of evidence. Since the optimized methods require specific amounts of DNA, it is extremely important to establish methods of quantifying the exact concentration of DNA, but also to determine its origin, that is, whether the DNA found is derived from a human source. Also, one of the key parameters for successful application of PCR is the purity of the sample. Specifically, the samples arriving for processing are very often contaminated with bacterial DNA or contain large amounts of so-called PCR inhibitors. Hence, the PCR method is the most used in the quantification of DNA within forensic DNA analysis, and it will be explained in more detail in the following section.

1.5.1 Quantitative RT-PCR Quantification Technology

The most recent improvement in the quantification of human DNA is the use of quantitative PCR reactions in a real-time (quantitative real-time PCR [QRT-PCR]) instrument. The real-time PCR scan is done on a Cepheid Smart Cycler (Figure 1.12). Currently, the most widely used commercial kits are Quantifiler® Human DNA Quantification Kit and Quantifiler® Y Human Male DNA Quantification Kit (Applied Biosystems 2014), Quantifiler® HP (Human Plus) DNA Quantification Kit, Quantifiler® Trio DNA Quantification Kit (Applied Biosystems 2017), The Investigator® Quantiplex Kit (Qiagen 2018a) and The Investigator® Quantiplex HYres Kit (Qiagen 2018b), and The Plexor® HY System (Promega Corporation 2017).

Both Quantifiler (Applied Biosystems) kits are based on TaqMan (5′ nuclease assay) technology with probes labeled with two different fluorescent dyes (Reporter dye attached on the 5′-end of probe and Quencher dye located on the 3′-end). If hybridization between the probe and DNA target occurs, a reporter dye will be released and start to fluoresce. This reaction will be used for detection of the presence of the DNA target. Also, IPC (internal PCR control) is included within each reaction to verify the reaction setup and/or to point the presence of PCR inhibitors.

The Plexor HY (Promega Corporation) is based on the interaction between two modified nucleotides to achieve quantitative PCR analysis. One of the PCR primers contains a modified nucleotide (iso-dC) linked to a fluorescent label at the 5′-end. The second PCR primer is unlabeled. The reaction mix includes deoxynucleotides and iso-dGTP modified with the quencher, which is incorporated opposite the iso-dC residue in the primer. The incorporation of the quencher-iso-dGTP at this position results in quenching of the fluorescent dye on the complementary strand and a reduction in fluorescence, which allows quantitation during amplification. IPC control is added to each reaction.

Finally, Qiagen Quantiplex kits (Qiagen) are based on the detection of amplification using "Scorpion" primers and fast PCR chemistry. Scorpion primers are bifunctional

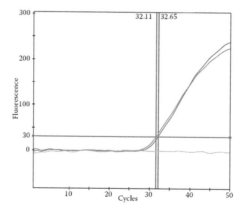

Figure 1.12 The real-time PCR scan done on a Cepheid Smart Cycler at the Department of Forensic Sciences, George Washington University, Washington, DC, shows two samples going through the cycle threshold (CT) between cycles 32 and 33, and a third negative sample showing a flat baseline that never rises to the CT cutoff height. (Used with permission from Department of Forensic Sciences, George Washington University, Washington DC.)

molecules containing a PCR primer covalently linked to a probe. The fluorophore in this probe interacts with a quencher, also incorporated into the probe, which reduces fluorescence. During PCR, when the probe binds to the PCR products, the fluorophore and quencher become separated. This leads to an increase in fluorescence. This kit contains reagents and a DNA polymerase for specific amplification of proprietary regions present on several autosomal chromosomes of the human genome.

As could be seen, different companies developed different methods, but with one main goal: to improve sensitivity and stability of the DNA quantification procedure.

1.6 Polymerase Chain Reaction

The first method that has been used in forensic DNA analysis was the analysis of RFLP. In this method, the double-stranded DNA is cleaved in the presence of a restriction endonuclease, after which DNA fragments are separated using electrophoresis on the agarose gel. However, RFLP as the method of choice within forensic DNA analysis was promptly substituted with new more suitable technology. PCR has literally revolutionized the field of molecular genetics and biology in general. The starting amount of DNA, which for decades has been a limiting factor in research, is now reduced to the existence of only one well-preserved molecule, and the arduous procedures of DNA isolation and purification have become simpler. Suddenly, whole new opportunities opened up before the researchers. There is almost no social activity that the technique has not directly or indirectly touched and changed forever. The main "culprits" for this situation are Kary Mullis and his colleagues from the Department of Human Genetics, Cetus Corporation. In 1985, the journal *Science* published the research study by R. Saiki, K. Mullis and H. Erlich, with the first description of in vitro amplification of specific DNA fragments, catalyzed by DNA polymerase, the enzyme isolated from *Escherichia coli* (Saiki et al. 1985). However, because the *E. coli* enzyme is inherently degraded by heating to denature the DNA, a new enzyme was needed. In 1988, the same team of scientists published the paper in which they introduced heat-stable DNA polymerase isolated from bacterium *Thermus aquaticus* (*Taq*), instead of the previously used *E. coli* polymerase (Saiki et al. 1988). Despite numerous modifications, various applications, and process automation, the basic aspects of this technique, established in 1985 and 1988, have not been significantly changed.

The basic premises of the PCR are as follows—in the native state DNA is a double helix consisting of two antiparallel polynucleotide chains that are interconnected by hydrogen bonds. The nucleotides in one polynucleotide chain are joined by a covalent bond between the sugar and the phosphate group of adjacent nucleotides, and hydrogen bonds that hold two polynucleotide chains together are always formed between complementary bases, that is, A (adenine) always binds to T (thymine) and C (cytosine) with G (guanine). Hydrogen bonds are weak and easily disrupted by heating (denaturation), whereas the covalent bonds persist; the cooling leads to a reestablishment of hydrogen bonds between complementary bases (renaturation).

Based on the preceding explanation, the basic model of PCR reaction consists of three principal steps: denaturation (separation of polynucleotide strands caused by heating to 95°C), hybridization (annealing of artificially synthesized DNA primers

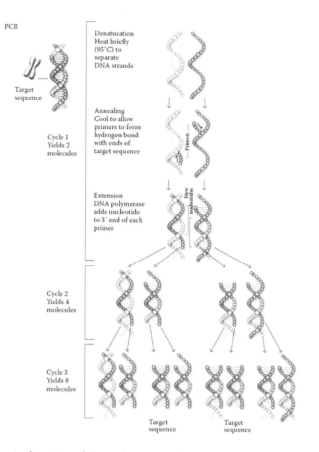

PCR

Target
sequence

Cycle 1
Yields 2
molecules

Cycle 2
Yields 4
molecules

Cycle 3
Yields 8
molecules

Denaturation
Heat briefly
(95°C) to
separate
DNA strands

Annealing
Cool to allow
primers to form
hydrogen bond
with ends of
target sequence

Extension
DNA polymerase
adds nucleotide
to 3′ end of each
primer

Target
sequence

Target
sequence

Figure 1.13 Schematic drawing of the polymerase chain reaction (PCR) process.

that can be fluorescently labeled and flank the DNA fragments to be amplified at 50°C–65°C), and chain elongation (binding of complementary bases on the free sites, in the presence of denatured matrix, DNA primers, and *Taq* polymerase activity at 72°C). Under ideal conditions and 100% amplification efficiency (which certainly do not exist in nature) by cycle 30, approximately a billion copies of the target area on the DNA template have been produced (Butler 2011).

Therefore, PCR is the method by which the fragment of DNA multiplies to produce billions of copies (depending on the number of reaction cycles) that are completely identical to the DNA fragment of interest (Figure 1.13). This technique has several key advantages over RFLP: it requires a smaller amount of DNA, it is faster (results are obtained in a matter of hours), and it does not involve radioactive isotopes (NRC 1996). However, there are two fundamental drawbacks in the use of this method: (1) because of the exceptional sensitivity, there is a possibility of contamination; and (2) many loci analyzed by PCR, especially STR molecular markers, have fewer alleles than minisatellite VNTR loci analyzed by RFLP. However, these two problems can be relatively easily overcome by conducting this method in strictly controlled conditions and using multiplex STR systems that, for example, nowadays allow the simultaneous analysis of up to 15 STR loci.

1.7 PCR Methods

1.7.1 Multiplex STR Systems

Like the RFLP system, the first tests for DNA analysis in forensic medicine were based on the variability of DNA sequences (sequential polymorphism). Researchers soon discovered another class of polymorphic loci with short repetitive blocks (two to nine nucleotide repeats). This class of loci is called STR or *microsatellites*; these are in contrast to systems such as D1S80, which are referred to as long tandem repeats or *minisatellites*. These molecular markers have already been discussed in the previous text, and it has been mentioned that one of their main advantages is the possibility of relatively simple, rapid, and simultaneous processing of more than 10 STR loci.

Almost all systems used in forensic testing have repeat units of 4 bp (tetranucleotide) or 5 bp (pentanucleotide), which occur from 5 to about 50 times, depending on the locus (Hammond et al. 1994; Alford et al. 1994). This type of molecular marker is very short (100–400 bp) and very useful in the analysis of degraded DNA, in contrast to the RFLP fragments whose length can vary from 500 to 12,000 bp. The advantage of these systems is the fact that the occurrence of so-called *stutter* bands is much less pronounced than in loci with one to three nucleotide repeats, making the interpretation of results much easier. The term *stutter* band is used for DNA fragments that are one repeat unit shorter than the expected band. Stutter occurs relatively frequently in tetranucleotide repeat sequences, as an amplicon that is one repeat unit shorter than expected allele. Stutter bands can make the interpretation of mixtures difficult when it is important to determine whether the minor fragment that appears in a mixture is a stutter band or belongs to a different source. As the number of nucleotides in a repeat increases, the formation of stutter bands decreases; thus, the frequency of stutter bands for pentanucleotide repeats in PowerPlex 16 is much lower (Primorac 2001). The United States has adopted 13 STR loci for a national database of convicted offenders (McEwen 1995). This database is called CODIS (Combined DNA Index System). This system was initially limited to convicted rapists and murderers, but was later extended to other crimes (Hares 2011). Thirteen CODIS loci were selected in a manner to coincide with other loci chosen by Forensic Science Services in Great Britain for its national database and organizations such as INTERPOL. Two facts are related with CODIS loci: none of the CODIS markers are linked within exons and no CODIS genotype is associated with known phenotypes (Katsanis and Wagner 2013). Genomic characterization of CODIS markers and phenotypic relevance of genomic regions of CODIS markers are shown in Tables 1.2 and 1.3, respectively (Katsanis and Wagner 2013).

The new STR typing systems have several advantages over the old ones. The first advantage, already mentioned, is that more loci can be amplified simultaneously (multiplex reaction). Other features important in development of STR multiplexes include discrete and distinguishable alleles, amplification of the locus should be robust, a high power of discrimination, an absence of genetic linkage with other loci being analyzed, and low levels of artifact formation during the amplification (Goodwin et al. 2011). Since one primer of each pair has a fluorescent tag, it is possible to differentiate different loci by both size and color after separation by electrophoresis (Buel et al. 1998).

Table 1.2   **Genomic Characterization of CODIS Markers**

	CODIS Marker	Cytogenetic Location	Intragenic or Distance from Nearest Gene	Included in Marshfield Human Genetic Linkage Maps	Number of SNPs (dbSNP Build 132) within 1 kbp
1	D18S51	18q21.33	Intron 1	Included	13
2	FGA	4q28	Intron 3		4
3	D21S11	21q21.1	>100 kb from nearest gene	Removed	7
4	D8S1179	8q24.13	>50 kb from nearest gene	Included	14
5	VWA[a]	12p13.31	Intron 40		27
6	D13S317	13q31.1	>100 kb from nearest gene	Included	10
7	D16S539	16q24.1	~10 kb from nearest gene	Removed	29
8	D7S820	7q21.11	Intron 1	Included	7
9	TH01	11p15.5	Intron 1		8
10	D3S1358	3p21.31	Intron 20		11
11	D5S818	5q23.2	>100 kb from nearest gene	Removed	8
12	CSF1PO	5q33.1	Intron 6		15
13	D2S1338	2q35	~20 kb from nearest gene	Included	11
14	D19S433	19q12	Intron 1	Included	10
15	D1S1656	1q42	Intron 6	Removed	22
16	D12S391[a]	12p13.2	~40 kb from nearest gene	Included	18
17	D2S441	2p14	~30 kb from nearest gene	Removed	13
18	D10S1248	10q26.3	~3 kb from nearest gene	Included	16
19	Penta E	15q26.2	within uncharacterized EST; ~50 kb from nearest gene		19
20	DYS391	Yq11.21	~5 kb from nearest gene		0
21	TPOX	2p25.3	Intron 10		33
22	D22S1045	22q12.3	Intron 4	Included	19
23	SE33	6q14	~30 kb from nearest gene, pseudogene		8
24	Penta D	21q22.3	Intron 4		6

Source: Katsamis, S.H., Duke Institute for Genome Sciences and Policy, Duke University, Durham, NC. With permission.

Note: Markers are shown in their relative rank according to Hares (2011).

[a] VWA and D12S391 are colocated on 12p13 within 6 Mbp.

Table 1.3 **Reported Phenotypic Relevance of Genomic Regions of CODIS Markers**

	CODIS Marker	Gene Name	Disorder(s) Caused by Gene Mutations	Number of Phenotypes Associated within 1 kbp	Predicted DNA Elements
1	D18S51	*BCL2* (B-cell CLL/lymphoma 2)	Leukemia/lymphoma, B-cell	11	ELAV1 binding site
2	FGA	*FGA* (fibrogen alpha chain)	Congenital afibrinogenemia; hereditary renal amyloidosis; dysfibrinogenemia (alpha type)	17	PABPC1 binding site
3	D21S11	None		1	None
4	D8S1179	None		17	None
5	VWA[a]	*VWF* (von Willebrand factor)	von Willebrand disease	12	ELAV1 binding site
6	D13S317	None		5	None
7	D16S539	None		8	None
8	D7S820	*SEMA3A* (sema domain, immunoglobulin domain, short basic domain, secreted (semaphoring) 3A)		8	CELF1, ELAV1, and PABPC1 binding site
9	TH01	*TH* (tyrosine hydroxylase)	Segawa syndrome, recessive	18	ELAVL1, PABPC 1, and SLBP binding site
10	D3S1358	*LARS2* (leucyl-tRNA synthetase 2, mitochondria)		15	None
11	D5S818	None		5	None
12	CSF1PO	*CSF1R* (colony stimulating factor 1 receptor)	Predisposition to myeloid malignancy	15	eGFP-GATA2 transcription factor; PABPC 1 binding site
13	D2S1338	None		9	None
14	D19S433	*C19orf2* (uncharacterized gene)		7	DNase I hypersensitivity site; SLBP binding site
15	D1S1656	*CAPN9* (calpain 9)		10	PABPC 1 binding site
16	D12S391[a]	None		6	None
17	D2S441	None		6	None
18	D10S1248	None		6	DNase I hypersensitivity site
19	Penta E	EST: BG210743 (uncharacterized EST)		8	None

(Continued)

Table 1.3 (Continued)   Reported Phenotypic Relevance of Genomic Regions of CODIS Markers

	CODIS Marker	Gene Name	Disorder(s) Caused by Gene Mutations	Number of Phenotypes Associated within 1 kbp	Predicted DNA Elements
20	DYS391	None		1	None
21	TPOX	*TPO* (thyroid peroxidase)	Thyroid dyshormonogenesis 2 A	5	PABPC 1 and SLBP binding site
22	D22S1045	*IL2RB* (interleukin 2 receptor, beta)		11	None
23	SE33	None		9	None
24	Penta D	*HSF2BP* (heat shock factor 2-binding protein)		6	PABPC 1 and SLBP binding site

Source: Katsamis, S.H., Duke Institute for Genome Sciences and Policy, Duke University, Durham, NC. With permission.

Note: Markers are shown in their relative rank according to Hares (2011).

[a] VWA and D12S391 are colocated on 12p13 within 6 Mbp.

The most common STR systems that are now used in routine forensic work are manufactured by three companies: Promega, Life Technologies, and Qiagen.

A detailed overview of basic PCR technology, STR-PCR technology, and their applications in forensic investigations can be found in supplementary literature (McEwen 1995; Schanfield 2000; Butler 2010). Here, we will only look at the most recent multiplex STR systems from three different companies.

1.7.2 PowerPlex® Fusion System

The PowerPlex® Fusion System is a 24-locus multiplex for human identification applications including forensic analysis, relationship testing, and research use. This five-color system allows coamplification and fluorescent detection of the 13 core CODIS (US) loci (CSF1PO, FGA, TH01, TPOX, vWA, D3S1358, D5S818, D7S820, D8S1179, D13S317, D16S539, D18S51, and D21S11), the 12 core European Standard Set loci (TH01, vWA, FGA, D21S11, D3S1358, D8S1179, D18S51, D10S1248, D22S1045, D2S441, D1S1656, and D12S391) and Amelogenin for gender determination (Figure 1.14). In addition, the male-specific DYS391 locus is included to identify null Y allele results for Amelogenin. The Penta D and Penta E loci are included to increase discrimination and allow searching of databases that include profiles with these Penta loci. Finally, the D2S1338 and D19S433 loci, which are popular loci included in a number of databases, were incorporated to further increase the power of discrimination. This extended panel of STR markers is intended to satisfy both CODIS and ESS (European Standard Set) recommendations (Promega Corporation 2020).

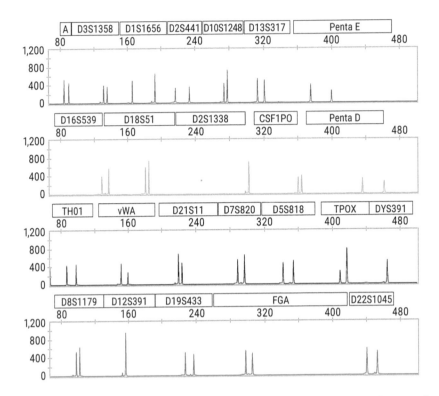

Figure 1.14 Electropherogram of The PowerPlex® Fusion System Promega Corporation. Note: Five-color detection system (five colors for loci and fifth color for sizing ladder, not shown). (From https://worldwide.promega.com/products/genetic-identity/genetic-identity-workflow/str -amplification/powerplex-fusion-system/?catNum=DC2402, accessed July 31, 2019.)

1.7.3 GlobalFiler® PCR Amplification Kit

GlobalFiler® PCR Amplification Kit is the first six-dye fluorescent system STR kit from Applied Biosystems that allows amplification of 24 loci Amelogenin, 21 autosomal STR loci (D3S1358, vWA, D16S539, CSF1PO, TPOX, D8S1179, D21S11, D18S51, D2S441, D19S433 TH01, FGA, D22S1045, D5S818, D13S317, D7S820, SE33, D10S1248, D1S1656, D12S391, D2S1338), one Y-STR (DYS391), and one Y indel. Detection of different loci is enabled using 6-FAM (blue), VIC (green), NED (yellow), TAZ (red) and SID (purple) dyes. The sizing of DNA fragments is done using GeneScan® using 600 LIZ size standard in orange dye spectrum. Amplification can be completed in approximately 80 minutes and full profiles can be obtained with 125 pg input DNA (Figure 1.15).

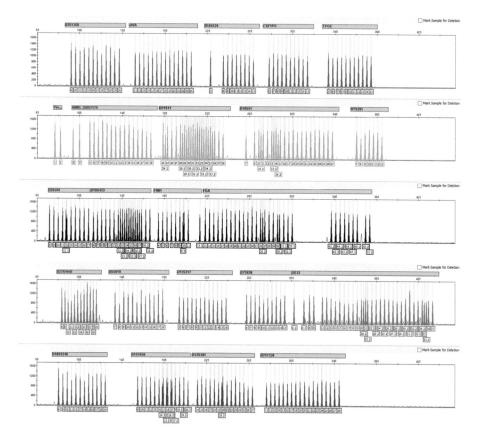

Figure 1.15 Allelic ladder of GlobalFiler® PCR Amplification Kit (Applied Biosystems). Note: Six-color detection system (five colors for loci and fifth color for sizing ladder, not shown). (From https://www.thermofisher.com/order/catalog/product/4476135?SID=srch-srp-4476135, accessed July 31, 2019.)

1.7.4 Investigator 24plex QS Kit

Qiagen, as the relatively new company in the field of the manufacturing of multiplex STR systems, released more than ten different human identification assays between 2011 and 2012. Investigator 24plex QS Kit could be recognized as the most recent approach from Qiagen produced for the forensic genetics community. This six-color system for human identification allows multiplex amplification of the CODIS (Combined DNA Index System) core loci, the ESS (European Standard Set) markers, SE33, DYS391, D2S1338, D19S433, and Amelogenin (Figure 1.16).

1.8 Detection of PCR Products

The next step after amplification is the detection of PCR products. It is conducted on one of the analytical machines, using different software to generate final genetic profiles. Detection, as the final phase of the process of determining the genetic identity of the person or biological evidence, is an automated procedure that is most susceptible to frequent modifications and innovations. These changes essentially follow the trend of progress in computational and fluorescent detection technology and aim for complete automation of the process. Because of the necessity that all laboratories should produce results that can be shared, all testing is done with standardized forensically validated kits. At this time, the shift from gel to capillary electrophoresis has been completed.

Unlike RFLP testing, where produced DNA fragments had significant measurement errors and limited reproducibility, STR-based testing produces measured DNA fragments, and the lengths of the fragments are "measured" by allelic ladders used like molecular rulers. The allele ladder contains the majority of commonly detected allelic variants at a given locus and allows detection and analysis of particular allelic variants. This makes the technology extremely useful in parentage as well as forensic testing. Because only the length of the fragment is at issue, and the DNA migration process of PCR-amplified DNA is quite different than restriction digested genomic DNA, many different detection methods have been used, with the initial technology using radioactivity and silver staining, followed by the development of fluorescent detection and the use of fluorescent-labeled primers. This was further enhanced by the development of charge-coupled detectors and computer software that could simultaneously unscramble multiple colors allowing for single reaction, single-read detection of multiple colors. In most cases, a number of STR loci are used, some of which match in size and position on the gel. To avoid misinterpretation, four—and recently, five—dyes are used to distinguish individual loci. The loci with considerably different molecular weights are labeled in the same color, and different colors are used for the loci whose allele ranges overlap. Since alleles from different loci emit fluorescent light at different wavelengths, genetic analyzers are able to distinguish each allele as they simultaneously pass through the detection system. Thus, after the data are collected and digitally processed, the resulting peaks do not represent individual bases, but rather allelic variants, which are indicators of how many repeats are present in a given fragment. Consequently, the term "sequencing," which is often used in laboratory jargon for this process, is inappropriate for this type of testing, which is normally referred to as "DNA profiling or genotyping."

Although some of the loci used forensically contain either sequence variation in the repeats or a combined pattern of repeat sequence variation and repeat order variation, these are not detected, such that for two individuals with the type D21S11 29, one could be homozygous for the same allele for repeat sequence variation, whereas another individual could have two different repeat sequence alleles. However, to determine this, each D21S11 allele would have to be sequenced. Although this would increase the information content, it is not practical in routine forensic testing. Sequencing of DNA has been a critical part of the evolution of molecular genetics tools. Although it has limited application forensically, it will become more important as sequencing technology improves and becomes less expensive.

Figure 1.16 Electropherogram of Investigator 24plex QS Kit (Qiagen). Note: Six-color detection system (five colors for loci and sixth color for sizing ladder, not shown). (From Qiagen, *Investigator 24plex QS Kit Handbook, 94*, 2021.)

1.8.1 Analytical Thresholds and Sensitivity for Forensic DNA Analysis

Heterozygous STR loci for forensic casework and database samples produce two peaks in the profile. In an ideal situation, the ratio between these two peaks will be 1:1 in terms of peak height and area (Goodwin et al. 2011). Unfortunately, it does not happen commonly and variations in peak height are common. However, in good-quality DNA extracts, the smaller peak is on average approximately 90% the size of the larger peak (Gill et al. 1997). During forensic DNA analysis when the quantity of analyzed DNA is too low, it may be difficult to distinguish true low-level peaks from technical artifacts, including noise. While working with low amounts of DNA, degraded DNA, or DNA mixtures, most forensic scientists experienced the "stochastic effect during PCR amplification of low-level DNA" because of a random primer banding in the low copy number milieu. In that case during the detection process, several outcomes are possible (Figure 1.17): *allele drop-in, allele drop-out, increased stutter production, locus drop-out,* and *heterozygote peak imbalance* (Butler 2011; National Forensic Science Technology Center 2013). Allele drop-in (additional allele is observed) could be of unknown origin or from a variety of intralaboratory sources including consumable items (Gill 2010) and personnel, whereas a single allelic drop-out produces false homozygous profiles, because of stochastic effects. On the other hand, stutter peaks are formed because of strand slippage during the extension of nascent DNA strand during PCR amplification (Schlotterer and Tautz 1992). Stutter peaks usually appear one repeat unit before true alleles but can occur one repeat unit after true alleles, and most of the stutters (di- and trinucleotide repeats are more prone to stutter than are tetra- and pentanucleotide repeats) are less than 15% of the main peak (Goodwin et al. 2011). Heterozygote peak imbalance also happens because of the stochastic PCR effect, where one of the alleles is amplified preferentially by chance. In order to distinguish a real peak from noise, it is necessary to establish a minimum distinguishable signal that may be considered a relative fluorescent unit (RFU) or analytical threshold (AT) (Bregu 2013). The RFU peak height mainly depends on the amount of DNA being analyzed. Recently, the Scientific Working Group on DNA Analysis Methods (SWGDAM) published Interpretation Guidelines for Autosomal STR Typing by Forensic DNA Testing Laboratories, where they underline that in general, nonallelic data such as stutter, non-template-dependent nucleotide addition, disassociated dye, and incomplete spectral separation are reproducible, whereas spikes and raised baselines are nonreproducible (Scientific Working Group on DNA Analysis Methods 2010). Bregu and his group published an excellent paper stating that determination of ATs should be widely implemented. However, they observed that when a substantial mass of DNA (>1 ng) was amplified the baseline noise

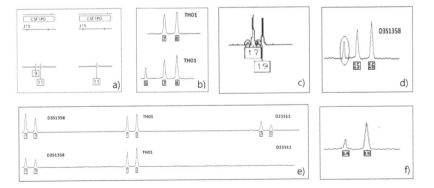

Figure 1.17 Stochastic effect during PCR amplification of low-level DNA: (a) allele drop-out; (b) allele drop-in; (c) regular stutter; (d) enlarged stutter; (e) locus dropout; (f) peak imbalance.

increased, and that the number and intensity of noise peaks increased with increasing injection times (Bregu et al. 2013).

1.8.2 Sequencing

DNA sequencing is the process of determining the order of bases within the DNA strand or fragment, and it represents one of the basic methods underlying molecular genetics. Knowing the DNA sequence is a precondition for any manipulation of the target segment of the hereditary material. For example, a computer search of sequences of all known restriction (endonuclease) sites can result in the creation of complete and accurate restriction maps.

Simplicity, rapidity, and a wide range of applicability promoted the automated sequencing analysis currently in use. The initial requirement for a successful automatic sequencing is the simultaneous separation of fragments using electrophoresis and the detection of labeled fragments. In automated detection, fluorescent DNA labeling has the advantage over other detection methods. However, no matter what generation of automated sequencing is used, the principles of each technique are based on the Sanger method.

The current generation of fluorescence techniques is based on the use of fluorescent markers linked to the chain-terminator ddNTPs, which implies a different labeling of each of the bases (C, G, T, A). Each of these four ddNTPs fluoresces in a different spectrum (Figure 1.18). The labeled ddNTP is incorporated into the DNA molecule by using *Taq* polymerase and has two functions: to stop further synthesis of the DNA chain and to bind fluorophore to the end of the molecules. The main advantage of this method is that the reaction does not have to be conducted in four separate tubes. Even with this method, the final detection occurs during electrophoresis, when labeled fragments pass through the detector's reading region. Using this method, 3–500 bp can be generated per lane in either a lane of an acrylamide gel or in a capillary. What is important to note is the fact that this approach to testing of hereditary material in forensic DNA analysis is primarily applied in the analysis of mtDNA and its hypervariable regions.

Automated sequencing is designed to sequence large pieces of DNA such as at the control regions of mtDNA where information is distributed throughout the region or if you are looking for specific substitutions (SNPs). As the number of AIMs and PIMs are developed and validated as forensic investigative tools for the majority of DNA profiles that do not match DNA database files, the need for rapid SNP-based testing have been increased, and commercial kits are being developed. Minisequencing multiplexes have the ability to amplify multiple regions in a single reaction, and then with a second reaction detect specific alleles (SNPs)

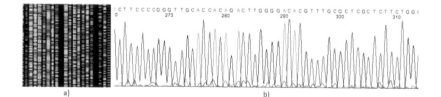

Figure 1.18 Results of autosomal sequencing. Each base on the gel is dyed differently. Information shown in the picture of the gel represent initial, raw data (a), which are converted to actual sequences by software processing (b). (From htp://www.insidestandards.com/dna -sequencing-standard-provides-measuring-stck/; https://seqcore.brcf.med.umich.edu/sites/defau lt/fles/html/interpret.html, accessed August 27, 2018.)

at these regions. The system that appears to be the most widely used and is compatible with capillary electrophoresis DNA analyzers currently in use is Life Technologies's SNAPSHOT™ single-base extension assay or "minisequencing" system. The system consists of multiplexes of unlabeled primers that amplify one or more loci. The amplicon is cleaned with ExoSap (a mixture of exonuclease and shrimp alkaline phosphatase that simultaneously destroys the old primers and spent nucleotides), and then added to a new reaction mixture that contains unlabeled primers for the targeted SNP and the four ddNTPs each labeled with a different fluorophore. The material is cleaned up with ExoSAP and detected on a standard capillary electrophoresis DNA analyzer. The SNP primers have been designed so that different loci have different-sized fragments, and each allele has a different color.

1.9 Massive Parallel Sequencing

The development of *next-generation sequencing* (NGS) technology (also called massively parallel sequencing [MPS]) has revolutionized genomics, which has in turn lead to the possibility of performing large-scale sequencing projects. By increasing the number of DNA fragments that can be analyzed independently of each other, enabled a high level of information flow, also called *high-throughput* technology. MPS is a high-throughput approach used for DNA sequencing of one million to several billion short reads per instrument run. The system uses parallelized platforms for sequencing of clonally amplified DNA templates. Commercially available NGS technologies differ in methods, such as sequencing by ligation, sequencing by hybridization, sequencing by synthesis, single-molecule DNA sequencing, nanopore sequencing, and multiplex polony sequencing. The greatest advantage of NGS systems is the possibility to give results in cases of highly degraded DNA or just traces of DNA that was obtained from a crime scene. MPS methods can be divided into short-read and long-read NGS methods, which depend on the instrument and chemistry used. The read length of a short-read method is between 100 and 600 bp, while the read length obtained with a long-read NGS method varies between 10,000 to 15,000 bp. Three major NGS platforms are the Genome Analyzer (Illumina/Solexa), ABI SOLiD System (Life Technologies), and 454 Genome Sequencer FLX (Roche) (Marjanović, Primorac, and Dogan 2018).

1.10 DNA Phenotyping

Recent trends in forensic DNA analysis strive to use SNP genotyping for the purpose of predicting the potential of some phenotypic characteristics such as skin, eye and hair color, and others, and it has been described as DNA phenotyping (FDP). The basic idea of FDP is based on the need to collect the largest possible number of phenotypic information about a person whose identity needs to be determined and to predict physical appearance from DNA. FDP may be of great importance in cases when identification by STR profiling is challenging. It can be used to narrow suspect lists, to help solve human remains cases and to generate leads in cases where there are no suspects or database hits.

This information can be extremely useful in situations where an unknown missing person for whom there is no starting information about possible relatives with whom to compare DNA needs to be identified. On the other hand, in cases in which only a biological trace but no additional data about the perpetrator (witness statements, etc.) is available, and the generated DNA profile is not in the DNA database, phenotyping could theoretically provide some baseline information about the observed phenotypic characteristics. It is important to note that the results of phenotypic analyses are not entered into the DNA database. Rather, they direct the investigation toward a relevant person so that their identity can be established via STR profiling (Marjanović, Primorac, and Dogan 2018). As noted before, the role of a forensic anthropologist is to determine the basic characteristics, such as race, gender, age, height, from human skeletal remains in order to help perform their individualization. Until recently, it was impossible to perform any prediction of many phenotypical characteristics such as eye, hair, and skin color on the basis of genetic analysis. The latest research has determined certain genetic markers that are correlated with the genetic assessment of these properties, and the forensic scientific community is working on establishing multiplex systems which incorporate SNPs that are positioned on the genes responsible for pigmentation (Marjanović, Primorac, and Dogan 2018).

1.11 Forensic Analysis of Plant DNA

Analysis of DNA isolated from plants has a primary role in connecting individuals with crime scene or linking the evidence to specific geographic locations, and, recently, it has become possible to monitor the movement of certain drugs from breeding sites to the places of consumption. The first forensic case that involved the analysis of DNA from plants, which demonstrated the importance and potential use of plant DNA testing, occurred in 1992. In Arizona, the body of a female was found under a Palo Verde tree. Lying close to the body, there was a pager that could be traced to its owner. During the investigation, police found a few seed pods from a Palo Verde tree in the suspect's truck. Police wanted to know if the seeds found in a truck could be associated with the Palo Verde tree next to the body establishing the suspect's tie to the crime scene, which the suspect denied. A study was made of DNA from Palo Verde trees in the area and they were all found to be different. The DNA from the seeds in the truck was identical to the DNA from the tree next to the body. This was the first use of plant DNA to link a suspect to a crime scene and was critical in finding the suspect guilty of murder (Miller Coyle et al. 2001).

Another area of interest is the possible identification of the origin of plant material. Marijuana plants propagate clonally so all descendants of a parent plant will have the same genetic profile, although there is variation among clones. Because of this, it is now possible to monitor the distribution of seeds and leaves, connecting paths of drug distribution, as well as drug dealers (Shirley et al. 2013). Surely, in the near future, it will be required to implement a number of population studies on different plants in order to accurately determine the occurrence of particular genes in specific types of plants within a certain geographic area, which will ultimately have a major impact on the acceptance of the results of plant DNA analysis in court (Zeller et al. 2001). This issue is so topical that it is separately discussed in more detail in Chapter 14.

1.12 Forensic Analysis of Animal DNA

Many studies have clearly shown that animal DNA has a unique genetic code, making it possible to distinguish individual animals even within the same species and subspecies (breed). Recently, a new SNP assay was developed in order to identify individual horses from urine samples that are submitted for postracing doping tests (Kakoi et al. 2013). It is often necessary to determine the origin of animal hairs found at a crime scene or on the clothes of a victim or suspect. Usually, it is dog or cat hair, or hair from other pets. So far, numerous research studies have been carried out that clearly showed the specificity of nuclear DNA or mtDNA isolated from various evidence of animal origin (Savolainen et al. 2000; Fridez et al. 1999; Padar et al. 2001; Roney et al. 2001; Raymond-Menotti 2001). This topic is discussed further in Chapter 12.

References

1992. 1991 report concerning recommendations of the DNA Commission of the International Society for Forensic Haemogenetics relating to the use of DNA polymorphism. *Vox Sang* 63:70–73.

Abaz, J., S.J. Walsh, J.M. Curran et al. 2002. Comparison of variables affecting the recovery of DNA from common drinking containers. *Forensic Sci Int* 126(3):233–240.

Alford, R.L., H.A. Hammond, I. Coto, and C.T. Caskey. 1994. Rapid and efficient resolution of parentage by amplification of short tandem repeats. *Am J Hum Genet* 55(1):190–195.

Alonso, A., S. Andjelinovic, P. Martin et al. 2001. DNA typing from skeletal remains: Evaluation of multiplex and megaplex STR systems on DNA isolated from bone and teeth samples. *Croat Med J* 42(3):260–266.

Andjelinović, Š., D. Sutlović, I. Erceg-Ivkošić et al. 2005. Twelve-year experience in identification of skeletal remains from mass graves. *Croat Med J* 46(4):530–539.

Andrews, R.M., I. Kubacka, P.F. Chinnery, R.N. Lightowlers, D.M. Turnbull, and N. Howell. 1999. Reanalysis and revision of the Cambridge reference sequence for human mitochondrial DNA. *Nat Genet* 23(2):147.

Applied Biosystems. 2017. *Quantifiler™ HP and Trio DNA Quantification Kits*. Waltham, MA: Thermo Fisher Scientific.

Applied Biosystems. 2014. *Quantifiler® Human DNA Quantification Kit and Quantifiler® Y Human Male DNA Quantification Kit User's Manual*. Carlsbad, CA: Life Technologies.

Barać, L., M. Peričić, I.M. Klarić et al. 2003. Y chromosomal heritage of Croatian population and its island isolates. *Eur J Hum Genet* 11(7):535–542.

Bregu, J., D. Conklin, E. Coronado, M. Terrill, W.R. Cotton, and M.C. Grgicak. 2013. Analytical thresholds and sensitivity: Establishing RFU thresholds for forensic DNA analysis. *J Forensic Sci* 58(1):120–129.

Brettell, T.A., J.M. Butler, and J.R. Almirall. 2011. Forensic science. *Anal Chem* 83(12):4539–4556.

Budowle, B., R. Chakraborty, A.M. Guisti, A.J. Eisenberg, and R.C. Allen. 1991. Analysis of the VNTR locus DIS80 by the PCR followed by high resolution PAGE. *Am J Hum Genet* 48(1):137–144.

Budowle, B., F.S. Baechtel, and C.T. Comey. 1992. Some considerations for use of AMP-FLPs for identity testing. In *Advances in Forensic Haemogenetics*, eds. C. Ritter and P.M. Schneider. New York: Springer Verlag, 11–17.

Buel, E., M.B. Schwartz, and M.J. LaFountain. 1998. Capillary electrophoresis STR analysis: Comparison to gel-based systems. *J Forensic Sci* 43(1):164–170.

Butler, J.M. 2010. *Fundamentals of Forensic DNA Typing*. San Diego, CA: Elsevier Academic Press.

Butler, J.M. 2011. *Advanced Topics in Forensic DNA Typing: Methodology*. San Diego, CA: Elsevier Academic Press.

Coble, M.D., and J.M. Butler. 2005. Characterization of new miniSTR loci to aid analysis of degraded DNA. *J Forensic Sci* 50(1):43–53.

Coble, M.D., R.S. Just, J.E. O'Callaghan et al. 2004. Single nucleotide polymorphisms over the entire mtDNA genome that increase the power of forensic testing in Caucasians. *Int J Legal Med* 118(3):137–146.

Cooper, G.M., and R.E. Hausman. 2004. *The Cell: A Molecular Approach*, 3rd edn. Washington, DC: ASM Press.

Davis, C., J. Ge, C. Sprecher et al. 2013. Prototype PowerPlex Y23 System: A concordance study. *Forensic Sci Int Genet* 7(1):204–208.

Dexheimer, P.J., and L. Cochella. 2020. MicroRNAs: From mechanism to organism. *Front Cell Dev Biol Jun* 3(8):409.

Doležel, J., J. Bartoš, H. Voglmayr, and J. Greilhuber. 2003. Nuclear DNA content and genome size of trout and human. *Cytom A* 51(2):127–128.

Džijan, S., G. Ćurić, D. Pavlinić, M. Marcikić, D. Primorac, and G. Lauc. 2009. Evaluation of the reliability of DNA typing in the process of identification of war victims in Croatia. *J Forensic Sci* 54(3):608–609.

Federal Bureau of Investigation. 1990. *The Application of Forensic DNA Testing to Solve Violent Crimes*. Washington, DC: US Department of Justice.

Fisher, A.B., and D.R. Fisher. 2012. *Techniques of Crime Scene Investigation*. Boca Raton, FL: Taylor and Francis.

Fridez, F., S. Rochat, and R. Coquoz. 1999. Individual identification of cats and dogs using mitochondrial DNA tandem repeats? *Sci Justice* 39(3):167–171.

Gabriel, N.M., D.C. Calloway, L.R. Rebecca, Š. Andjelinović, and D. Primorac. 2001. Population variation of human mitochondrial DNA hypervariable regions I and II in 105 Croatian individuals demonstrated by immobilized sequence specific oligonucleotide probe analysis. *Croat Med J* 42(3):328–335.

Giampaoli, S., A. Berti, F. Valeriani et al. 2012. Molecular identification of vaginal fluid by microbial signature. *Forensic Sci Int Genet* 6(5):559–564.

Gill, P., R. Sparkes, and C. Kimpton. 1997. Development of guidelines to designate alleles using an STR multiplex system. *Forensic Sci Int* 89(3):185–197.

Gill, P., D. Rowlands, G. Tully, I. Bastisch, T. Staples, and P. Scott. 2010. Manufacturer contamination on disposable plastic-ware and other reagents—An agreed position statement by ENFSI, SWGDAM and BSAG. *Forensic Sci Int Genet* 4(4):269–270.

Geberth, J.V. 2006. *Practical Homicide Investigation: Tactics, Procedures and Forensic Techniques*. Boca Raton, FL: Taylor and Francis.

Goodwin, W., A. Linacre, and S. Hadi. 2011. *An Introduction to Forensic Genetics* (2nd ed.). Hoboken, NJ: Wiley-Bleckwell.

Gornik, I., M. Marcikic, M. Kubat, D. Primorac, and G. Lauc. 2002. The identification of war victims by reverse paternity is associated with significant risks of false inclusion. *Int J Legal Med* 116(5):255–257.

Graur, D., and W.H. Li. 2000. *Fundamentals of Molecular Evolution*, 2nd edn. Sunderland, MA: Sinauer, 14, 274–275, 386–387, 392–394.

Gunn, A. 2006. *Essential Forensic Biology*. Chichester: John Willey & Sons.

Haas, C., E. Hanson, M.J. Anjos et al. 2013. RNA/DNA co-analysis from human saliva and semen stains: Results of a third collaborative EDNAP exercise. *Forensic Sci Int Genet* 7(2):230–239.

Hammond, H.A., L. Jin, Y. Hong, C.T. Caskey, and R. Chakraborty. 1994. Evaluation of 13 short tandem repeat loci for use in personal identification applications. *Am J Hum Genet* 55(1):175–189.

Hares, D.R. 2011. Expanding the CODIS core loci in the United States. *Forensic Sci Int Genet* 6(1):e52–e54.

Holland, M.M., D.L. Fisher, L.G. Mitchell et al. 1993. Mitochondrial DNA sequence analysis of human skeletal remains: Identification of remains from the Vietnam War. *J Forensic Sci* 38(3):542–553.

Holland, M.M., and T.J. Parsons. 1999. Mitochondrial DNA sequence analysis—Validation and use for forensic casework. *Forensic Sci Rev* 11(1):22–50.

Houck, M.M., and J.A. Siegel. 2006. *Fundamentals of Forensic Science*. London: Elsevier Academic Press.

Inman, K., and N. Rudin. 1997. *An Introduction to Forensic DNA Analysis*. New York: CRC Press.

International Human Genome Sequencing Consortium. 2001. Initial sequencing and analysis of the human genome. *Nature* 409(6822):860–921.

Irwin, J.A., J.L. Saunier, H. Niederstatter et al. 2009. Investigation of heteroplasmy in the human mitochondrial DNA control region: A synthesis of observations from more than 5000 global population samples. *J Mol Evol* 68(5):16–527.

Ivanov, P.L., M.J. Wadhams, R.K. Roby, M.M. Holland, V.W. Weedn, and T. Parsons. 1996. Mitochondrial DNA sequence heteroplasmy in the Grand Duke of Russia Georgij Romanov, establishes the authenticity of the remains of Tzar Nicholas II. *Nat Genet* 12(4):417–420.

Jeffreys, A.J., V. Wilson, and S.L. Thein. 1985a. Hypervariable "minisatellite" regions in human DNA. *Nature* 314(6006):67–73.

Jeffreys, A.J., S.L. Thein, and V. Wilson. 1985b. Individual specific "fingerprints" of human DNA. *Nature* 316(6023):76–79.

Jobling, M.A., E. Heyer, P. Dieltjes, and P. de Knijff. 1999. Y-chromosome-specific microsatellite mutation rates re-examined using a minisatellite, MSY1. *Hum Mol Genet* 8(11):2117–2120.

Jobling, M.A., M.E. Hurles, and C. Tyler-Smith. 2004. *Human Evolutionary Genetics: Origins, Peoples and Disease.* New York: Garland Science, Taylor & Francis Group.

Johnson, C.V., D. Primorac, D.M. McKinstry, D.W. Rowe, and J.B. Lawrence. 2000. Tracking COL1A1 RNA in osteogenesis imperfecta: Splice-defective transcripts initiate transport from the gene but are retained within the SC35 domain. *J Cell Biol* 150(3):417–432.

Kaida, D., T. Schneider-Poetsch, and M. Yoshida. 2012. Splicing in oncogenesis and tumor suppression. *Cancer Sci* 103(9):1611–1616.

Kakoi, H., I. Kijima-Suda, H. Gawahara et al. 2013. Individual identification of racehorses from urine samples using a 26-plex single-nucleotide polymorphism assay. *J Forensic Sci* 58(1):21–28.

Katsanis, H.S., and J.K. Wagner. 2013. Characterization of the standard and recommended CODIS markers. *J Forensic Sci* 58(Suppl. 1):S169–S171.

Kayser, M., and A. Sajantila. 2000. Mutations at Y-STR loci: Implications for paternity testing and forensic analysis. *Forensic Sci Int* 118(2–3):116–121.

Kelemen, O., P. Convertini, Z. Zhang et al. 2013. Function of alternative splicing. *Gene* 514(1):1–30.

Lee, H.C., C. Ladd, C.A. Scherczinger, and M.T. Bourke. 1998. Forensic applications of DNA typing: Part 2. Collection and preservation of evidence. *Am J For Med Pathol* 19:10–18.

Marjanovic, D., S. Fomarino, S. Montagna et al. 2005. The peopling of modern Bosnia-Herzegovina: Y-chromosome haplogroups in the three main ethnic groups. *Ann Hum Genet* 69(6):757–763.

Marjanović, D., A. Durmić-Pašić, N. Bakal et al. 2007. DNA identification of skeletal remains from the second world war mass graves uncovered in Slovenia. *Croat Med J* 48(4):513–519.

Marjanović, D., and D. Primorac. 2013. DNA variability and molecular markers in forensic genetics. In *Forensic Genetics: Theory and Application*, eds. D. Marjanović, D. Primorac, L.L. Bilela et al. Sarajevo: Lelo Publishing (Bosnia and Herzegovina's edition), 75–98.

Marjanović, D., N. Hadžić Metjahić, J. Čakar et al. 2015. Identification of human remains from the Second World War mass graves uncovered in Bosnia and Herzegovina. *Croat Med J* 56(3):257–262.

Marjanović, D., D. Primorac, and S. Dogan. 2018. *Forensic Genetics: Theory and Application.* Sarajevo: International Burch University.

Maroni, G. 2001. *Molecular and Genetic Analysis of Human Traits.* Malden, MA: Blackwell Science.

Mayr, W.R. 1993. Recommendations of the DNA Commission of the International Society for Forensic Haemogenetics relating to the use of PCR-based polymorphisms. *Vox Sang* 64(2):124–126.

McEwen, J.E. 1995. Forensic DNA data banking by state crime laboratories. *Am J Hum Genet* 56(6):1487–1492.

Miller Coyle, H., C. Ladd, T. Palmbach, and H.C. Lee. 2001. The green revolution: Botanical contributions to forensics and drug enforcement. *Croat Med J* 42(3):340–345.

Nakanishi, H., H. Shojo, T. Ohmori et al. 2013. Identification of feces by detection of Bacteroides genes. *Forensic Sci Int Genet* 7(1):176–179.

National Forensic Science Technology Center. 2013. Stochastic effects of LCN DNA analysis. http://www.nfstc.org/pdi/Subject09/pdi_s09_m01_03_b.htm (accessed March 25, 2013).

National Research Council, National Academy of Sciences. 1996. *The Evaluation of Forensic DNA Evidence.* Washington, DC: National Academy Press.

Neilson, J.R., and R. Sandberg. 2010. Heterogeneity in mammalian RNA 3' end formation. *Exp Cell Res* 316(8):1357–1364.

Padar, Z., B. Egyed, K. Kontadakis et al. 2001. An importance of canine identification in Hungarian forensic practice. In Proceedings of the Second European-American Intensive Course in Clinical and Forensic Genetics, eds. D. Primorac, I. Erceg, and A. Ivkosic, September 3–14, 2001. Dubrovnik and Zagreb: Studio HRG.

Parsons, T.J., and M.D. Coble. 2001. Increasing the forensic discrimination of mitochondrial DNA testing through analysis of the entire mitochondrial DNA genome. *Croat Med J* 42(3):304–309.

Pereira, R., I. Gomes, A. Amorim, and L. Gusmão. 2007. Genetic diversity of 10 X chromosome STRs in northern Portugal. *Int J Legal Med* 121(3):192–197.

Pinto, N., L. Gusmao, and A. Amorim. 2011. X-chromosome markers in kinship testing: A generalisation of the IBD approach identifying situations where their contribution is crucial. *Forensic Sci Int Genet* 5(1):27–32.

Prieto, L., C. Alves, B. Zimmermann et al. 2013. GHEP-ISFG proficiency test 2011: Paper challenge on evaluation of mitochondrial DNA results. *Forensic Sci Int Genet* 7(1):10–15.

Primorac, D., M.L. Stover, S.H. Clark, and D.W. Rowe. 1994. Molecular basis of nanomelia, a heritable chondrodystrophy of chicken. *Matrix Biol* 14(4):297–305.

Primorac, D., S. Andelinovic, M. Definis-Gojanovic et al. 1996. Identification of war victims from mass graves in Croatia and Bosnia and Herzegovina through the use of DNA typing and standards forensic methods. *J Forensic Sci* 41(5):891–894.

Primorac, D., C.V. Johnson, J.B. Lawrence et al. 1999. Premature termination codon in the aggrecan gene of nanomelia and its influence on mRNA transport and stability. *Croat Med J* 40(4):528–532.

Primorac, D., S.M. Schanfield, and D. Primorac. 2000. Application of forensic DNA in the legal system. *Croat Med J* 41:33–47.

Primorac, D. 2001. Validation of PP16 System on teeth and bone samples. In Proceedings of the Promega's STR Educational Forum, June 5–6. Zagreb. Promega Corporation.

Primorac, D. 2004. The role of DNA technology in identification of skeletal remains discovered in mass graves. *Forensic Sci Int* 146(Suppl):S163–S164.

Primorac, D., and F. Paić. 2004. DNA analysis in forensic medicine. In *Medical and Diagnostic Biochemistry in Clinical Practice*, eds. E. Topic, D. Primorac, and S. Jankovic. Zagreb: Medicinska Naklada, 365–384. (Croatian edition).

Primorac, D., and D. Marjanović. 2008. DNA analysis in forensic medicine and legal system. In *DNA Analysis in Forensic Medicine and Legal System*, ed. D. Primorac. Zagreb: Medicinska Naklada, 1–59. (Croatian edition).

Primorac, D. 2009. Human genome project-based applications in forensic sciences. *Croat Med J* 50(30):2005–2006.

Primorac, D., D. Marjanović, P. Rudan, R. Villems, and P. Underhill. 2011. Croatian genetic heritage: Y chromosome story. *Croat Med J* 52(3):225–234.

Primorac, D., G. Lauc, and I. Rudan. 2013. Predicting age from biological markers in forensic traces. *The Eight ISABS Conference on Forensic, Anthropologic and Medical Genetics and Mayo Clinic Lectures in Translational Medicine*, June 24–28 (Book of abstracts, p. 66). Split.

Prinz, M., and M. Sansone. 2001. Y chromosome-specific short tandem repeats in forensic casework. *Croat Med J* 42(3):288–291.

Promega Corporation. 2017. *Plexor˙ HY System for the Applied Biosystems 7500 and 7500 FAST RealTime PCR Systems – Technical Manual* (Revised 9/17). Madison, WI: Promega Corporation.

Promega Corporation. 2020. *PowerPlex® Fusion System for Use on the Applied Biosystems® Genetic Analyzers*. (Revised 7/20). Madison, WI: Promega Corporation.

Qiagen. 2018a. *Investigator˙ Quantiplex Handbook*. Hilden, Germany: Qiagen.

Qiagen. 2018b. *Investigator˙ Quantiplex HYres Handbook*. Hilden, Germany: Qiagen.

Qiagen. 2021. *Investigator˙ 24plex QS Kit Handbook*. Hilden, Germany: Qiagen.

Raymond-Menotti, M. 2001. Genetics of coat patterns in the domestic cat. In Proceedings of the Seventh ISABS Conference in Forensics, Anthropologic and Medical Genetics and Mayo Clinic Lectures in Translational Medicine, eds. D. Primorac, M. Schanfield, and S. Vuk-Pavlovic, June 20–24, 2011. Bol: ISABS.

Read, A., and D. Donnai. 2011. *New Clinical Genetics*. Banbury: Scion Publishing Ltd.

Report of the committee on the human mitochondrial genome. http://www.mitomap.org/report (accessed July 7, 2009).

Roney, C., A. Spriggs, C. Tsang, and J. Wetton. 2001. A DNA test for the identification of tiger bone. In: Proceedings of the Second European–American Intensive Course in Clinical and Forensic Genetics, eds. D. Primorac, I. Erceg, and A. Ivkošić, September 3–14, 2001. Dubrovnik and Zagreb: Studio HRG.

Saferstein, R. 2001. *Criminalistics: An Introduction to Forensic Science*, 7th edn. Hoboken, NJ: Prentice-Hall.

Saiki, R.K., S. Scharf, F. Falooona et al. 1985. Enzymatic amplification of beta-globin sequences and restriction site analysis for diagnosis of sickle cell anemia. *Science* 230(4372):1350–1354.

Saiki, R.K., D.H. Gelfand, S. Stoffel et al. 1988. Primer-directed enzymatic amplification of DNA with a thermostable DNA polymerase. *Science* 2239(4839):487–491.

Savolainen, P., L. Arvestad, and J. Lundeberg. 2000. A novel method for forensic DNA investigations: Repeat-type sequence analysis of tandemly repeated mtDNA in domestic dogs. *J Forensic Sci* 45(5):990–999.

Schanfield, M. 2000. DNA: PCR-STR. In *Encyclopedia of Forensic Sciences*, eds. J.A. Siegel, P.J. Saukko, and G.C. Knupfer. London: Academic Press, 526–535.

Schlotterer, C., and D. Tautz. 1992. Slippage synthesis of simple sequence DNA. *Nucleic Acids Res* 20(2):211–215.

Scientific Working Group on DNA Analysis Methods (SWGDAM). 2010. http://www.fbi.gov/about-us/lab/biometric-analysis/codis/swgdam.pdf (accessed March 25, 2013).

Semino, O., G. Passarino, J.P. Oefner et al. 2000. The genetic legacy of paleolithic *Homo sapiens sapiens* in extant Europeans: A Y chromosome perspective. *Science* 290(5494):155–159.

Shirley, N., L. Allgeier, T. LaNier, and H. Miller Coyle. 2013. Analysis of the NMI01 marker for a population database of cannabis seeds. *J Forensic Sci* 58(Suppl. 1):S176–S182.

Shows, T.B., P.J. McAlpine, C. Boucheix, F.S. Collins, P.M. Conneally, Frézal, et al. 1987. Guidelines for human gene nomenclature. An international system for human gene nomenclature (ISGN, 1987). *Cytogenetics and cell genetics* 46(1–4):11–28.

Škaro, V., C.D. Calloway, S.M. Stuart et al. 2011. Mitochondrial DNA polymorphisms in 312 individuals of Croatian population determined by 105 probe panel targeting 61 hypervariable and coding region sites. Abstract. Proceedings of the Seventh ISABS Conference in Forensic, Anthropologic and Medical Genetics and Mayo Clinic Lectures in Translational Medicine. http://www.isabs.hr/PDF/ISABS_2011_abstract_book_web.pdf (accessed February 24, 2013).

Slack, F.J. 2006. Regulatory RNAs and the demise of "junk" DIVA. *Genome Biol* 7(9):328.

Sprecher, C., B. Krenke, B. Amiott, D. Rabbach, and K. Grooms. 2000. The PowerPlex™ 16 system. *Profiles DNA* 4:3–6.

Stover, M.L., D. Primorac, S.C. Liu, M.B. McKinstry, and D.W. Rowe. 1994. Defective splicing of mRNA from one COL1A allele of type I collagen in nondeforming (Type I) osteogenesis imperfecta. *J Clin Invest* 92:1994–2002.

Szibor, R., M. Krawczak, S. Hering, J. Edelmann, E. Kuhlisch, and D. Krause. 2003. Use of X-linked markers for forensic purposes. *Int J Legal Med* 117(2):67–74.

Tamarin, H.R. 2002. *Principles of Genetics*. New York: McGraw-Hill.

Urquhart, A., C.P. Kimpton, T.J. Downes, and P. Gill. 1994. Variation in short tandem repeat sequences—A survey of twelve microsatellite loci for use as forensic identification markers. *Int J Leg Med* 107(1):13–20.

Van Kirk, M.E., A. McCary, and D. Podini. 2009. The molecular basis of microvariant STR alleles at the D21S11 locus using forensic DNA identification. In Proceedings of the American Academy of Forensic Sciences, Annual Scientific Meeting, February 16–21, Denver, CO: Abstract A134, 118.

Venter, J.C., M.D. Adams, E.W. Myers et al. 2001. The sequence of the human genome. *Science* 291(5507):1304–1351.

Wagner, J.K. 2013. Out with the "junk DNA" phrase. *J Forensic Sci* 58(1):292–294.

Wambaugh, J. 1989. *The Blooding*. New York: Bantam Books.

Willyard, C. 2018. New human gene tally reignites debate. *Nature* 558(7710):354–355.

Y Chromosome Consortium. 2002. A nomenclature system for the tree of human Y-chromosomal binary haplogroups. *Genome Res* 12(2):339–348.

Zeller, M., D.H. Wehner, and V. Hemleben. 2001. DNA fingerprinting of plants. In Proceedings of the Second European-American Intensive Course in Clinical and Forensic Genetics, eds. D. Primorac, I. Erceg, and A. Ivkošić, September 3–14, 2001. Dubrovnik and Zagreb: Studio HRG.

Forensic DNA Analysis and Statistics

2

MOSES S. SCHANFIELD, DRAGAN
PRIMORAC, AND DAMIR MARJANOVIC

Contents

2.1 Introduction

2.1.1 Genetic and Statistical Principles in Forensic Genetics

The distribution (segregation) of parental genotypes in offspring depends on the combination of alleles in the parents. Mendelian laws of inheritance determine the expected distribution of alleles in offspring of various matings. The principles of Mendelian inheritance are formulated in the form of three laws.

1. *Monohybrid crossing*: If a father who is homozygote (A,A) mates with a mother who is also homozygous (B,B), all offspring will be heterozygous, that is, A,B.
2. *Segregation*: If two heterozygous A,B parents mate, the following genotypes are possible: A,A, A,B, and B,B at a ratio of 1:2:1. This example shows that there is segregation of alleles at the same locus, such that on average 50% of gametes will carry the A allele, whereas the other 50% of gametes will carry the B allele.

DOI: 10.4324/9780429019944-3

3. *Independent assortment*: When individuals are segregating at more than one unlinked loci—that is, each locus segregates independently of the other—then, the segregation of alleles at one locus is independent of the segregation of alleles of the other loci during meiosis.

2.1.2 Principles of Parentage Testing

Based on Mendelian laws, the four rules for paternity/maternity testing are established:

1. A child, with the exception of a mutation, cannot have a marker (allele) that is not present in one of its parents (direct exclusion).
2. A child must inherit one marker (allele) from a pair of genetic markers from each parent (direct exclusion).
3. A child cannot have a pair of identical genetic markers, unless both parents have the same marker (indirect exclusion).
4. A child must have a genetic marker that is present as an identical pair in both parents (indirect exclusion).

2.1.3 Hardy–Weinberg Equilibrium

Gregor Mendel described the behavior of alleles in various matings. Similarly, in 1908, Hardy and Weinberg, independently, explained the behavior of alleles in the population. The Hardy-Weinberg equilibrium (HWE) describes the expected relationship between allele frequencies and genotype frequencies at a single locus (Weir 1993; Primorac et al. 2000). Thus, for a locus with two alleles A and B, in which A has a frequency of p and B has a frequency q, the HWE expected frequencies of genotypes from these allele frequencies is the binomial expansion of $p + q$, such that

$$(p+q)^2 = p^2 + 2pq + q^2 = 1.00$$

where

$$p + q = 1.00$$

where p^2 is the frequency of the homozygous A,A individuals; q^2 is the frequency of the homozygous B,B individuals; and $2pq$ is the frequency of the heterozygous A,B individuals.

This can be expanded to any number of alleles in the multinomial example. Thus, for three alleles, A, B, and C, with frequencies of p, q, and r, respectively, and $p + q + r = 1.000$. Then, the HWE expected values for the genotypes will be the multinomial expansion of the allele frequencies, such that the HWE expected distribution of $(p + q + r)^2$ will be

$$p^2 + 2pq + 2pr + q^2 + 2qr + r^2 = A,A + A,B + A,C + B,B + B,C + C,C$$

The HWE relates to the constancy of the frequency of genotypes over generations, and since allele frequencies are generated from the genotype frequencies, then the constancy of allele frequencies over generations is also implied. The HWE has several assumptions: that the population is very large, that individuals are mating at random, that the allele frequencies in males and females are the same, that there are nonoverlapping generations, and that there is no mutation and no migration (gene flow) into the population or natural selection. For human populations, many of these assumptions are or can be violated.

However, deviations from HWE may occur for several reasons:

1. Nonrandom formation of reproductive couples violates the assumption of random mating. In human populations, this can occur in two ways: *inbreeding* (mating of genetically related individuals) and *outbreeding* (mating of individuals known not to be related); only inbreeding violates HWE. Inbreeding—because the parents are closely related—can lead to a reduction in heterozygosity; in contrast, outbreeding increases heterozygosity in a parent population. *Inbreeding only affects genotype frequencies and not allele frequencies, whereas outbreeding affects all loci.*

2. A population that is not infinitely large can lead to *genetic drift*, which is a random change in allele frequency within smaller populations when the next generation is not a random sample of the genotypes in the previous generation. Under certain conditions, some alleles may entirely disappear (loss of alleles) or be present in every individual within that population (fixation). Thus, genetic drift will lead to changes in allele and genotype frequencies over time. Genetic drift affects individual loci.

3. Mutation or change in alleles, which may affect the allele frequency; however, it acts very slowly as most mutation rates are very low.

4. Gene flow between populations changes the frequency alleles and impacts the HWE until it reaches a new equilibrium. Continuous gene flow will alter the genetic structure by changing allele frequencies and genotype frequencies over generations. The effect of gene flow will depend on the differences in allele frequencies between the two mixing populations.

5. Natural selection will lead to increased frequency of the favorable alleles over time since children with the favored allele are more likely to survive than children with other alleles. Examples of G6PD deficiency or thalassemia are found in areas with malaria. Note: In adulthood, the genotype distribution will not be HWE; it will only be seen in newborns or before natural selection occurs.

6. Random mating also implies that the population is not substructured or divided into multiple breeding units that individually may be in HWE, but across the populations isolated by distance or physical barriers there will be an overall reduction in heterozygosity in the total population.

Although many of the parameters of HWE are often violated, it is very difficult to establish that a population is out of HWE. To do so, you would have to show significant deviations from HWE statistical tests at multiple loci. It is possible to indicate significant substructuring at the population over large distances and sometimes within a population that has had obvious recent mixing, in the absence of gene flow.

2.1.4 Linkage Equilibrium

Two loci close to each other on a chromosome will not recombine because they are physically too close together for sister chromosomes to exchange DNA. This is referred to as genetic linkage. The further apart two loci are, the higher the likelihood of recombination. At its maximum, recombination will occur in 50% of meiosis. Loci on different chromosomes are not physically linked, but through segregation will recombine 50% of the time. Linkage equilibrium denotes concepts by which loci on the same chromosome that are physically unlinked and loci on different chromosomes should behave as independent loci. This is the HWE equivalent of Mendel's law of independent assortment. *Linkage disequilibrium* can occur when a new mutation occurs in a region of DNA that was previously in linkage equilibrium, but now the mutation is associated with a single region. Until recombination moves the new mutation to all of the alternative sequences for that region, the region will be in linkage disequilibrium. The amount of time it will take the region to come into linkage equilibrium is inversely related to the genetic distance between the markers—that is, the closer together they are on the same DNA segment, the longer it will take to reach equilibrium. Regions of DNA that are in linkage disequilibrium are inherited in blocks referred to either as "haplogroups" or "haplotypes" depending on the number and type of markers that define the region. For markers on different chromosomes, the maximum rate of recombination due to segregation is 0.5. *Gametic disequilibrium* occurs when individuals from two different populations mix, and when alleles on different chromosomes have markedly different frequencies then certain unlinked alleles may occur together more often than predicted by chance because of the recent mixing. Gametic disequilibrium will disappear in a mixed population more quickly than linkage disequilibrium. Both types of equilibrium can be detected at the population level and at the segregation level in families. Because these states of disequilibrium violate the law in independent assortment, they have an impact on the calculation of match likelihood and parentage statistics.

2.2 DNA Evidence in Court

Some of the following describe in great detail DNA in the courtroom, ethics in the laboratory and the courtroom, DNA immigration and human trafficking, and DNA data banking. In many countries, presentation of scientific evidence belongs to the testimony of experts (expert witness). But the fact that the experts carried out analyses does not mean that such evidence will be automatically accepted. Under normal circumstances, it must be proven that the presented evidence material is reliable and informative. The Council of Europe and the Committee of Ministers, issued on February 10, 1992, during its 470th meeting, Recommendation No. 92 on the use of DNA analysis in the criminal justice system (Council of Europe 1992). Some of the most important guidelines are the following:

1. Samples collected for DNA analysis and the information derived from such analysis for the purpose of the investigation and prosecution of criminal offences must not be used for other purposes. However, where the individual from whom the samples have been taken so wishes, the information should be given to him.
2. Samples collected from living persons for DNA analysis for medical purposes, and the information derived from such samples, may not be used for the purposes of investigation and prosecution of criminal offences unless in circumstances laid down expressly by the domestic law.

3. Samples or other body tissues taken from individuals for DNA analysis should not be kept after the rendering of the final decision in the case for which they were used, unless it is necessary for purposes directly linked to those for which they were collected.

The FBI launched the National DNA Index System (NDIS) in 1998—along with the Combined DNA Index System (CODIS) software to manage the program—and since that time it has become the world's largest repository of known offender DNA records. CODIS includes a Convicted Offender Index (containing profiles of offenders submitted by states) and a Forensic Index (containing DNA profiles of evidence related to unsolved crimes). In partnership with local, state, and federal crime laboratories and law enforcement agencies, CODIS aided over 545,000 criminal investigations (FBI 2021).

Chris Asplen (personal communication) provides the following for U.S. courts: Samples or other body tissues taken from individuals for DNA analysis should not be kept after the rendering of the final decision in the case for which they were used, unless it is necessary for purposes directly linked to those for which they were collected. Sample retention in the United States is a state-specific policy particular to each state's legislation. However, federal legislation does require that all states have legislation requiring the destruction of samples and removal of profiles, subsequent to a finding of "not guilty" or the dropping of charges.

2.3 Forensic Identification

The aim of the forensic identification is to compare evidence (blood, body fluids, or tissue) with a victim or suspect. Within a given population, each person has a unique genetic record and, except in case of identical twins (identical twins, are genetically identical except for mutations that occur after the separation of the twins), DNA results will normally unambiguously link or exclude a suspect/victim from an item of evidence. When evidence and a possible donor are the same, then the match probability is calculated to provide additional weight to the evidence. For example, formerly, in sexual assault cases the ABO blood group system was used. If the victim has blood group O, and the evidence belongs to blood group A, then the suspect must have blood type A. If the person does have blood group A, then they cannot be excluded from the list of suspects. Is this information useful and does it serve as a proof in the legal sense? Given that approximately 40% of European populations belong to blood group A, such information is not too helpful. When it comes to DNA analysis, only a small portion of the population can be randomly matched. In determining kinship, the frequency of allele donors is an important value. For identification, statistics based on the expected frequency of genotypes are used. As mentioned earlier, the expected frequencies of genotypes are determined by HWE.

In forensic identification, two basic approaches to communication of results can be applied: the first involves informing the results in the form of frequency of the assessed DNA profile in a specific population (Marjanović et al. 2006). Thus, the final formulation may contain the fact that the frequency of the assessed DNA profile within the Croatian population is 1.33×10^{-9}. This is often referred to as the random match probability (RMP). This formulation clearly and accurately displays the results in mathematical terms. In contrast, a likelihood ratio (LR), which is simply the inverse of the RMP, denotes how many times more likely it is that the tested person contributed to the evidence sample than someone at random from the population. For example, if the RMP is 0.001, the LR says that the person tested is 1000 times as likely to be the donor of the sample as a random individual. Note: This is not the probability that the person tested is the donor. In this way, the common stand of the prosecution (evidence originates from a suspect) whose probability is 1, and the stand of the defense (evidence comes from someone else), where the probability for homozygotes is p^2, and for heterozygotes is $2pq$, are compared. In forensic genetics statistics, this concept is called a *likelihood ratio* (Butler 2010, Butler 2011). LR for one, say, heterozygous locus, equals $1/2pq$.

2.3.1 Correction for Substructuring

In the preceding example, only the HWE estimators of genotype frequencies were used to estimate the RMP and LR values. In the United States where forensic DNA testing was initially regulated by the DNA Advisory Board and later the U.S. Federal Bureau of Investigation, calculations were influenced by the results of the second National Research Council (NRC 1996). In that report, the NRC suggested that homozygous genotypes can significantly decrease the likelihood of an RMP. They suggested that a conservative approach was to increase the frequency of the homozygous genotype $\left(p_i^2 \right)$ by the correction for substructuring, such that the following formula applied:

$$\text{Corrected homozygous genotype frequency} = p_i^2 + \theta \times p_i \times (1 - p_i)$$

where θ is the degree of substructuring and p_i are the allele frequencies. For areas with large populations such as the United States, Europe, and Croatia, θ is set at 0.01, whereas for smaller endogamous populations such as Roma or religious isolates, θ is set at 0.03.

If the corrected value is used for the homozygous genotype, whereas the uncorrected value is for heterozygous genotypes, the following inequality will always hold, and the results will always be the most conservative value for the RMP:

$$\text{NRC2 inequality}: p_i^2 + \theta \times p_i \times (1 - p_i) + 2p_i p_j$$

$$> p_i^2 + \theta \times p_i \times (1 - p_i) + 2p_i p_j > 2\theta \times p_i \times (1 - p_i)$$

For the example in Table 2.1 (the RMP), if $\theta = 0.01$ is used, the RMP goes from 4.214×10^{-18} to 4.462×10^{-18} or a 5.6% decrease in the RMP, indicating that it is a more conservative number.

2.3.2 Individualization and Identification

Individualization is based on the fact that a particular person is characterized by that individual's genetic information. The example of that is identification through fingerprints, where it is possible to distinguish that a certain fingerprint belongs to the exact person. Even identical twins do not have identical fingerprints, although they carry the same genetic structure. Today, with the number of available markers, it is possible to achieve practically any degree of individualization. The degree of individualization required for

Table 2.1 Example of Forensic Identification of a Blood Stain by Analyzing 15 STR Loci

STR Locus	Known Sample of a Suspect	Question Blood Evidence	p_i	p_j	Formula	Frequency
D3S1358	16, 18	16, 18	0.265	0.155	$2p_ip_j$	0.0821
TH01	6, 9.3	6, 9.3	0.225	0.330	$2p_ip_j$	0.1485
D21S11	28, 32.2	28, 32.2	0.165	0.085	$2p_ip_j$	0.0281
D18S51	12, 17	12, 17	0.080	0.100	$2p_ip_j$	0.0160
PENTA E	7, 13	7, 13	0.190	0.155	$2p_ip_j$	0.0589
D5S818	12, 12	12, 12	0.340	0.340	p_i^2	0.1156
D13S317	12, 13	12, 13	0.270	0.075	$2p_ip_j$	0.0405
D7S820	10, 11	10, 11	0.290	0.185	$2p_ip_j$	0.1073
D16S539	11, 12	11, 12	0.340	0.280	$2p_ip_j$	0.1904
CSF1PO	11, 12	11, 12	0.245	0.345	$2p_ip_j$	0.1691
PENTA D	9, 9	9, 9	0.245	0.245	p_i^2	0.0600
VWA	16, 19	16, 19	0.205	0.065	$2p_ip_j$	0.0267
D8S1179	12, 15	12, 15	0.165	0.080	$2p_ip_j$	0.0264
TPOX	8, 8	8, 8	0.570	0.570	p_i^2	0.3249
FGA	21, 22	21, 22	0.155	0.190	$2p_ip_j$	0.0589
Amelogenin	X Y	X Y	/////	/////	/////	/////
Combined frequency (CF)						4.214×10^{-18}
Likelihood 1/CF						237,303,870,652,031,000

Note: p_i and p_j are the frequency of allele variants in Bosnia-Herzegovinian population (Marjanovic et al. 2006).

identification, if there is no international agreement, must be established at the local level. In order to do that, it is important to clearly determine the frequency of allele variation within the studied population, as well as its total size. The resulting probability in Table 2.1 is more than sufficient for both the Bosnia-Herzegovinian and the Croatian populations, but even much lower values obtained by DNA profiling of an individual at a smaller number of short tandem repeat (STR) loci in some cases may be sufficient for the Croatian population. Specifically, the probability of an accidental match of, for example, 1 in 23 billion Croats or population of Bosnia and Herzegovina is solid evidence that the sample originated from a common source. The definition of individualization is an administrative one. Such a rule originally proposed by the FBI was approximately 100 times the U.S. population, such that any value less than 1 in 30 billion was considered individualized, such that an expert could report that "to a reasonable degree of scientific knowledge the profile had to originate with the suspect or their identical twin" (Budowle et al. 2000).

2.3.3 Parentage Testing

The need for determining the identity of an individual person or parentage has been around since biblical days. Even King Solomon encountered this problem, when two women came before him, each claiming to be the true mother of a child and asking him to decide which one was the true mother. Realizing that he would not be able to resolve this based on their testimony, as only one could be speaking the truth, and trusting the fact that the real mother would do anything to save her child, Solomon suggested that the child should be cut in half and that each mother will be given a half. Solomon recognized the real mother as the woman who fiercely opposed the decision and assigned the child to her (Kastelan and Duda 1987). Today, modern forensic laboratories perform parentage testing in less painful and more efficient ways, thanks to the development of DNA technology.

The primary function of all forensic testing, regardless of whether it is parentage testing or personal identification, is to exclude the maximum number of individuals. In cases of parentage testing, this is done by identifying the obligate allele(s) and determining if the presumed or alleged parent also carries this allele(s) (in cases of disputed parentage, the disputed parent is always referred to as the *alleged parent*) (Schanfield 2000). The obligate allele is the allele that originated from the biological parent. For example, if we assume that the mother is AA and the child is AB, then the child's B allele had to arise from the biological father. In a case of disputed paternity, it is always assumed that the mother is truly the mother of the child, unless the test proves disagreement with Mendelian laws of inheritance. If the alleged father is homozygous (two B alleles) or heterozygous (one B allele) for the B allele, he cannot be excluded as the biological parent. If the alleged father does not have the B allele, that is, a child cannot have inherited it from him, he must be excluded. The exception to this rule is when there is a relatively common mutation, A to B. Therefore, it is generally accepted that in the case when at least three genetic inconsistencies are found with the child being tested, the alleged father is automatically excluded as the biological father of the child. If both the mother and child are A,B, then the obligate alleles are A and B, and either allele A or B could have come from the biological father. It is obvious that in this case there are two obligate alleles. Every alleged father who is the carrier of an A or B allele cannot be excluded. However, if the alleged father has C,C or any other non-A or non-B genotype, he can be excluded.

2.3.4 Paternity Index or Combined Paternity Index

If the alleged father is not excluded as a possible biological father, one method of evaluating the weight of the evidence is equal to the relative probability that the alleged father passed the obligate allele to the child versus an unrelated individual from the same population. The ratio between the relative probability that the alleged father passed the obligate allele(s) and the probability that a randomly selected unrelated man from the same population could pass the allele to the child is called the *paternity index* (PI). The PI is an LR or an "odds ratio," that is, the ratio of probabilities that the alleged father, and not another man from the population, passed the obligate allele. Mendelian laws determine probability or possibility that the alleged father can pass the gene. If the father is homozygous for obligate allele or has both obligate alleles, then the likelihood that he passed the obligate allele is 1.0 (2/2). If the father has only one copy of obligate allele or only one of the two obligate alleles, then the probability that he passed obligate allele(s) equals to 0.5 (1/2). The value of PI for a particular locus is equal to the probability that the alleged father can pass allele to the child, divided by the frequency of obligate allele(s). Therefore, PI will be $1/p$ or $0.5/p$, depending on the number of obligate alleles, which the alleged father carries. If the genotypes of mother and child suggest two obligate alleles, p will be calculated using the following formula: $p = p_1 + p_2$, and the PI will be

$$PI = 1/\left(p_1 + p_2\right) \text{ or } PI = 0.5/\left(p_1 + p_2\right)$$

PI is calculated for each locus tested.

The rules of probability theory determine how to combine multiple PIs to determine the combined probability of paternity. (Note: There are two rules for combining probabilities, the "and" rule and the "or" rule. The expected value of the heterozygote $2pq$ uses both rules. The probability [or likelihood] of getting a "p" and a "q" is the product of p and q, whereas the probability [or likelihood] of getting a p from the mother and q from the father, or a p from the father and a q from the mother, is the sum of the two p,q's.) To combine all of the individuals' paternity indices, each of the individual PIs is multiplied together. This calculation is called the *combined paternity index (CPI)*. The combined paternity index is the "likelihood ratio" or "odds ratio," i.e., the ratio of likelihood that the alleged father, and not some other randomly selected man from the population, passed the obligate alleles from all loci examined.

Rules for combining probabilities are quite simple. If you want to know the combined probability of the traits A and B and C, it is necessary to multiply the individual probabilities of A and B and C. Therefore, the likelihood that the alleged father passed loci 1*A, 2*C and locus 3*E to the child equals to the product of probabilities 1*A2*C3*E. Simply put, no matter how many loci are used for the analysis of contested paternity, the CPI is determined by multiplying all calculated PIs, i.e., the paternity indices calculated for each of the observed loci. In the beginning, when a relatively small number of loci were used for these purposes (three to five loci), the acceptable value of CPI, say, for the German judiciary was 1000, while in the United States this value must be at least 100. In other words, in these cases, CPI expresses the fact that the alleged father is 1000 or 100 times more likely to be the biological father than any other nonrelated man from the same population. Nowadays, in the use of, say, 20-plus STR loci represented in most commercial kits (PowerPlex Fusion 6C® [Promega], Globalfiler® [Thermal Fisher], or Investigator 24plex® [Qaigen]), it is nothing unusual for the value of the CPI in the standard paternity test (includes mother, child, and presumed father) to be in the range of several hundred thousand to up to several million, while in the case of the motherless paternity testing, this value is somewhat lower. It is important to note that fluctuations in these values primarily depend on the frequency of alleles that biological father passes to a child. Since many have difficulties in understanding and interpreting the likelihood ratios, an alternative way of showing data is based on a conversion of probability ratio in the probability of paternity.

2.3.5 Probability of Paternity

In the 18th century, the mathematician Bayes developed a theorem to assess the probability that an event occurred, even when this event cannot be directly measured. This is the basis of Bayesian statistics that is widely used, especially in areas such as genetic counseling for genetic risk assessment. The Bayesian formula to estimate the possibility for an event to happen is

$$\frac{X \times p}{X \times p + Y \times (1 - p)}$$

where X is the probability that some event will occur, Y is the probability that it will not occur, p is the prior probability that X will happen, and $1 - p$ is the prior probability that it will not happen.

Bayes's formula has several limitations. The formula is exhaustive, that is, it includes all possibilities of the event space. In the case of paternity testing, X is the combined probability that the alleged father passed all the obligate alleles, and consists of the product of all 0.5 and 1.0 values for each locus, and Y is the combined probability that a nonrelated individual in the population is the biological father and the product of frequencies of all obligate alleles. The prior probability is the probability that this event could have happened without the knowledge of the current results. In this case, it represents the probability that the alleged father is the biological father before any laboratory tests were performed. There are different ways to calculate the prior probability that the alleged father is also the biological father. It can be assumed that it is a laboratory inclusion rate that is approximately 70%. It can also be assumed that the prior probability is the number of males that may be taken into account within the time when fertilization occurred or, more precisely, men from the area but who are of reproductive age. However, it is customary to assume that it is the same probability that the alleged father is or is not the biological father. This is the so-called neutral prior probability. Therefore, the prior probability equals to 0.5. Some statisticians argue that the prior probability should be calculated by dividing the desired result by the total number of possible outcomes, for example, the prior probability to get "one" when throwing dice (half pair of dice) is equal to 1/6, whereas the probability of getting "heads" when tossing a coin is equal to 1/2. On the occasion of paternity testing, only two outcomes are possible: he is or he is not the biological father. The starting probability is equal to 1 for the two options, or 0.5, which means that it is possible that the "neutral starting probability" is the possible correct starting probability. If the value 0.5 is chosen as the initial probability, the formula reduces to the one that uses the CPI.

$$\text{Probability of paternity} \left(\text{if prior probability} = 0.5 \right) = \text{PP} = 1 / \left[1 + \left(1 / \text{CPI} \right) \right]$$

This formula can be further simplified to

$$\text{Probability of paternity} = \text{CPI} / \left(\text{CPI} + 1 \right)$$

All prosecutors, judges, and lawyers who read this formula will see clearly why it is not possible to obtain a score of 100%. In fact, no matter how large the CPI number is, it must always be divided by a number that is greater than 1 and so this ratio will never equal 1, and hence the probability of paternity, put in percentage, will never be 100%. Therefore, when the score is, say, 99.999999678998%, it is ridiculous to ask for absolute certainty. Unfortunately, however, DNA experts are often asked: "Are you 100% sure that the person concerned is the child's biological father?"

Thus, the probability of paternity is a possibility that the alleged father is the biological father of the child. Therefore, if a hypothetical CPI equals, say 6,477,698, then the probability that the alleged father is the biological father equals 0.999999846 or 99.9999846% [6,477,698/(6,477,698 + 1)]. In contrast, the probability that the alleged father is not the biological father is 0.000000154 or 0.0000154%.

The logical question that can be asked is, What is the lowest value of the probability of paternity, so that it can be fully accepted? This is normally set on a national level. In the US it was set at 99% in 2000 (Uniform Parentage Act 2000 and American Association of Blood Banks 2000), in Germany it is 99.999% (Poetsch et al. 2013), and in France it is banned (Irish Times 2009, Service Public 2020). Thus, a high percentage of probability is determined by using a larger number of molecular markers, which is nowadays seldom less than 20. A large number of laboratories suggest the analysis of additional molecular markers if they don't get these values, regardless of whether the father's identified allelic variants match with obligate alleles at all loci tested.

2.3.6 Random Man Not Excluded

PI and probability of paternity have a common assumption with Mendelian likelihood that the alleged father is also the biological father of the child. To assess the same data in a different way, it is possible to ask how much genetic information is present in the mother–child pair, that is, what is the discrimination power of a test at preventing the erroneous inclusion of an alleged father. This is similar to the concept of "the power of the test" in statistics. Ideally, the test should be strong and practical, and should exclude all falsely accused alleged fathers. Today, by the method of DNA analysis, approximately 30% of accused alleged fathers are excluded worldwide (Allen 1999).

The power of the paternity test can be determined via the so-called *random man not excluded* (RMNE). This statistical test can be compared with the calculation of a random match probability in the population, which is applied in forensic identification (see later). RMNE is the proportion of the population that can provide all obligate alleles and cannot be excluded. This term is used to express the frequency of individuals that cannot be excluded as potential donors of genetic material.

The formula for RMNE for one locus is

$$p^2 + 2p(1-p) \text{ or, simply,} 1-(1-p)^2$$

Combined random man not excluded (CRMNE) is analogous to the CPI, that is, it is equal to the product of individual values. The value of CRMNE is usually very low. However, it is easier to use the reciprocal value of the formula (1 – CRMNE), for which the term "probability of exclusion" is commonly used. But since it is possible to mix the concept of "probability of exclusion" with the notion of "probability of paternity," the less confusing term "exclusionary power" (EP) or "power of exclusion" (PE) is used. EP or PE is the probability of excluding a man falsely accused (1 – CRMNE). When an a priori probability of 0.5 is used, PE equals the Bayesian probability of paternity in the so-called nonexclusion model. To avoid confusion with the term "probability of paternity," which derives from the model of inheritance, one should use the term "nonexcluded probability of paternity" or "Weiner's probability of paternity" named for the scientist who first proposed the concept in 1976 (Weiner 1976). The term nonexcluded probability of paternity has several advantages over the standard probability of paternity (Li and Chakravarti 1988). The biggest advantage of this method is that it always increases by increasing the number of conducted tests and is completely analogous to the concept of population frequency or random match probability, which is used in forensic identifications. The probability of paternity (the term traditionally used for the probability of paternity with respect to the transfer of alleles) can be reduced if the alleged father is heterozygote and the mother and child are carrying the same heterozygous genotype containing two common alleles, where $p_1 + p_2$ is greater than 0.5.

Table 2.2 shows the general formulas used in paternity testing, whereas Table 2.3 shows an example of testing. The fourth column in Table 2.3 displays all of the obligate alleles of

Table 2.2 **Formulas for Calculating Paternity Indices**

Alleles of a Mother	Alleles of a Child	Alleles of a Father	PI
AB	AA	AA	1/a
BC or BB	AB	AA	1/a
AA	AA	AA	1/a
BB or BC	AB	AB or AC	1/2a
AB	AA	AB or AC	1/2a
AA	AA	AB	1/2a
AB	AB	AB	1/(a + b)
AB	AB	AA	1/(a + b)
AB	AB	AC	1/2(a + b)

Source: www.DNA-view.com.
Note: A, B, C—alleles; a, b, c—frequency of A, B, C alleles.

the alleged father. The column p denotes the frequency of each allele, which is based on the Croatian database (Projić et al. 2007). RMNE is calculated by the aforementioned formula. X indicates the probability that the alleged father can pass the obligate allele, and it is determined by the father's genotype (if the father has only one copy of obligate allele or only one of two obligate alleles, then the probability [X] that he passed the obligate allele to a child equals to 0.5). PI is calculated as X/p or, in the case that there are two obligate alleles at the same locus, as $X/(p_1 + p_2)$. From the combined values, shown at the bottom of the table, it is evident that the probability of the alleged father is almost 790 million times higher than of any other unrelated male in the Croatian population. This value indicates the probability of paternity of 99.99999987%. RMNE shows that the probability to exclude a falsely accused father is 99.99999973%. Thus, in this case, the probability of paternity based on the passage of alleles is approximate to the likelihood based on the method of nonexclusion.

Table 2.3 **Example of Complete Paternity Testing**

STR Locus	Mother	Child	Obligate Allele	Alleged Father	p(Y)	RMNE	X	PI
D8S1179	13, 14	11, 13	11	11, 13	0.0564	0.1096	0.5	8.8652
D21S11	30, 32.2	31, 32.2	31	30,2 31	0.0590	0.1145	0.5	8.4746
D7S820	10, 12	9, 10	9	9, 9	0.1795	0.3268	1	5.5710
CSF1PO	12, 12	10,12	10	10, 12	0.2538	0.4432	0.5	1.9701
D3S1358	15, 15	15, 18	18	15, 18	0.1487	0.2753	0.5	3.3625
TH01	6, 9.3	6, 8	8	6, 8	0.1308	0.2445	0.5	3.8226
D13S317	8, 10	10, 11	11	11, 11	0.3308	0.5522	1	3.0230
D16S539	9, 11	11, 13	13	9, 13	0.1718	0.3141	0.5	2.9104
D2S1338	17, 18	17, 23	23	23, 25	0.1026	0.1947	0.5	4.8733
D19S433	14.2, 16	13, 16	13	12, 13	0.2128	0.3803	0.5	2.3496
VWA	15, 16	16, 17	17	17, 17	0.2256	0.4003	1	4.4326
TPOX	8, 9	9, 9	9	8, 9	0.0846	0.1620	0.5	5.9102
D18S51	16, 17	15, 17	15	14, 15	0.1359	0.2533	0.5	3.6792
D5S818	11, 12	11, 11	11	11, 12	0.3256	0.5452	0.5	1.5356
FGA	20, 21	20, 25	25	20, 25	0.1000	0.1900	0.5	5.0000
Amelogenin	X X	X Y		X Y				
Combined								790, 169, 857
Power of exclusion	1 – CRMNE					2.67×10^{-9}	99.9999997%	
Likelihood	1/CRMNE					374, 531, 835		
Probability of paternity	CPI/CPI + 1					99.99999973%		99.9999987%

Notes: p(Y), frequency of each allele or each obligate allele based on Croatian database (Projic et al. 2007); RMNE, random man not excluded; X, the probability that alleged father can pass obligate allele; PI, paternity index.

2.3.7 Motherless Paternity Testing

Occasionally, it is necessary to conduct a paternity testing when the mother is unavailable for analysis. In Germany, these cases are labeled as motherless or deficient tests, because not all parties are present. In such cases, there is a considerable loss of information. The formula used for calculating the frequency of potential fathers in a given population is as follows:

$$RMNE = p^2 + q^2 + 2pq + 2p(1-p-q) + 2q(1-p-q) = (p+q)^2 + 2(p+q)(1-p-q)$$

Accordingly, the PI is calculated differently. If one parent is missing, a special formula for each individual phenotype of an alleged parent is used. Formulas taken from Brenner (1993) and further elaborated based on the data from www.DNA-view.com are shown in Table 2.4. To demonstrate the loss of information, as well as differences in the results, the paternity test, shown in Table 2.3, the CPI, and exclusion probability were recalculated without the data on the mother, using appropriate formulas from Table 2.4. The results are shown in Table 2.5. The first thing that one would notice is that the power of exclusion decreases approximately 20,525-fold, whereas the CPI drops only about 760-fold; this illustrates the differences between the two methods for calculating the probability of parentage. In these instances, the CPI tends to overestimate the likelihoods. Since the CPI method is used more frequently, the greater loss of information is not noticed. However, it is clear the RMNE method is the more conservative of the two.

Table 2.4 **Formulas for Calculating the Index of Paternity or Maternity in the Absence of the Other Parent**

Alleles of a Child	Alleles of a Tested Parent	PI
AA	AA	1/a
AB	AA	1/2a
AA	AB	1/2a
AB	AB	A + b/4ab
AB	AC	1/4a

Source: www.DNA-view.com,
Notes: A, B, C—alleles; a, b, c—frequency of A, B, C alleles.

Table 2.5   **Example of Motherless Paternity Testing**

STR loci	Child	Alleged Father	Obligate Alleles	*p*(Y)	RMNE	PI
D8S1179	11, 13	11, 13	11, 13	0.0564 0.3487	0.6461	5.1496
D21S11	31, 32.2	30.2, 31	31, 32.2	0.0590 0.1128	0.3141	4.2373
D7S820	9, 10	9, 9	9, 10	0.1795 0.2590	0.6847	2.7855
CSF1PO	10, 12	10, 12	10, 12	0.2538 0.3538	0.8460	1.6916
D3S1358	15, 18	15, 18	15, 18	0.1487 0.2538	0.6430	2.6663
TH01	6, 8	6, 8	6, 8	0.1308 0.2641	0.6338	2.8580
D13S317	10, 11	11, 11	10, 11	0.3308 0.0615	0.6307	1.5115
D16S539	11, 13	9, 13	11, 13	0.1718 0.2795	0.6990	1.4552
D2S1338	17, 23	23, 25	17, 23	0.1026 0.2154	0.5348	2.4366
D19S433	13, 16	12, 13	13, 16	0.2128 0.0641	0.4772	1.1748
VWA	16, 17	17, 17	16, 17	0.2256 0.2205	0.6932	2.2163
TPOX	9, 9	8, 9	9	0.0846 0.0846	0.3097	5.9102
D18S51	15,17	14, 15	15, 17	0.1359 0.0897	0.4003	1.8396
D5S818	11, 11	11, 12	11	0.3256 0.3256	0.8784	1.5356
FGA	20, 25	20, 25	20, 25	0.1000 0.1231	0.3965	2.5000
Amelogenin	X Y	X Y				
Combined					1.61×10^{-4}	826 693
Power of exclusion	1 – CRMNE				99.9839%	
Likelihood	1/CRMNE				6170	
Probability of paternity	CPI/CPI + 1				99.9838%	99.9998%

Notes: In the absence of the mother, there is no identification of the paternal alleles, thus the biological father could contribute either allele. p(Y), frequency of each allele or each obligate allele based on the Croatian database (Projic et al. 2007); PI, paternity index; RMNE, random man not excluded.

2.3.8 Effect of Mutations

One of the open, but very important, issues is the potential occurrence of mutations, which can lead to discrepancies between a DNA profile of the alleged father and a specified set of obligate alleles, on one but sometimes even on two molecular markers. This issue points to the fact that paternity/maternity may not be automatically rejected if the inconsistency in allele variants is observed on 1 out of 15 tested molecular markers. In this case, complex statistical forms are applied that take into account the frequency of mutations at specific loci (more information can be found on www.DNA-view.com).

After that the probability of paternity is calculated which, according to the experience of the majority of laboratories, in cases of standard paternity testing, as noted earlier, must exceed a value of 99.999% and in cases of motherless paternity testing a value of 99.99% (European standards). When this does not happen, and there is a likelihood of mutations, it is customary to request further testing on the X or Y related markers (depending on the child's gender). It is important to note that the more markers are used in analysis, the bigger the chance to detect the mutations at these loci. However, these cases are not too frequent and are treated differently in different countries. In most cases, only two exclusions are enough for a tested person to be automatically excluded as the biological parent of a child (*two-exclusion rule*) (Promega 1996). However, as more and more parentage cases are tested with increasing numbers of loci, it has been observed that two exclusions can occur by chance due to mutations, and it is necessary to carefully evaluate cases with only two exclusions, as most non-parents have an average of four to five exclusions (Promega 1996).

2.3.9 Maternity Testing

Sometimes, there is a need to find out who the biological mother of an abandoned or a deceased child is. In this case, unlike paternity testing, there is no knowledge about the child's parents. Therefore, as with any other forensic evidence, the genetic profile of the child is considered a proof. If a child is a heterozygote at a particular locus, then there are two obligate alleles. Conversely, when the child is a homozygote, there is only one obligate allele.

If there is no information about the mother, the formula for calculating *random female not excluded* (RFNE) is the same as the one used for computing RMNE. The mother is excluded from further analysis if she does not have either of the child's two alleles. But a mother cannot be excluded if she has only one of the child's two alleles. If this counts for all loci, the alleged mother cannot be excluded. *Combined random female not excluded* (CRFNE) is the product of all individual RFNEs and displays the percentage of women in the population that cannot be excluded at random or may be falsely included. The opposite is 1/CRFNE, showing the number of women who should be tested in order to reach the matching result. 1/CRFNE is also the Bayesian probability of maternity. Conversely, it is possible to calculate the index of maternity based on the passage of alleles, similar to the aforementioned PI, and convert it into Bayes's probability. So, in case of lack of information about the father, the maternity index is calculated by the formulas shown in Table 2.4.

The principle is completely identical to the principle in Table 2.5, only the allelic variants identified in the child's DNA profile are compared with allelic variants in the mother's DNA profile.

2.3.10 Parentage Testing with Mixed Populations

When calculating the parentage index, it is common to use the allele frequencies for the alleged father or mother depending on the question. If the alleged father or mother originates from the mixture of two populations, such as Croatian and sub-Saharan African, then the frequency of the obligate allele, if possible, should be the average of the two frequencies. On the other hand, if the alleged parent is Italian, and the only database available is the one in Croatia, it will not drastically affect the results as long as populations from the same geographical area are used.

In reality, almost all human populations are mixed to some extent. Alleles specific to Central Asia, Northern Asia, and Africa may be found in European populations, as there is historical evidence of admixture with the Mongols, the Avars, and African slaves throughout history. It is not required to make any special adjustments for these historical events, because they will be included in the database of a population. On the other hand, if the person that we are testing is a tourist, for example, from Puerto Rico, it would be very desirable, if possible, to use the database from that country. However, if a specific database is not available, the analysis may lead to certain errors. If the genetic structure of the population in question is known, the frequency can be calculated using the allele frequencies proportional to the corresponding subpopulation.

Precisely because of these reasons, journals such as *Forensic Science International*, *Forensic Science International: Genetics*, *Journal of Forensic Science*, *Internal Journal of Legal Medicine*, as well as *Croatian Medical Journal* often publish population data for a large number of the world's populations. This approach allows comparative analysis in order to investigate if there are statistically significant differences between the populations, and to enable the accurate calculation of certain forensic genetics parameters.

2.4 Identification of Human Body Remains

Identification of the human remains may be achieved by a number of methods, depending on the circumstances, as well as the condition of the human remains. If the decedent is recovered before decomposition, visual identification of the remains by a close relative is possible as is identification of the victim using scars or other physical marks (tattoos), fingerprints (if any), dental record or fractures, and medical implants as well as DNA analysis.

In wartime, identification of victims is often difficult, because a large number of bodies are often buried in mass graves, and *premortem* data are not always available as well (Gunby 1994). Development of new technologies for the analysis of genomic (autosomal as well as sex) and mitochondrial DNA provided forensic experts with new tools for identification. Furthermore, DNA analysis has become a standard method that is used for the confirmation/rejection of results of previous analyses, even in cases where the level of recognition and possibilities for body identification are relatively good. Because of the rapid decay of soft tissue of bodies in war or in mass catastrophes, forensic experts often can only perform analysis of DNA from the skeletal and dental remains for identification purposes (Gaensslen and Lee 1990). Several authors emphasize that DNA extraction methods that precede the amplification of DNA are of major importance for a higher percentage of successful identifications (Fisher et al. 1993; Ortner et al. 1992; Hochmeister et al. 1991; Burgi 1997). Furthermore, it is stated that the analysis of mtDNA is the best choice if working with highly degraded material. However, using modified standard procedures of DNA extraction and purifying the isolated DNA (repurification) with NaOH or other chemicals, it was found that the success rate of identification by genomic DNA may reach more than 96% (Andjelinović et al. 2005; Primorac 1999; Primorac et al. 1996; Keys et al. 1996; Alonso et al. 2001). It was also noted that the analysis of dental remains in comparison with the results of the analysis of long bones gives better results in 20%–30% of all cases analyzed (Primorac 1999).

2.4.1 Victim Identification Using Parental DNA

In the previous section, we discussed identification of the father or mother of a living person. Now we will talk about the identification of remains, which is also possible if either parents or children of a presumed victim are available for testing. Sometimes, in order to determine the identity of the victim, reconstruction is necessary. The first case involves a missing person who is assumed to be dead, but the body is yet to be found (e.g., in one's home were found evidence of blood, the person is missing, and the body has not been found). Do the blood stains found at the scene belong to the person who lived there and whose parents are available for testing? In the second case, the human remains are found in a mass grave or they are from the murder case, and cannot be identified in any other way than through genetic analysis.

In the process of identifying the remains, the calculation is slightly different from that in paternity testing. In this case, instead of determining the frequency of an obligate allele in a population of potential fathers or mothers (RMNE or RFNE), it is necessary to compute the likelihood of the allele in common in each of the parents. If we go back to the product rule for combining probabilities, the probability of finding a parent with allele A and the other parent with allele B is the frequency of potential parents with allele A multiplied by the frequency of potential parents with allele B. In practice, it is calculated by computing the RMNE/RFNE for each allele, which is then multiplied by the specific RMNE for the first allele multiplied by RFNE for the second allele to obtain a probability of two parents who have the same allele. This method is called *random parents not excluded* (RPNE). Therefore, the formula for calculating RPNE for any two alleles, regardless of whether they are similar or different, is as follows: $(p^2 + 2p[1 - p]) \times (q^2 + 2q[1 - q])$. For example, using the locus D3S1358 and results in an unidentified stain of D3S1358 15,17 with an incidence of 0.2425 and 0.1875 for each allele, the probability of finding two parents, one with allele *D3S1358 * 15* and the other with allele *D3S1358 *17*, would be

$$\left(0.2425^2 + 2 \times 0.2425 \times 0.7575\right) \times \left(0.1875^2 + 2 \times 0.1875 \times 0.8125\right) = 0.4262 \times 0.3398 = 0.1448.$$

Accordingly, 14.5% of parents in Croatia would not be excluded as possible parents. On the other hand, 85.5% of parents in Croatia would be excluded by this simple test. By multiplying all RPNEs for each locus, a combined RPNE or *combined parents not excluded* (CRPNE) is created, generating a relative frequency or proportion of the population of possible parents. Similarly, the probability of parents who are not excluded from the analysis is 1/CRPNE. Bayes's probability of parentage, utilizing the method of nonexclusion and a priori probability of 0.5, is 1 – CRPNE.

Identification of body remains found in Croatia and Bosnia was carried out before the STR technique became available (Primorac et al. 1996). In this case, the population of potential parents is computed for each allele (column *p*), which means that the alleles shown in the RMNE column were multiplied by each other and the product was RPNE for each locus (Table 2.6). At that time the data were based on an analysis of DQA1 and Polymarker, which are less informative than STR (Keys et al. 1996; Gill and Evett 1995; Budowle et al. 1999), and the values are lower than those shown in previous tables. In this case, only 0.4% of couples in Croatia could be possible parents of the found victims, or 99.62% of couples would be excluded from further analysis. This means that the chances were 260 to 1 that some of the couples are parents of the found victims, or that there is a 99.62% probability to determine the parentage by the method of nonexclusion.

Table 2.6 Identification of Remains of Child Using Parents

Locus	Mother	Bone	Father	Obligate Alleles	p	RMNE	RPNE
LDLR	B	A,B	A	A	0.410	0.6519	
				B	0.590	0.8319	0.5423
GYPA	A,B	A,B	A,B	A	0.560	0.8064	
				B	0.440	0.6864	0.5535
HBGG	A	A,B	B	A	0.530	0.7791	
				B	0.470	0.7191	0.5603
D7S8	A	A,B	A,B	A	0.653	0.8796	
				B	0.347	0.5736	0.5045
GC	A,C	A,C	A,C	A	0.279	0.4802	
				C	0.595	0.8360	0.4014
DQA1	2,4	3,4	2,3	3	0.105	0.1990	
				4	0.340	0.5644	0.1123
Combined							0.0038
Likelihood of parenthood							261
Probability of parenthood							99.62%

2.4.2 Victim Identification Using Child's DNA

If the missing person is a father whose children are available, identification is usually carried out following the procedure described earlier. In Table 2.7, data are presented for the identification of NN, who was assumed to be the father of the child. The circumstances are more favorable if the mother is also available for analysis. The testing results of mother and child indicate that 99.9969% of unrelated males would be excluded; however, the remains of NN were not excluded. Calculating the SI and PI suggests that it is approximately 45,000 times more likely that the NN is the father of a boy than another unrelated man, or more precisely,

Table 2.7 Identification of Remains Using a Child of the Decedent

Loci	Wife	Child	Obligate Allele	Remains NN	p	RMNE	X	SI
D3S1358	14,16	15,16	15	15,17	0.2425	0.4262	0.50	2.062
VWA	15,16	16,19	19	19	0.0900	0.1719	1.00	11.111
FGA	21,22	21,22	21	21,22	0.2125			
			22		0.1575	0.6031	1.00	2.703
THO1	7,8	7,8	7	8,9	0.1724	FBI		
			8		0.1150	0.4922	0.50	1.740
TPOX	11	10,11	10	9,10	0.0625	0.1211	0.50	8.000
CSF1PO	10,12	10	10	10,11	0.2750	0.4744	0.50	1.818
D5S818	11,12	10,12	10	10,12	0.0775	0.1490	0.50	6.452
D13S317	8,12	8,11	11	11,12	0.3350	0.5578	0.50	1.493
D7S820	8,11	8	8	8,10	0.1650	0.3028	0.50	3.030
AMEL	X,X	X,Y		X,Y				
Combined						3.14×10^{-5}		45,720
Exclusion power						99.9969%		
Likelihood						31,812		
Probability of paternity						99.99781%		

the probability that NN is the boy's father is 99.9978%. The method of nonexclusion exhibited the probability of almost 32,000 to 1, that is, the probability of paternity is 99.998%. The two methods coincide with each other, and it is fairly certain that the remains found are of those of the boy's father. It is easy to notice that this table is based on the use of only nine STR loci and on the frequencies that are not entirely a realistic presentation of the situation. Such an approach was deliberately designed to show how, despite all drawbacks that were being solved along the way, strong results were achieved during the past 15 years in the DNA identification project, and to warn that even when there is only a partial profile (because it is not uncommon that, out of a total of 20 loci examined, only 9 to 10 loci can be profiled on the skeletal remains), a statistically valid identification can be accomplished.

Identification of war victims in Croatia was one of the greatest scientific challenges, not only for Croatians, but also for the world's forensic scientists. The experiences gained through this project, as well as in a similar project that was implemented in Bosnia and Herzegovina, under the coordination of the International Commission on Missing Persons, have found global applications. One very interesting project, whose first phase has been completed, was the identification of victims from a mass grave in Slovenia dated to the end of World War II (Marjanovic et al. 2007; Marjanovic et al. 2015). The success of this analysis gives us hope that in the near future, this method will allow for the identification of victims of that conflict, who died during World War II and in the postwar years.

2.4.3 Parentage Testing versus Forensic Identification

In normal parentage testing, the frequency of possible parents (mothers or fathers) is determined by the frequency of RM/FNE or population of potential allele donors. In forensic identification, the frequency of potential phenotype (genotype) donors is relevant (Schanfield 2000). Table 2.8 shows a comparison of the results of parentage testing [RMNE $= (p + q)^2 + 2(p + q) (1 - p - q)$] and the results of forensic identification ($2pq$). In this case, the paternity donor population (RMNE) is approximately seven times bigger than the frequency obtained from forensic identification. Thus, there is almost an entire order of magnitude difference at one locus between the two methods of identification.

The estimates shown in Table 2.8 are based on the heterozygous phenotype. If the results were for a homozygote, p^2 would serve for the assessment of HWE, assuming there is no significant deviation from HWE due to population substructuring or some other event. The U.S. National Research Council recommends that the substructuring correction factor should be taken into account for all populations, even those in which substructuring was not observed, such as Croatia. In this case, to assess the HWE, the formula $p^2 + p(1 - p) \times (0.01)$ is used (NRC 1996). The example in Table 2.9 shows that in that case for a homozygote sample at D3S1358 (15, 15), the calculation for the expected match likelihood would be $0.2425^2 + 2 \times 0.2425 \times$

Table 2.8 **Comparison of Calculating Population Frequencies at One Locus in the Case of Paternity Testing versus Forensic Identification**

	RMNE (Paternity)	Forensic Evidence
Allele	D3S1358 15, 17	D3S1358 15, 17
Frequency	$(p + q)2 + 2(p + q) (1 - p - q)$	$2pq$
	0.6751	0.0909

Notes: $p(15) = 0.2425$, $q(17) = 0.1875$.

Table 2.9 **Comparison of Calculating Population Frequencies at One Locus in the Case of Paternity Testing versus Forensic Identification with U.S. National Research Council Recommendations**

	RMNE (Paternity)	Forensic Evidence
Allele	D3S1358 15, 15	D3S1358 15, 15
Frequency	$P^2 + 2p(1 - p)$	$P^2 + 2p(1 - p) \, (0.01)$
	0.4262	0.0625

Note: $p(15) = 0.2425$.

$0.7575 \times 0.01 = 0.0625$, because the RMP for a homozygous person is $p^2 + 2p(1 - p)$ 0.01. Note: The substructuring value of (θ or F) for large populations is taken to be 0.01; for small or inbred populations, such as Native Americans or Roma, the value of 0.03 is used.

It is clear that it is much easier to obtain very low match likelihoods for forensic evidence rather than a parentage situation. The final step in forensic identification is to unite any individual measures in order to obtain the frequencies of the multilocus genotype or profile. This is achieved by multiplying the frequencies of the genotypes of each locus.

References

Allen, R. 1999. *The American Association of Blood Banks (AABB) Annual Meeting. Paternity Testing, Specialist Interested Group.* Annual Report. San Francisco, CA: AABB.

Alonso, A., Š. Andjelinović, P. Martin et al. 2001. DNA typing from skeletal remains: Evaluation of multiplex and megaplex systems on DNA isolated from bone and teeth samples. *Croat Med J* 42(3):260–266.

American Association of Blood Banks 53rd annual meeting. November 4-8, 2000. Washington DC, USA. Abstracts. Transfusion. 2000 Oct;40(10 Suppl):1S-175S.

Andjelinović, Š., D. Sutlović, I. Erceg-Ivkošić et al. 2005. Twelve-year experience in identification of skeletal remains from mass graves. *Croat Med J* 46(4):530–539.

Brenner, C. 1993. A note on paternity computation in cases lacking a mother. *Transfusion* 33(1):51–54.

Budowle, B., T.R. Moretti, A.L. Baumstark et al. 1999. Population data on the thirteen CODIS core short tandem repeat loci in African Americans, U.S. Caucasians, Hispanics, Bahamians, Jamaicans, and Trinidadians. *J Forensic Sci* 44(6):1277–1286.

Budowle, B., R. Chakraborty, G. Carmody et al. 2000. Source attribution of a forensic DNA profile. *Forensic Sci Commun* 2(3) https://www.researchgate.net/publication/267683506_Source_Attribution_of_a_Forensic_DNA_Profile_Forensic_Science_Communications_July_2000.

Burgi, S.B., ed. 1997. *First European–American Intensive Course in PCR Based Clinical and Forensic Testing. Laboratory Manual.* September 23–October 3 1997. Split: Laboratory for Clinical and Forensic Genetics.

Butler, J.M. 2010. *Fundamentals of Forensic DNA Typing.* San Diego, CA: Elsevier Academic Press.

Butler, J.M. 2011. *Advanced Topics in Forensic DNA Typing: Methodology.* San Diego, CA: Elsevier Academic Press.

Council of Europe. 1992. Committee of Minister's. Recommendation no. R (92). 1 on the Use of Analysis of Deoxyribonucleic Acid (DNA) within the Framework of the Criminal Justice System. 470th meeting of the Minister's Deputies.

FBI. 2021.The FBI's Combined DNA Index System (CODIS) Hits Major Milestone. https://www.fbi.gov/news/press-releases/press-releases/the-fbis-combined-dna-index-system-codis-hits-major-milestone (accessed November 28, 2022)

Fisher, D.L., M.M. Holland, L. Mitchell et al. 1993. Extraction, evaluation and amplification of DNA from decalcified and un-decalcified United States civil war bone. *J Forensic Sci* 38(1):60–68.

Gaensslen, R.E., and C.H. Lee. 1990. Genetic markers in human bone tissue. *Forensic Sci Rev* 2:126–146.

Gill, P., and I. Evett. 1995. Population genetics of short tandem repeat (STR) loci. In *Human Identification: The Use of DNA Markers*, ed. B. Weir. Dordrecht: Kluwer Publishing, 69–87.

Gunby, P. 1994. Medical team seeks to identify human remains from mass graves of war in former Yugoslavia. *JAMA* 272(23):1804–1806.

Hochmeister, M.N., B. Budowle, U.V. Borer et al. 1991. Typing of deoxyribonucleic acid (DNA) extracted from compact bone from human remains. *J Forensic Sci* 36(6):1649–1661.

Irish Times. 2009. French men's insecurity over paternity of offspring creating a society of doubt .https://www.irishtimes.com/news/french-men-s-insecurity-over-paternity-of-offspring-cre-ating-a-society-of-doubt-1.773569 (accessed May 10, 2020)

Kaštelan, J., and B. Duda. 1987. *Bible—The Old and New Testaments*. Zagreb: Kršćanska sadašnjost (Croatian edition).

Keys, K.M., B. Budowle, Š. Andjelinović et al. 1996. Northern and Southern Croatian population data on seven PCR-based loci. *Forensic Sci Int* 81(2–3):191–199.

Li, C.C., and A. Chakravarti. 1988. An expository review of two methods of calculating a paternity probability. *Am J Hum Genet* 43(2):197–205.

Marjanović, D., N. Bakal, N. Pojskić et al. 2006. Allele frequencies for 15 short tandem repeat loci in a representative sample of Bosnians and Herzegovinians. *Forensic Sci Int* 156(1):79–81.

Marjanović, D., A. Durmić-Pasić, N. Bakal et al. 2007. DNA Identification of skeletal remains from the Second World War mass graves uncovered in Slovenia—First results. *Croat Med J* 48:520–527.

Marjanović, D., N. Hadžić Metjahić, J. Čakar, M. Džehverović, S. Dogan, E. Ferić et. al. 2015. Identification of human remains from the Second World War mass graves uncovered in Bosnia and Herzegovina. *Croat Med J* 56(3), 257–262.

NRC. 1996. *The evaluation of forensic DNA evidence, Committee on Forensic DNA: an Update.* Washington, DC: National Research Council, National Academy Press.

Ortner, D.J., N. Tuross, and A.I. Stix. 1992. New approach to the study of disease in archeological new world populations. *Hum Biol* 64(3):337–360.

Poetsch, M., A. Preusse-Prange, T. Schwark, and N. von Wurmb-Schwark (2013). The new guidelines for paternity analysis in Germany-how many STR loci are necessary when investigating duo cases? *Int J Legal Med* 127(4), 731–734.

Primorac, D., Š. Andjelinović, M. Definis-Gojanović et al. 1996. Identification of war victims from mass graves in Croatia, Bosnia and Herzegovina by the use of standard forensic methods and DNA typing. *J Forensic Sci* 41:891–894.

Primorac, D. 1999. Identification of human remains from mass graves found in Croatia and Bosnia and Herzegovina. In Proceedings of the 10th International Symposium on Human Identification. September 29–October 2, 1999. Orlando, FL and Madison, WI: Promega Corporation.

Primorac, D., S.M. Schanfield, and D. Primorac. 2000. Application of forensic DNA in the legal system. *Croat Med J* 41:33–47.

Promega statistics workshop. 1996. September 16–18, 1996. Scottsdale, AZ.

Projić, P., V. Škaro, I. Šamija et al. 2007. Allele frequencies for 15 short tandem repeat loci in representative sample of Croatian population. *Croat Med J* 48(4):473–477.

Schanfield, M. 2000. DNA: Parentage testing. In *Encyclopedia of Forensic Sciences*, eds. J.A. Siegel, P.J. Saukko, and G.C. Knupfer. London: Academic Press, 504–515.

Service Public. 2020. In what context can a paternity test be carried out? https://www.service-pub-lic.fr/particuliers/vosdroits/F14042?lang=en (accessed November 28, 2022)

Weiner, A.S. 1976. Likelihood of parentage. In *Paternity Testing in Blood Grouping*, ed. L.N. Sussman. Springfield, VA: Charles C. Thomas, 124–131.

Weir, B.S. 1993. Population genetics of DNA profiles. *J Forensic Sci Soc* 33(4):218–225.

Forensic Aspects of mtDNA Analysis

3

MITCHELL M. HOLLAND
AND WALTHER PARSON

Contents

3.1 Mitochondrion and mtGenome Structure

Mitochondria, organelles found in the cytoplasm of most cell types, contain a second intracellular DNA genome. According to the widely accepted endosymbiotic theory of mitochondrial origin, mitochondria were derived from α-proteobacteria that lived approximately 2 billion years ago inside pre-eukaryotic cells resembling ancient protists (Gray 1992; Lang et al. 1997; Gray et al. 2004). Over the course of evolutionary time, the endosymbiont lost its ability to survive outside the eukaryotic cell, and portions of its associated DNA were retained (Andersson et al. 1998; Dolezal et al. 2006). The endosymbiont became the mitochondrion, and its genome became mitochondrial DNA (mtDNA). The mitochondrion continues to play a vital role in the production of cellular energy by means of oxidative phosphorylation and the synthesis of ATP molecules. Gene products of the mitochondrial genome (mtGenome) work together with the products of hundreds of genes located in the nuclear DNA (nDNA) genome that are necessary for organelle function and, therefore, are directly linked to the survival of the cell.

The first complete mtGenome sequences reported in the literature came, perhaps surprisingly, from mice and humans (Bibb et al. 1981; Anderson et al. 1981). As expected, the mouse and human genomes are quite similar in size and structure, with identical gene organization. The human mtGenome is a circular molecule of approximately 16,568 nucleotides in length (Andrews et al. 1999); revised from the original 16,569 nucleotides (Anderson et al. 1981) and referred to as the revised Cambridge reference sequence (rCRS). A noncoding portion of the mtGenome (nucleotides 16024-16568 and 1-576) is referred to as the control region (CR), given the presence of start sites for replication and transcription. The CR and coding region contain considerable information content for identification and phylogenetic purposes.

DOI: 10.4324/9780429019944-4

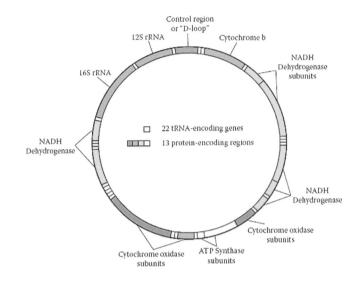

Figure 3.1 All portions of the mtGenome not colored blue represent coding information. A total of 37 genes are identified, including 22 for transport RNAs (tRNAs), 2 for ribosomal RNAs (rRNAs; the 12S and 16S subunits), and 13 genes for enzymes in the respiratory chain involved in the process of oxidative phosphorylation and ATP production. The portion colored in blue represents the control region (CR), where the initiation of replication and transcription can be found. (From Wikimedia Commons.)

A schematic representation of the human mtGenome can be found in Figure 3.1. Detailed information about the mtGenome is available on the internet in a variety of general and specialized databases (for example, MITOMAP; Lupi et al. 2010). The complementary strands of mtDNA sequence are quite different in their composition. The "heavy strand" is rich in purine nucleosides (adenosine and guanosine), whereas the "light strand" is subsequently rich in pyrimidine nucleosides (thymidine and cytidine). The coding region encompasses approximately 93% of the genome, with gene sequences densely arranged. There are 37 genes found in the coding region: 22 genes for transport RNAs (tRNAs), two genes for ribosomal RNAs (rRNAs; the 12S and 16S subunits), and 13 genes for enzymes in the respiratory chain involved in the process of oxidative phosphorylation and ATP production. Function, mutation, hereditary disorders, and population variability related to the mtGenome sequence have been studied in great detail (Arnheim and Cortopassi 1992; Wallace 2010; Lodeiro et al. 2010; Irwin et al. 2011; Federico et al. 2012; Murphy et al. 2016; Stenton and Prokisch 2020). Specific deletions, duplications, and point mutations can lead to a variety of pathophysiologic syndromes. In addition, somatic deletions and mutations that are acquired and accumulated during the life of an individual are held to be one of the most important causes for degeneration of cells and tissues in the aging process (Arnheim and Cortopassi 1992). For example, work on patients exposed to nucleoside analogue antiretroviral drugs to combat HIV has supported these assertions by connecting the progressive accumulation of somatic mtDNA mutations to premature aging in these patients (Payne et al. 2011).

The CR accounts for approximately 7% of the mtGenome and contains a displacement loop (D-loop), as during the initiation of transcription and replication DNA in the CR forms a three-stranded loop structure (Brown and Clayton 2006). The 1122 base pair (bp) CR includes promoters for transcription of polycystronic messages of the heavy and light strands, along with the initiation of heavy strand replication, highlighting the fact that although no gene sequences exist in the CR, it still contains vital regulatory elements for function. Subsequent initiation of light strand replication occurs at one or more sites in the coding region, from a primary initiation site or from multiple alternative sites. Given the reduced evolutionary pressures on the CR, mtDNA profiles in this region are quite variable between individuals in the general population.

The numbering system of the circular mtGenome begins in the middle of the CR (i.e., position 1) and proceeds in a clockwise fashion until it reunites with position 1 at nucleotide position 16568 of the rCRS. The CR is typically defined as the sequence between nucleotide position (np) 16024 (immediately following the tRNA for proline) and position 576 (immediately before the tRNA for phenylalanine), having crossed over position 1 in the genome's numbering system. There are two hypervariable (HV) segments located in the CR that are densely polymorphic. Hypervariable region 1 (HV1) begins at nucleotide position 16024 and ends at 16365, and hypervariable region 2 (HV2) starts at 73 and ends at 340. Because of the high levels of variability within HV1 and HV2, these two segments have most often been exploited for forensic purposes (Holland and Parsons 1999; Melton et al. 2012; Forsythe et al. 2020). There are two variable regions (VRs) of DNA sequence that span the remaining portions of the CR: VR1 encompasses sequence positions 16366–72 and VR2 encompasses sequence positions 341–576 (Figure 3.2). These regions can often be

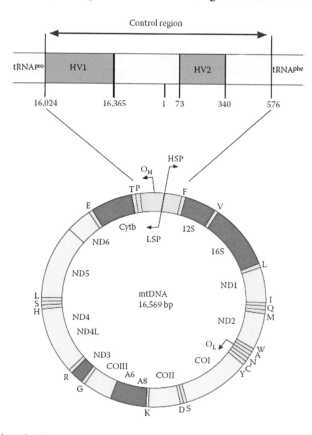

Figure 3.2 Mitochondrial DNA control region (CR). There are two hypervariable (HV) segments located in the CR that contain densely distributed polymorphic content. Hypervariable region 1 (HV1) begins at nucleotide position 16024 and ends at 16365, and hypervariable region 2 (HV2) starts at 73 and ends at 340. There are two variable regions (VRs) that span the remaining portions of the control region; i.e., VR1 encompasses sequence positions 16366–72, and VR2 encompasses sequence positions 341–576. (From Holland and Parsons 1999.)

used to distinguish between individuals with the same HV sequences (Lutz et al. 2000). Lastly, coding region sequences can be used to further differentiate between individuals with the same CR sequence (Coble et al. 2006; Parsons and Coble 2001).

3.2 mtDNA Copy Number

There are numerous mitochondria located inside a single eukaryotic cell, and inside a single mitochondrion there are multiple copies of the mtGenome resulting in a range of 100 to 10,000 copies per cell (Wai et al. 2010). The total number is typically dependent on the cell's energy requirements. For example, the average somatic cell may contain 200 to more than 1700 copies of mtDNA, whereas the higher energy requirements of a mature oocyte require >150,000 copies (Bogenhagen and Clayton 1974; Robin and Wong 1988; Piko and Matsumoto 1976; Michaels et al. 1982). Regardless of the precise number, the elevated copies typically enable the isolation of suitable quantities of mtDNA for sequence analysis from poor-quality forensic samples. In contrast to the two copies of each autosomal locus in a single cell, or the single copy of each Y chromosome locus in males, this phenomenon allows mtDNA analysis to remain accessible when nDNA testing has failed. As an example, given that >200 pg of nDNA is typically desired for short tandem repeat (STR) analysis in forensic casework, this translates into ~30 cells worth of DNA, assuming 6.6 pg of nDNA per cell. Within the DNA recovered from these 30 cells, there are ~60 copies of STR alleles (two copies per locus), but ~6000 to ~51,000 copies of mtDNA (assuming a per cell copy number of 200–1700). To further enhance the sensitivity of mtDNA testing, the analysis typically involves a higher number of polymerase chain reaction (PCR) cycles for amplification; for example, 38 cycles versus 28–30 cycles for STR amplification. Therefore, one can reliably obtain an mtDNA sequencing result from DNA samples containing femto-gram (subpicogram) levels of nDNA. Historically, the types of evidence where this has most often been applied is for the analysis of hair shafts and older, poorly preserved bones and teeth (Hopgood et al. 1992; Melton et al. 2005; Fisher et al. 1993; Ivanov et al. 1996; Ginther et al. 1992; Melton et al. 2012; Amorim et al. 2019). In recent years, the development of new DNA extraction techniques has made it possible to generate nDNA testing results on a higher percentage of older bones, and to successfully perform mtDNA analysis on the most highly degraded samples (Loreille et al. 2007). However, to date, nDNA analysis has remained an elusive and ineffective technique when analyzing hair shafts (Muller et al. 2007), unless scientists are able to combine hair material from multiple hairs or take samplings on the proximal end of the hair shaft near or including the root (Amory et al. 2007). It is quite possible that this will change in the near future as new methodologies are applied (Brandhagen et al. 2018).

3.3 mtDNA Inheritance

Relatives within the same maternal lineage will typically share the same mtDNA sequence; for example, all siblings, male and female, typically have the same mtDNA sequence as their mother. At first glance, the reason for this mode of inheritance appears to be simple.

The mature ovum has >150,000 copies of mtDNA, whereas the sperm cell entering the egg has only a few copies; the number of copies found in the midsection of the sperm tail ranges from ~2 to 100 copies (Chen et al. 1995; Wai et al. 2010). In addition, there is evidence for a mechanism that recognizes and eliminates copies of paternal mtDNA as they enter the oocyte (Kaneda et al. 1995; Sutovsky et al. 2000). A small number of studies have revealed the potential presence of paternal copies of human mtDNA in mature adults (Gyllensten et al. 1991; Schwartz and Vissing 2002), including an unsubstantiated report of pervasive paternal inheritance in multiple families (Luo et al. 2018; Lutz-Bonengel and Parson 2019). The Schwartz study was not confirmed in a second laboratory, was associated with a disease state, and the phenomenon has not been observed since, so it is difficult to know whether the findings are unique (Bandelt et al. 2005). Overall, the literature clearly supports the contention that for practical purposes of mtDNA analysis in forensic investigations, it is correct to assume that mtDNA will typically be inherited through the maternal line. Support for this assertion has been documented, including by researchers who were not able to identify paternal mtDNA in the somatic cells of offspring by means of routine methods used in forensic laboratories (Parsons et al. 1997). However, with the emergence of quantitative PCR methods and deep DNA sequencing, the validity of the paradigm that mtDNA is "strictly" maternal inheritance could be challenged in the future (Van Wormhoudt et al. 2011). The ability for these technologies to query pools of sequence and assess each copy of mtDNA from a cell or cluster of cells may allow researchers to determine whether paternal copies of mtDNA persist in adult humans, albeit at a low level. More important, these methods may allow researchers to determine whether copies of paternal mtDNA can be found in female germ cells.

3.4 Massively Parallel Sequencing

As of mid-2021, more than 150 peer-reviewed journal articles had been published that used some form of massively parallel sequencing (MPS) to study mtDNA variation for forensic applications. In addition, more than 1000 articles had been published on MPS analysis of mtDNA in all scientific disciplines, including anthropologic and medical genetics. These data reflect a growing interest in the use of MPS analysis to sequence mtDNA, and an increasing interest in mitochondrial genetics as it relates to human migration and health. This section of the chapter will not cover the origins or details of MPS technologies (Bruijns et al. 2018), nor the association to anthropology or medicine-oriented studies, but instead will focus on the forensic applications associated with mtDNA analysis.

The first studies of mtDNA analysis using an MPS approach were proof of concept for mixtures (Holland et al. 2011) and skeletal remains identification cases (Loreille et al. 2011). Soon thereafter, publications emerged that evaluated different sequencing chemistries and instruments for MPS analysis of the CR and the entire mtGenome (Parson et al. 2013; Mikkelsen et al. 2014; King et al. 2014a; McElhoe et al. 2014; Chaitanya et al. 2015), including concordance studies (Peck et al. 2016). Along the way, software alignment tools were developed to ensure that mtDNA MPS data is properly aligned and the resulting haplotypes reflect phylogenetically correct SNP (single-nucleotide polymorphism) and INDEL (insertion/deletion polymorphism) assignments (King et al. 2014b; Holland et al. 2017; Sturk-Andreaggi et al. 2017), with the latter two available as commercial products (GeneMarker™ HTS, SoftGenetics, Inc., State College, Pennsylvania; and CLC Genomics

Workbench, QIAGEN, Hilden, Germany, respectively). Standard reference material is also available for MPS mtDNA analysis (Riman et al. 2017), allowing forensic laboratories to assess the quality of their data.

More recent studies have begun to address the validation of methods for implementation in forensic casework (Peck et al. 2018; Hollard et al. 2019; Brandhagen et al. 2020; Cihlar et al. 2020; Holt et al. 2021) and on newly developed kits, which now allow for rapid analysis of mtGenome sequence from all sample types (Strobl et al. 2018; Woerner et al. 2018; Pereira et al. 2018; Holland et al. 2019; Cuenca et al. 2020). In addition, methods involving capture approaches (Marshall et al. 2017; Eduardoff et al. 2017; Shih et al. 2018; Gorden et al. 2021) and for the analysis of mtDNA mixtures (Kim et al. 2015; Vohr et al. 2017; Churchhill et al. 2018; Crysup et al. 2021; Smart et al. 2021) have been developed and are now available when faced with the most challenging sample types (for example, old, degraded skeletal remains), and when mixtures are encountered.

Hair shafts have been a typical sample type in forensic mtDNA casework (see Section 3.9). Therefore, considerable work has been done to evaluate MPS methods to analyze hairs (Parson et al. 2015; Desmyter et al. 2016) and to study the drift of mtDNA heteroplasmy in hairs (Gallimore et al. 2018). Overall, the forensic community is well positioned to transition into MPS mtDNA analysis, which will allow for greater levels of discrimination potential when performing mtDNA analysis, will encourage more laboratories to adopt the testing method, and will put those laboratories in position for the analysis of SNP and STR data when using an MPS approach.

3.5 Alignment, Nomenclature, and Databasing of mtDNA Profiles

The results of mtDNA sequence analysis are typically a string of nucleotides, or bases (FASTA files for MPS analysis), that are conveniently reported relative to the 16,568 bp rCRS (Andrews et al. 1999; NC012920) of the first sequenced human mtGenome (Anderson et al. 1981). The rCRS-coded haplotype is less complex and easier to report and compare to other mitohaplotypes than the full sequence. The rCRS stems from a Westeurasian sample (England) and, in fact, represents a very young lineage of the human mtDNA phylogeny. Thus, when mtDNA data profiles are presented relative to the rCRS, the resulting haplotype is in most cases older than the rCRS.

Naming conventions for substitutions (SNPs) relative to the rCRS follow the International Union of Pure and Applied Chemistry (IUPAC) code: identifying the rCRS base, followed by the nucleotide position (np) and the variant base. For example, the transition relative to the rCRS at np 152 is called T152C. The rCRS base can be skipped and the SNP designated as 152C. In population genetics, the mere mention of the position would indicate the transition (152 = T152C), whereas a transversion would be explicitly stated (152G). Mixtures of multiple bases at a position are called with the IUPAC codes. For example, 152Y indicates a mixture of T and C at that position and 152B would be a mixture of A, G, and T. The letter N is used to designate any base at a given position.

The context of SNP origin can be a complex issue. For example, the variant 263G does not mean that this is derived to the rCRS, as one may suggest, thinking in medical genetic terms of wildtype versus mutation constellations. The 263G SNP simply means that there is a difference at that position to the rCRS without any mutational direction or context. In fact, the rCRS is different from the majority of mtDNA sequences worldwide, which is why almost all other sequences have 263G in their haplotype.

Insertions are typically indicated 5 prime (medical genetics) or 3 prime (forensic and population genetics) of a particular base or homopolymeric region. For example, 309.1C or 315.1C represents tracts of 8 or 6 cytosines in regions where the rCRS has only 7 or 5 cytosines, indicating a cytosine insertion after np 309 and 315, respectively. Deletions are noted with "del" or "–" at the deleted position; for example, 249del or 249– for a deletion at np 249. When possible, the use of "D" for deletion should be avoided, as "D" is already used as an IUPAC code for a mixture of C, G, and T. The alignment of INDELs relative to the rCRS can sometimes pose a problem, including when length heteroplasmy is observed (Sturk-Andreaggi et al. 2020), as there can be more than one way to align the nps. For example, within the segment of 16184 to 16194, the two transitions at 16188 and 16189 (C16188T T16189C) refer to the same nucleotide sequence as 16188del 16193.1C. Both haplotypes, 16188T 16189C and 16188del 16193.1C, correspond to the same nucleotide sequence, but obviously appear to be different to an inexperienced practitioner. One can easily imagine that alignment differences may lead to confusion, wrongful exclusion of sequences that derive from the same (or related) individual, and database searches that fail to identify matches. For the latter, this would lead to an underestimation of the probability of a given haplotype and, thus, an overestimation of the weight of the evidence, which would put the suspect in a forensic case at a disadvantage. In order to avoid this, alignment-free search engine software has been developed (Röck et al. 2011; Huber et al. 2018). The concept of phylogenetic alignment and examples can be found in Bandelt and Parson (2008).

It is clear that alignment relative to the rCRS becomes increasingly difficult with (geographic and/or temporal) distance of the sequence from the rCRS. This is particularly true for INDELs. It has been shown that strict, most parsimonious alignment relative to the rCRS leads to jumping alignment that can put a sequence at a larger distance than the actual mutational events would suggest (see examples in Bandelt and Parson, 2008). In forensic practice, that could mean that the sequence of another tissue of the same individual (or from a close relative) could be put at further distance than it would when considering the mutational difference alone. For example, the haplotype 16183C 16188T 16189C 16193.1C represents a sequence identical to 16183del in a range between 16183 and 16194, although the former uses four differences relative to the rCRS and the latter only one. Therefore, the forensic community has decided to use phylogenetic alignment, or alignment to the next evolutionary neighbor (Bandelt and Parson 2008), rather than alignment to a fixed reference sequence, while the haplotype is still reported relative to the rCRS (Parson et al. 2014; SWGDAM Guidelines 2019). A possible limitation of the phylogenetic approach would be an unforeseen alignment change with newly observed mtDNA variation. However, practical experience with large datasets shows that this is expected to happen infrequently (Huber et al. 2018). In order to find the phylogenetic alignment of a given sequence, the EMPOP database (https://empop.online) can be used. EMPOP accepts any alignment or FASTA-like string of the query sequence and turns it into the phylogenetic representation of the haplotype according to phylogenetic principles.

In the early 1990s, small mtDNA databases were developed and maintained by individual laboratories dealing with mtDNA cases and comprised of samples from local populations. With the understanding of phylogenetic signatures of mtDNA sequences, and the relevance of regional subsets of mtDNA databases (for example, Bobillo et al. 2010), it has become clear that the quality of an mtDNA database search strongly correlates to the size of the database and the representativeness of its content. A centralized mtDNA databasing effort was undertaken by the EDNAP (European DNA Profiling) Group. They realized in the early 2000s that quality control is a key element for legally defensible mtDNA databases. The resulting product, EMPOP, increased the quality of mtDNA datasets by introducing sound laboratory protocols for mtDNA sequence generation (for example, Parson and Bandelt 2007) and phylogenetic tools (quasi-median networks) to control and monitor the quality of new datasets (Parson and Dür 2007). The current EMPOP release (v4/R13) holds 48,572 haplotypes from dozens of populations worldwide; 4289 mtGenome, 38,361 CR, and 43,573 HV1/HV2 sequences. While the majority of haplotypes only cover regions of the CR, entire mtGenome sequences will become increasingly available with the ongoing rapid production of MPS data.

3.6 mtDNA Heteroplasmy

Heteroplasmy is defined as a mixture of more than one mtGenome sequence within a cell or between cells of a single individual. As they accumulate, these heteroplasmic variants can lead to disease and can have a profound impact on the aging process (Arnheim and Cortopassi 1992; Wallace 2010; Payne et al. 2011; Nissanka and Moraes 2020). In a forensic context, heteroplasmic variants can provide additional layers of discrimination potential (Ivanov et al. 1996) and have become a routine part of the interpretation process in casework (Melton 2004; Melton et al. 2005; Melton et al. 2012). Therefore, because the presence of heteroplasmy can impact the interpretation of mtDNA comparisons between evidentiary and known samples encountered in forensic casework, it is important to understand the biological basis of heteroplasmy, how often it is observed in relation to the sequence detection method being used, and how best to interpret the results.

It is possible, and indeed very likely, that heteroplasmy exists in each individual. The γ-polymerase used for replicating the mtGenome is quite faithful (Longley et al. 2001), so the likely source of mutation is from DNA damage induced by reactive oxygen species (ROSs) generated during oxidative phosphorylation, followed by the actions of a less-than-adequate repair system (Yakes and VanHouten 1997). The repair of ROS damage occurs primarily through the base excision repair pathway (Driggers et al. 1993), and through homologous recombination and nonhomologous end joining pathways (Sykora et al. 2011). While this is an effective process, cells struggle to keep up with the rate of ROS damage in the mitochondrion. It is this constant challenge of combating mtDNA damage, and the failure ultimately to do so, that allows for the accumulation of heteroplasmic variants over the lifetime of an individual.

In addition to single-nucleotide heteroplasmy (SNH), there is considerable length heteroplasmy (LH) observed in the mtGenome, especially within long stretches of GC (guanine–cytosine) base pairs (Forster et al. 2010). LH is typically caused by slippage of the replicative polymerase, similar to the type of slippage that occurs at STR loci to create stutter artifacts. Two primary regions of the CR are hot spots for LH: 16,184–16,193 in HV1 when there is a T to C transition at np 16,189; and 303–309 in HV2, especially when there is a C to T transition at np 310. When the number of homopolymeric nucleotides reaches eight or more, γ-polymerase struggles to faithfully replicate these stretches of DNA and will create a set of length variants. The resulting family of length variants will produce uninterpretable and out-of-reading-frame data when performing conventional, Sanger-type sequencing (STS). The introduction of MPS analysis has helped to address this problem, especially when using chemistries that allow for sequential base incorporation; for example, the chemistry from Illumina, Inc. However, interpretation of LH remains challenging (Sturk-Andreaggi et al. 2020).

A genetic bottleneck theory was proposed in the early 1980s to explain the rapid segregation of heteroplasmic sequence variants in Holstein cows, providing a mechanism for the transmission of heteroplasmy from cell to cell or from mother to child (Hauswirth and Laipis 1982; Hauswirth et al. 1984). The theory purported a reduction in the number of mtDNA molecules during the development of germ cells. Only a small number of mtDNA molecules pass through the bottleneck and are subsequently replicated to >150,000 in the mature ovum. Researchers have attempted to identify where in the developmental process the bottleneck occurs and elucidate the mechanism of the bottleneck (Jenuth et al. 1996; Cree et al. 2008; Cao et al. 2009). Reports by Jenuth and Cree support a reduction mtDNA copy number in primordial germ cells, while Cao supports a theory that the bottleneck occurs through the selection of specific replicating units, potentially in conjunction with protein-rich structures called nucleoids (Satoh and Kuroiwa 1991; Kanki et al. 2004; Kukat et al. 2011; Lodeiro et al. 2012). Regardless, of the precise mechanism, the narrow bottleneck provides for considerable drift in heteroplasmic variants across the germline (Rebolledo-Jaramillo et al. 2014; Holland et al. 2018). In addition, bottlenecks exist during fetal tissue development that results in drift across and within tissues, including hair which exhibits higher rates of drift than blood and buccal cells (Naue et al. 2015; Gallimore et al. 2018; Wilton et al. 2018; Barrett et al. 2020).

When occurring in germ cells, de novo mutations in mtDNA are the basis for potential changes in lineage-based CR haplotypes over evolutionary time. However, complete fixation of these changes is most certainly preceded by a heteroplasmic state, which may persist for many generations. The frequency of observing heteroplasmy in humans, detected by conventional STS, ranges from 2% to 8% of the population, including examples of SNH at more than two sites within the CR (Melton 2004; Irwin et al. 2009). This relatively low rate is primarily due to the lack of sensitivity of STS in detecting minor sequence variants. At best, the Sanger method can detect minor variants at a ratio of 1:10 to 1:20, or 5%–10% of the major component. Low-level variants are masked by the predominant sequence, and even higher-level variants are not routinely reported by operating forensic laboratories. Deep sequencing MPS has emerged to allow for the detection and, more important, the resolution of low-level (<2%) heteroplasmic variants (Holland et al. 2011; Just et al. 2015; Holland et al. 2018; Gallimore et al. 2018; González et al. 2020). Rates of heteroplasmy in the population range from 30% to 60% when applying an MPS approach, and depending on whether the CR or entire mtGenome (for example, Holland et al. 2018) is being evaluated. Damage of mtDNA can impact the interpretation of heteroplasmy (Rathbun et al. 2017; Holland et al. 2019; Holland et al. 2021), but can often be repaired prior to analysis (Gorden et al. 2018). Overall, it is important to properly interpret heteroplasmy, while avoiding the reporting of mixed sites associated with system or instrument noise (McElhoe and Holland 2020), and poor alignment strategies (Just et al. 2014).

3.7 Nuclear mtDNA Segments

Nuclear mtDNA segments (NUMTs) are known as transpositions of mtDNA into the nuclear genome in eukaryotic cells (Lopez et al. 1994). A systematic study has demonstrated the breadth of NUMTs (Dayama et al. 2014) that were in the past often observed by coincidence and have been collected in databases (for example, MITOMAP; www.mito-map.org). The Dayama study revealed 141 yet unobserved NUMTs in the 1000 Genomes data, including one full and multiple near-full mtGenomes. The exact process of exchange is not fully understood (Hazkani-Covo et al. 2010), but it has been demonstrated that mtDNA copies can appear in higher copy number than just single instances; for example, 38-76X in the cat nuclear genome (Lopez et al. 1996). This has significant consequences in PCR-based molecular genetic applications, as mtDNA-specific PCR primers that span a NUMT can lead to co-amplification of the nuclear target. The chances of picking up NUMTs increase with reduced targeted amplicon size (Santibanez-Koref et al. 2019), which is particularly the case for midi-amplicon (Berger and Parson 2009) and mini-amplicon (Eichmann and Parson 2008) approaches used in forensic casework to successfully amplify severely degraded mtDNA. In such cases, the resulting sequence represents a mixture of the genuine mtDNA and the NUMTs. Most NUMTs are easily identified as they represent mutational patterns that are unknown to the mtDNA phylogeny (see MITOMAP) and can therefore be subtracted from the mixture (Ring et al. 2018; Smart et al. 2019; Marshall and Parson 2021). Also, it is important to note that mtDNA-specific quantification and use of the appropriate amount of mtDNA copies for PCR (1000–5000) mitigate the co-amplification of NUMTs. Repetitive NUMTs may not be avoidable in PCR-based approaches, and thus, need to be dealt with during analysis and interpretation of data. It is also known that NUMTs can consist of relatively recent transpositions from the cytoplasm (Lutz-Bonengel et al. 2021) that resemble modern extant lineages and thus appear as genuine mixtures of mtDNA. Such constellations could pose a problem when interpreting forensic evidence, but fortunately are rare occurrences.

3.8 Application of mtDNA Analysis to Forensic Cases

The analysis of mtDNA sequence has been used by forensic laboratories since the early 1990s to help exonerate individuals who were falsely convicted of crimes, to assist the trier of fact in criminal investigations, and to support the identification of military personnel and historical figures. Common sample types include hair shafts and old skeletal remains, as the high copy number mtGenome is an attractive alternative when STR analysis fails. Hair shafts are routinely recovered in criminal investigations, but in the United States are often sent to the FBI (Federal Bureau of Investigation), regional crime laboratories, or private laboratories for mtDNA analysis (see Melton et al. 2005 and Melton et al. 2012 for a review). The first criminal case in the United States that used mtDNA evidence was the 1996 *State of Tennessee v. Paul Ware* (Davis 1998), in which mtDNA analysis was used to associate hairs found in a rape victim's throat and on a bed sheet to Ware. Laboratories around the world continue to use mtDNA analysis to help solve crime and to provide evidence for exoneration in cases of false indictment or conviction.

Interestingly, the vast majority of DNA in a hair shaft is from the nuclear genome (Brandhagen et al. 2018). Unfortunately, the nDNA is highly fragmented, so conventional STR analysis is not an option. This is due primarily to the action of a nuclease that preferentially digests nDNA following cell death and initiation of the keratinization process of the hair. In the future, methods may be developed to test the small fragments of nDNA (average of 65 bp) in hair shafts through the analysis of identity SNPs. However, Brandhagen et al. (2018) suggest that the majority of hairs will most likely still require mtDNA testing.

The identification of skeletal remains associated with individuals of historical interest (for example, Gill et al. 1994; Ivanov et al. 1996; Coble et al. 2009; King et al. 2014; Bodner et al. 2015; Marshall et al. 2020a), mass fatality incidents (for example, Biesecker et al. 2005; Buś et al. 2019; Ambers et al. 2020), and military personnel (Holland et al. 1993; Palo et al. 2007; Piccinini et al. 2010; Ossowski et al. 2016) has routinely involved mtDNA testing. Laboratory procedures often involve specialized DNA extraction methods (for example, Loreille et al. 2007; Dabney et al. 2013) to recover minute quantities of highly degraded DNA. All procedures performed on human remains or on trace evidence must be performed in a laboratory separated from reference samples, and reference samples should be analyzed after the analysis of evidence, when possible, to minimize the chance of contamination. The PCR product generated from extracted DNA is subjected to MPS or STS. Once data have been confirmed (by analysis of the complementary strands of mtDNA sequence and, in most cases, through duplicate analysis), the results are compared to the rCRS (Andrews et al. 1999), and the differences are listed to generate a mitohaplotype (see Section 3.6). The list of differences becomes the mtDNA "profile" associated with the biological sample. Only then are different sample profiles compared to determine whether matches have occurred. In the case of a clear mismatch, a person can be eliminated as a potential donor of the sample or as a potential relative of the person who is being identified. Inconclusive results are possible when a single sequence difference exists or with LH, as the disparity can be the result of a severe shift in heteroplasmic variant ratios. In the case of a match, the profile must be compared to a population database and the probability of a coincidental match calculated (see Sections 3.6 and 3.10).

3.9 Genetic Variability and Random Match Probabilities

The strengths and limitations of conventional mtDNA testing have been well documented (Holland and Parsons 1999; Melton et al. 2012; Forsythe et al. 2020). Increasing the discrimination potential of mtDNA analysis has been an ongoing desire of the forensic community. These efforts have included increasing the size (and quality) of the search database (EMPOP), incorporating heteroplasmy in the interpretation process, and expanding the scope of analysis to the coding region (Lutz et al. 2000; Parsons and Coble 2001; Coble et al. 2006). However, STS presents practitioners with a difficult hurdle to overcome when attempting to sequence the entire mtGenome from forensically challenged samples (Lyons et al. 2013). With the introduction of MPS analysis, routine testing of the mtGenome is now available (McElhoe et al. 2014; Strobl et al. 2018; Pereira et al. 2018; Woerner et al. 2018), allowing for resolution of low-level heteroplasmy and the frequent differentiation of maternal relatives (Holland et al. 2018). Therefore, MPS enables practitioners to harness the maximum discrimination potential of the testing method. With that said, it is important to note that policy and legal questions will need to be addressed when reporting information associated with the mtDNA coding region (Scudder et al. 2018; Marshall et al. 2020b).

One of the first published examples of providing weight estimates associated with an mtDNA match was the identification of Nicholas Romanov, the last Russian tsar (Gill et al. 1994; Ivanov et al. 1996; Coble et al. 2009). The presence of heteroplasmy in the Romanov family allowed for the calculation of a likelihood ratio (LR) that took both the haplotype frequency and probability of observing heteroplasmy into consideration. The

LR for identity when based strictly on the haplotype of the skeletal remains was calculated as 150, and the LR for the presence of a heteroplasmic sequence shared by two brothers (Nicholas and Georgij Romanov) was calculated as 2500. Assuming the two events are independent of one another, the combined LR was 375,000 if the remains were, in fact, those of Nicholas Romanov versus a random person in the population. Recalculation of the LR using recently determined empirical mutation rates for the mtGenome (Rebolledo-Jaramillo et al. 2014), rates and number of sites of heteroplasmy across the CR (data from the Holland laboratory), along with an updated database search using the 41,385 HV1/HV2 EMPOP profiles (v4/R12), pushes the value above one million. Therefore, it is clear that reporting of heteroplasmy can significantly increase the discrimination potential of mtDNA testing.

The significance of a match in mtDNA analysis depends on the circumstances of the case and on the profiles in question. Definitive results of identification are possible only in cases with a closed population, for example, a traffic accident in which it is sufficient to identify individuals from a list of people involved in the accident. If a mitohaplotype match is made between a reference sample from a maternal relative and one of the investigated samples, and all other individuals have differing mitohaplotypes, a qualified positive identification can be established. Such certainty does not rely on calculation of absolute probabilities, but on the fact that the sample in question could only have originated from a small number of people, and the sample profile matches only one individual in the group.

Most forensic cases do not involve a closed group of individuals, and thus must consider the general population as a potential source of the sample. In this case, a match cannot be considered as a definitive finding and a statistical weight must be calculated. Two approaches are used to calculate a statistic. The Clopper–Pearson method (Clopper and Pearson 1934) is used when observations have been made in a database search and can be estimated using the following equation, where N is the database size, x is the number of observations of the mitohaplotype in the database, k is 0 to x observations, and p is the mitohaplotype frequency when x or fewer observations are expected 5% of the time, assuming a 95% upper confidence limit. A binominal distribution is solved for p through sequential iterations, requiring a computer algorithm to generate the statistic.

$$\sum_{k=0}^{x} \binom{N}{k} p_0^k \left(1-p_0\right)^{N-k} = 0.05$$

Using the following equation, a confidence limit from zero proportion is used when no observations are made in a database search, where p is the estimated frequency or probability of the profile in the general population, α is set at 0.05 for a 95% confidence level, and N is the database size:

$$p = 1 - \alpha^{1/N}$$

For example, in a database of 41,385 HV1/HV2 sequences, a case profile not observed in the database will have an estimated frequency of 7.24×10^{-5}. This means that, with 95% confidence, 99.99276% of the population can be excluded as potential donors of the case sample; or 1 in ~13,812 individuals in the general population will have the mitohaplotype. This corresponds to a *maximum* match probability and is not an estimate of the actual match probability, which in a great majority of cases would be a considerably lower value.

As the size of the search database increases, and the forensic community moves to mtGenome MPS analysis, including the interpretation of heteroplasmy, the probability estimates of matching mitohaplotypes will continue to improve. Therefore, it is expected that the global application of mtDNA analysis will increase in the future.

References

Ambers, A., M.M. Bus, J.L. King et al. 2020. Forensic genetic investigation of human skeletal remains recovered from the La Belle shipwreck. *Forensic Sci Int* 306:110050.

Amorim, A., T. Fernandes, and N. Taveira. 2019. Mitochondrial DNA in human identification: A review. *PeerJ* 7:e7314.

Amory, S., C. Keyser, E. Crubezy, and B. Ludes. 2007. STR typing for ancient DNA extracted from hair shafts of Siberian mummies. *Forensic Sci Int* 166(2–3):218–229.

Anderson, S., A.T. Bankier, B.G. Barrell et al. 1981. Sequence and organization of the human mitochondrial genome. *Nature* 290(5806):457–465.

Andersson, S.G., A. Zomorodipour, J.O. Andersson et al. 1998. The genome sequence of *Rickettsia prowazekii* and the origin of mitochondria. *Nature* 396(6707):133–140.

Andrews, R.M., I. Kubacka, P.F. Chinnery et al. 1999. Reanalysis and revision of the Cambridge reference sequence for human mitochondrial DNA. *Nat Genet* 23(2):147.

Arnheim, N., and G. Cortopassi. 1992. Deleterious mitochondrial DNA mutations accumulate in aging human tissues. *Mutat Res* 275(3–6):157–167.

Bandelt, H.-J., Q.-P. Kong, W. Parson, and A. Salas. 2005. More evidence for non-maternal inheritance of mitochondrial DNA? *J Med Genet* 42(12):957–960.

Bandelt, H.-J., and W. Parson. 2008. Consistent treatment of length variants in the human mtDNA control region: A reappraisal. *Int J Legal Med* 122(1):11–21.

Barrett, A., B. Arbeithuber, A. Zaidi et al. 2020. Pronounced somatic bottleneck in mitochondrial DNA of human hair. *Phil Trans R Soc B Biol Sci* 375(1790):20190175.

Berger, C., and W. Parson. 2009. Mini-midi-mito: Adapting the amplification and sequencing strategy of mtDNA to degradation state of crime scene samples. *Forensic Sci Int Genet* 3(3):149–153.

Bibb, M.J., R.A. Van Etten, C.T. Wright et al. 1981. Sequence and gene organization of mouse mitochondrial DNA. *Cell* 26(2 Pt 2):167–180.

Biesecker, L.G., J.E. Baily-Wilson, J. Ballantyne et al. 2005. DNA identification after the 9/11 World Trade Center attack. *Science* 310(5751):1122–1123.

Bobillo, M.C., B. Zimmermann, A. Sala et al. 2010. Amerindian mitochondrial DNA haplogroups predominate in the population of Argentina: Towards a first nationwide forensic mitochondrial DNA sequence database. *Int J Legal Med* 124(4):263–268.

Bodner, M., A. Iuvaro, C. Strobl et al. 2015. Helena, the hidden beauty: Resolving the most common west Eurasian mtDNA control region haplotype by massively parallel sequencing an Italian population sample. *Forensic Sci Int Genet* 15:21–26.

Bogenhagen, D., and D.A. Clayton. 1974. The number of mitochondrial deoxyribonucleic acid genomes in mouse L and human HeLa cells: Quantitative isolation of mitochondrial deoxyribonucleic acid. *J Biol Chem* 249(24):7991–7995.

Brandhagen, M.D., O. Loreille, and J. Irwin. 2018. Fragmented nuclear DNA is the predominant genetic material in human hair shafts. *Genes* 9(12):640–660.

Brandhagen, M.D., R.S. Just, and J.A. Irwin. 2020. Validation of NGS for mitochondrial DNA casework at the FBI laboratory. *Forensic Sci Int Genet* 44:102151.

Brown, T.A., and D.A. Clayton. 2006. Genesis and wanderings: Origins and migrations in asymmetrically replicating mitochondrial DNA. *Cell Cycle* 5(9):917–921.

Bruijns, B., R. Tiggelaar, and H. Gardeniers. 2018. Massively parallel sequencing techniques for forensics: A review. *Electrophoresis* 39(21):2642–2654.

Buś, M.M., M. Lembring, A. Kjellström et al. 2019. Mitochondrial DNA analysis of a Viking age mass grave in Sweden. *Forensic Sci Int Genet* 42:268–274.

Cao, L., H. Shitara, M. Sugimoto et al. 2009. New evidence confirms that the mitochondrial bottle-neck is generated without reduction of mitochondrial DNA content in early primordial cells in mice. *PLOS Genet* 5(12):e1000756.

Chaitanya, L., A. Ralf, M. van Oven et al. 2015. Simultaneous whole mitochondrial genome sequencing with short overlapping amplicons suitable for degraded DNA using the Ion Torrent Personal Genome Machine. *Hum Mutat* 36(12):1236–1247.

Chen, X., R. Prosser, S. Simonetti et al. 1995. Rearranged mitochondrial genomes are present in human oocytes. *Am J Hum Genet* 57(2):239–.247.

Churchill, J.D., M. Stoljarova, J.L. King, and B. Budowle. 2018. Massively parallel sequencing-enabled mixture analysis of mitochondrial DNA samples. *Int J Legal Med* 132(5):1263–1272.

Cihlar, J.C., C. Amory, R. Lagac et al. 2020. Developmental validation of a MPSWorkflow with a PCR-based short amplicon whole mitochondrial genome panel. *Genes* 11(11):1345.

Clopper, C.J., and E.S. Pearson. 1934. The use of confidence or fiducial limits illustrated in the case of the binomial. *Biometrika* 26(4):404–413.

Coble, M.D., P.M. Vallone, R.S. Just et al. 2006. Effective strategies for forensic analysis in the mito-chondrial DNA coding region. *Int J Legal Med* 120(1):27–32.

Coble, M.D., O.M. Loreille, M.J. Wadhams et al. 2009. Mystery solved: The identification of the two missing Romanov children using DNA analysis. *PLOS ONE* 4(3):4838–4846.

Cree, L.M., D.C. Samuels, S.C. de Sousa Lopes et al. 2008. A reduction of mitochondrial DNA molecules during embryogenesis explains the rapid segregation of genotypes. *Nat Genet* 40(2):249–254.

Crysup, B., A.E. Woerner, J.L. King, and B. Budowle. 2021. Graph algorithms for mixture interpre-tation. *Genes* 12(2):185.

Cuenca, D., J. Battaglia, M. Halsing, and S. Sheehan. 2020. Mitochondrial sequencing of miss-ing persons DNA casework by implementing Thermo Fisher's Precision ID mtDNA whole genome assay. *Genes* 11(11):1303.

Dabney, J., M. Knapp, I. Glocke et al. 2013. Complete mitochondrial genome sequence of a Middle Pleistocene cave bear reconstructed from ultrashort DNA fragments. *Proc Natl Acad Sci U S A* 110(39):15758–15763.

Davis, C.L. 1998. Mitochondrial DNA: State of Tennessee v. Ware. *Profiles DNA*:6–7.

Dayama, G., S.B. Emery, J.M. Kidd, and R.E. Mills. 2014. The genomic landscape of polymorphic human nuclear mitochondrial insertions. *Nucleic Acids Res* 42(20):12640–12649.

Desmyter, S., M. Bodner, G. Huber et al. 2016. Hairy matters: MtDNA quantity and sequence varia-tion along and among human head hairs. *Forensic Sci Int Genet* 25:1–9.

Dolezal, P., V. Likic, J. Tachezy, and T. Lithgow. 2006. Evolution of the molecular machines for protein import into mitochondria. *Science* 313(5785):314–318.

Driggers, W.J., S.P. LeDoux, and G.L. Wilson. 1993. Repair of oxidative damage within the mito-chondrial DNA RINr 38 cells. *J Biol Chem* 268(29):22042–22045.

Eduardoff, M., C. Xavier, C. Strobl, A. Casas-Vargas, and W. Parson. 2017. Optimized mtDNA control region primer extension capture analysis for forensically relevant samples and highly compromised mtDNA of different age and origin. *Genes* 8(10):237–253.

Eichmann, C., and W. Parson. 2008. "Mitominis": Multiplex PCR analysis of reduced size ampli-cons for compound sequence analysis of the entire mtDNA control region in highly degraded samples. *Int J Legal Med* 122(5):385–388.

Federico, A., E. Cardaioli, P. DaPozzo et al. 2012. Mitochondria, oxidative stress and neurodegen-eration. *J Neurol Sci*, in press.

Fisher, D.L., M.M. Holland, L. Mitchell et al. 1993. Extraction, evaluation, and amplification of DNA from decalcified and undecalcified United States Civil War bone. *J Forensic Sci* 38(1):60–68.

Forster, L., P. Forster, S.M.R. Gurney et al. 2010. Evaluating length heteroplasmy in human mito-chondrial DNA control region. *Int J Legal Med* 124(2):133–142.

Forsythe, B., L. Melia, and S.A. Harbison. 2020. Methods for the analysis of mitochondrial DNA. *Wiley Interdiscip Rev Forensic Sci* 3(1):e1388.

Gallimore, J.M., J.A. McElhoe, and M.M. Holland. 2018. Assessing heteroplasmic variant drift in the mtDNA control region of human hairs using an MPS approach. *Forensic Sci Int Genet* 32:7–17.

Gill, P., P.L. Ivanov, C. Kimpton et al. 1994. Identification of the remains of the Romanov family by DNA analysis. *Nat Genet* 6(2):130–135.

Ginther, C., L. Issel-Tarver, and M.C. King. 1992. Identifying individuals by sequencing mitochondrial DNA from teeth. *Nat Genet* 2(2):135–138.

González, M. del M., A. Ramos, M.P. Aluja, and C. Santos. 2020. Sensitivity of mitochondrial DNA heteroplasmy detection using next generation sequencing. *Mitochondrion* 50:88–93.

Gorden, E.M., K. Sturk-Andreaggi, and C. Marshall. 2018. Repair of DNA damage caused by cytosine deamination in mitochondrial DNA of forensic case samples. *Forensic Sci Int Genet* 34:257–264.

Gorden, E.M., K. Sturk-Andreaggi, and C. Marshall. 2021. Capture enrichment and massively parallel sequencing for human identification. *Forensic Sci Int Genet* 53:102496.

Gray, M.W. 1992. The endosymbiont hypothesis revisited. *Int Rev Cytol* 141:233–357.

Gray, M.W., B.F. Lang, and G. Burger. 2004. Mitochondria of protists. *Annu Rev Genet* 38:477–524.

Gyllensten, U., D. Wharton, A. Josefsson, and A.C. Wilson. 1991. Paternal inheritance of mitochondrial DNA in mice. *Nature* 352(6332):255–257.

Hauswirth, W.W., and P.J. Laipis. 1982. Mitochondrial DNA polymorphism in a maternal lineage of Holstein cows. *Proc Natl Acad Sci U S A* 79(15):4686–4690.

Hauswirth, W.W., M.J. Van de Walle, P.J. Laipis, and P.D. Olivo. 1984. Heterogeneous mitochondrial DNA D-loop sequence in bovine tissue. *Cell* 37(3):1001–1007.

Hazkani-Covo, E., R.M. Zeller, and W. Martin. 2010. Molecular poltergeists: Mitochondrial DNA copies (numts) in sequence nuclear genomes. *PLOS Genet* 6(2):e1000834.

Holland, C.A., J.A. McElhoe, S. Gaston-Sanchez, and M.M. Holland. 2021. Damage patterns observed in mtDNA control region MPS data for a range of template concentrations and when using different amplification approaches. *Int J Legal Med* 135(1):91–106.

Holland, M.M., D.L. Fisher, L.G. Mitchell et al. 1993. Mitochondrial DNA sequence analysis of human skeletal remains: Identification of remains from the Vietnam War. *J Forensic Sci* 38(3):542–553.

Holland, M.M., and T.J. Parsons. 1999. Mitochondrial DNA sequence analysis — Validation and use for forensic casework. *Forensic Sci Rev* 11(1):21–50.

Holland, M.M., M.R. McQuillan, and K.A. O'Hanlon. 2011. Second generation sequencing allows for mtDNA mixture deconvolution and high resolution detection of heteroplasmy. *Croat Med J* 52(3):299–313.

Holland, M.M., E. Pack, and J.A. McElhoe. 2017. Evaluation of GeneMarker® HTS for improved alignment of mtDNA MPS data, haplotype determination, and heteroplasmy assessment. *Forensic Sci Int Genet* 28:90–98.

Holland, M.M., K.D. Makova, and J.A. McElhoe. 2018. Deep-coverage MPS analysis of heteroplasmic variants within the mtGenome allows for frequent differentiation of maternal relatives. *Genes* 9(3):124–145.

Holland, M.M., R.M. Bonds, C.A. Holland, and J.A. McElhoe. 2019. Recovery of mtDNA from unfired metallic ammunition components with an assessment of sequence profile quality and DNA damage through MPS analysis. *Forensic Sci Int Genet* 39:86–96.

Hollard, C., L. Ausset, Y. Chantrel et al. 2019. Automation and developmental validation of the ForenSeq™ DNA signature preparation kit for high-throughput analysis in forensic laboratories. *Forensic Sci Int Genet* 40:37–45.

Holt, C.L., K.M. Stephens, P. Walichiewicz et al. 2021. Human mitochondrial control region and mtGenome: Design and forensic validation of NGS multiplexes, sequencing and analysis software. *Genes* 12(4):599.

Hopgood, R., K.M. Sullivan, and P. Gill. 1992. Strategies for automated sequencing of human mitochondrial DNA directly from PCR products. *BioTechniques* 13(1):82–92.

Huber, N., W. Parson, and A. Dür. 2018. Next generation database search algorithm for forensic mitogenome analysis. *Forensic Sci Int Genet* 37:204–214.

Irwin, J.A., J.L. Saunier, H. Niederstatter et al. 2009. Investigation of heteroplasmy in the human mitochondrial DNA control region: A synthesis of observations from more than 5000 global population samples. *J Mol Evol* 68(5):516–527.

Irwin, J.A., W. Parson, M.D. Coble, and R.S. Just. 2011. mtGenome reference population databases and the future of forensic mtDNA analysis. *Forensic Sci Int Genet* 5(3):222–225.

Ivanov, P.L., M.J. Wadhams, R.K. Roby et al. 1996. Mitochondrial DNA sequence heteroplasmy in the Grand Duke of Russia Georgij Romanov establishes the authenticity of the remains of Tsar Nicholas II. *Nat Genet* 12(4):417–420.

Jenuth, J.P., A.C. Peterson, K. Fu, and E.A. Shoubridge. 1996. Random genetic drift in the female germline explains the rapid segregation of mammalian mitochondrial DNA. *Nat Genet* 14(2):146–151.

Just, R.S., J.A. Irwin, and W. Parson. 2014. Questioning the prevalence and reliability of human mitochondrial DNA heteroplasmy from massively parallel sequencing data. *Proc Natl Acad Sci* 111(43):E4546–E4547.

Just, R.S., J.A. Irwin, and W. Parson. 2015. Mitochondrial DNA heteroplasmy in the emerging field of massively parallel sequencing. *Forensic Sci Int Genet* 18:131–139.

Kaneda, H., J. Hayashi, S. Takahama et al. 1995. Elimination of paternal mitochondrial DNA in intraspecific crosses during early mouse embryogenesis. *Proc Natl Acad Sci* 92(10):4542–4546.

Kanki, T., K. Ohgaki, M. Gaspari et al. 2004. Architectural role of TFAM in maintenance of human mitochondrial DNA. *Mol Cell Biol* 24(22):9823–9834.

Kim, H., H.A. Erlich, and C.D. Calloway. 2015. Analysis of mixtures using next generation sequencing of mitochondrial DNA hypervariable regions. *Croat Med J* 56(3):208–217.

King, J.L., B.L. LaRue, N.M. Novroski et al. 2014a. High-quality and high-throughput massively parallel sequencing of the human mitochondrial genome using the Illumina MiSeq. *Forensic Sci Int Genet* 12:128–135.

King, J.L., A. Sanjantila, and B. Budowle. 2014b. mitoSAVE: Mitochondrial sequence analysis of variants in Excel. *Forensic Sci Int Genet* 12:122–125.

King, T.E., G.G. Fortes, P. Balaresque et al. 2014. Identification of the remains of King Richard III. *Nat Commun* 5:5631–5638.

Kukat, C., C.A. Wurm, H. Spahr et al. 2011. Super-resolution microscopy reveals that mammalian mitochondrial nucleoids have a uniform size and frequently contain a single copy of mtDNA. *Proc Natl Acad Sci* 108(33):13534–13539.

Lang, B.F., G. Burger, C.J. O'Kelly et al. 1997. An ancestral mitochondrial DNA resembling a eubacterial genome in miniature. *Nature* 387(6632):493–497.

Lodeiro, M.F., A.U. Uchida, J.J. Arnold et al. 2010. Identification of multiple rate-limiting steps during the human mitochondrial transcription cycle in vitro. *J Biol Chem* 285(21):16387–16402.

Lodeiro, M.F., A. Uchida, M. Bestwick et al. 2012. Transcription from the second heavy-strand promoter of human mtDNA is repressed by transcription factor A in vitro. *Proc Natl Acad Sci* 109(17):6513–6518.

Longley, M.J., D. Nguyen, T.A. Kunkel, and W.C. Copeland. 2001. The fidelity of human DNA polymerase gamma with and without exonucleolytic proofreading and the p55 accessory subunit. *J Biol Chem* 276(42):38555–38562.

Lopez, J.V., N. Yuhki, W. Modi et al. 1994. Numt, a recent transfer and tandem amplification of mitochondrial DNA in the nuclear genome of the domestic cat. *J Mol Evol* 39:174–190.

Lopez, J.V., S. Cevario, and S.J. O'Brien. 1996. Complete nucleotide sequences of the domestic cat (Felis catus) mitochondrial genome and a transposed mtDNA tandem repeat (Numt) in the nuclear genome. *Genomics* 33(2):229–246.

Loreille, O.M., T.M. Diegoli, J.A. Irwin, M.D. Coble, and T.J. Parsons. 2007. High efficiency DNA extraction from bone by total demineralization. *Forensic Sci Int Genet* 1(2):191–195.

Loreille, O., H. Koshinsky, V.Y. Fofanov, and J.A. Irwin. 2011. Application of next generation sequencing technologies to the identification of highly degraded unknown soldiers' remains. *Forensic Sci Int Genet Suppl Ser* 3(1):e540–e541.

Luo, S., C.A. Valencia, J. Zhang et al. 2018. Biparental inheritance of mitochondrial DNA in humans. *Proc Natl Acad Sci U S A* 115(51):13039–13044.

Lupi, R., P.D. de Meo, E. Picardi et al. 2010. MitoZoa: A curated mitochondrial genome database of metazoans for comparative genomics studies. *Mitochondrion* 10(2):192–199.

Lutz, S., H. Wittig, H.J. Weisser et al. 2000. Is it possible to differentiate mtDNA by means of HVIII in samples that cannot be distinguished by sequencing the HVI and HVII regions? *Forensic Sci Int* 113(1–3):97–101.

Lutz-Bonengel, S., and W. Parson. 2019. No further evidence for paternal leakage of mitochondrial DNA in human yet. *Proc Natl Acad Sci U S A* 116(6):1821–1822.

Lutz-Bonengel, S., H. Niederstätter, J. Naue et al. 2021. Evidence for multi-copy Mega-NUMTs in the human genome. *Nucleic Acids Res* 49(3):1517–1531.

Lyons, E.A., M.K. Scheible, K. Sturk-Andreaggi, J.A. Irwin, and R.S. Just. 2013. A high-throughput Sanger strategy for human mitochondrial genome sequencing. *BMC Genomics* 14:881–896.

Marshall, C., K. Sturk-Andreaggi, J. Daniels-Higginbotham et al. 2017. Performance evaluation of a mitogenome capture and Illumina sequencing protocol using non-probative, case-type skeletal samples: Implications for the use of a positive control in a next-generation sequencing procedure. *Forensic Sci Int Genet* 31:198–206.

Marshall, C., K. Sturk-Andreaggi, E.M. Gorden et al. 2020a. A forensic genomics approach for the identification of Sister Marija Crucifiksa Kozulic. *Genes* 11(10):e1140.

Marshall, C., K. Sturk-Andreaggi, J.D. Ring, A. Dür, and W. Parson. 2020b. Pathogenic variant filtering for mitochondrial genome haplotype reporting. *Genes* 11(10):1–10.

Marshall, C., and W. Parson. 2021. Interpreting NUMTs in forensic genetics: Seeing the forest for the trees. *Forensic Sci Int Genet* 53:102497.

McElhoe, J., M. Holland, K. Makova et al. 2014. Development and assessment of an optimized next-generation DNA sequencing approach for the mtgenome using the Illumina MiSeq. *Forensic Sci Int Genet* 13:20–29.

McElhoe, J.A., and M.M. Holland. 2020. Characterization of background noise in MiSeq MPS data when sequencing human mitochondrial DNA from various sample sources and library preparation methods. *Mitochondrion* 52:40–55.

Melton, T. 2004. Mitochondrial DNA heteroplasmy. *Forensic Sci Rev* 16(1):1–20.

Melton, T., G. Dimick, B. Higgins, L. Lindstrom, and K. Nelson. 2005. Forensic mitochondrial DNA analysis of 691 casework hairs. *J Forensic Sci* 50(1):73–80.

Melton, T., C. Holland, and M. Holland. 2012. Forensic mitochondrial DNA—Current practice and future potential. *Forensic Sci Rev* 10:101–122.

Michaels, G.S., W.W. Hauswirth, and P.J. Laipis. 1982. Mitochondrial DNA copy number in bovine oocytes and somatic cells. *Dev Biol* 94(1):246–251.

Mikkelsen, M., R. Frank-Hansen, A.J. Hansen, and N. Morling. 2014. Massively parallel Pyrosequencing of the mitochondrial genome with the 454 methodology in forensic genetics. *Forensic Sci Int Genet* 12:30–37.

Mitochondrial DNA search database. Available at: http://empop.online/ (accessed on June 30, 2019).

MITOMAP. A human mitochondrial genome database. Available at: http://www.mitomap.org/ (accessed on June 30, 2019).

Muller, K., R. Klein, E. Miltner, and P. Wiegand. 2007. Improved STR typing of telogen hair root and hair shaft DNA. *Electrophoresis* 28(16):2835–2842.

Murphy, E., H. Ardehali, R.S. Balaban et al. 2016. Mitochondrial function, biology, and role in disease: A scientific statement from the American Heart Association. *Circ Res* 118(12):1960–1991.

Naue, J., S. Horer, T. Sanger et al. 2015. Evidence for frequent and tissue-specific sequence heteroplasmy in human mitochondrial DNA. *Mitochondrion* 20:82–94.

Nissanka, N., and C.T. Moraes. 2020. Mitochondrial DNA heteroplasmy in disease and targeted nuclease-based therapeutic approaches. *EMBO Rep* 21(3):e49612.

Ossowski, A., M. Diepenbroek, T. Kupiec et al. 2016. Genetic identification of communist crimes' victims (1944–1956) based on the analysis of one of many mass graves discovered on the Powazki Military Cemetery in Warsaw, Poland. *J Forensic Sci* 61(6):1450–1455.

Palo, J.U., M. Hedman, N. Söderholm, and A. Sajantila. 2007. Repatriation and identification of the Finnish World War II soldiers. *Croat Med J* 48(4):528–535.

Parson, W., and H.-J. Bandelt. 2007. Extending guidelines for mtDNA typing of population data in forensic science. *Forensic Sci Int Genet* 1:13–19.

Parson, W., and A. Dür. 2007. EMPOP: A forensic mtDNA database. *Forensic Sci Int Genet* 1(2):88–92.

Parson, W., C. Strobl, G. Huber et al. 2013. Evaluation of next generation mtGenome sequencing using the Ion Torrent Personal Genome Machine (PGM). *Forensic Sci Int Genet* 7(5):543–549.

Parson, W., L. Gusmao, D.R. Hares et al. 2014. DNA Commission of the International Society for Forensic Genetics: Revised and extended guidelines for mitochondrial DNA typing. *Forensic Sci Int Genet* 13:134–142.

Parson, W., G. Huber, L. Moreno et al. 2015. Massively parallel sequencing of complete mitochondrial genomes from hair shaft samples. *Forensic Sci Int Genet* 15:8–15.

Parsons, T.J., D.S. Muniec, K. Sullivan et al. 1997. A high observed substitution rate in the human mitochondrial DNA control region. *Nat Genet* 15(4):363–368.

Parsons, T.J., and M.D. Coble. 2001. Increasing the forensic discrimination of mitochondrial DNA testing through analysis of the entire mitochondrial DNA genome. *Croat Med J* 43(3):304–309.

Payne, B.A., I.J. Wilson, C.A. Hateley et al. 2011. Mitochondrial aging is accelerated by anti-retroviral therapy through the clonal expansion of mtDNA mutations. *Nat Genet* 43(8):806–810.

Peck, M.P., M.D. Brandhagen, C. Marshall et al. 2016. Concordance and reproducibility of a next generation mtGenome sequencing method for high-quality samples using the Illumina MiSeq. *Forensic Sci Int Genet* 24:103–111.

Peck, M.P., K. Sturk-Andreaggi, J.T. Thomas et al. 2018. Developmental validation of a Nextera XT mitogenome Illumina MiSeq sequencing method for high-quality samples. *Forensic Sci Int Genet* 34:25–36.

Pereira, V., A. Longobardi, and C. Børsting. 2018. Sequencing of mitochondrial genomes using the Precision ID mtDNA whole genome panel. *Electrophoresis* 39(21):2766–2775.

Piccinini, A., S. Coco, W.Parson et al. 2010. World War one Italian and Austrian soldier identification project: DNA results of the first case. *Forensic Sci Int Genet* 4(5):329–333.

Piko, L., and L. Matsumoto. 1976. Number of mitochondria and some properties mitochondrial DNA in the mouse egg. *Dev Biol* 49(1):1–10.

Rathbun, M.M., J.A. McElhoe, W. Parson, and M.M. Holland. 2017. Considering DNA damage when interpreting mtDNA heteroplasmy in deep sequencing data. *Forensic Sci Int Genet* 26:1–11.

Rebolledo-Jaramillo, B., M.S.W. Su, N. Stoler et al. 2014. Maternal age effect and severe germline bottleneck in the inheritance of human mitochondrial DNA. *Proc Natl Acad Sci U S A* 111(43):15474–15479.

Riman, S., K.M. Kiesler, L.A. Borsuk, and P.M. Vallone. 2017. Characterization of NIST human mitochondrial DNA SRM-2392 and SRM-2392-I standard reference materials by next generation sequencing. *Forensic Sci Int Genet* 29:181–192.

Ring, J.D., K. Sturk-Andreaggi, M.A. Peck, and C. Marshall. 2018. Bioinformatic removal of NUMT-associated variants in mitotiling next-generation sequencing data from whole blood samples. *Electrophoresis* 39(21):2785–2797.

Robin, E.D., and R. Wong. 1988. Mitochondrial DNA molecules and virtual number of mitochondria per cell in mammalian cells. *J Cell Physiol* 136(3):507–513.

Röck, A., J. Irwin, A. Dür, T. Parsons, and W. Parson. 2011. SAM: String-based sequence search algorithm for mitochondrial DNA database queries. *Forensic Sci Int Genet* 5(2):126–132.

Santibanez-Koref, M., H. Griffin, D.M. Turnbull et al. 2019. Assessing mitochondrial heteroplasmy using next generation sequencing: A note of caution. *Mitochondrion* 46:302–306.

Satoh, M., and T. Kuroiwa. 1991. Organization of multiple nucleoids and DNA molecules in mitochondria of a human cell. *Exp Cell Res* 196(1):137–140.

Schwartz, M., and J. Vissing. 2002. Paternal inheritance of mitochondrial DNA. *N Engl J Med* 347(8):576–580.

Scudder, N., D. McNeven, S.F. Kelty, S.J. Walsh, and J. Robertson. 2018. Massively parallel sequencing and the emergence of forensic genomics: Defining the policy and legal issues for law enforcement. *Sci Justice* 58(2):153–158.

Shih, S.Y., N. Bose, A.B.R. Gonçalves, H.A. Erlich, and C.D. Calloway. 2018. Applications of probe capture enrichment next generation sequencing for whole mitochondrial genome and 426 nuclear SNPs for forensically challenging samples. *Genes* 9(1):49–68.

Smart, U., B. Budowle, A. Ambers et al. 2019. A novel phylogenetic approach for de novo discovery of putative nuclear mitochondrial (pNumt) haplotypes. *Forensic Sci Int Genet* 43:102146.

Smart, U., J.C. Cihlar, S.N. Mandape et al. 2021. A continuous statistical phasing framework for the analysis of forensic mitochondrial DNA mixtures. *Genes* 12(2):128.

Stenton, S.L., and H. Prokisch. 2020. Genetics of mitochondrial disease: Identifying mutations to help diagnosis. *EBiomedicine* 56:102784.

Strobl, C., M. Eduardoff, M.M. Bus, M. Allen, and W. Parson. 2018. Evaluation of the precision ID whole mtDNA genome panel for forensic analyses. *Forensic Sci Int Genet* 35:21–25.

Sturk-Andreaggi, K., M.A. Peck, C. Boysen et al. 2017. AQME: A forensic mitochondrial DNA analysis tool for next-generation sequencing data. *Forensic Sci Int Genet* 31:189–197.

Sturk-Andreaggi, K., W. Parson, M. Allen, and C. Marshall. 2020. Impact of the sequencing method on the detection and interpretation of mitochondrial DNA length heteroplasmy. *Forensic Sci Int Genet* 44:102205.

Sutovsky, P., R.D. Moreno, J. Ramalho-Santos et al. 2000. Ubiquitinated sperm mitochondria, selective proteolysis, and the regulation of mitochondrial inheritance in mammalian embryos. *Biol Reprod* 63(2):582–590.

SWGDAM Guidelines. 2019. Interpretation guidelines for mitochondrial DNA analysis by forensic DNA testing laboratories. Approved 04/23/2019.

Sykora, P., D.M. Wilson III, and V.A. Bohr. 2011. Repair of persistent strand breaks in the mitochondrial genome. *Mech Ageing Dev* 133(4):169–175.

Van Wormhoudt, A., V. Roussel, G. Courtois, and S. Huchette. 2011. Mitochondrial DNA introgression in the European abalone *Haliotis tuberculata tuberculata*: Evidence for experimental mtDNA paternal inheritance and a natural hybrid sequence. *Mar Biotechnol (NY)* 13(3):563–574.

Vohr, S.H., R. Gordon, J.M. Eizenga et al. 2017. A phylogenetic approach for haplotype analysis of sequence data from complex mitochondrial mixtures. *Forensic Sci Int Genet* 30:93–105.

Wai, T., A. Ao, X. Zhang et al. 2010. The role of mitochondrial DNA copy number in mammalian fertility. *Biol Reprod* 83(1):52–62.

Wallace, D.C. 2010. Mitochondrial DNA mutations in disease and aging. *Environ Mol Mutagen* 51(5):440–450.

Wilton, P.R., A. Zaidi, K. Makova, and R. Nielsen. 2018. A population phylogenetic view of mitochondrial heteroplasmy. *Genetics* 208(3):1261–1274.

Woerner, A.E., A. Ambers, F. Wendt et al. 2018. Evaluation of the precision ID mtDNA whole genome panel on two massively parallel sequencing systems. *Forensic Sci Int Genet* 36:213–224.

Yakes, F.M., and B. VanHouten. 1997. Mitochondrial DNA damage is more extensive and persists longer than nuclear DNA damage in human cells following oxidative stress. *Proc Natl Acad Sci* 94(2):514–519.

Y Chromosome in Forensic Science

4

MANFRED KAYSER AND
KAYE N. BALLANTYNE

Contents

4.1 Introduction

The Y chromosome has long been regarded as the poor cousin of the human genome. At only ~60 MB in size, the Y chromosome is the second smallest human chromosome (after chromosome 21) and contains the lowest number of genes. In contrast to all other human chromosomes, it is extremely rich in repetitive DNA sequences of all kinds. Only the euchromatic region, covering about half of the Y, is genetically active, consisting of Y-specific single-copy regions, Y-specific repetitive regions, and X-Y homologous regions, as well as the repetitive centromere region of the Y. The heterochromatic part consists of only repetitive sequences and does not harbor any genes. Unlike any other human chromosome, most of the Y chromosome (~95%) does not undergo homologous recombination during meiosis (termed the non-recombining portion of the Y, or NRY). Recombination with homologous regions on the X chromosome only occurs at the pseudoautosomal regions, located at the tips of the Y (Tilford et al. 2001). The most important gene functions on the Y are those involved in male sex determination (e.g., *SRY* gene) and spermatogenesis (e.g., *AZF*), although genes with other functions are also found, often with homologue partners on the X chromosome (such as *AMELY/AMELX*). The latter indicates the shared evolutionary history of Y and X going back to a homologue pair of autosomes in early mammalian history. The relative dearth of coding genes, combined with the Y's largely haploid nature, has resulted in this chromosome displaying genetic features and encoding human evolutionary history unlike any other human chromosome, some of which are suitable for forensic applications.

In order to fully utilize the human Y chromosome for forensic purposes, it is necessary to understand precisely what makes it such a unique chromosome. The usual absence

DOI: 10.4324/9780429019944-5

of Y chromosomes in females allows the use of the Y chromosome as a marker for human sex identification, which can add helpful information in forensic investigations. The strict male-specific inheritance of the NRY provides opportunities to specifically analyze DNA components of a crime scene sample that were provided by males only, and differentiate them from those provided by females, which can be highly important in mixed stain analysis in forensics such as in cases of sexual assault. At the same time, recombination-free inheritance from fathers to sons, combined with low to moderate mutation rates of most NRY-DNA polymorphisms, means that male relatives usually share the same NRY polymorphisms. This feature has both advantages and disadvantages for forensic applications of Y-chromosome DNA. Disadvantages come in the way that conclusions from Y-chromosome DNA analysis usually cannot be made on an individual level, as desired in forensic investigation. This is because in the event of a matching Y-DNA profile between samples from a suspect and a crime scene, the hypothesis that either the suspect or, alternatively, any of his paternal male relatives has left the crime scene sample have the same estimated probability (but see later for a potential solution). Advantages are that because of shared Y-DNA profiles between male relatives, a close paternal male relative of a deceased alleged father can be used to replace the father in paternity testing of a male offspring using Y-DNA analysis in deficiency cases where autosomal DNA profiling often is not informative. The same principle can also be used in disaster victim identification (DVI) of males using close or distant paternal male relatives in cases where autosomal DNA profiling fails. This principle also is the basis for using Y-chromosome DNA for familial search such as in DNA dragnets/mass screenings to trace unknown male perpetrators in cases where no autosomal DNA-profile match has initially been obtained. Here, the identification of participating male relatives of the unknown male perpetrator via Y-DNA-profile matching can provide investigative leads to trace the nonparticipating male perpetrator and identify him as a crime scene sample donor via autosomal DNA-profile matching.

The haploid nature of NRY also leads to the Y chromosome having a lower effective population size than the autosomes, with four copies of autosomal loci relative to each Y locus (Jobling and Tyler-Smith 2003). This lower effective population size results in the Y chromosome displaying the lowest genetic diversity of any chromosome (Group 2001). As a consequence of the lower effective population size, Y polymorphisms can be more strongly affected by genetic drift or population-level events such as bottlenecks or founder effects than autosomal loci. In addition, the asymmetrical spread of distinct polymorphisms is aided by the patrilineal transmission of the Y mirroring certain cultural practices, such as patrilocality (where males retain their familial lands, with females relocating) or polygyny (low numbers of males having the highest reproductive success) (Oota et al. 2001; Seielstad, Minch, and Cavalli-Sforza 1998). These features in part explain the relatively strong geographic information content provided with some Y-chromosomal DNA polymorphisms, as further outlined later.

4.2 Sex Determination

The use of the Y chromosome for male sex determination in forensic applications started about 40 years ago when luminescence microscopy was applied for detecting Y chromosomes in cells from cadaver material (Radam and Strauch 1973). In the late 1980s to early 1990s, specific Y-chromosome DNA sequences were used for this purpose (Ebensperger,

Studer, and Epplen 1989; Fattorinil et al. 1991). However, analyzing only Y-specific DNA for the purpose of male sex determination is semi-optimal, as the absence of the signal in principle can mean either the presence of female material or a negative result due to technical reasons. Therefore, systems have been developed that take advantage of the homologous nature of human X and Y chromosomes, targeting sites that display sequences with length polymorphisms between the copies. Since the early 1990s (Akane et al. 1992), the amelogenin system has been used for human sex determination in forensics and other applications such as in paleogenetics, and is part of many commercial kits for human identification. The most often used PCR primers amplify a 112 bp Y fragment together with a 106 bp X fragment (Sullivan et al. 1993). Observing in a DNA sample a fragment of 106 bp indicates the presence of DNA from a female with two X copies of the same length, whereas two fragments of 106 bp and 112 bp indicate DNA from a male. However, this system is not free of error, as it was observed that some men can carry a Y-chromosome deletion, which includes the *AMELY* gene locus, and consequently appear as females in the test results. Although the frequency of the respective deletion is low (<1%) in many geographic regions such as Europe (Mitchell et al. 2006; Steinlechner et al. 2002), it can be as high as 3% in some populations such as from India or Sri Lanka (Chang, Burgoyne, and Both 2003; Thangaraj, Reddy, and Singh 2002). To make DNA-based sex determination more reliable, there are proposals to combine the *AMELY/AMELX* system with other X-Y differential markers or with Y-specific markers that are more distant to the *AMELY* region (Santos, Pandya, and Tyler-Smith 1998). This has already occurred with the release of the PowerPlex Fusion Kit (Promega) containing both *AMELY/AMELX* and the male-specific Y-STR DYS391, as well as the GlobalFiler Kit (Life Technologies) containing *AMELY/AMELX*, DYS391, and a Y-chromosomal insertion/deletion (indel) marker.

4.3 Paternal Lineage Differentiation and Identification

In principle, any NRY-DNA marker with a low or medium mutation rate is suitable for characterizing groups of male relatives belonging to the same paternal lineage, especially when multiple markers are combined to create compound haplotypes. However, if the mutation rate of an NRY marker is too low, as it is, for instance, for Y-chromosomal single nucleotide polymorphisms (Y-SNPs) with a mutation rate per site per generation of about 10^{-8} (Xue et al. 2009), the markers will not be practically useful for forensic applications (besides biogeographic ancestry inference, see later). Although such a low mutation rate ensures that all males carrying a particular Y-SNP mutation can be linked back to a common ancestor, the time back to the most common ancestor is expected to be long when using Y-SNPs, especially if the Y-SNP is frequent enough. Consequently, close but also very distantly related males (so distant that it usually escapes family knowledge) will carry such Y-SNP mutations, and as such the level of male lineage identification is extremely low. Therefore, more polymorphic NRY markers, i.e., those with a higher underlying mutation rate such as Y-STRs that have an average mutation rate about 100,000 times higher than Y-SNPs (Goedbloed et al. 2009), are the preferred choice for male lineage differentiation for forensic purposes either alone as usually applied or in combination with Y-SNPs (but see later for a different forensic application of Y-SNPs).

The introduction of Y-STRs for paternal lineage identification was relatively straightforward for forensic biology, as they are biologically and analytically similar to autosomal STRs, although they are haploid rather than diploid. However, in the early days of forensic

STR analysis, the knowledge about Y-STR markers was lagging far behind that of autosomal STRs. This was because autosomal STRs were mainly identified in systematic studies with considerable funding, aiming to provide polymorphic markers for gene mapping purposes and disease gene identification. However, because of the non-recombining nature of most of the Y chromosome, the principle of linkage mapping does not work in practice; hence, the Y chromosome was left out in the search for STRs. It took until the early 1990s that the first human polymorphic male-specific Y-STR, DYS19, was identified (Roewer and Epplen 1992b) and was immediately applied to a sexual assault case where it provided an exclusion constellation (Roewer and Epplen 1992a). The coming years saw only a very minor increase in Y-STR markers, so that in the late 1990s less than 20 Y-STRs were known. In a milestone study published in 1997, 14 Y-STRs were analyzed in a multicenter approach involving many colleagues from the forensic genetic community interested in the future application of Y-STRs to forensic casework. The considerable population data generated resulted in the recommendation of seven Y-STRs for forensic application, the so-called minimal haplotype (MH), and three extra Y-STRs for supplementation (Kayser et al. 1997). This publication marks the beginning of Y-STR implementation in forensic case work, so that today after only 16 years Y-STR profiling has become a routine application in most forensic laboratories worldwide. Although about 30 additional Y-STRs were identified in the following years, it took until 2004 that a systematic search for polymorphic STRs on the Y-chromosome was published, which provided 166 new useful Y-STRs (Kayser et al. 2004) as valuable resources for forensic and other applications. The International Society of Forensic Genetics (ISFG) has issued guidelines on the use of Y-STRs for forensic purposes (Gill et al. 2001; Gusmao et al. 2006).

Since their introduction to forensic science, Y-STRs have been used for one main purpose—to identify male lineages for identifying and excluding male suspects (Roewer 2009; Kayser 2007). Y-STR analysis is particularly useful in DNA mixtures where female cells are present in substantially higher quantities than the male contribution. This is most commonly observed in sexual assault cases, where differential lysis to enrich male DNA from semen over female DNA from epithelial cells can help. Often this still results in a mixed autosomal STR profile not allowing the separation of the male perpetrator, despite the knowledge of the STR profile of the female victim from reference DNA analysis. Also in sexual assault cases involving sperm-negative samples, where differential DNA extraction cannot be applied, Y-STR profiles can be obtained, e.g., in ~45% of cases where autosomal STR profiling was unsuccessful (Olofsson et al. 2011). Moreover, any male–female mixed sample (such as blood/saliva or skin/skin mixtures) can benefit from the male-specific amplification (Dekairelle and Hoste 2001; Sibille et al. 2002).

In principle, paternal lineages can also be identified with Y-SNPs on the level of Y haplogroups. However, the 100,000 lower mutation rates of Y-SNPs compared to Y-STRs means that a Y haplogroup goes back to a common ancestry in the more distant past, hence, defining a more ancient Y lineage. In contrast, Y-STR haplotypes define a paternal lineage in the recent past, i.e., males belonging to an (extended) paternal family. Therefore, paternal lineages in forensic applications are typically characterized by Y-STRs based on haplotypes (see for a different forensic use of Y-SNPs later).

4.3.1 Y-STR Markers in Forensic Genetics

Commercial multiplexes such as AmpF/STR® Yfiler™ (Applied Biosystems; Figure 4.1) (Mulero et al. 2006) and PowerPlex Y® (Promega) (Krenke et al. 2005) provide highly sensitive

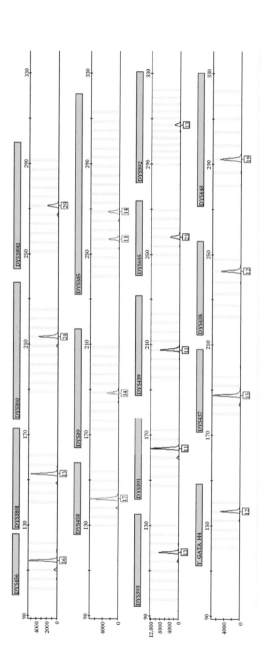

Figure 4.1 A Y-STR electropherogram. Representative 17-locus Y-STR profile that was generated with the AmpFLSTR® Yfiler® PCR Amplification Kit (Applied Biosystem/Life Technologies).

Table 4.1 Information of the Most Widely Forensically Applied Y-STR Markers

Marker	Repeat Motif	Alleles	Mutation Rate (95% Credible Interval; Number of Meiotic Transfers Investigated)	Gene Diversity
DYS19*,#,$	$(TAGA)_3(TAGG)_1(TAGA)_{6-16}$	9-19	2.6×10^{-3} (1.6×10^{-3} – 3.9×10^{-3}; 11,900)	0.758
DYS389I*,#,$	$(TCTG)_3(TCTA)_{6-14}$	9-17	2.4×10^{-3} (1.1×10^{-3} – 4.0×10^{-3}; 10,103)	0.691
DYS389II*,#,$	$(TCTG)_{4-5}(TCTA)_{10-14}$ $N_{28}(TCTG)_3 (TCTA)_{6-14}$	24-36	3.0×10^{-3} (1.8×10^{-3} – 4.5×10^{-3}; 10,079)	0.646
DYS390*,#,$	$(TCTG)_8(TCTA)_{9-14} (TCTG)_1$ $(TCTG)_4$	17-29	1.9×10^{-3} (8.0×10^{-4} – 3.2×10^{-3}; 11,385)	0.774
DYS391*,#,$	$(TCTG)_3(TCTA)_{6-15}$	5-16	2.8×10^{-3} (1.6×10^{-3} – 4.1×10^{-3}; 11,336)	0.474
DYS392*,#,$	$(TAT)_{4-20}$	4-20	4.0×10^{-4} (1.0×10^{-5} – 1.1×10^{-3}; 11,268)	0.669
DYS393*,#,$	$(AGAT)_{7-18}$	7-18	9.0×10^{-4} (3.0×10^{-4} – 1.7×10^{-3}; 10,079)	0.676
DYS385a/b*,#,$	$(AAGG)_4N_{14}(AAAG)_3$ $N_{12}(AAAG)_3N_{29}$ $(AAGG)_{6-7}(GAAA)_{7-23}$	6-28	2.0×10^{-3} (1.3×10^{-3} – 2.9×10^{-3}; 19,108 joined analysis)	0.968
DYS438#,$	$(TTTTC)_{7-16}$	7-18	5.0×10^{-4} (1.0×10^{-4} – 1.3×10^{-3}; 6,947)	0.664
DYS439#,$	$(GATA)_3N_{32}(GATA)_{5-19}$	5-19	5.5×10^{-3} (3.5×10^{-3} – 7.9×10^{-3}; 6,908)	0.699
DYS437#,$	$(TCTA)_{4-12}(TCTG)_2(TCTA)_4$	10-18	1.1×10^{-3} (30×10^{-4} – 2.3×10^{-3}; 6,919)	0.506
DYS448$	$(AGAGAT)_{11-13}N_{42}(AGAGAT)_{8-9}$	14-24	2.0×10^{-4} (2.0×10^{-5} – 8.0×10^{-4}; 3,531)	0.748
DYS456$	$(AGAT)_{11-23}$	5-23	4.3×10^{-3} (1.7×10^{-3} – 9.5×10^{-3}; 3,384)	0.597
DYS458$	$(GAAA)_{11-24}$	11-24	6.5×10^{-3} (2.3×10^{-3} – 1.3×10^{-2}; 3,382)	0.795
DYS635$	$(TCTA)_4(TGTA)_2(TCTA)_2(TGTA)_2(TCTA)_2(TATG)_{0-2}$ $(TCTA)_{4-17}$	16-30	3.7×10^{-3} (1.5×10^{-3} – 6.6×10^{-3}; 4,349)	0.791
Y-GATA- H4$	$(TAGA)_3N_{12}(TAGG)_3(TAGA)_{8-15}$ $N_{22}(TAGA)_4$	8-15.1	2.9×10^{-3} (1.3×10^{-3} – 5.5×10^{-3}; 4,534)	0.614

Notes: Mutation rates are median rates from Bayesian approach using summarized family data (Goedbloed et al. 2009), while global diversity values are calculated from the HGDP-CEPH panel of 604 males (Ballantyne et al. 2012.).

* Minimal haplotype (MH); # included in PowerPlex® Y kit (Promega); $ included in AmpFLSTR® Yfiler® PCR Amplification Kit (Applied Biosystems/Life Technologies).

methods of amplifying 12 or 17 overlapping markers, respectively (Table 4.1), with template requirements of only 0.25–0.5 ng, although full Y-STR profiles can reliably be obtained from only 125 pg of DNA (Mulero et al. 2006; Krenke et al. 2005; Gross et al. 2008; Sturk et al. 2009). Analytically, the methodology is identical to that used for autosomal STRs—highly multiplexed single-reaction PCRs followed by capillary electrophoresis, and semi-automatic software-supported allele scoring are used to genotype samples (Figure 4.1). Loci range in size from 90 to 330 bp, allowing amplification in high-quality or moderately degraded DNA samples (Table 4.1). Both Yfiler and PowerPlex Y are able to specifically amplify Y-STRs in the presence of overwhelming quantities of female DNA, with reported successful amplification at 1:2000 ratios of male to female DNA (Krenke et al. 2005; Mulero et al. 2006). Full male profiles are obtainable from 1:10 mixtures of male DNA, with partial profiles routinely seen from 1:20 and lower (Gross et al. 2008; Mulero et al. 2006). Y-STR results can be obtained from sperm-negative (as measured with the Sperm Hy-liter, microscopic evaluation, or prostate-specific antigen detection) sexual assault samples (Dekairelle and Hoste 2001; Sibille et al. 2002; Olofsson et al. 2011) or in fingernail scrapings that show a single female profile with autosomal STR typing (Malsom et al. 2009). The full set of 17 core Y-STRs provides high levels of discrimination between paternal lineages in outbred populations, with haplotype resolution (a measure of the ability of the set of markers to discriminate between unrelated males) reaching 0.989 in Europeans, 0.889 in sub-Saharan Africans, and 0.905 in East Asians (Table 4.1) (Ballantyne et al. 2012). The high discrimination power ensures a high probability of differentiating between male lineages in a population. However, the high discrimination power is not seen in all populations (but see later for a solution).

Although commercially available Y-STR kits provide high haplotype resolution in many populations, in certain populations they show reduced diversity. Examples of such populations that have experienced population bottlenecks or sex-biased migration are Finns, Xhosa, and Polynesians (D'Amato et al. 2010; Hedman et al. 2011; Kayser, Brauer, et al. 2000). In addition, as the number of males in Y-STR frequency databases expands, the number of unrelated individuals sharing haplotypes is growing. Thus, to ensure that the evidentiary value of Y-STRs remains high, it is necessary to expand the core set of Y-STR loci, in the same manner as the autosomal core set has already been increased in Europe and the United States. Most of the currently used Y-STRs, with their low- to mid-range diversities, were selected from a limited panel of known markers, as only 30 Y-STRs were available in early 2000. When creating new STR panels, the key to successful discrimination between unrelated individuals is the selection of sufficient numbers of markers which show high levels of diversity within the population of interest. The inclusion of highly diverse markers within the haplotype ensures that few males will carry identical haplotypes by chance (referred to as identity-by-state, IBS), rather than by a shared origin (Identity-by-Descent, IBD).

The Y-STRs showing the highest levels of diversity within and between worldwide populations generally share key characteristics that generate increased mutation rates and, therefore, increased allelic diversity within the locus. These molecular features include the number of repeats within a locus (the higher the number of repeats, the greater the mutation rate), the complexity of the STR sequence (the more complex a sequence in terms of numbers of repeat blocks the greater the rate), and the repeat size (mutation rate decreasing as repeat size increases from tri- to tetra- to pentanucleotide repeats) (Ballantyne et al. 2010). Furthermore, the multicopy status of Y-STRs usually provides enhanced value in differentiating between haplotypes (Ballantyne et al. 2012). The improved knowledge regarding both the number and characteristics of diverse Y-STRs has allowed the selection of candidate loci to improve current Y-STR testing capabilities. There are now mutation rates and sequence data available for 186 Y-STRs (Ballantyne et al. 2010; Goedbloed et al. 2009), and limited population data for ~110 of these (D'Amato, Benjeddou, and Davison 2009; D'Amato et al. 2010; Ehler, Marvan, and Vanek 2010; Hanson, Berdos, and Ballantyne 2006; Hedman et al. 2011; Leat et al. 2007; Lessig et al. 2009; Lim et al. 2007; Maybruck et al. 2009; Redd et al. 2002; Rodig et al. 2008; Xu et al. 2010; Geppert, Edelmann, and Lessig 2009; Hanson and

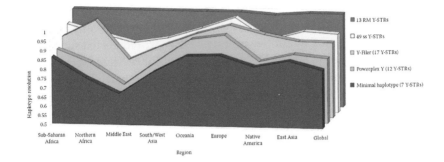

Figure 4.2 Increasing haplotype resolution in the global HGDP-CEPH panel with differing Y-STR sets. Only the RM Y-STRs provide near-complete resolution in all global populations. (Data from Ballantyne et al. 2012; Vermeulen et al. 2009.)

Ballantyne 2007a; Palha et al. 2012). To date, various combinations of 41 distinct Y-STR markers have been proposed to either supplement or replace the current set (for a review of additional Y-STRs in forensics, see Ballantyne and Kayser, 2012). These panels of 7–21 markers can increase the power of discrimination by 1.25% in American Africans and Europeans (Hanson and Ballantyne 2007b), 11.1% in South Africans (D'Amato, Bajic, and Davison 2011), and 28% in Finns (Hedman et al. 2011). The panel of novel Y-STRs with the highest increase in resolution on a global scale (Figure 4.2) is a set of 13 Y-STRs characterized by high mutation rates (1×10^{-2} or higher), called rapidly mutating (RM) Y-STRs, as will be further described later (Ballantyne et al. 2010; Ballantyne et al. 2012).

In 2012, the first next-generation commercial Y-STR kit was released that took advantage of the more recent scientific developments in the field of Y-STRs. The PowerPlex Y23 kit (Promega) contains all 17 Yfiler Y-STRs, plus an additional six Y-STR loci. These six Y-STRs—DYS481, DYS549, DYS533, DYS643, DYS570, and DYS576 (the latter two representing RM Y-STRs, see later)—were shown in a previous worldwide population study to be most informative, together with the commonly used Y-STRs, in discriminating paternal lineages out of a large number of simple, single-copy Y-STRs tested (Vermeulen et al. 2009). As expected, PPY23 provides additional discrimination power as demonstrated already for US populations (Davis et al. 2013), and this new kit comes with improved sensitivity and PCR inhibition resistance (Thompson et al. 2013). More recently, the AmpF/STR® Yfiler Plus PCR Amplification Kit (Life Technologies/Thermo Fisher Scientific) has been released (Gopinath et al. 2016), which currently represents the commercial Y-STR kit with the largest number of markers included. The Yfiler Plus kit targets all of the 17 Yfiler Y-STRs, 4 of the 6 Y-STRs additionally added in the PowerPlex Y23 kit (DYS481, DYS533, DYS570, DYS576), 1 additional highly polymorphic Y-STR (DYS460), and 5 additional RM Y-STRs (DYF387S1a/b, DYS449, DYS518, and DYS627) (Ballantyne et al. 2010) (for RM Y-STRs, see later), totaling 27 Y-STRs. In their developmental validation study, the company's authors reported increased discrimination of male lineages, improved performance in inhibited samples, improved balance in male–female mixed samples, and faster time to results (Gopinath et al. 2016).

4.3.2 Forensic Interpretation of Y-STR-Profile Matches

Although Y-STRs may be similar analytically to autosomal STRs, the interpretation of Y-STR profiles differs in several key aspects from that of autosomal STR profiles. Allele calling is generally less complex than for autosomal STRs, due to the haploid status. This also simplifies the separation of the haplotypes in mixed samples, with mixtures containing two to four males being relatively easily genotyped, provided there are measurable differences in relative amounts of all the contributors (Cerri et al. 2003; Parson et al. 2001; Prinz et al. 1997). However, some of the current Y-STRs (such as DYS385) have several male-specific copies, and cases with multicopy Y-STR status were observed for almost all markers that usually are present in single-copy (see www.yhrd.org for details). Although multicopy Y-STRs can be difficult to interpret in some situations, such as mixed samples, the multiple copies give greater diversity within the marker and have been shown to be the most informative markers in differentiating between haplotypes (Ballantyne et al. 2012).

When it comes to the way of estimating the strength of evidence, Y-STRs are conceptionally different from autosomal STRs in two ways. First, as the NRY is inherited intact from father to son, any statement of nonexclusion of a suspect from being the donor of a crime stain must also include all the suspects' paternal relatives in the nonexclusion (de Knijff 2003). In the absence of mutations, relatives separated by as many as 20 generations have been shown to share identical 17 locus haplotypes, and the nonexclusion of these tens or hundreds of male relatives must be conveyed in the evidentiary statement (Ballantyne et al. 2012). Second, due to the complete genetic linkage of all NRY markers, Y-STR frequencies have to be collected and used for statistical interpretation on the level of complete haplotypes instead of single loci. As such, the statistical interpretation of Y-STRs does not use the product rule with multiplication of the individual allele frequencies, but is instead most commonly done by estimating the frequency of the entire haplotype within the population of interest using representative databases.

Because compound haplotypes are much more variable than single independent STRs ever can be, the population databases used to derive Y-STR haplotype frequency estimates have to be much larger than for autosomal STRs. The larger the number of individuals in the frequency database, the more accurate the frequency estimate of common haplotypes will be, and the greater the chance of observing rare haplotypes. There are currently three Y-STR haplotype frequency databases available that were created for forensic purposes: YHRD, US Y-STR, and the Yfiler Haplotype Database. As of July 2019, the largest Y-STR haplotype frequency database, YHRD (www.yhrd.org), contains 285,406 minimal haplotypes, 244,777 PowerPlex Y haplotypes, 225,098 Yfiler haplotypes, 62,737 PowerPlex Y23 haplotypes, and 56,114 Yfiler Plus haplotypes.

When the case-observed Y-STR haplotype is contained in the database, the frequency of the haplotype can be calculated from the database as x observations in a sample of N haplotypes (Gill et al. 2001). As this is an estimate from a sample, rather than the entire population, upper and lower bounds of the estimate should be presented with the frequency. Although the confidence interval was previously calculated using the normal approximation, there has been a shift to using the binomial Clopper–Pearson confidence interval calculation as being more conservative for rare haplotypes (Buckleton, Krawczak, and Weir 2011). Despite the large sizes of current databases, it is estimated that 95% of all 17-locus Y-STR profiles will not be represented (Butler 2011). As a frequency estimate cannot be obtained for these singleton haplotypes not observed in the database before, there are several options for estimating evidentiary value. A 95% confidence interval, approximately equal to 3/N, may be used as a conservative estimate of the upper bound of the frequency (Buckleton, Krawczak, and Weir 2011). An alternative approach using the number of singletons already existing in the database has been proposed as a method to estimate the probability of an innocent suspect matching a previously unobserved haplotype

(Brenner 2010). Although promising, this approach is alleged to be anticonservative in its estimation of frequencies, and has yet to be adopted (Buckleton, Krawczak, and Weir 2011). Other methods are frequency surveying (Roewer et al. 2000; Willuweit et al. 2011) and the methods based on coalescent theory (Meyer Andersen et al. 2013). Thus far, none of these methods have reached the consensus status of being universally accepted and applied, which clearly marks a disadvantage of the forensic use of Y-STRs compared with autosomal STRs.

A general problem of currently unknown dimension is posed by the fact that all currently available Y-STR haplotype frequency databases for forensic use, including YHRD, were generated from unrelated individuals only. Thus, in principle, the derived frequency estimates are underestimating the true frequencies in the population where close and distant relatives are often living in the same population. Without empirical data available it is difficult to know how problematic this is for final conclusions in forensic case work, but it can be expected that the difference between estimated frequencies from such databases and the true frequencies are larger in rural areas and are smaller in metropolitan areas. Ideally, Y-STR haplotype frequency databases should be established from randomly chosen men, including related and unrelated individuals, to reflect the amount of male population substructure in a region.

At present, guidelines on Y-STR interpretation allow for the combination of Y-STR haplotype frequency estimates and autosomal STR match probabilities (Gusmao et al. 2006; Walsh, Redd, and Hammer 2008; Gill et al. 2001). However, there has been some criticism of this, due to the different assumptions used in each weight-of-evidence calculation (Amorim 2008). In particular, the frequencies of current Y-STR haplotypes do not exclude an individual's paternal relatives, while the autosomal probabilities generally do. This leads to each calculation addressing different, and exclusive, hypotheses, leading to joint probabilities being factually incorrect. To counter these arguments, it has been stressed that the nonexclusion of relatives is a case-specific problem, and may be discounted in some instances. Alternatively, any combined likelihood ratio should include the caveat that it may not be relevant for excluding relatives of the suspect (Buckleton, Krawczak, and Weir 2011).

4.4 Paternal Male-Relative Differentiation and Identification

The strongest limitation of the Y-STR sets widely used in forensics is that, in principle, conclusions cannot be drawn to a single individual because paternal male relatives of a suspect usually share the same Y-STR haplotype. This is because the mutation rates of the Y-STR loci involved are relatively low (10^{-4} to 10^{-3}, see Table 4.1; a mutation rate of 10^{-3} means 1 mutation every 1000 generations/meiosis per single locus; Goedbloed et al. 2009). However, a large Y-STR mutation rate study that analyzed almost 190 Y-STRs in close to 2000 DNA-confirmed father–son pairs (Ballantyne et al. 2010) previously identified 13 Y-STRs with higher mutation rates of $>10^{-2}$ referred to as rapidly mutating (RM) Y-STRs (Table 4.2).

Empirical evidence for male-relative differentiation with this set of RM Y-STRs was obtained in several studies with increasing number of relative pairs tested (Ballantyne et al. 2010; Ballantyne et al. 2012; Ballantyne et al. 2014). In the currently most comprehensive study (Adnan et al. 2016), this set of 13 RM Y-STRs differentiated father–sons (one

Table 4.2 Currently Available Rapidly Mutating (RM) Y-STR Markers

Marker	Copy Number	Repeat Motif	Mutation Rate	Diversity (Global)
DYS449[+]	1	$(TTCT)_{13-19}N_{22}(TTCT)_3 N_{12}(TTCT)_{13-19}$	1.22×10^{-2} (7.54×10^{-3} $- 1.85 \times 10^{-2}$, 1617)	0.88
DYS518[+]	1	$(AAAG)_3(GAAG)_1(AAAG)_{14-22}(GGAG)_1(AAAG)_4N_6(AAAG)_{11-19} N_{27}(AAGG)_4$	1.84×10^{-2} (1.25×10^{-2} $- 2.60 \times 10^{-2}$, 1556)	0.87
DYS526 a/b	2	$(CCTT)_{10-17}$ (a) $(CCCT)_3N_{20}(CTTT)_{11-17}(CCTT)_{6-10}$ $N_{113}(CCTT)_{10-17}$ (b)	(a) 2.72×10^{-3} ($9.52 \times 10^{-4} - 5.97 \times 10^{-3}$, 1716) (b) 1.25×10^{-2} ($7.88 \times 10^{-3} - 1.87 \times 10^{-2}$, 1651)	0.88
DYS547	1	$(CCTT)_{9-13}T(CTTC)_{4-5} N_{56} (TTTC)_{10-22}N_{10}(CCTT)_4(TCTC)_1(T$ $TTC)_{9-16}N_{14}(TTTC)_3$	2.36×10^{-2} (1.70×10^{-2} $- 3.18 \times 10^{-2}$, 1679)	0.87
DYS570[*,+]	1	$(TTTC)_{14-24}$	1.24×10^{-2} ($7.52 \times 10^{-3} - 1.91 \times 10^{-2}$, 1426)	0.83
DYS576[*,+]	1	$(AAAG)_{13-22}$	1.43×10^{-2} (9.41×10^{-3} $- 2.07 \times 10^{-2}$, 1727)	0.83
DYS612	1	$(CCT)_5(CTT)_1(TCT)_4(CCT)_1 (TCT)_{19-31}$	1.45×10^{-2} (9.61×10^{-3} $- 2.09 \times 10^{-2}$, 1767)	0.85
DYS626	1	$(GAAA)_{14-23}N_{24}(GAAA)_3 N_6$ $(GAAA)_5(AAA)_1(GAAA)_{2-3}$ $(GAAG)_1(GAAA)_3$	1.22×10^{-2} (7.70×10^{-3} $- 1.82 \times 10^{-2}$, 1689)	0.85
DYS627[+]	1	$(AGAA)_3N_{16}(AGAG)_3(AAAG)_{12-24}N_{81}(AAGG)_3$	1.23×10^{-2} (7.80×10^{-3} $- 1.81 \times 10^{-2}$, 1766)	0.85
DYF387S1[+]	2	$(AAAG)_3(GTAG)_1(GAAG)_4N_{16}$ $(GAAG)_9(AAAG)_{13}$	1.59×10^{-2} (1.08×10^{-2} $- 2.24 \times 10^{-2}$, 1804)	0.95
DYF399S1	3	$(GAAA)_3N_{7-8}(GAAA)_{10-23}$	7.73×10^{-2} (6.51×10^{-2} $- 9.09 \times 10^{-2}$, 1794)	0.99
DYF403S1a/b	4	$(TTCT)_{10-17}N_{2-3}(TTCT)_{3-17}$ (a) $(TTCT)_{12}N_2(TTCT)_8(TTCC)_9(TTCT)_{14}$ $N_2(TTCT)_3$ (b)	(a) 3.10×10^{-2} ($2.30 \times 10^{-2} - 4.07 \times 10^{-2}$, 1504) (b) 1.19×10^{-2} ($7.05 \times 10^{-3} - 1.86 \times 10^{-2}$, 1402)	0.89 (a)/0.99 (b)
DYF404S1	2	$(TTTC)_{10-20}N_{42}(TTTC)_3$	1.25×10^{-2} (7.92×10^{-3} $- 1.84 \times 10^{-2}$, 1739)	0.92

Notes: Mutation rate estimates are from family data (Ballantyne et al. 2010), while global diversity values are calculated from the HGDP-CEPH panel (Ballantyne et al. 2012.).

* Included in the PowerPlex® Y23 kit (Promega); [+] included in the AmpFLSTR® Yfiler® Plus kit (Applied Biosystems/Life Technologies).

meiosis apart) in 26.5% of the 2806 pairs tested, relatives two meioses apart (brothers, grandfather–grandson) in 45.7% of the 560 pairs, relatives three meioses apart in 53.8% of the 318 pairs, relatives four meioses apart in 62% of the pairs, with increasing differentiating rates in relatives with more meiosis apart. Male-relative differentiation rates were drastically lower based on Yfiler with 4.9%, 11.2%, 16.7%, and 24.8% for relative pairs separates by one to four meioses, respectively. Hence, the set of 13 RM Y-STRs carries a remarkable value for differentiating paternal male relatives and thus individualizing males by means

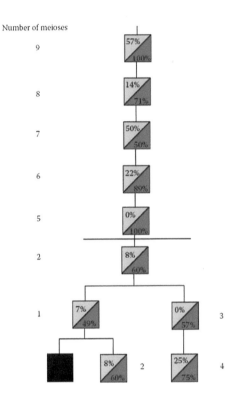

Figure 4.3 Observed rates of male relative differentiation with Yfiler (above the diagonal) and RM Y-STRs (below the diagonal), with respect to the level of relationship with the proband (black square). Full data and sample descriptions can be found elsewhere (Ballantyne et al. 2012).

of Y-chromosome analysis (Figure 4.3). In general, RM Y-STRs should not only be applied to those forensic cases with an a priori hypothesis of male relatives being involved, but moreover to all forensic cases where Yfiler analysis has revealed a match between suspect and crime scene materials to test for involvement of paternal males relatives (Ballantyne et al. 2012).

In the absence of commercial kits for the complete set of currently known RM Y-STRs, the 13 RM Y-STRs can be analyzed with three multiplex reactions (Ballantyne et al. 2012); companies are encouraged to develop and provide a commercial RM Y-STR product for forensic applications. Further, the AmpF*l*STR® Yfiler Plus Kit (Thermo Fisher Scientific) includes seven RM Y-STRs, whereas the PowerPlex Y23 kit (Promega) includes two (see earlier).

As indicated earlier (and see Figure 4.2), RM Y-STRs are also increasing male lineage differentiation when applied to unrelated males, as convincingly demonstrated in a worldwide multicenter study (Ballantyne et al. 2014). As shown by this study, haplotype diversity of all regional population groups as well as globally was higher based on the 13 RM Y-STRs compared to the 17 Yfiler Y-STRs. For instance, among the total of 7784 individuals tested in this study, 7714 different RM Y-STR haplotypes were observed, whereas 9075 different Yfiler haplotypes were found. Moreover, from the many unrelated individuals that shared Yfiler haplotypes with individuals from other populations around the world, almost all were differentiated based on RM Y-STR haplotypes (Ballantyne et al. 2014), which demonstrates Yfiler haplotype identity was by state and not by descent.

However, male-relative differentiation rates achieved with the currently known 13 RM Y-STRs are still improvable, especially for close relatives. To further improve male-relative differentiation rates, more RM Y-STR markers are needed, as are currently identified by one of the authors (MK).

4.5 Paternity Testing, DVI, and Familial Search

The use of the human Y chromosome in paternity testing of male offspring goes back more than 50 years ago where for the first time whole Y-chromosome length differences were applied to conclude an exclusion from paternity (Nuzzo, Caviezel, and de Carli 1966). The true value of applying NRY-DNA polymorphisms to paternity/family testing lies in the ability to replace unavailable persons such as deceased alleged fathers with paternal relatives (Kayser et al. 1998; Junge et al. 2006). In particular, such deficiency cases can only be solved with autosomal DNA analysis in cases where both parents of the deceased alleged father are available for DNA analysis, which is often not the case. Deficiency paternity cases involving male offspring can also be addressed with NRY-DNA analysis when the paternity issue occurred many generations ago, if paternal male relatives of both the alleged father and the male child are available for analysis. Obviously, such cases are impossible to solve with autosomal DNA analysis due to recombination issues. The most famous such case is that of US president Thomas Jefferson (1743–1826) who was speculated to have sired Eston Hemings Jefferson (born 1808), the son of his African American slave Sally Hemings (1773–1835) (Foster et al. 1998). NRY-DNA analysis of modern direct male descendants of Eston Hemings Jefferson and of several modern direct male descendants of Field Jefferson, the brother of Thomas Jefferson's father, revealed complete identity in the haplogroup characterized by several Y-SNPs and in 11 Y-STRs. This result is consistent with Thomas Jefferson's fathering of Easton Hemings Jefferson. However, given the nature of NRY-DNA, any one of Thomas Jefferson's contemporary paternal relatives, including his brother Randolph, could have been the true father with the same probability as Thomas Jefferson himself.

One of the key requirements to use NRY markers for paternity testing in deficiency cases involving male offspring, and to reconstruct more comprehensive paternal family relationships, is that the markers applied to display a sufficiently low mutation rate, since mutations usually complicate the statistical interpretations. By now, a relatively large amount of family-based mutation rate data has been gathered for the particular Y-STR markers used in forensics (Goedbloed et al. 2009) (Table 4.1). For the seven Y-STRs of the minimal haplotype and DYS385 more than 10,000 male meiotic transfers have been studied at each locus resulting in somewhat reliable locus-specific mutation rate estimates between 4×10^{-4} (DYS392) and 3×10^{-3} (DYS389II). Since other forensically used Y-STRs were introduced to the community later, the number of meiotic transfers investigated is smaller, hence the mutation rate estimates are somewhat less reliable, such as about 7000 meioses for DYS437, -38, and -39 with estimated mutation rates between 5×10^{-4} (DYS438) and 5×10^{-3} (DYS439) and about 3500-4000 meiotic transfers for the remaining Yfiler markers with estimated mutation rates between 2×10^{-4} (DYS448) and 6.5×10^{-3} (DYS458). The average mutation rate of all 17 Y-STRs commonly used in forensics (Table 4.1) was estimated as 2.2×10^{-3} based on more than 135,000 meiotic transfers (Goedbloed et al. 2009). Such relatively low rates established from large amounts of family data practically mean that in most cases of Y-STR applications to paternity testing involving these markers, a true biological father will show the same alleles as his true biological son.

If allelic differences are observed between a son and his putative father (or replacing paternal relative), they need to be considered in the paternity probability estimation (Rolf et al. 2001). It was previously recommended that an exclusion from paternity should be based on exclusion constellations at the minimum of three Y-STR loci, requiring the analysis of a sufficiently large number (≥9) of Y-STRs (Kayser and Sajantila 2001). This conclusion was based on the observation of two DNA-confirmed father–son pairs that both showed mutation at two out of nine Y-STRs tested (Kayser, Roewer, et al. 2000). Expectedly, this knowledge, and thus the recommendation, may be biased by the relatively small number of father–son pairs investigated at several Y-STRs in parallel. Indeed, a larger study involving close to 2000 father–son pairs and analyzing 17 Y-STRs (Yfiler) found one DNA-confirmed father–son pair with mutations at three Y-STRs (Goedbloed et al. 2009). Hence, based on the previous recommendation, this individual would have been excluded as the true father if only Y-STR data were available. However, not only did autosomal DNA analysis of the trio provided clear evidence in favor of paternity, but extended analysis of 157 Y-STRs showed no further allelic differences, clearly confirming the paternity (Ballantyne and Kayser 2012). Such an observation led us to recommend the use of probabilistic estimates of the likelihood of observing X number of mutations in a given meiotic transfer, where the locus-specific mutation rates of the (Y) STRs applied should be considered (as well as the number of meiotic transfers) (Ballantyne et al. 2012). A further solution for the issue of mutations in paternity testing of male offspring would be to use Y-STRs with an even lower mutation rate than the markers currently in use. Many of such slowly mutating (SM) yet polymorphic Y-STRs are available from the previously published Y-STR mutation survey (Ballantyne et al. 2010), although no specific SM Y-STR set has yet been developed for paternity testing.

As with paternity testing, Y-STRs are also useful for disaster victim identification (DVI) in cases where all other means of investigation, including autosomal STR (or SNP) profiling, were not informative. In particular, Y-STRs are suitable in DVI cases of males where only distant paternal male relatives are available for DNA testing that usually cannot be linked via autosomal STR profiling due to occurred recombination events. This can be important in disaster events involving entire families (hence all close relatives) such as airplane disasters to/from holiday destinations or disasters involving the destruction of residential areas including holiday resorts, such as the 2004 tsunami disaster in Southeast Asia.

Y-STR profiling can also be highly useful for familial search such as via DNA dragnets or mass screenings, if legally allowed, in cases with unknown suspects (no initial STR-profile match) and where the perpetrator escapes from voluntary participation. In familial search, the identity of Y-STR haplotypes or strong similarity of autosomal STR profile is used to highlight biological relatives of the (typically nonparticipating) unknown perpetrator among all participants of the mass screening (or a criminal offender DNA database). However, in contrast to autosomal STRs that are only able to highlight very close relatives (one to two generations apart), Y-STR allows identifying close and distant relatives of a nonparticipating perpetrator as they typically all share the same standard Y-STR haplotype defined by Y-STRs with moderate to low mutation rates (Table 4.1). The success of Y-STR–based dragnets has been demonstrated in practice in several cases (Dettlaff-Kakol and Pawlowski 2002; Huang et al. 2011). For instance, Y-STR–based mass screening was used in 2012 to finally solve the Marianne Vaatstra murder case in the Netherlands that happened in 1999. Although in this case the true perpetrator did participate in the Y-STR dragnet together with several thousands of other inhabitants of the area where the case happened, including several of his close and distant male relatives, he would also have been traced via his participating male relatives in case he had not participated (Kayser 2017). Expectedly, especially in rural areas, and also confronted with in the Vaatstra case (Kayser 2017), standard Y-STRs can unveil haplotype matches of several participating men all belonging to the perpetrator's paternal family related in a closely or distantly manner. Without further testing, however, it is impossible to differentiate between close relatives, who in principle are more likely to provide a lead to the nonparticipating unknown perpetrator, and distant relatives, who are less useful in finding the unknown perpetrator. The application of additional Y-STRs, particularly RM Y-STRs, as applied in the Vaatstra case (Kayser 2017), can help here. They allow separating out the distant relative, who will more likely show mutations the more distantly related they are to the unknown perpetrator, being left with those relatives without mutations, who most likely represent close relatives suitable to provide investigative leads. A detailed description of the Vaatstra case, which initiated two changes of the Dutch forensic DNA legislation, and how it was solved by means of Y-STRs is provided elsewhere (Kayser 2017).

4.6 Paternal Geographic Origin Inferences

4.6.1 Paternal Ancestry from Y-STR Haplotypes

Although most Y-STR haplotypes do not carry any geographic information, some particular ones indeed do. Direct geographic inference from Y-STR haplotypes is possible in cases where a Y-SNP haplogroup that contains geographic information harbors a relatively frequent Y-STR haplotype, for instance, because of population effects such as bottlenecks or founders. Examples of particular Y-STR haplotypes are DYS19-DYS389I-DYS389 II-DYS390-DYS391-DYS392-DYS393-DYS385a,b 17-13-30-25-10-11-13-10,14 indicating an Eastern European origin due to association with haplogroup R1a; haplotype 14-13-29-24-11-13-13-11,14 indicating a Western European origin due to association with haplogroup R1b; haplotype 13-13-30-24-10-11-13-16,18 indicating a Southwestern European origin due to association with haplogroup E3b; haplotype 15-13-31-21-10-11-13-16,17 indicating a sub-Saharan African origin due to association with haplogroup E*(xM35); and haplotype 15-12-28-24-10-13-13-12,16 indicating an origin in East/Southeast Asia or Oceania due to association with haplogroup O3. The reader is invited to search the named haplotypes in YHRD.org to see their geographic distribution and grasp the geographic value attached. However, as explained earlier, Y-SNPs are usually the better choice for inferring paternal geographic origins compared with Y-STRs.

4.6.2 Paternal Ancestry from Y-SNP Haplogroups

Although currently used Y-STRs are suitable for differentiating paternal lineages, their mutation rates provide strong limitation to their applicability for longer time-scale analyses including paternal ancestry inference, with some exceptions as mentioned earlier. Because of their about 100,000× lower mutation rate, Y-SNPs are therefore the marker of choice to

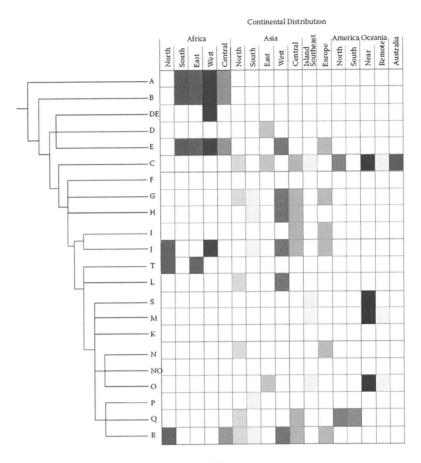

Figure 4.4 Y-chromosome phylogenetic tree of the 20 major haplogroups, with associated continental distributions. Haplogroups without distribution information are either found in all regions outside of Africa in low frequencies (F, K) or insufficient data exists (haplogroup NO).

illuminate paternal aspects of human demographic, including migration, history that also allow paternal ancestry inference. In the past, the Y Chromosome Consortium (YCC) Phylogenetic tree listed ~600 Y-SNPs defining 316 haplogroups, with each being defined by 1–25 characteristic Y-SNPs (Karafet et al. 2008). Across the globe, there are 20 main branches of the Y-chromosome phylogeny referred to as major haplogroups and labelled A-T (Figure 4.4) (Karafet et al. 2008; Y Chromosome Consortium 2002). The formation and spread of each major haplogroup reflect the establishment and expansion of major population groups and can give an indication of the time scales and the route of major migration events. At the same time major Y haplogroups can reveal information about paternal geographic origins. For example, the oldest and most variable haplogroups are the African haplogroups A and B, which is in line with the Out of Africa hypothesis of modern human origins. Distributions of other major haplogroups have been used as evidence for the spread of populations out of Africa and major migration routes throughout the globe such as haplogroup J in Northern African, European, and Central/Southern Asian populations, or haplogroup Q within the Americas (Chiaroni, Underhill, and Cavalli-Sforza 2009).

Consequently, at least some major haplogroups can be used directly to infer the geographic region of paternal genetic origin such as A and B indicating a sub-Saharan African origin, H indicating an South, Central, West Asian (and Roma) origin, and M indicating an origin in Oceania. However, most major haplogroups do not allow geographic inferences as they include subhaplogroups of diverse geographic origins; hence, additional Y-SNP typing is needed for determining the specific subhaplogroup to infer geographic information correctly. For instance, the major haplogroup C is found throughout North Asia, Southeast Asia, Central Asia, Oceania and Australia, and in Native Americans. However, certain C subhaplogroups specify particular geographic subregions such as C1 (M8) for Japan, C2 (M208) for Near Oceania, C2a1 (P33) for Remote Oceania, C3b (P39) for Native Americans, C4 (M347) for Australia, and C5 (M356) for South Asia. With some haplogroups, the specific geographic information can only be derived after high-resolution Y-SNP typing and subhaplogrouping, such as for the major haplogroup E which occurs in Europe, Africa, and West Asia. However, certain E subhaplogroups indicate particular geographic subregions such as E2 (M75) for sub-Saharan Africa, E1b1b1a1b (V13) for Europe, E1b1b1b1 (M81) for Northern Africa/Europe, and E1b1b1c1 (M34) for Middle East/Anatolia/Northern Africa/Europe. Another example of a major haplogroup with various derived subhaplogroups carrying different geographic signatures is the major haplogroup R such as R2a (M124) for South Asia, R1b1c (V88) for Africa/Middle East, R1a1a1g (M458) for (Eastern) Europe, and R1a1a1f (M434) for West Asia. These examples demonstrate that geographic inferences from NRY-DNA analysis must be accompanied by complete knowledge about the (sub)haplogroup in question and its geographic distribution from appropriate reference data, which do not yet fully exist for all subhaplogroups.

The predominant use for Y-SNPs in forensic science will most likely be for inference of paternal geographic ancestry in search for unknown suspects, as detailed earlier. However, the success rate will be strongly dependent on the population in question and the Y haplogroup of the individual, as indicated earlier. Therefore, when designing Y-SNP multiplex tools for paternal geographic ancestry inferences, it is necessary to account for the distributions of all subhaplogroups within each major clade in case they reveal different geographic information. Previously, a hierarchical Y-SNP-typing strategy has been described to screen for all major worldwide haplogroups, genotyping 28 SNPs for 29 branches of the phylogenetic tree (van Oven, Ralf, and Kayser 2011). This tool provides a method to rapidly determine the major haplogroup of a sample. Some of the revealed haplogroups already allow a broad geographic inference such as A and B for sub-Saharan Africa, H for South, Central, West Asia, and P for South Asia, while most major haplogroups identified are shared between larger geographic regions and additional Y-SNP typing is needed to accurately focus on the correct subhaplogroups with the respective geographic information associated. To provide finer resolution, Y-SNP multiplexes have also been developed for European-specific haplogroups (Brion et al. 2005), East African (Gomes et al. 2010), and South American (Geppert et al. 2011), to list just a few examples.

Y-SNPs can also be useful for highlighting admixture within individuals and populations. The former is of high relevance when performing forensic casework in regions of the world where genetic admixture between different continental groups is known to have played a substantial role in human population history. For instance, Y-SNP haplogroup data usually show high proportions of European ancestry in many urban populations in South America believed to be of mainly European ancestry (Corach et al. 2010). However, mtDNA and suitable autosomal markers analyzed in the same individuals revealed much less European and much more Native American and African admixture components (Corach et al. 2010). This usually is explained by sex-biased genetic admixture starting with the European occupation of South America dominated by males and continuing with the transfer of African slaves. Furthermore, this is similarly important in any region of the world where different continental groups are nowadays living next to each other due to more recent migrations, such as in many metropolitan areas in Europe or North America, because admixture is possible and cannot be excluded when investigating unknown samples. For instance, Y-SNP haplogrouping in self-declared US African Americans revealed considerably high proportions of European admixture, and such proportions were much smaller when detected with suitable mtDNA and autosomal DNA markers in the same individuals (Lao et al. 2010; Vallone and Butler 2004). An example for very distant admixture detected with Y-SNP analysis is that of an indigenous British male with a rare East Yorkshire surname carrying a haplogroup A1 Y-chromosome indicating an African origin (King et al. 2007). Some other men with the same rare surname also carried the African A1 Y-chromosomes suggesting that the A1 lineage need not have been a founding type at the time of surname establishment in Britain about 700 years ago or before (King et al. 2007). Clearly, an admixture event that old would not be manifested in any appearance traits anymore and is unlikely to be known from family records, but can still be detected with Y-SNP haplogroup analysis. These examples illustrate on one hand how powerful in general Y-SNPs are for geographic origin inference, but on the other hand demonstrate that solely analyzing Y-SNPs to infer the geographic origin of an individual can be misleading in cases of admixture involving ancestors from different geographic regions. Hence, Y-SNP analysis should always be combined with suitable mtDNA and autosomal markers for accurate inference of a person's geographic origin from DNA (Corach et al. 2010; Lao et al. 2010). Whereas autosomal ancestry-informative markers are most sensitive to reveal recent admixture, non-recombining Y-SNPs (and mtDNA SNPs) are able to detect recent as well as ancient admixture and additionally allow inferring the parental side (paternal versus maternal) through which an admixture was introduced.

On a further note, the ability to detect admixture with Y-SNPs may be useful for creating population-specific Y-STR haplotype frequency databases. The National Research Council recommended the use of separate databases for major population groups within a population of heterogeneous ancestry to ensure that frequency estimates were not biased for rare alleles or haplotypes of minor population groups (National Research 1996). However, in admixed societies, it is likely that some individuals who self-declare as one group may carry a Y chromosome of another group as a result of paternal admixture at any time as described earlier. By genotyping Y-SNPs for these individuals, it would be possible to obtain accurate frequency estimates for the parental population for Y-STR databases (Hammer et al. 2006).

With the advent of next-generation sequencing (NGS) technologies, which in the forensic world is typically referred to as massively parallel sequencing (MPS), the number of identified Y-SNPs and the Y-haplogroups they identify exploded. The most advanced currently available Y trees include tens of thousands of Y-SNPs that define many thousands of Y haplogroups (Jobling and Tyler-Smith 2017) compared to the few hundreds identified in pre-MPS times. By utilizing targeted MPS technology it has been demonstrated that hundreds of Y-SNPs can now be analyzed simultaneously allowing the inference of hundreds of Y-haplogroups (Ralf et al. 2015; Ralf et al. 2019). With these MPS tools, Y haplogroup inference from forensic DNA samples of low quality and quantity is now possible on a high level of resolution, which is allowing detailed paternal ancestry inference as demonstrated (Ralf et al. 2019). However, for many of the recently identified Y-SNPs, some of which are included in recent MPS tools, population data are currently unavailable. Future population application of the introduced Y-SNP MPS tools shall deliver the population data needed to allow the full utilization of these MP tools for paternal ancestry inference.

4.6.3 Y-SNP-Typing Technologies in Forensics

There is a wide range of technologies and platforms available to genotype SNPs rapidly and reliably depending on the degree of sample as well as marker throughput from singleplex assays suitable for parallel genotyping of a larger number of samples such as with TaqMan to parallel genotyping of up to one million SNPs in single samples that can be upgraded to large numbers of samples as well (Ding and Jin 2009). However, a small selection is usually applied in forensic laboratories (Sobrino, Brion, and Carracedo 2005), mostly concentrating on low throughput approaches that can make use of already existing equipment such as DNA analyzers. The SNP-typing platform commonly used by the forensic community is SNaPshot from Applied Biosystems, a single-base extension assay. An initial PCR amplifies a short segment (50–100 bp) around the target SNP, followed by a second PCR, where a fluorescently labelled dideoxy nucleotide, complementary to the SNP of interest, is added to a single primer annealed immediately adjacent to the SNP. Multiplexes of 20–30 SNPs can be developed but become more demanding in their design the more SNPs that are added. Nonoverlap of extension fragment lengths during capillary electrophoresis can be achieved by added nonhuman sequence "tails" to the end of the extension primer. There are SNaPshot-based assays for continental-level resolution of Y-SNP haplogroups (Figure 4.5) (van Oven, Ralf, and Kayser 2011) as well as for Y-SNPs informative within continental levels such as, for example, Europe (Brion et al. 2005), East Africa (Gomes et al. 2010), and South America (Geppert et al. 2011).

Alternatively, the concept of allele-specific hybridization is sometimes being used in forensics to distinguish between SNP alleles by utilizing two specific probes differing in sequence at only the target site. A signal will only be generated if the oligonucleotide probe matches the target (and thus the SNP) exactly. Attaching different color fluorescent dyes to each allele-specific probe allows the genotype to be determined. Such probes are generally detected using real-time PCR methods (such as TaqMan [Applied Biosystems], LightCycler [Roche], or Molecular Beacon probes [Invitrogen]), which limit the multiplexing ability. As the PCR products are not separated by size, as with capillary electrophoresis, the number of SNPs that can be analyzed in a single assay is limited by the number of dyes available (at present, five colors simultaneously, or two SNPs). Another limitation of this approach is that accurate allele calling requires reference DNA samples of known genotypes, which are not always easily available. Other methods, although only rarely applied in forensic laboratories for SNP typing, include direct DNA sequencing, restriction enzyme digestion, or MALDI-TOF mass spectrometry (Sequenom)-based technologies.

One of the major disadvantages of the Y-SNP technologies used in forensics, because of their ability to cope with compromised DNA, is their limited multiplexing capacity. For instance, the SNaPshot technology allows not more than 20–30 SNPs to be analyzed simultaneously per each SNaPshot multiplex. If more Y-SNPs are required, more SNaPshot assays need to be developed that multiply the consumption of evidence DNA. This provides major limitations to the level of Y haplogrouping and paternal ancestry inference available with the methods. On the other hand, forensic applications of Y haplogrouping including paternal ancestry inference would benefit from high-resolution Y haplogrouping, which

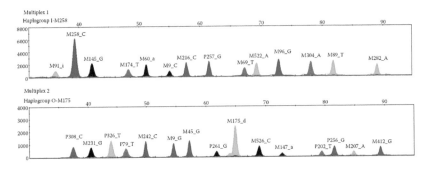

Figure 4.5 Electropherogram for an individual belonging to Y-chromosome haplogroup K2-P308 obtained with "Multiplex 2" (van Oven, Ralf, and Kayser 2011). For each peak, the marker name and the observed allele is indicated.

is not possible with the previous technologies. As with the discovery of Y-SNPs and Y haplogroups, MPS is also revolutionizing the way Y-SNPs are analyzed including in the forensic context, as targeted MPS methods allow multiplex amplification and sequencing of hundreds of Y-SNPs simultaneously (Ralf et al. 2015; Ralf et al. 2019). Because of the short sequence length of most currently available MPS technologies, they are generally suitable for analyzing compromised DNA, as confronted with in forensic analysis. As recently demonstrated, over 800 Y-SNPs allowing to infer almost 650 Y haplogroups can now be analyzed simultaneously (Ralf et al. 2019). A first developmental validation study of this Y-SNP MPS tool demonstrated robustness and reliability (Ralf et al. 2019). Moreover, a complete data analysis pipeline from the sequence reads generated by the MPS machines up to the final Y haplogroup including quality control has been developed and made available with this MPS tool (Ralf et al. 2019). The future is expected to see more and likely larger targeted MPS tools for Y-chromosomal (and other) SNPs to benefit forensic applications.

Acknowledgments

We thank Mannis van Oven and Arwin Ralf for providing Figure 4.5. We are grateful to the numerous colleagues who published on the Y chromosome and their application to forensic science, whose work we had the privilege to partly summarize here. Manfred Kayser was supported by funding from the Netherlands Forensic Institute (NFI) and by a grant from the Netherlands Genomics Initiative (NGI)/Netherlands Organization for Scientific Research (NWO) within the framework of the Forensic Genomics Consortium Netherlands (FGCN).

References

Adnan, A., A. Ralf, A. Rakha, N. Kousouri, and M. Kayser. 2016. Improving empirical evidence on differentiating closely related men with RM Y-STRs: A comprehensive pedigree study from Pakistan, *Forensic Sci. Int. Genet.*, 25: 45–51.

Akane, A., S. Seki, H. Shiono, H. Nakamura, M. Hasegawa, M. Kagawa, K. Matsubara, Y. Nakahori, S. Nagafuchi, and Y. Nakagome. 1992. Sex determination of forensic samples by dual PCR amplification of an X-Y homologous gene, *Forensic Sci. Int.*, 52(2): 143–48.

Amorim, A. 2008. A cautionary note on the evaluation of genetic evidence from uniparentally transmitted markers, *Forensic Sci. Int. Genet.*, 2(4): 376–78.

Andersen, M.M., A. Caliebe, A. Jochens, S. Willuweit, and M. Krawczak. 2013. Estimating trace-suspect match probabilities for singleton Y-STR haplotypes using coalescent theory. *Forensic Sci. Int. Genet.*, 7(2), 264–271.

Ballantyne, K.N., M. Goedbloed, R. Fang, O. Schaap, O. Lao, A. Wollstein, Y. Choi, K. van Duijn, M. Vermeulen, S. Brauer, et al. 2010. Mutability of Y-chromosomal microsatellites: Rates, characteristics, molecular bases, and forensic implications, *Am. J. Hum. Genet.*, 87(3): 341–53.

Ballantyne, K.N., V. Keerl, A. Wollstein, Y. Choi, S.B. Zuniga, A. Ralf, M. Vermeulen, P. de Knijff, and M. Kayser. 2012. A new future of forensic Y-chromosome analysis: Rapidly mutating Y-STRs for differentiating male relatives and paternal lineages, *Forensic Sci. Int. Genet.*, 6(2):208–218.

Ballantyne, K.N., and M. Kayser. 2012. Additional Y-STRs in forensics: Why, which and when, *Forensic Sci. Rev.*, 24(1): 63–78.

Ballantyne, K.N., A. Ralf, R. Aboukhalid, N.M. Achakzai, M.J. Anjos, Q. Ayub, J. Balazic, J. Ballantyne, D.J. Ballard, B. Berger, et al. 2014. Toward male individualization with rapidly mutating y-chromosomal short tandem repeats, *Hum. Mutat.*, 35(8): 1021–32.

Brenner, C.H. 2010. Fundamental problem of forensic mathematics - The evidential value of a rare haplotype, *Forensic Sci. Int. Genet.*, 4(5): 281–91.

Brion, M., B. Sobrino, A. Blanco-Verea, M.V. Lareu, and A. Carracedo. 2005. Hierarchical analysis of 30 Y-chromosome SNPs in European populations, *Int. J. Legal Med.* 119(1):10–15.

Buckleton, J.S., M. Krawczak, and B.S. Weir. 2011. The interpretation of lineage markers in forensic DNA testing, *Forensic Sci. Int. Genet.*, 5(2): 78–83.

Butler, J.M. 2011. *Advanced Topics in Forensic DNA Typing: Methodology*. Waltham, MA: Academic Press.

Cerri, N., U. Ricci, I. Sani, A. Verzeletti, and F. De Ferrari. 2003. Mixed stains from sexual assault cases: Autosomal or Y-chromosome short tandem repeats? *Croat. Med. J.*, 44(3): 289–92.

Chang, Y.M., L.A. Burgoyne, and K. Both. 2003. Higher failures of amelogenin sex test in an Indian population group, *J. Forensic Sci.*, 48(6): 1306–13.

Chiaroni, J., P.A. Underhill, and L.L. Cavalli-Sforza. 2009. Y chromosome diversity, human expansion, drift, and cultural evolution, *Proc. Natl Acad. Sci. U. S. A.*, 106(48): 20174–79.

Corach, D., O. Lao, C. Bobillo, K. Van der Gaag, S.B. Zuniga, M. Vermeulen, K. van Duijn, M. Goedbloed, P.M. Vallone, W. Parson, P. De Knijff, and M. Kayser. 2010. Inferring continental ancestry of Argentineans from autosomal, Y-chromosomal and mitochondrial DNA, *Ann. Hum. Genet.*, 74(1): 65–76.

D'Amato, M.E., M. Benjeddou, and S. Davison. 2009. Evaluation of 21 Y-STRs for population and forensic studies, *Forensic Sci. Int. Genet. Suppl. S.*, 2(1): 446–47.

D'Amato, M.E., L. Ehrenreich, K. Cloete, M. Benjeddou, and S. Davison. 2010. Characterization of the highly discriminatory loci DYS449, DYS481, DYS518, DYS612, DYS626, DYS644 and DYS710, *Forensic Sci. Int. Genet.*, 4(2): 104–10.

D'Amato, M.E., V.B. Bajic, and S. Davison. 2011. Design and validation of a highly discriminatory 10-locus Y-chromosome STR multiplex system, *Forensic Sci. Int. Genet.*, 5(2): 122–5.

Davis, C., J. Ge, C. Sprecher, A. Chidambaram, J. Thompson, M. Ewing, P. Fulmer, D. Rabbach, D. Storts, and B. Budowle. 2013. Prototype PowerPlex Y23 System: A concordance study, *Forensic Sci. Int. Genet.* 7(1): 204–208.

de Knijff, P. 2003. Son, give up your gun: Presenting Y-STR results in court, *Profiles DNA*, 7: 3–5.

Dekairelle, A.F., and B. Hoste. 2001. Application of a Y-STR pentaplex PCR (DYS19, DYS389I and II, DYS390 and DYS393) to sexual assault cases, *Forensic Sci. Int.*, 118(2–3): 122–25.

Dettlaff-Kakol, A., and R. Pawlowski. 2002. First Polish DNA "manhunt" – An application of Y-chromosome STRs, *Int. J. Legal Med.*, 116(5): 289–91.

Ding, C., and S. Jin. 2009. High-throughput methods for SNP genotyping, *Methods Mol. Biol.*, 578: 245–54.

Ebensperger, C., R. Studer, and J.T. Epplen. 1989. Specific amplification of the ZFY gene to screen sex in man, *Hum. Genet.*, 82(3): 289–90.

Ehler, E., R. Marvan, and D. Vanek. 2010. Evaluation of 14 Y-chromosomal short tandem repeat haplotype with focus on DYS449, DYS456, and DYS458: Czech population sample, *Croat. Med. J.*, 51(1): 54–60.

Fattorinil, P., S. Cacció, S. Gustincich, J. Wolfe, B.M. Altamura, and G. Graziosil. 1991. Sex determination and species exclusion in forensic samples with probe cY97, *Int. J. Legal Med.*, 104(5): 247–50.

Foster, E.A., M. Jobling, P.G. Taylor, P. Donnelly, P. De Knijff, R. Mieremet, T. Zerjal, and C. Tyler-Smith. 1998. Jefferson fathered slaves last child, *Nature*, 6706: 27–8.

Geppert, M., M. Baeta, C. Nunez, B. Martinez-Jarreta, S. Zweynert, O.W. Vacas Cruz, F. Gonzalez-Andrade, J. Gonzalez-Solorzano, M. Nagy, and L. Roewer. 2011. Hierarchical Y-SNP assay to study the hidden diversity and phylogenetic relationship of native populations in South America, *Forensic Sci. Int. Genet.*, 5(2): 100–04.

Geppert, M., J. Edelmann, and R. Lessig. 2009. The Y-chromosomal STRs DYS481, DYS570, DYS576 and DYS643, *Leg. Med. (Tokyo)*, 11: S109–SS10.

Gill, P., C.H. Brenner, B. Brinkmann, B. Budowle, A. Carracedo, M.A. Jobling, P. De Knijff, M. Kayser, M. Krawczak, W.R. Mayr, et al. 2001. DNA commission of the international society of forensic genetics: Recommendations on forensic analysis using Y-chromosome STRs, *Forensic Sci. Int.*, 124(1): 5–10.

Goedbloed, M., M. Vermeulen, R.N. Fang, M. Lembring, A. Wollstein, K. Ballantyne, O. Lao, S. Brauer, C. Kruger, L. Roewer, et al. 2009. Comprehensive mutation analysis of 17 Y-chromosomal short tandem repeat polymorphisms included in the AmpFlSTR Yfiler PCR amplification kit, *Int. J. Legal Med.*, 123(6): 471–82.

Gomes, V., P. Sanchez-Diz, A. Amorim, A. Carracedo, and L. Gusmao. 2010. Digging deeper into East African human Y chromosome lineages, *Hum. Genet.*, 127(5): 603–13.

Gopinath, S., C. Zhong, V. Nguyen, J. Ge, R.E. Lagace, M.L. Short, and J.J. Mulero. 2016. Developmental validation of the Yfiler((R)) Plus PCR amplification Kit: An enhanced Y-STR multiplex for casework and database applications, *Forensic Sci. Int. Genet.*, 24: 164–75.

Gross, A.M., A.A. Liberty, M.M. Ulland, and J.K. Kuriger. 2008. Internal validation of the AmpFlSTR Yfiler amplification kit for use in forensic casework, *J. Forensic Sci.*, 53(1): 125–34.

Group, International SNP Map Working. 2001. A map of human genome sequence variation containing 1.42 million single nucleotide polymorphisms, *Nature*, 409(6822): 928–33.

Gusmao, L., J.M. Butler, A. Carracedo, P. Gill, M. Kayser, W.R. Mayr, N. Morling, M. Prinz, L. Roewer, C. Tyler-Smith, and P.M. Schneider. 2006. DNA commission of the international society of forensic genetics (ISFG): An update of the recommendations on the use of Y-STRs in forensic analysis, *Forensic Sci. Int.*, 157(2–3): 187–97.

Hammer, M.F., V.F. Chamberlain, V.F. Kearney, D. Stover, G. Zhang, T. Karafet, B. Walsh, and A.J. Redd. 2006. Population structure of Y chromosome SNP haplogroups in the United States and forensic implications for constructing Y chromosome STR databases, *Forensic Sci. Int.*, 164(1): 45–55.

Hanson, E., and J. Ballantyne. 2007a. Population data for 48 'non-core' Y chromosome STR loci, *Leg. Med. (Tokyo)*, 9(4): 221–31.

Hanson, E.K., and J. Ballantyne. 2007b. An ultra-high discrimination Y chromosome short tandem repeat multiplex DNA typing system, *PLOS ONE*, 2(8): e688.

Hanson, E.K., P.N. Berdos, and J. Ballantyne. 2006. Testing and evaluation of 43 'noncore' Y chromosome markers for forensic casework applications, *J. Forensic Sci.*, 51(6): 1298–314.

Hedman, M., A.M. Neuvonen, A. Sajantila, and J.U. Palo. 2011. Dissecting the Finnish male uniformity: The value of additional Y-STR loci, *Forensic Sci. Int. Genet.*, 5(3): 199–201.

Huang, D., S. Shi, C. Zhu, S. Yi, W. Ma, H. Wang, and H. Li. 2011. Y-haplotype screening of local patrilineages followed by autosomal STR typing can detect likely perpetrators in some populations, *J. Forensic Sci.*, 56(5): 1340–42.

Jobling, M.A., and C. Tyler-Smith. 2017. Human Y-chromosome variation in the genome-sequencing era, *Nat. Rev. Genet.*, 18(8): 485–97.

Jobling, M.A., and C. Tyler-Smith. 2003. The human Y chromosome: An evolutionary marker comes of age, *Nat. Rev. Genet.*, 4(8): 598–612.

Junge, A., B. Brinkmann, R. Fimmers, and B. Madea. 2006. Mutations or exclusion: An unusual case in paternity testing, *Int. J. Legal Med.*, 120(6): 360–63.

Karafet, T.M., F.L. Mendez, M.B. Meilerman, P.A. Underhill, S.L. Zegura, and M.F. Hammer. 2008. New binary polymorphisms reshape and increase resolution of the human Y chromosomal haplogroup tree, *Genome Res.*, 18(5): 830–38.

Kayser, M., A. Caglia, D. Corach, N. Fretwell, C. Gehrig, G. Graziosi, F. Heidorn, S. Herrmann, B. Herzog, M. Hidding, et al. 1997. Evaluation of Y-chromosomal STRs: A multicenter study, *Int. J. Legal Med.*, 110(3): 125–33, 41–9.

Kayser, M., C. Krüger, M. Nagy, G. Geserick, and L. Roewer. 1998. Y-chromosomal DNA-analysis in paternity testing: Experiences and recommendations. In B. Olaisen and B. Brinkmann (Eds.), *Progress in Forensic Genetics*. Amsterdam: Elsevier.494–496.

Kayser, M., S. Brauer, G. Weiss, P.A. Underhill, L. Roewer, W. Schiefenhovel, and M. Stoneking. 2000. Melanesian origin of Polynesian Y chromosomes, *Curr. Biol.*, 10(20): 1237–46.

Kayser, M., L. Roewer, M. Hedman, L. Henke, J. Henke, S. Brauer, C. Kruger, M. Krawczak, M. Nagy, T. Dobosz, R. Szibor, P. De Knijff, M. Stoneking, and A. Sajantila. 2000. Characteristics and frequency of germline mutations at microsatellite loci from the human Y chromosome, as revealed by direct observation in father/son pairs, *Am. J. Hum. Genet.*, 66(5): 1580–88.

Kayser, M., and A. Sajantila. 2001. Mutations at Y-STR loci: Implications for paternity testing and forensic analysis, *Forensic Sci. Int.*, 118(2–3): 116–21.

Kayser, M., R. Kittler, A. Erler, M. Hedman, A.C. Lee, A. Mohyuddin, S.Q. Mehdi, Z. Rosser, M. Stoneking, M. Jobling, A. Sajantila, and C. Tyler-Smith. 2004. A comprehensive survey of human Y-chromosomal microsatellites, *Am. J. Hum. Genet.*, 74(6): 1183–97.

Kayser, M. 2007. Uni-parental markers in human identity testing including forensic DNA analysis, *Biotechniques*, 43 (suppl): xv–xxi.

Kayser, M. 2017. Forensic use of Y-chromosome DNA: A general overview, *Hum. Genet.*, 136(5): 621–35.

King, T.E., E.J. Parkin, G. Swinfield, F. Cruciani, R. Scozzari, A. Rosa, S.K. Lim, Y. Xue, C. Tyler-Smith, and M.A. Jobling. 2007. Africans in Yorkshire? The deepest-rooting clade of the Y phylogeny within an English genealogy, *Eur. J. Hum. Genet.*, 15(3): 288–93.

Krenke, B.E., L. Viculis, M.L. Richard, M. Prinz, S.C. Milne, C. Ladd, A.M. Gross, T. Gornall, J.R. Frappier, A.J. Eisenberg, C. Barna, X.G. Aranda, M.S. Adamowicz, and B. Budowle. 2005. Validation of male-specific, 12-locus fluorescent short tandem repeat (STR) multiplex, *Forensic Sci. Int.*, 151(1): 111–24.

Lao, O., P.M. Vallone, M.D. Coble, T.M. Diegoli, M. Van Oven, K. Van der Gaag, J. Pijpe, P. De Knijff, and M. Kayser. 2010. Evaluating self-declared ancestry of U.S. Americans with autosomal, Y-chromosomal and mitochondrial DNA, *Hum. Mutat.*, 31(12): E1875–EE93.

Leat, N., L. Ehrenreich, M. Benjeddou, K. Cloete, and S. Davison. 2007. Properties of novel and widely studied Y-STR loci in three South African populations, *Forensic Sci. Int.*, 168(2–3): 154–61.

Lessig, R., J. Edelmann, J. Dressler, and M. Krawczak. 2009. Haplotyping of Y-chromosomal short tandem repeats DYS481, DYS570, DYS576 and DYS643 in three Baltic populations, *Forensic Sci. Int. Genet. Suppl. S.*, 2(1): 429–30.

Lim, S.K., Y. Xue, E.J. Parkin, and C. Tyler-Smith. 2007. Variation of 52 new Y-STR loci in the Y Chromosome Consortium worldwide panel of 76 diverse individuals, *Int. J. Legal Med.*, 121(2): 124–7.

Malsom, S., N. Flanagan, C. McAlister, and L. Dixon. 2009. The prevalence of mixed DNA profiles in fingernail samples taken from couples who co-habit using autosomal and Y-STRs, *Forensic Sci. Int. Genet.*, 3(2): 57–62.

Maybruck, J.L., E. Hanson, J. Ballantyne, B. Budowle, and P.A. Fuerst. 2009. A comparative analysis of two different sets of Y-chromosome short tandem repeats (Y-STRs) on a common population panel, *Forensic Sci. Int. Genet.*, 4(1): 11–20.

Mitchell, R.J., M. Kreskas, E. Baxter, L. Buffalino, and R.A. van Oorschot. 2006. An investigation of sequence deletions of amelogenin (AMELY), a Y-chromosome locus commonly used for gender determination, *Ann. Hum. Biol.*, 33(2): 227–40.

Mulero, J.J., C.W. Chang, L.M. Calandro, R.L. Green, Y. Li, C.L. Johnson, and L.K. Hennessy. 2006. Development and validation of the AmpFlSTR Yfiler PCR amplification kit: A male specific, single amplification 17 Y-STR multiplex system, *J. Forensic Sci.*, 51(1): 64–75.

National Research Council. 1996. The evaluation of forensic DNA evidence. In *Committee on DNA Forensic Science: An Update*. Washington, DC: The National Academies Press. 89–124.

Nuzzo, F., F. Caviezel, and L. de Carli. 1966. Y chromosome and exclusion of paternity, *Lancet*, 2(7457): 260–62.

Olofsson, J., H.S. Mogensen, B.B. Hjort, and N. Morling. 2011. Evaluation of Y-STR analyses of sperm cell negative vaginal samples, *Forensic Sci. Int. Genet. Suppl. S.*, 3(1): e141–42.

Oota, H., W. Settheetham-Ishida, D. Tiwawech, T. Ishida, and M. Stoneking. 2001. Human mtDNA and Y-chromosome variation is correlated with matrilocal versus patrilocal residence, *Nat. Genet.*, 29(1): 20–21.

Palha, T., E. Ribeiro-Rodrigues, A. Ribeiro-dos-Santos, and S. Santos. 2012. Fourteen short tandem repeat loci Y chromosome haplotypes: Genetic analysis in populations from Northern Brazil, *Forensic Sci. Int. Genet.* 6(3): 413–418.

Parson, W., H. Niederstatter, S. Kochl, M. Steinlechner, and B. Berger. 2001. When autosomal short tandem repeats fail: Optimized primer and reaction design for Y-chromosome short tandem repeat analysis in forensic casework, *Croat. Med. J.*, 42(3): 285–7.

Prinz, M., K. Boll, H.J. Baum, and B. Shaler. 1997. Multiplexing of Y chromosome specific STRs and performance for mixed samples, *Forensic Sci. Int.*, 85(3): 209–18.

Radam, G., and H. Strauch. 1973. Lumineszenzmikroskopischer Nachweis des Y Chromosoms in Knochenmarkszellen – Eine neue Methode zur Geschlechterkennung an Leichenmaterial, *Krim. Forensische Wiss.*, 6: 149–51.

Ralf, A., M. van Oven, D. Montiel Gonzalez, P. de Knijff, K. van der Beek, S. Wootton, R. Lagace, and M. Kayser. 2019. Forensic Y-SNP analysis beyond SNaPshot: High-resolution Y-chromosomal haplogrouping from low quality and quantity DNA using Ion AmpliSeq and targeted massively parallel sequencing, *Forensic Sci. Int. Genet.*, 41: 93–106.

Ralf, A., M. van Oven, K. Zhong, and M. Kayser. 2015. Simultaneous analysis of hundreds of Y-chromosomal SNPs for high-resolution paternal lineage classification using targeted semiconductor sequencing, *Hum. Mutat.*, 36(1): 151–9.

Redd, A.J., A.B. Agellon, V.A. Kearney, V.A. Contreras, T. Karafet, H. Park, P. De Knijff, J.M. Butler, and M.F. Hammer. 2002. Forensic value of 14 novel STRs on the human Y chromosome, *Forensic Sci. Int.*, 130(2–3): 97–111.

Rodig, H., L. Roewer, A.M. Gross, T. Richter, P. De Knijff, M. Kayser, and W. Brabetz. 2008. Evaluation of haplotype discrimination capacity of 35 Y-chromosomal short tandem repeat loci, *Forensic Sci. Int.*, 174(2–3): 182–88.

Roewer, L., and J.T. Epplen. 1992a. Rapid and sensitive typing of forensic stains by PCR amplification of polymorphic simple repeat sequences in case work, *Forensic Sci. Int.*, 53(2): 163–71.

Roewer, L., J. Arnemann, N.K. Spurr, K.H. Grzeschik, and J.T. Epplen. 1992b. Simple repeat sequences on the human Y chromosome are equally polymorphic as their autosomal counterparts, *Hum. Genet.*, 89(4): 389–94.

Roewer, L., M. Kayser, P. de Knijff, K. Anslinger, A. Betz, A. Caglia, D. Corach, S. Furedi, L. Henke, M. Hidding, et al. 2000. A new method for the evaluation of matches in non-recombining genomes: Application to Y-chromosomal short tandem repeat (STR) haplotypes in European males, *Forensic Sci. Int.*, 114(1): 31–43.

Roewer, L. 2009. Y chromosome STR typing in crime casework, *Forensic Sci. Med. Pathol.*, 5(2): 77–84.

Rolf, B., W. Keil, B. Brinkmann, L. Roewer, and R. Fimmers. 2001. Paternity testing using Y-STR haplotypes: Assigning a probability for paternity in cases of mutations, *Int. J. Legal Med.*, 115(1): 12–5.

Santos, F.R., A. Pandya, and C. Tyler-Smith. 1998. Reliability of DNA-based sex tests, *Nat. Genet.*, 18(2): 103.

Seielstad, M.T., F. Minch, and L.L. Cavalli-Sforza. 1998. Genetic evidence for a higher female migration rate in humans, *Nat. Genet.*, 20(3): 278–80.

Sibille, I., C. Duverneuil, G. Lorin de Grandmaison, K. Guerrouache, F. Teissiere, M. Durigon, and P. de Mazancourt. 2002. Y-STR DNA amplification as biological evidence in sexually assaulted female victims with no cytological detection of spermatozoa, *Forensic Sci. Int.*, 125(2–3): 212–16.

Sobrino, B., M. Brion, and A. Carracedo. 2005. SNPs in forensic genetics: A review on SNP typing methodologies, *Forensic Sci. Int.*, 154(2–3): 181–94.

Steinlechner, M., B. Berger, H. Niederstätter, and W. Parson. 2002. Rare failures in the amelogenin sex test, *Int. J. Legal Med.*, 116(2): 117–20.

Sturk, K.A., M.D. Coble, S.M. Barritt, and J.A. Irwin. 2009. Evaluation of modified Yfiler amplification strategy for compromised samples, *Croat. Med. J.*, 50(3): 228–38.

Sullivan, K.M., A. Mannucci, C.P. Kimpton, and P. Gill. 1993. A rapid and quantitative DNA sex test: Fluorescence-based PCR analysis of X-Y homologous gene amelogenin, *BioTechniques*, 15(4): 636–38, 40–41.

Thangaraj, K., A.G. Reddy, and L. Singh. 2002. Is the amelogenin gene reliable for gender identification in forensic casework and prenatal diagnosis? *Int. J. Legal Med.*, 116(2): 121–23.

Thompson, J.M., M.M. Ewing, W.E. Frank, J.J. Pogemiller, C.A. Nolde, D.J. Koehler, A.M. Shaffer, D.R. Rabbach, P.M. Fulmer, C.J. Sprecher, and D.R. Storts. 2013. Developmental validation of the PowerPlex Y23 System: A single multiplex Y-STR analysis system for casework and database samples, *Forensic Sci. Int. Genet.* 7(2): 240–250.

Tilford, C.A., T. Kuroda-Kawaguchi, H. Skaletsky, S. Rozen, L.G. Brown, M. Rosenberg, J.D. McPherson, K. Wylie, M. Sekhon, T.A. Kucaba, R.H. Waterson, and D.C. Page. 2001. A physical map of the human Y chromosome, *Nature*, 409(6822): 943–45.

Vallone, P.M., and J.M. Butler. 2004. Y-SNP typing of U.S. African American and Caucasian samples using allele-specific hybridization and primer extension, *J. Forensic Sci.*, 49(4): 723–32.

van Oven, M., A. Ralf, and M. Kayser. 2011. An efficient multiplex genotyping approach for detecting the major worldwide human Y-chromosome haplogroups, *Int. J. Legal Med.* 125(6): 879–885.

Vermeulen, M., A. Wollstein, K. van der Gaag, O. Lao, Y. Xue, Q. Wang, L. Roewer, H. Knoblauch, C. Tyler-Smith, P. de Knijff, and M. Kayser. 2009. Improving global and regional resolution of male lineage differentiation by simple single-copy Y-chromosomal short tandem repeat polymorphisms, *Forensic Sci. Int. Genet.*, 3(4): 205–13.

Walsh, B., A.J. Redd, and M.F. Hammer. 2008. Joint match probabilities for Y chromosomal and autosomal markers, *Forensic Sci. Int.*, 174(2–3): 234–38.

Willuweit, S., A. Caliebe, M.M. Andersen, and L. Roewer. 2011. Y-STR frequency surveying method: A critical reappraisal, *Forensic Sci. Int. Genet.*, 5(2): 84–90.

Xu, Z., H. Sun, Y. Yu, Y. Jin, X. Meng, D. Sun, J. Bai, F. Chen, and S. Fu. 2010. Diversity of five novel Y-STR loci and their application in studies of north Chinese populations, *J. Genet.*, 89(1): 29–36.

Xue, Y., Q. Wang, Q. Long, B.L. Ng, H. Swerdlow, J. Burton, C. Skuce, R. Taylor, Z. Abdellah, Y. Zhao, et al. 2009. Human Y chromosome base-substitution mutation rate measured by direct sequencing in a deep-rooting pedigree, *Curr. Biol.*, 19(17): 1453–57.

Y Chromosome Consortium. 2002. A nomenclature system for the tree of human Y-chromosomal binary haplogroups, *Genome Res.*, 12(2): 339–48.

Forensic Application of X Chromosome STRs

5

MOSES S. SCHANFIELD AND
MICHAEL D. COBLE

Contents

5.1 Introduction

Loci on the X chromosome in mammals and other organisms with XX/XY/XO have the same pattern of inheritance as all genes in haplodiploid organisms such as bees, ants, and wasps. In both types of organisms, there is an initial difference in allele frequencies in the two sexes, with ultimate convergence on an equilibrium frequency after several generations. For our purposes, we will assume that females are homogametic XX and the males will be heterogametic XY in this case.

For a locus with two alleles, there are six mating types and five types of offspring, as listed in Table 5.1. The table uses allele frequencies and is similar to the one found in Li (1976, p. 134). Females receive one gamete from their mother and one from their father, while males only receive an X chromosome from their mothers. The alleles present in females is p_f and q_f and for males p_m and q_m. Thus, the frequencies of the homozygous and heterozygous genotypes in the diploid females will be for A_1A_1, A_1A_2, and A_2A_2, respectively, $p_f p_m$, $(p_f q_m + p_m q_f)$, and $q_f q_m$. The frequency of the hemizygous genotypes A_1Y and A_2Y in males will be p_m and q_m, respectively.

The proof of the above is easily shown with the $A_1 A_1$ genotype total in females. Where

$$A_1A_1 \text{ Total} = p_f p_m^2 + \tfrac{1}{2} p_f q_m p_m + \tfrac{1}{2} p_m^2 q_f$$

$$= p_m [p_f p_m + \tfrac{1}{2}(p_f q_m + p_m q_f)]$$

$$= p_f p_m$$

DOI: 10.4324/9780429019944-6

Table 5.1 Equilibrium Mating Types and Frequencies for a Sex-Linked or Haplodiploid Locus

Mother	Father	Freq	Daughters			Sons	
			A_1A_1	A_1A_2	A_2A_2	A_1Y	A_2Y
A_1A_1	A_1Y	$p_f p_m^2$	$p_f p_m^2$	0	0	$p_f p_m^2$	0
A_1A_1	A_2Y	$p_f p_m q_m$	0	$p_f p_m q_m$	0	$p_f p_m q_m$	0
A_1A_2	A_1Y	$p_f q_m P_m +$ $P_m^2 q_f$	$1/2 p_f q_m P_m +$ $1/2 P_m^2 q_f$	$1/2 p_f q_m P_m +$ $1/2 P_m^2 q_f$	0	$1/2 p_f q_m P_m +$ $1/2 P_m^2 q_f$	$1/2 p_f q_m P_m +$ $1/2 P_m^2 q_f$
A_1A_2	A_2Y	$p_f q_m^2 +$ $p_m q_f q_m$	0 $1/2 p_m q_f q_m$	$1/2 p_f q_m^2 +$ $1/2 p_m q_f p_m$	$1/2 p_f$ $q_m^2 +$	$1/2 p_f q_m^2 +$ $1/2 p_m q_f p_m$	$1/2 p_f q_m^2 +$ $1/2 p_m q_f p_m$
A_2A_2	A_1Y	$q_f q_m p_m$	0	$q_f q_m p_m$	0	0	$q_f q_m p_m$
A_2A_2	A_2Y	$q_f q_m^2$	0	0	$q_f q_m^2$	0	$q_f q_m^2$
Total		1	$p_f p_m$	$p_f q_m + p_m q_f$	$q_f q_m$	p_f	q_f

Note $[p_f p_m + \frac{1}{2} (p_f q_m + p_m q_f)]$ is the allele frequency for $p_f p = P + \frac{1}{2} H$, thus $[p_f p_m + \frac{1}{2} (p_f q_m + p_m q_f)] = p_f$

Thus, after one generation of random mating for a sex-linked trait, the alleles are in Hardy–Weinberg equilibrium (HWE). However, the allele frequencies will have changed slightly, depending on the magnitude of the differences.

5.1.1 Changes in Allele Frequencies Over Time

The frequency of A_1 in the next generation will be

$$p'_f = p_f p_m + \frac{1}{2} (p_f q_m + p_m q_f)$$
$$= \frac{1}{2} p_f p_m + \frac{1}{2} p_f q_m + \frac{1}{2} p_m q_f + \frac{1}{2} p_f p_m$$
$$= \frac{1}{2} p_f (p_m + q_m) + \frac{1}{2} p_m (q_f + p_f)$$

since $p_m + q_m = 1$ and $p_f + q_f = 1$

$$= \frac{1}{2}(p_f + p_m)$$

However, since the X chromosome only originates with the mother, sons can only have their mothers X chromosome and thus X allele, thus the frequency of the A_1 and A_2 in the next generation will always be the frequency of the A_1 and A_2 in the previous maternal generation. Thus,

$$p'_m = p_f$$

Using the frequencies in the two sexes the point estimations for the allele frequencies are going to be

$$p_f = P_f + \frac{1}{2} H_f$$
$$p_m = P_m$$

As there are a total of three X chromosomes between males (X) and females (XX), assuming equal numbers of males and females, then the mean allelic frequency will be

$$\bar{p}\ 2/3\ p_f + 1/3\ p_m$$

Though the distribution of genotypes is expected to be in HWE, we have seen that the frequencies are not the same, and that they will converge over time. The preceding equation is easily rearranged such that

$$p_m = 3\ \bar{p}\ - 2\ p_f$$

(Hint: multiply by 3 and rearrange the components.)

Using the formula from earlier for $p'_f = \frac{1}{2}(p_f + p_m)$, substituting the above into this formula

$$
\begin{aligned}
p'_f &= \frac{1}{2}\ (p_f + p_m)\\
&= \frac{1}{2}\ (p_f + 3\ \bar{p}\ - 2\ p_f)\\
&= \frac{1}{2}\ p_f + 3/2\ \bar{p}\ - 2/2\ p_f\\
&= 3/2\ \bar{p}\ - \frac{1}{2}p_f\\
&= \frac{1}{2}\ (3\ \bar{p}\ - p_f)
\end{aligned}
$$

One can convert the female frequency in the daughter generation to a component of the mean allele frequency and the female allele frequency in the maternal generation.

Similarly, it can be shown what the relationship is between the daughter's allele frequency in the next generation and the mean allele frequency by rearranging $p'_f = \frac{1}{2}(3 - p_f)$:

$$
\begin{aligned}
p'f &= \frac{1}{2}(3\ \bar{p}\ - p_f)\\
p'_f &= 3/2\ \bar{p}\ - 1/2\ p_f\\
p'_f &= (2/2\ \bar{p}\ + \frac{1}{2}\ \bar{p}\ ) - \frac{1}{2}p_f\\
p'_f - 2/2\ \bar{p}\ &= \frac{1}{2}\ \bar{p}\ - \frac{1}{2}\ p_f\\
p'_f - \bar{p}\ &= \frac{1}{2}\ \bar{p}\ (-\ p_f)\ \text{to make the sides symmetrical}\\
p'_f - \bar{p}\ &= -\frac{1}{2}\ (p_f - \bar{p}\ )
\end{aligned}
$$

Thus, based on this, it is clear that the deviation in each succeeding generation in females will be half that of the previous generation. The deviation in males will also be halved, however since $p'_m = q_f$ there will be a one-generation lag. This is represented in Figure 5.1.

Using the preceding formulas, we can calculate the changes in allele frequencies in males and females in succeeding generations as well as the average allele frequency in the population and the difference in allele frequency between males and females. This data is presented in Table 5.2.

As we can see in the Figure 5.1, the oscillations in allele frequency converge on the mean allele frequency relatively quickly. The rate of convergence can be determined by looking at the change in the deviations from the average allele frequency p. Thus if $d_0 = (p_f - \bar{p}\ )$ represents the difference between the mean frequency of p and the female frequency, then in the next generation at time $d_1 = p'_f - \bar{p} = -\frac{1}{2}\ (p_f - \bar{p}\ )$ or

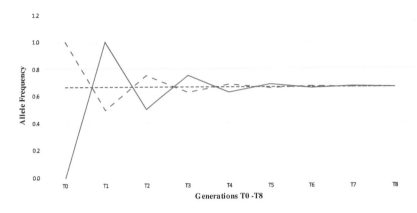

Figure 5.1 The change in allele frequencies for female (p_f) and male (p_m), and mean (p) allelic frequencies for an X-linked locus or a locus in a haplodiploid organism, where $p_f = 1.0$ and $p_m = 0.0$ in the first generation.

Table 5.2   **Distribution of Changes in Male and Female Allele Frequencies over Time**

Generation	Male	Female	*P*	d_t
0	0	1	0.667	0.333
1	1	0.5	0.667	−0.167
2	0.5	0.75	0.667	0.083
3	0.75	0.625	0.667	−0.042
4	0.625	0.687	0.667	0.021
5	0.687	0.656	0.667	-0.01
6	0.656	0.672	0.667	0.005
7	0.672	0.664	0.667	-0.003
8	0.664	0.668	0.667	0.001
9	0.668	0.666	0.667	−0.0007
10	0.666	0.667	0.667	0.0003

Allele frequencies of artificial population are respectively 0.000 and 1.000 in males and females. Frequencies are compared to the population average allele frequency () and the difference in allele frequencies between males and females (*dt*).

$$d_1 = -\tfrac{1}{2}d_0$$

Then given the relationship stated earlier, $d_1 = -\tfrac{1}{2}d_0$. This process can be repeated over and over. This repetitive process becomes a *recursive* process. We will see this process in the future. To estimate the deviation in the any number of generations the generalized formula becomes

$$d_t = (-\tfrac{1}{2})^t d_0$$

After six generations the deviation will be only 0.0039. Thus, after six or seven generations the deviations from the mean allele frequency should be negligible. However, until that occurs, as seen earlier in the generalized case in which males and females have differing frequencies, deviations from HWE will occur. As we saw previously, they are generally not large, however, the formulas are given next:

$$P_f - \bar{p}^2 = -1/9 \ (4p_f - p_m)(p_f - p_m)$$
$$H_f - 2\overline{pq} = 1/9 \ (4p_f - p_m)(p_f - p_m) + (4q_f - q_m)(q_f - q_m)$$
$$Q_f - \bar{q}^2 = -1/9 \ (4q_f - q_m)(q_f - q_m)$$

The point estimates for males and females can be estimated from the genotype arrays the same as autosomal loci. Thus, the estimated frequencies for A_1 males and females are

$$p_m = N_1 Y/N$$
$$p_f = (N_{11} + \tfrac{1}{2} N_{12})/N$$

Hedrick (2005) presented calculations for sex-linked alleles for males and females using p and q. In the general case, using the alleles p_i, the formulas for allele frequencies in males and females are

$$p_{im} = N_{im}/N_m$$
$$p_{if} = (N_{iif} + 1/2N_{ijf})/N_f$$

Hedrich (2005) goes on to state that the average frequency is $\bar{p} = (2/3p_f + 1/3p_m)$. *Though this is the average frequency of the sex-linked allele, it is not the population estimate from the sample.* The population point estimate for the sample would be

$$p_i = (N_i Y + 2N_{ii} + \Sigma N_{ij})/(N_m + 2N_f)$$

Note: The only time the average and total sample frequencies will be the same is when the number of males and females is equal.

5.1.2 Random Match

As stated previously, for our purposes we will assume that females are homogametic XX and the males will be heterogametic XY in this case. Thus, females are diploid, while males are haploid. This ultimately complicates the calculation of random match probability (RMP) for males and females. For female profiles, the calculation of random match likelihoods is the same as for any other diploid locus with heterozygotes equal to $2p_i p_j$ and homozygotes p_i^2, with or without the substructuring correction. In contrast, hemizygous males will always be estimated by the allele frequency p_i.

Unlike the autosomal markers used forensically and in parentage testing which are presumed to be unlinked and scattered throughout the genome, all of the X-linked makers are on a single chromosome. The X chromosome has traditionally been broken into four linkage groups. The four linkage groups and associated short tandem repeats (STRs) are found in Table 5.3. Closely linked markers often are in linkage disequilibrium (LD). LD indicates that the markers found within LD units are not randomly distributed. Depending on the degree of LD, there may be an effect on the RMP, but there will be an obvious effect on parentage calculations. In parentage testing one can select either the most informative

Table 5.3   Distribution of XSTR Loci in Four Sets of Multiplexes for Use in Forensic Testing

Linkage Group	Map	Qiagen Investigator Argus 8	Qiagen Investigator Argus 12	GenePhile X-Plex	Diegoli and Coble (2010)
1	Xpter-22.2			DXS6807	
1	Xp22.4 (22.1)			DXS9902	DXS9902
1	Xp22.3		DXS10148		
1	Xp22.32	DXS10135	DXS10135		
1	Xp22.31	DXS8378	DXS8378	DXS8378	DXS8378
1	Xp22.1				DXS6795
2	Xq11.2	DXS7132	DXS7132	DXS7132	DXS7132
2	Xq12	DXS10074	DXS10074		
2	Xq12		DXS10079		
2	Xq21.1			DXS9898	
2	Xq21.1				DXS6803
2	Xq21.2			DXS6809	
2	Xq21.3			DXS6789	DXS6789
2	Xq22			DXS7424	DXS7424
2	Xq22			DXS101	DXS101
2	Xq23(24)			GATA172D05	GATA172D05
2	Xq24				DXS7130
2	Xq25				GATA165B12
3	Xq26.2		DXS10103		
3	Xq26.2	HPRTB	HPRTB	HPRTB	HPRTB
3	Xq26.2	DXS10101	DXS10101		
4	Xq27				GATA31E08
4	Xq28				DXS10147
4	Xq28			DXS8377	
4	Xq28	DXS10134	DXS10134		
4	Xq28	DXS7423	DXS7423	DXS7423	DXS7423
4	Xq28		DXS10146		

Notes: Map locations taken from publications; disagreements are indicated in parenthesis. Loci in bold are in significant linkage disequilibrium and require haplotypes calculations.

STR or obtain haplotype frequencies for the markers in LD. The total RMP will be the product of the respective individual loci, or individual loci and haplotypes if there are linkage groups included.

    At the present time there are a limited number of kits commercially available. The three that the authors are aware of are Qaigen Argus Investigator X-8 and X-12 kits from Biotype Diagnostic GmbH (Dresden, Germany; distributed by Qaigen), and GenePhile X-Plex from GenePhile Bioscience Co., Ltd. (Taipei, Taiwan). These commercially available kits are single multiplexes with 8, 12, and 13 loci, respectively, which have had analysis of linkage disequilibrium performed. The Argus Investigator kits were extensively studied by Tilmar et al. (2008) and further discussed by Machado and Medina-Acosta (2009). The markers in each of the four linkage groups on the

X chromosome are in marked linkage disequilibrium and require the generation of two or three locus haplotypes for each of the linkage groups. In contrast, Hwa et al. (2009) did not find significant linkage disequilibrium across the 13 loci in the multiplex they evaluated, allowing them the simply multiply the individual loci. Diegoli and Coble (2010) developed two mini-multiplexes for forensic applications at the US Armed Forces DNA Identification Laboratory supported by National Institute of Justice (NIJ) grant funding. They found no linkage disequilibrium among the 15 loci in Europeans, African Americans, Southwestern Hispanic Americans (SW Hispanics), and US Asians. The location of the loci and those with marked linkage disequilibrium from the kits can be found in Table 5.3.

Table 5.4 is an example of paternity using the Diegoli and Coble (2010) XSTR multiplexes. Because they are not in linkage disequilibrium, the individual probabilities can be multiplied. As these frequencies would be the same for paternity testing or forensic identification for a male, this could also serve as an example of a forensic identification.

The requirements regarding linkage equilibrium are specified both in the United States and Europe by standards set by the DNA Advisory Board's Quality Assurance Standards 8.1.3.2 (DNA Advisory Board, 2000), and the International Society for Forensic Haemogenetics (ISFH), which specified that independent assortment = linkage independence be maintained before the product rule can be applied (ISFH, 1992).

Table 5.4 Paternity Case Using XSTR Loci from a US European Family and European Allele Frequencies

Sample Name	Mother 002-M	Child 002-C	AF 002-F	Obligate Allele	Frequency	RMNE	KI
SRY			SRY+				
DXS8378	11,12	9,11	9	9	0.017	0.017	58.824
DXS9902	10,10	10,10	10	10	0.305	0.305	3.279
DXS6795	9,11	11,11	11	11	0.161	0.161	6.211
DXS7132	12,14	12,15	15	15	0.229	0.229	4.367
DXS6803	11,13	13,13	13	13	0.093	0.093	10.753
DXS6789	20,20	20,21	21	21	0.22	0.22	4.545
DXS7424	15,15	15,16	16	16	0.136	0.136	7.353
DXS101	24,25	25,25	25	25	0.119	0.119	8.403
GATA172D05	6,11	6,12	12	12	0.025	0.025	40
DXS7130	13,16.3	13,15.3	15.3	15.3	0.237	0.237	4.219
GATA165B12	9,10	10,10	10	10	0.297	0.297	3.367
HPRTB	11,11	11,12	12	12	0.263	0.263	3.802
GATA31E08	12,14	12,12	12	12	0.246	0.246	4.065
DXS10147	9,9	8,9	8	8	0.407	0.407	2.457
DXS7423	16,16	15,16	15	15	0.356	0.356	2.809
Combined						6.142E-11	1.628E+10

Probability of Paternity = 99.9999999999%

Source: Data from Diegoli et al. (2014).

Note: Using 15 XSTR loci the results are highly informative.

5.2 Mutation Rates

A review of original data and a literature review of mutations in XSTRs is found in Diegoli et al., 2014). Mutations pose little problem in forensic evidence analysis, as there is a low likelihood of a mutation creating differences between tissues. In contrast, for parentage testing the presence of mutation poses a problem in the interpretation of the results of intergenerational tests, artificially creating exclusions. (See Chapter 2 for a discussion of this.) It is well known that the mutation rate between variable number tandem repeat (VNTR) polymorphism and single-nucleotide polymorphism (SNP) differ by orders of magnitude, mostly because the mechanisms differ by length polymorphism. The question is whether the detectable mutation rates for XSTRs are the same as nuclear STRs. The study by Diegoli et al. (2014) looked at approximately 21,000 meiosis in three populations—US Europeans, African Americans, and US SW Hispanics. Diegoli et al. (2014) compared their data to some literature on mutation in nuclear chromosome STRs. A larger summary can be found in AABB (2004), which was the last report of the AABB committee on parentage testing and covered 22 loci and 252,000 meioses. The data in the AABB parentage testing report compared mutation data on restriction fragment length polymorphisms (RFLPs) and STRs. Though the mutation rate was higher at RFLP loci (mu = 0.0023) than STR loci (mu = 0.0011), it was not significantly higher (surveysar.com/ztest.htm). Similarly, the difference reported in Diegoli et al. (2014) was not significant. In contrast, for nuclear STRs, the mutation rate was significantly higher in males than in females (M/F $\approx$ 3.31). However, the observed values for M/F XSTRs, though in the correct direction, were not significant. These results could be sample-size dependent. Thus, in conclusion, there is no convincing evidence that XSTRs have the same abnormal mutation rate as non-XSTRs, though this certainly could be subject to additional studies.

5.2.1 Frequency of Zero Mutation Rates

Among the 15 XSTR loci in the female African American samples, the loci had no mutations. Is this a significant result of the lack of mutations for either African Americans, or between males and females? Testing the XSTR data by chi-square indicates this observation is not significant. To test the difference between males and females with no mutations in a larger dataset, the frequency of no mutations in males and females in the AABB (2004) non-XSTRs was tested for 41 loci in 3717 female meioses (3 loci) and 19,721 male meioses (9 loci) with zero mutations. These results were also nonsignificant, suggesting that a lack of mutations at a specific locus and population is simply a sampling artifact and, with enough testing, mutations will be found in all populations and loci.

5.3 Exchange and Compatibility of Data

The nomenclature for DNA markers was established in 1986 (Shows et al., 1987) and has been modified as new types of markers have been added. Unfortunately, not all contributors have followed a uniform nomenclature. Table 5.3 lists the 27 loci found in various kits. Most of these follow the Shows et al. (1987) nomenclature. Most of the XSTR loci are anonymous polymorphisms that follow the nomenclature D (DNA polymorphism), X (X chromosome), S (single copy), and number (order in which polymorphism was assigned). In contrast, the nomenclature for a polymorphism in a structural gene uses the abbreviation for the gene. In this case the tetranucleotide repeat at Hypoxanthine-guanine phosphoribosyltransferase (HPRT) is used. Unfortunately, the use of markers such as GATA172D05, GATA165B12, and GATA31E08 are not part of the system. They reflect a series of tetranucleotide polymorphism (GATA) but provide no information on chromosome location. In a sense, these are like Penta D and Penta E autosomal STR from the Promega kits. Ideally, all DNA markers should be sent to the organization that classifies them.

In any case, identification of alleles whether by capillary electrophoresis (CE) or massive parallel sequencing (MPS) is done by comparison to allele ladders or known reference sequences. Even though different manufacturers will have different lengths, the classification of alleles should be accurate.

5.4 The Anthropological Genetics of XSTRs

After completion of the mutation study (Diegoli et al., 2014), a graduate student decided to investigate the anthropological usefulness of the XSTRs tested (Schanfield et al., 2013). To that end the AFDIL multiplexes were tested at the George Washington University (GWU) Department of Forensic Science, DNA Laboratory, using samples tested with the AFDIL multiplexes to create allele ladders. The samples consisted of the independent samples used in the mutation study (132 Europeans, 155 SW Hispanics, and 118 African Americans). The samples originated from cases of disputed paternity originally provided by the Analytical Genetic Testing Center (AGTC; Denver, Colorado) and originally tested at AFDIL: West Africans ($N = 30$, from Ghana and Nigeria), Ethiopians ($N = 34$), Southeast Asians ($N = 32$, primarily from Vietnam, Laos, Cambodia, and Thailand), and Northeast Asians ($N = 34$, from several Siberian populations). The African and Southeast Asian samples originated from AGTC immigration cases, and the Siberian samples were anthropological studies performed at AGTC. The latter samples were tested at the GWU Department of Forensic Science.

The loci studied are presented in Table 5.3 (Diegoli and Coble, 2010). The anthropological study yielded 144 alleles from the 15 loci. Six of the loci were in three possible linkage disequilibrium clusters: DXS8378 and DXS9902 (LG1, P22.5/P22.4), DXS7428 and DXS101 (LG2, Q22/Q22), and DXS10147 and DXS7423 (LG2, Q28/Q28). The results of testing these three pairs for LD indicated no statistically significant LD was detected (LG1, $p = 0.052$; LG2, $p = 0.377$; and LG4, $p = 0.133$). Anthropologically, evolutionarily LD will tend to slow recombination, while mutations increase differentiation. The big question is whether these loci differ among populations. Are the differences large enough to have significant information content (Fst), and are there unique alleles to indicate unique evolutionary events?

5.4.1 Allele Frequency Differences

The allele frequencies used in the anthropological genetics study are found in Tables 5.1A–5.15A (see the Appendix at the end of the chapter). The distributions of alleles in the seven populations indicated that there were no significant deviations from HWE, and no significant frequency differences between males and females. However, there were some statistically significant differences observed among the seven populations. There were 15 unique alleles out of 144 alleles (10.4%) distributed as follows: Africa ($N = 10$), Europeans ($N =$

3), and SW Hispanics ($N = 2$). Note: SW Hispanics are a mixture of Native Americans and Europeans, with a small amount of African contribution (Schanfield, unpublished). Population genetic studies of XSTRs in Native American populations are greatly needed and should be a focus for future research funding.

Fst is an established method for measuring variation at a locus across populations. In 1978, Sewell Wright published an interpretation of Fst results. Values of Fst less than 0.05 indicate little genetic differentiation among populations; Fst values of 0.05–0.15 indicate moderate genetic differentiation among populations; and Fst values of 0.15–0.25 indicate great genetic differentiation among populations. Six of the 15 XSTR loci had Fst values less than 0.05, while 9 of the 15 XSTR loci had an Fst value greater than 0.05. None of the 15 XSTR loci had an Fst value greater than 0.15. The test of significance of Wright's Fst is chi-square ($\chi^2 = 2NFst$) distributed with one degree of freedom. However, in this case a correction has to be made for 15 tests, as multiple tests were performed on the same dataset. This correction changes the significance level from 0.05 to 0.003. Five of the nine values had a probability greater than 0.003, three values were less than 0.003 (significant), and one was marginal at 0.0035. Thus, three or four of the Fst values were significant and ranged from 8.87 to 12.125. The data indicates that 60% of the 15 XSTRs studied had significant moderate genetic differentiation, while the rest of the nine had moderate genetic differentiation that was not significant.

Since there were 144 alleles in the 15 loci, principal components analysis (PCA) was used as a data reduction tool (Townend, 2002a). This yielded six principal components that accounted for 100% of the data variation. Cluster analysis was used to generate an unrooted dendrogram showing the relationship of the populations (Townend, 2002b). The dendrogram shown in Figure 5.2 was generated using Ward's method, on squared Euclidean distances, using the six PCA vectors as synthetic genes (Statistica Dataminer, Statsoft; Tulsa, Oklahoma).

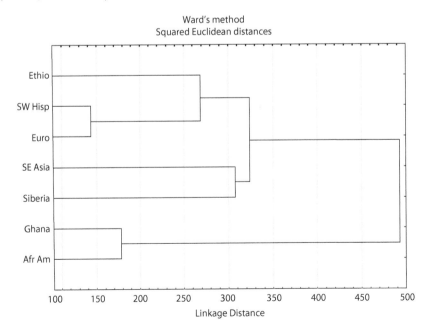

Figure 5.2 An unrooted dendrogram of cluster analysis of seven populations, using Ward's method on six PCA vectors as synthetic alleles generated from 15 XSTR loci and 122 alleles.

Figure 5.2 is of interest in that the linkage distances are quite long, which is unusual for STRs. Further, the major separation is between Africa and the rest of the world, with the exception of Ethiopia. For SNP markers, the similar examinations of populations of African versus non-African gave a linkage distance of 6.7 compared to almost 500 using the same software, but with PCA from 15 XSTRs. The position of Ethiopia as closer to Europeans than Africans is not unusual. Using other markers indicates significant Indo-European admixture in Ethiopians, probably created by back migration (Schanfield, unpublished data).

5.5 Future Directions

Uniparental markers on mtDNA (Chapter 3) and the Y chromosome (Chapter 4) are widely used in forensic testing and parentage testing. However, the haplodiploid X markers have limited applications in forensic and parentage testing. The primary use is in inheritance cases. For example, with autosomal STRs, a grandmother only shares on average one-fourth of her autosomal DNA with her granddaughter. Kinship statistics for this type of analysis tends to provide weak evidence of a grandmother–granddaughter relationship. However, if the granddaughter is the child of the grandmother's son, then the grandmother and grand-daughter will share one-half of their XSTRs. This is because the single X chromosome of the son is transmitted entirely from his mother, to him, to his female child without any recombination. Thus, 12 XSTR markers could provide a much stronger statistical associa-tion between the grandmother and the granddaughter than 20 autosomal STRs.

Future applications of XSTR markers could be useful for familial searching cases where females in the offender database provide a kinship index (KI) KI above a designated threshold for further investigation. Typically, laboratories utilize Y-chromosomal markers to limit the list of candidate leads. However, since females do not carry a Y chromosome, XSTRs may find some utility to reduce the number of investigative leads from a larger number of candidates. XSTRs may also have some potential utility in anthropological genetics, but more research will be required.

Acknowledgments

The authors would like to acknowledge the research and collaboration with Dr. Toni Diegoli and Diane Tiesma.

Appendix: XSTR World Distribution

Table 5.1A   XSTR World Distribution LG1 DXS9902

Population	European	SW Hisp	Afr-Amer	W Africa	Ethiopia	SE Asia	NE Asia
N	132	155	118	30.000	34.000	32.000	34.000
*6							
*7							
*8	0.000	0.000	0.093	0.100	0.000	0.000	0.000
*9	0.053	0.019	0.068	0.100	0.088	0.000	0.000
*10	0.303	0.355	0.305	0.333	0.235	0.219	0.412
*11	0.379	0.458	0.314	0.300	0.265	0.438	0.441
*11.1	0.030	0.019	0.008	0.167	0.000	0.063	0.000
*11.3							
*12	0.227	0.142	0.195	0.000	0.412	0.281	0.147
*12.3							
*13	0.008	0.006	0.017	0.000	0.000	0.000	0.000
*13.3							
*14							
*14.3							
*15							
*15.3							
*16							
*16.3							
*17							
*17.3							
*18							
*19							
*20							
*21							
*22							
*23							
*24							
*25							
*26							
*27							
*28							
*29							
*30							
*31							
He	0.709	0.643	0.757	0.751	0.697	0.678	0.614

Table 5.2A XSTR World Distribution LG1 DXS8378

Population	European	SW Hisp	Afr-Amer	Ghana	Ethiopian	SE Asia	NE Asia
N	132	155	118	30.000	34.000	32.000	34.000
*6	0.000	0.000	0.000	0.000	0.000	0.000	0.000
*7	0.000	0.000	0.000	0.000	0.000	0.000	0.000
*8	0.015	0.000	0.000	0.000	0.000	0.000	0.000
*9	0.023	0.026	0.017	0.000	0.029	0.000	0.118
*10	0.409	0.471	0.246	0.300	0.471	0.438	0.176
*11	0.303	0.303	0.390	0.300	0.235	0.438	0.235
*11.1							
*11.3							
*12	0.227	0.155	0.314	0.367	0.235	0.063	0.324
*12.3							
*13	0.023	0.045	0.034	0.033	0.029	0.063	0.147
*13.3							
*14							
*14.3							
*15							
*15.3							
*16							
*16.3							
*17							
*17.3							
*18							
*19							
*20							
*21							
*22							
*23							
*24							
*25							
*26							
*27							
*28							
*29							
*30							
*31							
He	0.688	0.660	0.688	0.684	0.666	0.609	0.773

Table 5.3A   **XSTR World Distribution LG1 DXS6795**

Population	European	SW Hisp	Afr-Amer	Ghana	Ethiopian	SE Asia	NE Asia
N	132	155	118	30.000	34.000	32.000	34.000
*6							
*7	0.000	0.000	0.000	0.000	0.029	0.000	0.000
*8							
*9	0.311	0.129	0.127	0.167	0.206	0.156	0.059
*10	0.038	0.110	0.322	0.267	0.235	0.438	0.000
*11	0.417	0.258	0.161	0.033	0.235	0.344	0.294
*11.1							
*11.3							
*12	0.015	0.181	0.076	0.033	0.029	0.031	0.059
*12.3							
*13	0.212	0.284	0.102	0.033	0.147	0.000	0.588
*13.3							
*14	0.008	0.013	0.025	0.000	0.088	0.031	0.000
*14.3							
*15	0.000	0.019	0.161	0.400	0.029	0.000	0.000
*15.3							
*16	0.000	0.006	0.017	0.067	0.000	0.000	0.000
*16.3							
*17	0.000	0.000	0.008	0.000	0.000	0.000	0.000
*17.3							
*18							
*19							
*20							
*21							
*22							
*23							
*24							
*25							
*26							
*27							
*28							
*29							
*30							
*31							
He	0.683	0.791	0.811	0.733	0.815	0.664	0.561

Table 5.4A **XSTR World Distribution LG2 DXS7132**

Population	European	SW Hisp	Afr-Amer	Ghana	Ethiopian	SE Asia	NE Asia
N	132	155	118	30.000	34.000	32.000	34.000
*6							
*7							
*8							
*9							
*10	0.000	0.006	0.000	0.000	0.000	0.000	0.000
*11	0.015	0.000	0.034	0.000	0.029	0.000	0.059
*11.1							
*11.3							
*12	0.061	0.090	0.110	0.200	0.059	0.000	0.059
*12.3							
*13	0.273	0.290	0.203	0.267	0.235	0.156	0.235
*13.3							
*14	0.432	0.342	0.331	0.267	0.265	0.281	0.412
*14.3							
*15	0.167		0.229	0.233	0.294	0.281	0.176
*15.3							
*16	0.045	0.200	0.059	0.033	0.088	0.281	0.029
*16.3	0.000	0.052	0.000	0.000	0.029	0.000	0.000
*17	0.008	0.013	0.025	0.000	0.000	0.000	0.029
*17.3							
*18	0.000	0.006	0.008	0.000	0.000	0.000	0.000
*19							
*20							
*21							
*22							
*23							
*24							
*25							
*26							
*27							
*28							
*29							
*30							
*31							
He	0.705349	0.747721	0.779517	0.762222	0.775087	0.738281	0.735294

Table 5.5A   **XSTR World Distribution LG2 DXS6803**

Population	European	SW Hisp	Afr-Amer	Ghana	Ethiopian	SE Asia	NE Asia
N	132	155	118	30.000	34.000	32.000	34.000
*6							
*7							
*8	0.000	0.000	0.042	0.000	0.000	0.000	0.000
*9	0.000	0.006	0.034	0.067	0.000	0.000	0.000
*10	0.030	0.006	0.059	0.067	0.118	0.000	0.000
*11	0.220	0.271	0.432	0.433	0.176	0.188	0.030
*11.1							
*11.3	0.023	0.058	0.008	0.000	0.118	0.125	0.152
*12	0.258	0.310	0.203	0.267	0.235	0.125	0.121
*12.3	0.061	0.155	0.068	0.067	0.059	0.313	0.303
*13	0.242	0.077	0.093	0.033	0.118	0.063	0.030
*13.3	0.152	0.110	0.017	0.067	0.176	0.188	0.303
*14	0.000	0.000	0.042	0.000	0.000	0.000	0.000
*14.3	0.008	0.006	0.000	0.000	0.000	0.000	0.061
*15							
*15.3	0.008	0.008	0.008	0.008	0.008	0.008	0.008
*16							
*16.3							
*17							
*17.3							
*18							
*19							
*20							
*21							
*22							
*23							
*24							
*25							
*26							
*27							
*28							
*29							
*30							
*31							
He	0.79844	0.78518	0.74993	0.72222	0.83737	0.79688	0.77319

Table 5.6A   XSTR World Distribution LG2 DXS6789

Population	European	SW Hisp	Afr-Amer	Ghana	Ethiopian	SE Asia	NE Asia
N	132	155	118	30.000	34.000	32.000	34.000
*6							
*7							
*8							
*9							
*10							
*11							
*11.1							
*11.3							
*12							
*12.3							
*13							
*13.3							
*14	0.015	0.026	0.000	0.033	0.000	0.000	0.000
*14.3							
*15	0.045	0.026	0.263	0.333	0.029	0.219	0.029
*15.3							
*16	0.015	0.039	0.161	0.067	0.000	0.406	0.235
*16.3							
*17	0.000	0.000	0.000	0.033	0.000	0.031	0.000
*17.3							
*18	0.000	0.000	0.025	0.133	0.000	0.000	0.000
*19	0.038	0.052	0.059	0.033	0.088	0.125	0.000
*20	0.364	0.394	0.178	0.067	0.412	0.125	0.471
*21	0.318	0.284	0.220	0.200	0.147	0.000	0.147
*22	0.136	0.090	0.076	0.067	0.235	0.031	0.118
*23	0.061	0.071	0.017	0.033	0.088	0.063	0.000
*24							
*25							
*26							
*27							
*28							
*29							
*30							
*31							
He	0.7403	0.74585	0.81456	0.81333	0.73702	0.75	0.68685

Table 5.7A   **XSTR World Distribution LG2 DXS7424**

Population	European	SW Hisp	Afr-Amer	Ghana	Ethiopian	SE Asia	NE Asia
N	132	155	118	30.000	34.000	32.000	34.000
*6							
*7							
*8							
*9							
*10	0.015	0.006	0.017	0.033	0.000	0.000	0.000
*11	0.008	0.006	0.127	0.067	0.000	0.000	0.000
*11.1							
*11.3							
*12	0.030	0.039	0.042	0.033	0.029	0.000	0.000
*12.3							
*13	0.053	0.097	0.237	0.300	0.059	0.156	0.412
*13.3							
*14	0.189	0.200	0.237	0.267	0.235	0.063	0.088
*14.3							
*15	0.311	0.168	0.178	0.167	0.294	0.156	0.147
*15.3							
*16	0.273	0.297	0.136	0.100	0.324	0.594	0.176
*16.3							
*17	0.106	0.148	0.025	0.033	0.029	0.031	0.059
*17.3							
*18	0.015	0.032	0.000	0.000	0.029	0.000	0.118
*19							
*20	0.000	0.006	0.000	0.000	0.000	0.000	0.000
*21							
*22							
*23							
*24							
*25							
*26							
*27							
*28							
*29							
*30							
*31							
He	0.77778	0.80974	0.81844	0.79333	0.7474	0.59375	0.7526

Table 5.8A **XSTR World Distribution LG2 DXS101**

Population	European	SW Hisp	Afr-Amer	Ghana	Ethiopian	SE Asia	NE Asia
N	132	155	118	30.000	34.000	32.000	34.000
*6							
*7							
*8							
*9							
*10							
*11							
*11.1							
*11.3							
*12							
*12.3							
*13							
*13.3							
*14							
*14.3							
*15	0.015	0.026	0.000	0.000	0.000	0.000	0.000
*15.3							
*16	0.000	0.000	0.008	0.000	0.000	0.000	0.000
*16.3							
*17	0.000	0.000	0.017	0.000	0.000	0.000	0.000
*17.3							
*18	0.061	0.097	0.034	0.033	0.000	0.000	0.000
*19	0.045	0.045	0.085	0.100	0.088	0.000	0.000
*20	0.015	0.000	0.034	0.067	0.000	0.000	0.000
*21	0.008	0.019	0.169	0.200	0.029	0.000	0.000
*22	0.045	0.013	0.068	0.067	0.029	0.000	0.029
*23	0.076	0.058	0.051	0.033	0.118	0.156	0.088
*24	0.212	0.271	0.068	0.033	0.118	0.563	0.059
*25	0.174	0.129	0.119	0.067	0.118	0.094	0.382
*26	0.152	0.213	0.093	0.100	0.206	0.125	0.265
*27	0.121	0.090	0.169	0.133	0.029	0.000	0.147
*28	0.061	0.019	0.034	0.067	0.206	0.031	0.029
*29	0.000	0.013	0.034	0.100	0.059	0.000	0.000
*30	0.008	0.000	0.017	0.000	0.000	0.031	0.000
*31	0.008	0.006	0.000	0.000	0.000	0.000	0.000
He	0.869146	0.839875	0.895576	0.891111	0.859862	0.632813	0.749135

Table 5.9A   **XSTR World Distribution LG2 GATA172D05**

Population	European	SW Hisp	Afr-Amer	Ghana	Ethiopian	SE Asia	NE Asia
N	132	155	118	30.000	34.000	32.000	34.000
*6	0.144	0.103	0.186	0.233	0.294	0.125	0.000
*7	0.000	0.000	0.059	0.067	0.029	0.063	0.000
*8	0.197	0.123	0.186	0.167	0.088	0.250	0.059
*9	0.015	0.071	0.246	0.333	0.265	0.156	0.059
*10	0.311	0.323	0.127	0.167	0.265	0.188	0.324
*11	0.250	0.271	0.169	0.033	0.029	0.156	0.529
*11.1							
*11.3							
*12	0.083	0.110	0.025	0.000	0.029	0.063	0.029
*12.3							
*13							
*13.3							
*14							
*14.3							
*15							
*15.3							
*16							
*16.3							
*17							
*17.3							
*18							
*19							
*20							
*21							
*22							
*23							
*24							
*25							
*26							
*27							
*28							
*29							
*30							
*31							
He	0.77433	0.77977	0.8210284	0.77333	0.76298	0.83008	0.60727

Table 5.10A **XSTR World Distribution LG2 DXS7130**

Population	European	SW Hisp	Afr-Amer	Ghana	Ethiopian	SE Asia	NE Asia
N	132	155	118	30.000	34.000	32.000	34.000
*6							
*7							
*8							
*9	0.000	0.000	0.008	0.000	0.000	0.000	0.000
*10	0.008	0.006	0.034	0.033	0.000	0.000	0.000
*11	0.030	0.058	0.051	0.100	0.029	0.156	0.000
*11.1							
*11.3							
*12	0.129	0.194	0.237	0.167	0.088	0.000	0.059
*12.3							
*13	0.023	0.052	0.178	0.200	0.235	0.094	0.000
*13.3	0.076	0.019	0.017	0.000	0.088	0.000	0.000
*14	0.000	0.026	0.042	0.033	0.000	0.031	0.000
*14.3	0.197	0.239	0.136	0.200	0.059	0.000	0.412
*15							
*15.3	0.417	0.323	0.237	0.200	0.265	0.594	0.382
*16							
*16.3	0.106	0.077	0.051	0.067	0.235	0.125	0.147
*17							
*17.3	0.015	0.006	0.008	0.000	0.000	0.000	0.000
*18							
*19							
*20							
*21							
*22							
*23							
*24							
*25							
*26							
*27							
*28							
*29							
*30							
*31							
He	0.752296	0.788345	0.828785	0.835556	0.799308	0.597656	0.65917

Table 5.11A   **XSTR World Distribution LG2 GATA165B12**

Population	European	SW Hisp	Afr-Amer	Ghana	Ethiopian	SE Asia	NE Asia
N	132	155	118	30.000	34.000	32.000	34.000
*6							
*7							
*8	0.008	0.006	0.042	0.000	0.059	0.000	0.000
*9	0.326	0.232	0.102	0.100	0.176	0.313	0.206
*10	0.326	0.484	0.297	0.267	0.088	0.563	0.706
*11	0.318	0.258	0.449	0.433	0.676	0.125	0.059
*11.1							
*11.3							
*12	0.023	0.019	0.110	0.200	0.000	0.000	0.029
*12.3							
*13							
*13.3							
*14							
*14.3							
*15							
*15.3							
*16							
*16.3							
*17							
*17.3							
*18							
*19							
*20							
*21							
*22							
*23							
*24							
*25							
*26							
*27							
*28							
*29							
*30							
*31							
He	0.68595	0.64491	0.68601	0.69111	0.5	0.57031	0.45502

Table 5.12A **XSTR World Distribution LG3 HPRTB**

Population	European	SW Hisp	Afr-Amer	Ghana	Ethiopian	SE Asia	NE Asia
N	132	155	118	30	34	32	34
*6							
*7							
*8							
*9	0.000	0.000	0.042	0.033	0.118	0.000	0.000
*10	0.000	0.006	0.017	0.067	0.000	0.000	0.000
*11	0.114	0.058	0.051	0.133	0.029	0.250	0.000
*11.1	0.341	0.265	0.263	0.267	0.353	0.125	0.324
*11.3	0.311	0.394	0.246	0.267	0.382	0.250	0.353
*12	0.167	0.194	0.237	0.200	0.059	0.375	0.235
*12.3	0.061	0.058	0.127	0.033	0.059	0.000	0.088
*13	0.008	0.026	0.017	0.000	0.000	0.000	0.000
*13.3							
*14							
*14.3							
*15							
*15.3							
*16							
*16.3							
*17							
*17.3							
*18							
*19							
*20							
*21							
*22							
*23							
*24							
*25							
*26							
*27							
*28							
*29							
*30							
*31							
He	0.743	0.730	0.793	0.793	0.708	0.719	0.708

Table 5.13A   **XSTR World Distribution LG4 GATA31E08**

Population	European	SW Hisp	Afr-Amer	Ghana	Ethiopian	SE Asia	NE Asia
N	132	155	118	30.000	34.000	32.000	34.000
*6							
*7	0.000	0.019	0.017	0.000	0.000	0.000	0.000
*8	0.000	0.000	0.042	0.067	0.000	0.031	0.000
*9	0.212	0.181	0.178	0.033	0.324	0.063	0.000
*10	0.015	0.052	0.136	0.267	0.000	0.000	0.000
*11	0.205	0.168	0.059	0.100	0.088	0.219	0.647
*11.1							
*11.3							
*12	0.220	0.342	0.246	0.233	0.353	0.281	0.088
*12.3							
*13	0.220	0.194	0.246	0.167	0.235	0.313	0.235
*13.3							
*14	0.091	0.045	0.051	0.067	0.000	0.031	0.029
*14.3							
*15	0.023	0.000	0.025	0.033	0.000	0.063	0.000
*15.3							
*16	0.008	0.000	0.000	0.033	0.000	0.000	0.000
*16.3							
*17	0.008	0.000	0.000	0.000	0.000	0.000	0.000
*17.3							
*18							
*19							
*20							
*21							
*22							
*23							
*24							
*25							
*26							
*27							
*28							
*29							
*30							
*31							
He	0.80751	0.77977	0.82031	0.82444	0.70761	0.76563	0.5173

Table 5.14A **XSTR World Distribution LG4 DXS10147**

Population	European	SW Hisp	Afr-Amer	Ghana	Ethiopian	SE Asia	NE Asia
N	132	155	118	30.000	34.000	32.000	34.000
*6	0.212	0.329	0.119	0.100	0.294	0.250	0.559
*7	0.038	0.052	0.263	0.267	0.000	0.250	0.000
*8	0.318	0.413	0.407	0.400	0.382	0.344	0.294
*9	0.424	0.200	0.195	0.133	0.265	0.156	0.147
*10	0.008	0.006	0.017	0.100	0.059	0.000	0.000
*11							
*11.1							
*11.3							
*12							
*12.3							
*13							
*13.3							
*14							
*14.3							
*15							
*15.3							
*16							
*16.3							
*17							
*17.3							
*18							
*19							
*20							
*21							
*22							
*23							
*24							
*25							
*26							
*27							
*28							
*29							
*30							
*31							
He	0.67229	0.67854	0.713157	0.73111	0.69377	0.73242	0.57958

Table 5.15A   **XSTR World Distribution LG4 DXS7423**

Population	European	SW Hisp	Afr-Amer	Ghana	Ethiopian	SE Asia	NE Asia
N	132	155	118	30.000	34.000	32.000	34.000
*6							
*7							
*8							
*9							
*10							
*11							
*11.1							
*11.3							
*12	0.000	0.006	0.000	0.000	0.000	0.000	0.000
*12.3	0.083	0.045	0.034	0.067	0.000	0.000	0.029
*13	0.258	0.323	0.458	0.600	0.324	0.250	0.059
*13.3	0.439	0.426	0.356	0.233	0.559	0.750	0.882
*14	0.159	0.097	0.144	0.100	0.118	0.000	0.029
*14.3	0.053	0.103	0.008	0.000	0.000	0.000	0.000
*15	0.008	0.000	0.000	0.000	0.000	0.000	0.000
*15.3							
*16							
*16.3							
*17							
*17.3							
*18							
*19							
*20							
*21							
*22							
*23							
*24							
*25							
*26							
*27							
*28							
*29							
*30							
*31							
He	0.70546	0.69253	0.64191	0.57111	0.5692	0.375	0.21626

References

AABB. 2004. 54 pp. Annual report summary for testing in 2003. *Prepared Parentage Test Stand Program Unit.*

Diegoli T.M., and M.D. Coble. 2010. Development and characterization of two mini-X chromosomal short tandem repeat multiplexes. *For Sci Int: Genet doi 10.1016/j.fsigen.2010.08.019* (online).

Diegoli T.M. 2014. In: *Forensic DNA Application: An Interdisciplinary Perspective.* Primorac D and Schanfield M (eds). Forensic application of X chromosome STRs. CRC Press, pp135–169.

Diegoli T.M., A. Linacre, M. Schanfield, and M.D. Coble. 2014. Mutation rates of 15 X chromosomal short tandem repeat markers. *Int J Leg Med 128*(4):579–587.

DNA Advisory Board 2000. *Quality Assurance Standards for Forensic DNA Testing Laboratories. For Sci Comm 2*(3).

Hedrick P.W. 2005. *Genetics of Populations* (3rd ed). Jones and Bartlett Publ.

Hwa H.L., Y.Y. Chang, J.C.I. Lee, H.Y. Yin, Y.H. Chen, L.H. Tseng, Y.N. Su, and T.M. Ko. 2009. Thirteen X-chromosomal short tandem repeat loci multiplex data from Taiwanese. *Int J Legal Med 123*(3):263–269.

ISFH 1992. 1991 Report Concerning Recommendations of the DNA Commission of the International Society for Forensic Haemogenetics Relating to the Use of DNA Polymorphisms. *For Sci Int 52*(2), pp 125–130.

Li C.C. 1976. *First Course in Population Genetics.* Boxwood Press, (p 631).

Machado F.B., and E. Medina-Acosta. 2009. Genetic map of human X-linked microsatellites used in forensic practice *For Sci Int Genet 3*:202-204.

Schanfield M.S., D. Tiesma, T. Diegoli, M. Coble, and M. Crawford. 2013. Preliminary report on the anthropology of 15 X STR loci. Program of the 82nd Annual Meeting of the American Association of Physical Anthropologists:36 (Abst).

Shows T.B., P.J. McAlpine, C. Bouchix, et al. 1987. Guidelines for human gene nomenclature. An international system for human gene nomenclature (ISGN, 1987). *Cytogenet Cell Genet 46*(1–4):11–28.

Tillmar A.O., P. Mostad, T. Egeland, B. Lindblom, G. Holmlund, and Montelius. 2008. *Analysis of Linkage and Linkage Equilibrium for 8 X-STR Markers. For Sci Int Genet 3*:37-41.

Townend J. 2002a. In: *Principal Components Analysis*, Chapter 14. Practical statistics for environmental and biological scientists. John Wiley & Sons, Ltd, pp 205–220.

Townend J. 2002b. In: *Cluster Analysis*, Chapter 15. Practical statistics for environmental and biological scientists. John Wiley & Sons, Ltd, pp 221–228.

Wright S. 1978. *Evolution and the Genetics of Populations, Vol. 4.* Univ Chicago Press.

Increasing the Efficiency of Typing Challenged Forensic Biological Samples

6

MAGDALENA M. BUŚ AND
BRUCE BUDOWLE

Contents

6.1 Introduction

Each compromised DNA sample may present different challenges that will affect downstream analyses. These challenges include DNA strand fragmentation (i.e., fragments <200 base pairs [bp] in length), reduced number of DNA copies (low template number or low copy number [LCN]; Budowle et al., 2001; Gill, 2001; Gill et al., 2000), single base damage, and the presence of inhibitors or contaminants. Additionally, due to environmental insults, highly degraded samples may be prone to producing sequence errors that can compromise the interpretation of results. These factors, either individually or in combination, impact DNA analyses. They can affect the success of polymerase chain reaction (PCR) amplification and/or present challenges for the interpretation of the result and data.

Diverse factors can influence the degree of DNA integrity in a sample. These factors include but are not limited to temperature, humidity, the activity of fungi and microorganisms, the type of environment (e.g., soil, water), and exposure to damaging factors (Jakubowska et al., 2012). Therefore, every bone sample, even collected from the same burial site, may display varying levels of DNA degradation and inhibition. Moreover, the type and concentration of coextracted inhibitors are not necessarily predictable.

Several methods have been developed to maximize the recovery of DNA from aged and degraded bone material. The techniques can be applied before PCR, during amplification, after PCR, or in combination. The choice of the DNA extraction method can impact

DOI: 10.4324/9780429019944-7

DNA recovery and copurification of inhibitors. Nevertheless, even the most seemingly efficient DNA extraction procedures may not overcome the damages inflicted on the DNA. Another crucial step in the success of data generation is PCR amplification. Since PCR generates copies of targeted sites in the genome, improvement of DNA amplification can enable typing of low template samples. Removal of inhibitors from DNA extracts, use of alternative polymerases, the addition of strand stabilizers and polymerase enhancers, and use of additives such as bovine serum albumin can enable PCR of challenged samples (Hochmeister et al., 1991).

6.2 Pre-PCR Improvement Strategies

DNA extracts from bone may contain compounds that have an inhibitory effect on PCR amplification. Various PCR inhibitors can be coextracted with DNA from bone, such as tannic, humic, and fulvic acids (Schrader et al., 2012; Tuross, 1994), all commonly soil-derived and therefore may be present in skeletal remains that were buried. The ubiquitous inhibitors in bone are calcium and collagen (Eilert and Foran, 2009; Opel et al., 2010). Another inhibitor that can be coextracted with DNA from skeletal remains is indigo. Indigo is a dye used for some types of fabrics that may become mingled with human remains during decomposition. Additionally, chemical reagents, for example, EDTA, used for DNA extraction may have an inhibitory effect on PCR amplification if not sufficiently removed during extraction (Schrader et al., 2012). Inhibitors can reduce or prevent amplification by interacting with PCR components, particularly the DNA polymerase or by binding to the DNA molecules. For example, tannic acid and calcium interact with the polymerase, whereas humic acid inhibits the PCR through binding to DNA and thus limits the amount of available templates for amplification (Opel et al., 2010).

Several strategies have been developed to overcome the adverse effect of inhibitors on PCR amplification. These approaches involve additional purification of extracts by precipitation (Hanni et al., 1995), silica columns (Faber et al., 2013), gel filtration (Miller, 2001), silica-coated magnetic beads (Nagy et al., 2005), filtration devices (Doran and Foran, 2014), or dedicated kits (Hu et al., 2015). All these methods are effective in the removal of various types of inhibitors to some degree; however, as with any method in which more sample manipulations are performed, substantial loss in DNA recovery has been observed (Doran and Foran, 2014). Therefore, additional purification methods can reduce the total number of copies available for amplification. Thus, a balance is struck between purification of inhibitors and amount of recoverable DNA. Additionally, low molecular weight or highly fragmented DNA can be lost during extraction and subsequent additional purification steps, thus reducing the available template for PCR.

An alternative treatment to overcome inhibition is dilution of DNA extracts such that the concentration of inhibitor does not impede the PCR. Usually, inhibitors display an inhibitory effect at certain concentration levels (Opel et al., 2010). Therefore, dilution of extracts will reduce the concentration of inhibitors while limiting sample manipulation. A drawback of this approach is that some samples are already in low copy number, and further dilution may reduce the number of copies in a sample below levels that will enable detection, even with PCR amplification. The number, type, and concentration of inhibitors may vary among samples; therefore, it is good practice to test different dilutions of DNA extracts. Testing serial extract dilutions—1:10, 1:20, 1:50, etc.—can allow for a balance between sufficient copy number and reduced concentration of inhibitors to promote DNA amplification. It is possible that one sample will amplify effectively after a 1:10 dilution, whereas another one may require 1:20 dilution to overcome the effects of inhibition. Dilution can be an effective approach to reduce the inhibitory effects of amplification of mitochondrial DNA (mtDNA). There are more copies of mtDNA in a cell (several hundred to a thousand) compared with nuclear DNA (two copies for nuclear DNA, except for gametes, which contain only one copy per cell). Therefore, dilution of a DNA sample may still yield sufficient mtDNA for downstream analyses. Another limitation of the dilution method is that it adds more labor and time to the analysis.

Dilution of samples to a level of genomic DNA that is considered as low copy number (Budowle et al., 2009, Gill, 2001, Gill et al., 2000) can influence the results obtained by DNA typing. Most massively parallel sequencing (MPS) or short tandem repeat (STR) kits for human identification suggest an optimum amount of input DNA of 500 pg to 1 ng for effective PCR amplification with manageable levels of stochastic effects (e.g., ForenSeq™ DNA Signature Kit, Verogen; Precision ID Panels, Thermo Fisher Scientific; GlobalFiler™, Thermo Fisher Scientific; PowerPlex® Fusion 6C System, Promega Corporation). Highly fragmented DNA can be considered as an LCN sample if a template molecule is fragmented to a degree that is not long enough for PCR amplification in which no or only a few copies can be generated. Moreover, these small fragments tie up reagents dedicated to amplification of the sufficient length target molecules and thus reduce PCR efficiency. Therefore, DNA profiles obtained from such samples may be problematic to interpret and additional efforts may be sought, often fruitless, to yield DNA data for attribution purposes (e.g., several independent amplifications and sequencing or capillary electrophoresis runs).

A number of DNA typing procedures involve determining the amount of DNA recovered from samples so that proper amounts of DNA are placed into the PCR and to limit needless sample consumption. Commercially available real-time PCR methods and quantification kits, such as the Quantifiler™ Trio DNA Quantification Kit (Thermo Fisher Scientific) or the PowerQuant® System (Promega Corporation), enable quantification of the DNA recovered from a sample down to picogram levels. For an LCN DNA extract, the sensitivity of the real-time PCR system can be increased by using a combination of buffers that substantially improve the detection of nucleic acids (Ahmad and Ghasemi, 2007). Additionally, the concentration of DNA and the level of degradation can be measured based on the presence of short and longer target DNA fragments. An ancillary benefit of these kits is that they contain internal controls that can indicate inhibition (see manuals for the Quantifiler Trio DNA Quantification Kit, and PowerQuant System). When a sample does not yield any amplified products, the underlying reasons may be inhibition and/or insufficient DNA template (or both). Lack of amplification of the internal control suggests that additional purification or sample dilution may be needed to reduce inhibitory effects. However, this method does not determine the particular types of inhibitors that were coextracted with the DNA. Nevertheless, this approach can facilitate determination whether inhibition is present in DNA extracts. If no sign of an inhibitory effect has been detected and no DNA is detected, it likely indicates that there are too few template molecules for PCR. In this situation, the DNA samples can be concentrated, which can effectively increase the DNA quantity that can be placed into a PCR. Then, this high-volume, pooled extract can be concentrated by using dedicated columns such as Amicon Ultra-0.5 mL Centrifugal Filters or Microcon® Centrifugal Filters (Millipore Sigma). DNA loss may be encountered (Noren et al., 2013) when concentrating a sample; however, pooling of several DNA extracts may yield a sufficient quantity of DNA for PCR.

6.3 PCR Improvement Strategies

6.3.1 Polymerases

Inhibitors may not always be effectively removed or reduced from DNA sample extracts (Glocke and Meyer, 2017). Additionally, fragmentation of challenged DNA samples may drastically reduce the PCR yield. However, there are processes related to PCR that may allow amplification of such challenged samples. Several strategies can be applied to increase PCR efficiency, which include the choice of polymerase or a blend of enzymes, addition of amplification enhancers, and reduction of amplicon sizes.

The polymerase is the enzyme that copies the template strands. Various polymerases are available as a separate enzyme product or as ready-to-use master mixes containing buffer, deoxynucleotide triphosphates (dNTPs), and $MgCl_2$. However, challenged samples display a complex set of characteristics that can prevent amplification, even in the presence of a well-validated and widely used enzyme, e.g., AmpliTaq Gold (Abu Al-Soud and Radstrom, 1998; Eilert and Foran, 2009). Each polymerase exhibits different characteristics (Terpe, 2013) and, thus, may perform differently in the presence of inhibitory substances. Therefore, new polymerases are continuously being evaluated for challenged samples (Nilsson et al., 2016) or a mixture of polymerases with diverse characteristics is used for PCR amplification (Hedman et al., 2011). It has been shown that Pfu polymerase is resistant to humic and fulvic acids (Matheson et al., 2010), whereas Tth polymerase demonstrates resistance to inhibition from skeletal remains (Eilert and Foran, 2009). However, when both of these polymerases were tested along with other polymerases on ancient DNA, Ex Taq HS, FastStart Taq, and PicoMaxx HF enzymes showed higher efficiency in amplification of highly degraded samples (Kim et al., 2015). KAPA2G Robust (KAPA Biosystems) is resistant to the inhibitor ammonium nitrate, while the KAPA3G (KAPA Biosystems) Plant enzyme has the highest efficiency in amplifying degraded DNA from old buried bone material (Nilsson et al., 2016). A blend of polymerases has been proposed to be more versatile in overcoming the effects of inhibition (Hedman et al., 2011). Replacing AmpliTaq Gold DNA polymerase in the AmpFlSTR SGM Plus kit with a mix of ExTaq Hot Start and PicoMaxx High Fidelity increased the yield of the number of STR alleles detected from challenged samples. The efficiency of PCR amplification also can be improved by reduction of the PCR volume and/or by use of twice the concentration of Taq polymerase (McNevin et al., 2015). These approaches offer higher sensitivity and better chance to obtain results from LCN DNA samples. Reducing the PCR volume does not actually increase yield, but it does increase the concentration of PCR product. With a higher concentration of PCR product, more product is available for downstream detection of genetic markers. Volume reduction and increased concentration of the polymerase can be advantageous when DNA analyses are performed on limited samples, e.g., a single piece of bone or a tooth.

Finding an efficient buffer system that interacts well with the polymerase can increase PCR yield. The buffer used in the PCR predominantly contains regular additives, such as salts like magnesium chloride, which have a twofold effect: (1) influencing the activity of the polymerase, and (2) neutralizing the negative-charged strands of DNA (primers and template DNA) so that they may anneal to make copies during PCR. However, these standard PCR components may not be effective when challenged samples are amplified. Therefore, using PCR enhancers may be added to increase the efficiency of amplification in a cost-effective way.

6.3.2 PCR Enhancers

Additional optimization may be necessary when inhibition is interrupting the PCR, the annealing temperatures of the primers differ, CG-rich templates are being amplified, and DNA markers are amplified in a PCR multiplex. Substances such as DMSO (dimethyl sulfoxide), betaine, formamide, BSA (bovine serum albumin), TMAC (tetramethyl ammonium chloride), glycol, and Tween 20 have been shown to effectively improve the yield of PCR amplification by chemical interaction with inhibitors or DNA structure. For example, GC-rich regions pose problems for PCR amplification because of secondary structure formation and different hybridization (or annealing) temperatures which may interfere with primer extension. Such enhancers as DMSO, formamide, and betaine effectively improve amplification of DNA sequences with high GC content (Comey et al., 1991; Sarkar et al., 1990; Henke et al., 1997; Jensen et al., 2010). Betaine at a final concentration of 1.25 mol/L improved amplification efficiency of LCN DNA samples (Marshall et al., 2015). Similarly, glycerol and Tween 20 improved the specificity of a PCR (Varadaraj and Skinner, 1994). TMAC enhances the specificity of primer annealing (Chevet et al., 1995). BSA significantly enhances PCR amplification of highly degraded DNA extract from old samples (Hochmeister et al., 1991; Paabo et al., 1988). Rohland and Hofreiter (2007) found that BSA helped to overcome the inhibitory effects in ancient DNA extracts. The addition of BSA at a concentration of 0.15 mg/mL considerably improved PCR amplification of DNA extracted from a Viking-age mass grave (Bus et al., 2019). Since degraded DNA from human remains usually exhibits complex characteristics that impact PCR amplification, a mixture of enhancers that display different properties may be effective in increasing the yield of the PCR (Ralser et al., 2006). It has been shown that BSA used in combination with DMSO or formamide increases PCR efficiency more so than each enhancer alone (Farell and Alexandre, 2012). It is important to optimize the concentration of each enhancer. DMSO (1%–10%) is recommended for use in a PCR; however, adding >2% may result in inhibition of Taq polymerase. Similarly, the recommended final reaction concentration of Tween-20 is in the range of 0.1%–1%, and the use of a concentration above 1% may result in PCR inhibition. Final reaction concentrations of formamide range from 1.25% to 10%, and for betaine the ranges are from 0.5 M to 2.5 M. For comprehensive PCR troubleshooting, see Lorenz (2012).

6.3.3 Reduction of the Size of Amplicons

Aged DNA samples (including those samples exposed to substantial environmental insults) often are characterized by extensive DNA degradation manifested as a reduction in average fragment sizes. DNA fragmentation severely reduces the efficiency of PCR, especially if primers (short pieces of DNA that target the site of interest and initiate copying of the target) are designed to amplify fragments of greater lengths than those present in a DNA extract. In such samples, PCR amplification may yield partial profiles or no amplification results. Therefore, a strategy that can improve data recovery from highly degraded samples is to redesign the primers so that they are more closely spaced, thus reducing amplicon sizes. This approach has been used for high discriminatory power markers such as autosomal STRs. Mitochondrial DNA, because of its greater copy number and protection within the mitochondrion, is less susceptible to degradation than nuclear DNA (Bogenhagen and Clayton, 1974; Miller et al., 2003; Satoh and Kuroiwa, 1991). Hence, amplification of 200–300 bp long fragments may be possible in aged samples where nuclear DNA does not provide any results. However, since nuclear DNA provides a higher discrimination power than mitochondrial DNA, reduction in amplicon sizes is sought to enhance amplification yield. In standard commercially available forensic STR kits, the fragment lengths for each marker vary from approximately 100 bp to 450 bp. It has been observed that the amplification of smaller-sized STRs yields a higher typing success rate from degraded samples (Coble and Butler, 2005; Wiegand and Kleiber, 2001). Therefore, shortening the amplicon size is highly recommended for typing DNA from remains that have been exposed to harsh environmental conditions (Zietkiewicz et al., 2012). Mini-STR kits are commercially available for CE analysis, such as the AmpFLSTR™ MiniFiler™ PCR Amplification Kit (Thermo Fisher Scientific) with all amplicon sizes less than 270 bp. Similarly, in the ForenSeq™ Signature Prep Kit (Verogen) designed for MPS, the maximum amplicon length of most autosomal STRs is less than 250 bp. Another system that can be used to improve the success of retrieving genetic information from degraded samples is InnoTyper® 21 (InnoGenomics Technologies). The InnoTyper kit contains 20 bi-allelic INNULs (insertion or null) markers and amelogenin with amplicons sizes from approximately 60 to 125 bp. Therefore, the assay is highly sensitive for analysis of degraded DNA. The system has been successfully used in typing analyses of highly degraded samples and rootless hair shafts (Brown et al., 2017; Grisedale et al., 2018). While STRs are the primary genetic markers for human identification, single-nucleotide polymorphisms (SNPs) may be better suited for typing highly degraded DNA. The PCR primers can be placed closer in proximity to the SNP position than those designed for STRs. Therefore, in theory, fragments as short as 50 bp in length can be amplified, which increases the chances of generating results from fragmented DNA. Recently, reverse complement PCR (RC-PCR) was shown to be able to analyze highly degraded DNA, i.e., fragments no more than 50 bases in length (Kieser et al., 2020). This technique enables amplification and library preparation for sequencing in a single, closed-tube assay. Kieser et al. (2020) showed that with a 27-human identity SNP panel that highly degraded DNA of only 50 bases in length could be analyzed. This method may be an effective alternative in the analysis of highly degraded DNA.

6.3.4 PCR Cycles

The analyses of samples with a low amount of DNA template will pose several problems in obtaining data and subsequent data interpretation. The most frequently observed drawbacks of STR analysis of low DNA input samples are heterozygote imbalance, an increased level of allele dropout (or extreme heterozygote imbalance), and relatively higher stutter peaks. When low template DNA is sequenced, usually higher background, low read depth, and increased sequence errors are observed. Increasing the number of PCR cycles has been applied to overcome some of these limitations.

Producers of commercial STR kits establish the optimal number of PCR cycles during developmental validation. Each cycle of PCR in theory doubles the number of target molecules (although in practice the PCR is not so efficient). Therefore, additional cycles can allow for detection of PCR product that may not be revealed under a standard number of cycles. For the GlobalFiler kit, 29 PCR cycles are recommended when 1 ng of DNA is amplified and 30 cycles are suggested for 500 pg of DNA. In the GlobalFiler Express kit, 25, 26, and 27 PCR cycles are recommended for direct amplification from blood collected on treated paper, and 26, 27, and 28 PCR cycles are indicated for buccal samples. The manufacturer of the PowerPlex Fusion 6C System recommends 29 cycles for amplification of 1 ng of DNA. While results can be obtained with less than the recommended optimum input amounts, DNA extracts from human remains frequently yield <100 pg total DNA. Kloosterman and Kersbergen (2003) reanalyzed a variety of samples, including remains from a 10-year-old mass grave, that produced no or partial STR profiles after using a 28 cycle-amplification protocol. By increasing the PCR cycle number to 34 the samples yielded mainly complete profiles. Hedell et al. (2015) detected complete STR profiles for 42 and 84 pg of input DNA by applying 32 PCR cycles. For 8 and 17 pg of DNA, the allele dropout decreased from 100% to 75% and 20%, respectively, compared with an amplification employing 30 cycles. Harrel et al. (2019) tested the effect of additional PCR cycles on DNA extracted from challenged skeletal samples using the GlobalFiler DNA Amplification Kit (Thermo Fisher Scientific). They observed significant improvement in recovery by STR profile completeness with <120 pg of input DNA amplified up to 32 cycles. However, increasing the number of PCR cycles may result in lower quality of obtained data, for instance, the presence of increased stutter, artifacts, and sequence errors. Additionally, with the increased sensitivity is a concomitant risk of contamination (Gill et al., 2000).

Similarly, an increase number of cycles can improve obtaining results for MPS. For example, for Precision ID SNP panels (Thermo Fisher Scientific), 21 cycles are recommended for sequencing 1 ng DNA, and additional 1 to 5 cycles for when <1 ng DNA is amplified. For Precision ID mtDNA panels (Thermo Fisher Scientific), 21 cycles are recommended for 0.1 ng (approximately 2900 mtDNA copies), and additional 1 to 5 cycles are optional when <0.1 ng genomic DNA is available for amplification. The manufacturer, however, emphasizes that input DNA is based on nuclear genomic DNA quantification and that the actual number of mtDNA copies may vary among samples and sample types (e.g., bone, blood, saliva).

6.3.5 Whole Genome Amplification

Whole genome amplification (WGA) is a methodology that in theory amplifies all the DNA in a sample, not just targeted regions as described earlier. After this comprehensive WGA, standard PCR that targets markers of interest is performed. If no amplification results are observed after standard or improved PCR amplification methods or the results are complex and difficult to interpret, WGA can be an alternative solution to increase the number of amplifiable templates. The amplified whole genome sample can be subsequently used for STR, SNP, or mitochondrial DNA typing (Maciejewska et al., 2013). Various techniques and commercially available WGA kits have been proposed. The approaches include degenerate oligonucleotide-primed PCR (Arneson et al., 2008a; Telenius et al., 1992), primer extension preamplification PCR (Arneson et al., 2008b), multiple displacement amplification (Spits et al., 2006), restriction and circularization-aided rolling circle amplification (Wang et al., 2004), blunt-end ligation-mediated WGA (Li et al., 2006), REPLI-g (Qiagen), Amplil™ WGA (Menarini Silicon Biosystems), PicoPLEX DNA-Seq (Picoseq) (Takara Bio), DOPlify

WGA (PerkinElmer), the GenomePlex™ WGA commercial kit (Takara Bio), and the GenomiPhi™ Amplification kit (GE Healthcare). Despite the methodological differences in the various protocols, all approaches are focused on the generation of a high copy number that covers the entire genome, and some scientists profess to be able to amplify DNA even from a single cell. These methods were used for amplification of degraded forensic DNA (Ambers et al., 2016; Maciejewska et al., 2013) as well as formalin-fixed paraffin-embedded (FFPE) samples (Mendez et al., 2017). Ballantyne et al. (2007) demonstrated that amplification was significantly improved after WGA treatment, and STR loci that were not typed previously were detected after the treatment. However, an increase in artifacts, such as stutter and amplification biases between alleles in specific STR loci, was observed in many samples. Most of the WGA methods are designed to amplify a small amount of good-quality DNA. Therefore, the method may not always be effective, and exaggerated stochastic effects may be observed when highly degraded DNA from human remains is amplified (Ambers et al., 2016).

6.3.6 DNA Damage Repair

After cellular death, the enzymatic repair system does not maintain its function. Consequently, the genome is exposed to various insults that disturb its stability and integrity (see the review by Alaeddini et al., 2010). These factors include a harsh environment with high humidity as well as microorganisms that catalyze the damage to cellular structure. However, under favorable conditions, for example, at low temperature and dry environment, degradation processes can be considerably slowed. Therefore, the level of DNA degradation in human remains will depend substantially on conditions in which the samples were maintained prior to recovery. DNA degradation will be manifested as fragmentation, blocking lesions, or miscoding lesions (Dabney et al., 2013). A common type of damage is intermolecular cross-links such as Maillard products. Maillard products, which can prevent PCR amplification, are formed by condensation reactions between sugars and primary amino groups in proteins and nucleic acids (Paabo et al., 2004).

Various forms of degradation can prevent PCR amplification and subsequent typing or sequencing, or present challenges with data interpretation. There are enzymes that can recognize and repair damage in DNA and can be used to correct some types of damage in DNA, thereby reducing the negative impact of degradation on sequencing results. DNA repair with these enzymes should be performed prior to PCR amplification or before MPS library preparation. For example, uracil–DNA–glycosylase (UNG) and endonuclease VIII efficiently remove uracil residues from ancient DNA (Briggs et al., 2010). Multienzyme cocktails are commercially available, e.g., the NEBNext FFPE DNA Repair Mix (New England BioLabs), which contain enzymes to repair FFPE DNA damage such as deamination of cytosine converting to uracil, nicks and gaps, oxidized bases, and blocked 3′-ends (see the NEBNext FFPE DNA Repair Mix application overview). It has been shown that cytosine deamination was significantly reduced after mitochondrial DNA repair with this kit (Gorden et al., 2018). Another kit, PreCR® Repair Mix (New England BioLabs), is designed to reduce abasic sites, nicks, thymidine dimers, blocked 3′-ends, oxidized guanine, oxidized pyrimidines, and deaminated cytosine (see PreCR Repair Mix product information). The modified PreCR Repair Mix protocol was used to repair damage caused by UV light exposure (Diegoli et al., 2012). The study demonstrated an increase in allelic peak heights and recovery of STR alleles from artificially damaged DNA after using the repair mix. However promising, the repair assays may be limited when significantly damaged samples are analyzed (Ambers et al., 2014).

6.4 Post-PCR Approaches

6.4.1 Cleanup of Post-Amplification Products

Cleanup of PCR products can substantially improve obtaining results from challenged samples. The purpose of the cleanup procedure is to remove remaining dNTPs, primers, salts, and other impurities that can interfere with the subsequent Sanger sequencing, CE, or MPS. Several methods can be used for the cleanup procedure such as column-based purification, e.g., the QIAquick PCR Purification Kit (Qiagen) or the SpinPrep™ PCR Clean-up Kit (Millipore Sigma); enzymatic cleanup, e.g., ExoSAP-IT™ (Thermo Fisher Scientific); ethanol precipitation; or magnetic bead purification. The column-based purification kits are designed to purify PCR fragments ranging from 100 bp to 10–12 kilobases (kb) in length by removing polymerases, dNTPs, salts, and primers. Filtration of the amplicons removes ions that compete with DNA injected into the capillary (Budowle et al., 2001). Smith and Ballantyne (2007) observed that post-PCR purification with a MinElute column (Qiagen) before CE analysis enhanced the recovery of full STR profiles with as little 20 pg of input DNA without increasing PCR cycles. This cleanup approach with standard cycle numbers may result in lower stutter ratios than increased cycle number methods. However, a loss of DNA copies was observed after using the column-based purification method (Kemp et al., 2014). Another method of purification is ExoSAP-IT, which utilizes two hydrolytic enzymes—exonuclease I and shrimp alkaline phosphatase—to remove the unwanted dNTPs and primers. It has been shown that ExoSAP-IT can be more effective than column-based purification as well as being suited for highly degraded DNA from skeletal remains (Dugan et al., 2002). ExoSAP-IT was used for cleanup of PCR products from DNA extracted from Viking-age bone samples improving Sanger sequencing of mitochondrial DNA (Bus et al., 2019). The same samples cleaned with ethanol exhibited more problems with data interpretation (data not published). The phenol cleanup is an inexpensive and efficient method for removal of most salts and small organic molecules; however, it is a time-consuming method, not completely evaporated phenol may interfere with subsequent sequencing, and it is a hazardous chemical.

6.4.2 Increased CE Injection Time

Specific injection settings are employed to place amplified PCR products in a CE. The amount of DNA injected in a CE system depends mostly on the applied voltage and injection time. A positive voltage is used to move the DNA into the capillary (Butler et al., 2004; Ulfelder and McCord, 1996). For example, in the ABI 3130 and 3130xl genetic analyzers (Thermo Fisher Scientific), the recommended injection voltage is up to 15 kV and the injection time is 1–600 sec (see Maintenance, Troubleshooting, and Reference Guide, 3130 and 3130xl genetic analyzers). DNA in samples in which negative ions are present can be poorly injected into capillaries. These other negative ions are more efficiently injected into a capillary than negatively charged DNA markers (because of a higher charge-to-mass ratio). An efficient post-PCR cleanup method, such as centrifugation with desalting columns (e.g.

Bio-Spin® and Micro Bio-Spin™ desalting columns [Bio-Rad] or Amicon Ultra-0.5 mL Centrifugal Filters [Millipore Sigma]), can be used to remove competing ions and increase DNA typing signals. It also has been shown that anion-exchange chromatography can be highly efficient at desalting, compared with other desalting methods such as ethanol precipitation, dialysis/ultrafiltration, or purification with commercial kits (Manduzio et al., 2010).

Hedell et al. (2015) analyzed LCN DNA using 5, 10, and 20 sec for CE injection with 3 kV. They found that 32 and 33 PCR cycles combined with 10 sec CE injection were beneficial for STR allele detection from challenged samples. The same positive correlation between increasing the CE injection settings from 3 kV/10 sec to 9 kV/15 sec and increased recovery of STR alleles was observed by Westen et al. (2009). At the higher injection settings of 9 kV/15 sec injection, improvement in genotyping of the minor component in a 10:1 mixture was observed (Westen et al., 2009). Amplification of these mixtures with increased PCR cycles resulted in overloaded profiles. Therefore, the authors recommended instead boosting the CE injection conditions for recovery of STR profiles from low DNA input.

6.5 Conclusion

Different factors can influence the quantity and structure of DNA residing in environmentally exposed forensic samples and aged human remains. Each DNA sample may display a different degree of fragmentation, single base damage patterns, various inhibitors and inhibitor concentrations, and numbers of amplifiable molecules. Therefore, the best practice in the analysis of challenged samples will be, where possible, to attempt several different methods for improving the chances of obtaining interpretable results. Cleaning up inhibitors and impurities, followed by the use of an effective polymerase(s), addition of PCR enhancers, and/or increasing the number of PCR cycles and the CE injection time may be beneficial. Additionally, WGA or DNA repair can be options for increasing typing success. Even with a validated series of protocols further modifications may be required for some samples. There are myriad insults that affect DNA residing in, for example, bones; therefore, not all possible situations have been addressed by studies to date. Additionally, validation and establishment of robust methodologies are worth the effort since these actions can reduce time and costs of analyses at later stages of working with challenged samples. While the establishment of efficient protocols may require time and can be laborious, the effort may be worthwhile and be the difference between obtaining a result or not having any data.

References

Abu Al-Soud, W. and P. Radstrom. 1998. Capacity of nine thermostable DNA polymerases to mediate DNA amplification in the presence of PCR-inhibiting samples. *Appl Environ Microbiol* 64(10), 3748–3753.

Ahmad, A. and J. Ghasemi. 2007. New buffers to improve the quantitative real-time polymerase chain reaction. *Biosci Biotechnol Biochem* 71(8), 1970–1978.

Alaeddini, R., S. J. Walsh, and A. Abbas. 2010. Forensic implications of genetic analyses from degraded DNA—A review. *Forensic Sci Int Genet* 4(3), 148–157.

Ambers, A., M. Turnbough, R. Benjamin, H. Gill-King, J. King, A. Sajantila, and B. Budowle. 2016. Modified DOP-PCR for improved STR typing of degraded DNA from human skeletal remains and bloodstains. *Leg Med (Tokyo)* 18, 7–12.

Ambers, A., M. Turnbough, R. Benjamin, J. King, and B. Budowle. 2014. Assessment of the role of DNA repair in damaged forensic samples. *Int J Legal Med* 128(6), 913–921.

Arneson, N., S. Hughes, R. Houlston, and S. Done. 2008a. Whole-genome amplification by degenerate oligonucleotide primed PCR (DOP-PCR). *CSH Protoc* 2008, pdb prot4919.

Arneson, N., S. Hughes, R. Houlston, and S. Done. 2008b. Whole-genome amplification by improved primer extension preamplification PCR (I-PEP-PCR). *CSH Protoc* 2008, pdb prot4921.

Ballantyne, K. N., R. A. H. Van Oorschot, and R. J. Mitchell. 2007. Comparison of two whole genome amplification methods for STR genotyping of LCN and degraded DNA samples. *Forensic Sci Int* 166(1), 35–41.

Bogenhagen, D. and D. A. Clayton. 1974. The number of mitochondrial deoxyribonucleic acid genomes in mouse L and human HeLa cells. Quantitative isolation of mitochondrial deoxyribonucleic acid. *J Biol Chem* 249(24), 7991–7995.

Briggs, A. W., U. Stenzel, M. Meyer, J. Krause, M. Kircher, and S. Paabo. 2010. Removal of deaminated cytosines and detection of in vivo methylation in ancient DNA. *Nucleic Acids Res* 38(6):e87.

Brown, H., R. Thompson, G. Murphy, D. Peters, B. La Rue, J. King, et al. 2017. Development and validation of a novel multiplexed DNA analysis system, InnoTyper((R)) 21. *Forensic Sci Int Genet* 29, 80–99.

Budowle, B., A. J. Eisenberg, and A. Van Daal. 2009. Validity of low copy number typing and applications to forensic science. *Croat Med J* 50(3), 207–217.

Budowle, B., D. L. Hobson, J. B. Smerick, and J. A. L. Smith. 2001. Low copy number - Consideration and caution. In: *Twelfth International Symposium on Human Identification 2001*, http://www.promega.com/ussymp12proc/default.htm.

Bus, M. M., M. Lembring, A. Kjellstrom, C. Strobl, B. Zimmermann, W. Parson, M. Allen. 2019. Mitochondrial DNA analysis of a Viking age mass grave in Sweden. *Forensic Sci Int Genet* 42, 268–274.

Butler, J. M., E. Buel, F. Crivellente, and B. R. Mccord. 2004. Forensic DNA typing by capillary electrophoresis using the ABI Prism 310 and 3100 genetic analyzers for STR analysis. *Electrophoresis* 25(10–11), 1397–1412.

Chevet, E., G. Lemaitre, and M. D. Katinka. 1995. Low concentrations of tetramethylammonium chloride increase yield and specificity of Pcr. *Nucleic Acids Res* 23(16), 3343–3344.

Coble, M. D. and J. M. Butler. 2005. Characterization of new MiniSTR loci to aid analysis of degraded DNA. *J Forensic Sci* 50(1), 43–53.

Comey, C. T., J. M. Jung, and B. Budowle. 1991. Use of formamide to improve amplification of HLA DQ alpha sequences. *BioTechniques* 10(1), 60–61.

Dabney, J., M. Meyer, and S. Paabo. 2013. Ancient DNA damage. *Cold Spring Harb Perspect Biol* 5(7):a012567.

Diegoli, T. M., M. Farr, C. Cromartie, M. D. Coble, and T. W. Bille. 2012. An optimized protocol for forensic application of the PreCR (TM) Repair Mix to multiplex STR amplification of UV-damaged DNA. *Forensic Sci Int Genet* 6(4), 498–503.

Doran, A. E. and D. R. Foran. 2014. Assessment and mitigation of DNA loss utilizing centrifugal filtration devices. *Forensic Sci Int Genet* 13, 187–190.

Dugan, K. A., H. S. Lawrence, D. R. Hares, C. L. Fisher, and B. Budowle. 2002. An improved method for post-PCR purification for mtDNA sequence analysis. *J Forensic Sci* 47(4), 811–818.

Eilert, K. D. and D. R. Foran. 2009. Polymerase resistance to polymerase chain reaction inhibitors in bone. *J Forensic Sci* 54(5), 1001–1007.

Faber, K. L., E. C. Person, and W. R. Hudlow. 2013. PCR inhibitor removal using the NucleoSpin (R) DNA clean-up XS kit. *Forensic Sci Int Genet* 7(1), 209–213.

Farell, E. M. and G. Alexandre. 2012. Bovine serum albumin further enhances the effects of organic solvents on increased yield of polymerase chain reaction of GC-rich templates. *BMC Res Notes* 5, 257.

Gill, P. 2001. Application of low copy number DNA profiling. *Croat Med J* 42(3), 229–232.

Gill, P., J. Whitaker, C. Flaxman, N. Brown, and J. Buckleton. 2000. An investigation of the rigor of interpretation rules for STRs derived from less than 100 pg of DNA. *Forensic Sci Int* 112(1), 17–40.

Glocke, I. and M. Meyer. 2017. Extending the spectrum of DNA sequences retrieved from ancient bones and teeth. *Genome Res* 27(7), 1230–1237.

Gorden, E. M., K. Sturk-Andreaggi, and C. Marshall. 2018. Repair of DNA damage caused by cytosine deamination in mitochondrial DNA of forensic case samples. *Forensic Sci Int Genet* 34, 257–264.

Grisedale, K. S., G. M. Murphy, H. Brown, M. R. Wilson, and S. K. Sinha. 2018. Successful nuclear DNA profiling of rootless hair shafts: A novel approach. *Int J Legal Med* 132(1), 107–115.

Hanni, C., T. Brousseau, V. Laudet, and D. Stehelin. 1995. Isopropanol precipitation removes Pcr inhibitors from ancient bone extracts. *Nucleic Acids Res* 23(5), 881–882.

Harrel, M., D. Gangitano, and S. Hughes-Stamm. 2019. The effects of extra PCR cycles when amplifying skeletal samples with the GlobalFiler (R) PCR amplification kit. *Int J Legal Med* 133(3), 745–750.

Hedell, R., C. Dufva, R. Ansell, P. Mostad, and J. Hedman. 2015. Enhanced low-template DNA analysis conditions and investigation of allele dropout patterns. *Forensic Sci Int Genet* 14, 61–75.

Hedman, J., C. Dufva, L. Norén, C. Ansell, L. Albinsson, and R. Ansell. 2011. Applying a PCR inhibitor tolerant DNA polymerase blend in forensic DNA profiling. *Forensic Sci Int Genet Suppl S* 3(1), e349–e350.

Henke, W., K. Herdel, K. Jung, D. Schnorr, and S. A. Loening. 1997. Betaine improves the PCR amplification of GC-rich DNA sequences. *Nucleic Acids Res* 25(19), 3957–3958.

Hochmeister, M. N., B. Budowle, U. V. Borer, U. Eggmann, C. T. Comey, and R. Dirnhofer. 1991. Typing of deoxyribonucleic acid (DNA) extracted from compact bone from human remains. *J Forensic Sci* 36(6), 1649–1661.

Hu, Q. Q., Y. X. Liu, S. H. Yi, and D. X. Huang. 2015. A comparison of four methods for PCR inhibitor removal. *Forensic Sci Int Genet* 16, 94–97.

Jakubowska, J., A. Maciejewska, and R. Pawlowski. 2012. Comparison of three methods of DNA extraction from human bones with different degrees of degradation. *Int J Legal Med* 126(1), 173–178.

Jensen, M. A., M. Fukushima, and R. W. Davis. 2010. DMSO and betaine greatly improve amplification of GC-rich constructs in de novo synthesis. *PLOS ONE* 5(6), e11024.

Kemp, B. M., M. Winters, C. Monroe, and J. L. Barta. 2014. How much DNA is lost? Measuring DNA loss of short-tandem-repeat length fragments targeted by the PowerPlex 16 (R) system using the Qiagen MinElute purification kit. *Hum Biol* 86(4), 313–329.

Kieser, R. E., M. M. Bus, J. L. King, W. Van Der Vliet, J. Theelen, and B. Budowle. 2020. Reverse complement PCR: A novel one-step PCR system for typing highly degraded DNA for human identification. *Forensic Sci Int Genet* 44:102201.

Kim, K., M. Bazarragchaa, C. H. Brenner, B. S. Choi, and K. Y. Kim. 2015. Extensive evaluation of DNA polymerase performance for highly degraded human DNA samples. *Forensic Sci Int* 251, 171–178.

Kloosterman, A. D. and P. Kersbergen. 2003. Efficacy and limits of genotyping low copy number (LCN) DNA samples by multiplex PCR of STR loci. *J Soc Biol* 197(4), 351–359.

Li, J., L. Harris, H. Mamon, M. H. Kulke, W. H. Liu, P. Zhu, and M. G. Makrigiorgos. 2006. Whole genome amplification of plasma-circulating DNA enables expanded screening for allelic imbalance in plasma. *J Mol Diagn* 8(1), 22–30.

Lorenz, T. C. 2012. Polymerase chain reaction: Basic protocol plus troubleshooting and optimization strategies. *Jove J Vis Exp*. 63, 3998.

Maciejewska, A., J. Jakubowska, and R. Pawlowski. 2013. Whole genome amplification of degraded and nondegraded DNA for forensic purposes. *Int J Legal Med* 127(2), 309–319.

Manduzio, H., A. Martelet, E. Ezan, and F. Fenaille. 2010. Comparison of approaches for purifying and desalting polymerase chain reaction products prior to electrospray ionization mass spectrometry. *Anal Biochem* 398(2), 272–274.

Marshall, P. L., J. L. King, and B. Budowle. 2015. Utility of amplification enhancers in low copy number DNA analysis. *Int J Legal Med* 129(1), 43–52.

Matheson, C. D., C. Gurney, N. Esau, and R. Lehto. 2010. Assessing PCR inhibition from humic substances. *Open Enzym Inhib J* 3(1), 38–45.

Mcnevin, D., J. Edson, J. Robertson, and J. J. Austin. 2015. Reduced reaction volumes and increased Taq DNA polymerase concentration improve STR profiling outcomes from a real-world low template DNA source: Telogen hairs. *Forensic Sci Med Pathol* 11(3), 326–338.

Mendez, P., L. T. Fang, D. M. Jablons, and I. J. Kim. 2017. Systematic comparison of two whole-genome amplification methods for targeted next-generation sequencing using frozen and FFPE normal and cancer tissues. *Sci Rep* 7(1), 4055.

Miller, D. N. 2001. Evaluation of gel filtration resins for the removal of PCR-inhibitory substances from soils and sediments. *J Microbiol Methods* 44(1), 49–58.

Miller, F. J., F. L. Rosenfeldt, C. Zhang, A. W. Linnane, and P. Nagley. 2003. Precise determination of mitochondrial DNA copy number in human skeletal and cardiac muscle by a PCR-based assay: Lack of change of copy number with age. *Nucleic Acids Res* 31(11), e61.

Nagy, M., P. Otremba, C. Kruger, S. Bergner-Greiner, P. Anders, B. Henske, et al. 2005. Optimization and validation of a fully automated silica-coated magnetic beads purification technology in forensics. *Forensic Sci Int* 152(1), 13–22.

Nilsson, M., J. Granemo, M. M. Bus, M. Havsjo, and M. Allen. 2016. Comparison of DNA polymerases for improved forensic analysis of challenging samples. *Forensic Sci Int Genet* 24, 55–59.

Noren, L., R. Hedell, R. Ansell, and J. Hedman. 2013. Purification of crime scene DNA extracts using centrifugal filter devices. *Investig Genet* 4(1), 8.

Opel, K. L., D. Chung, and B. R. McCord. 2010. A study of PCR inhibition mechanisms using real time PCR. *J Forensic Sci* 55(1), 25–33.

Paabo, S., J. A. Gifford, and A. C. Wilson. 1988. Mitochondrial DNA sequences from a 7000-year old brain. *Nucleic Acids Res* 16(20), 9775–9787.

Paabo, S., H. Poinar, D. Serre, V. Jaenicke-Despres, J. Hebler, N. Rohland, et al. 2004. Genetic analyses from ancient DNA. *Annu Rev Genet* 38, 645–679.

Ralser, M., R. Querfurth, H. J. Warnatz, H. Lehrach, M. L. Yaspo, and S. Krobitsch. 2006. An efficient and economic enhancer mix for PCR. *Biochem Biophys Res Commun* 347(3), 747–751.

Rohland, N. and M. Hofreiter. 2007. Comparison and optimization of ancient DNA extraction. *BioTechniques* 42(3), 343–352.

Sarkar, G., S. Kapelner, and S. S. Sommer. 1990. Formamide can dramatically improve the specificity of PCR. *Nucleic Acids Res* 18(24), 7465.

Satoh, M. and T. Kuroiwa. 1991. Organization of multiple nucleoids and DNA molecules in mitochondria of a human cell. *Exp Cell Res* 196(1), 137–140.

Schrader, C., A. Schielke, L. Ellerbroek, and R. Johne. 2012. PCR inhibitors - Occurrence, properties and removal. *J Appl Microbiol* 113(5), 1014–1026.

Smith, P. J. and J. Ballantyne. 2007. Simplified low-copy-number DNA analysis by post-PCR purification. *J Forensic Sci* 52(4), 820–829.

Spits, C., C. Le Caignec, M. De Rycke, L. Van Haute, A. Van Steirteghem, I. Liebaers, and K. Sermon. 2006. Whole-genome multiple displacement amplification from single cells. *Nat Protoc* 1(4), 1965–1970.

Telenius, H., N. P. Carter, C. E. Bebb, M. Nordenskjold, B. A. Ponder, and A. Tunnacliffe. 1992. Degenerate oligonucleotide-primed PCR: General amplification of target DNA by a single degenerate primer. *Genomics* 13(3), 718–725.

Terpe, K. 2013. Overview of thermostable DNA polymerases for classical PCR applications: From molecular and biochemical fundamentals to commercial systems. *Appl Microbiol Biotechnol* 97(24), 10243–10254.

Tuross, N. 1994. The biochemistry of ancient DNA in bone. *Experientia* 50(6), 530–535.

Ulfelder, K. J. and B. R. Mccord. 1996. Capillary electrophoresis of DNA. In: *Handbook of Capillary Electrophoresis* (Landers, J., ed.), CRC Press, New York, 347–378.

Varadaraj, K. and D. M. Skinner. 1994. Denaturants or cosolvents improve the specificity of PCR amplification of a G + C-rich DNA using genetically engineered DNA polymerases. *Gene* 140(1), 1–5.

Wang, G., E. Maher, C. Brennan, L. Chin, C. Leo, M. Kaur, et al. 2004. DNA amplification method tolerant to sample degradation. *Genome Res* 14(11), 2357–2366.

Westen, A. A., J. H. A. Nagel, C. C. G. Benschop, N. E. C. Weiler, B. J. De Jong, and T. Sijen. 2009. Higher capillary electrophoresis injection settings as an efficient approach to increase the sensitivity of STR typing. *J Forensic Sci* 54(3), 591–598.

Wiegand, P. and M. Kleiber. 2001. Less is more - Length reduction of STR amplicons using redesigned primers. *Int J Legal Med* 114(4–5), 285–287.

Zietkiewicz, E., M. Witt, P. Daca, J. Zebracka-Gala, M. Goniewicz, B. Jarzab, and M. Witt. 2012. Current genetic methodologies in the identification of disaster victims and in forensic analysis. *J Appl Genet* 53(1), 41–60.

Mixtures and Probabilistic Genotyping

7

MICHAEL COBLE, TODD BILLE,
AND DANIELE PODINI

Contents

7.1 Introduction

A mixed biological sample can be defined as a sample with two or more contributors. Interpretation of mixtures has been a part of forensic biology since the initial use of genetic markers. ABO typing and isozymes (e.g., PEPA and PGM) were genetic markers analyzed prior to DNA analysis. The limited number of alleles for these markers made the interpretation of mixtures relatively simple, yet fairly uninformative. Eventually, forensic biology moved away from protein/enzyme analysis toward the analysis of DNA. These markers contain more alleles and, therefore, significantly more possible genotypes. The interpretation of mixtures has evolved as the methods used (restriction fragment length polymorphism [RFLP] and polymerase chain reaction [PCR]-based methods such as DQA1, DQPM, D1S80, and short tandem repeats [STRs]), the sensitivity of the analysis, and the types of samples analyzed have changed over the years. The interpretation through this time, however, still generally follows the basic outline provided by Clayton et al. (1998):

1. Identify the presence of a mixture.
2. Identify artifacts from true alleles.
3. Determine the number of contributors.
4. Determine the approximate ratio of the components.
5. Determine the possible contributing genotypes.
6. Compare the reference sample to the possible genotypes.

While this list of steps appears to be straightforward and simple, challenges exist for each step in the process. Budowle et al. (2009) describe other features that should be considered when interpreting a mixed DNA profile. Various bodies have convened over the years to

DOI: 10.4324/9780429019944-8

provide guidance for the interpretation of mixtures. These include the Scientific Working Group on DNA Analysis Methods (SWGDAM 2017), the International Society for Forensic Genetics (Gill et al. 2006), UK Forensic Science Regulator (DNA Mixture Guidance 2018), and the German Stain Commission (Schneider et al. 2008). The National Research Council (NRC II 1996), National Academy of Sciences (NAS 2009), and the President's Council of Advisors on Science and Technology (PCAST 2016) have also weighed in on the topic.

The steps described by Clayton et al. (1998) assume that the profile is suitable for interpretation. This determination is made based on the quality and quantity of information present in the profile. For most interpretation approaches, one critical factor is whether the number of contributors (NOC) can be reasonably assumed. In essence, step 3 of the process has become step 1. If there are too few alleles, there may be insufficient information for the analyst to make the determination. If there are too many alleles, then this may also confound the determination. The presence of very low-level contributors, stochastic effects, artifacts such as stutter (step 2), and other factors also affect this decision. Therefore, the determination if a profile is suitable for comparison will depend on the interpretation approach used by the laboratory and, if the laboratory uses probabilistic genotyping, the sophistication of the models the software uses. The approaches for determining the NOC (step 3) have evolved from simple maximum allele counting to artificial intelligence and machine learning. A more in-depth discussion of the NOC determination is later in this chapter.

Once a profile has been determined to be interpretable, the first step in mixture interpretation is, obviously, identifying if the profile being examined consists of DNA from more than one person. This would appear to be the easiest step in the process, but even this "easy" step can be challenging. Under the assumption that an individual can have at most two alleles per locus, then any locus containing three alleles is an indication of the presence of a mixture. However, tri-allelic loci have been observed, albeit rarely. In addition, it may be difficult to differentiate a true allelic peak and elevated stutter, as noted in the second step by Clayton et al. (1998). A stutter peak may exceed expectations due to contributions from an additional contributor with an overlapping allele or it could be a stochastic event.

Another indication of a mixture is one or more loci with heterozygous balance outside of expectations. Kelly et al. (2012) describe the expected peak height balance between sister alleles of a heterozygous pair based on the average peak height (APH) of the alleles. As the average peak height decreases, the range of expected imbalance increases. Possible sister alleles with an observed imbalance exceeding expectations may be the result of a mutation in the primer sequence, a stochastic event, or the presence of at least one other contributor. Multiple loci with elevated imbalance are an indication of a mixture.

Finally, should peaks below the analytical threshold be considered? An apparent single-source profile with allelic peaks at 2,000 relative fluorescence units (RFU) is not affected by the possible presence of an additional contributor below the analytical threshold (assuming the thresholds are between 20 and 75 RFU). But what about an apparent single-source profile with the majority of alleles just above the analytical threshold and multiple peaks with the expected peak morphology and located within allelic bins just below the analytical threshold? While these peaks cannot be confidently considered alleles, their presence may be an indication of at least one additional contributor, and the additive effects due to allele sharing with the profile above the analytical threshold may affect heterozygous balance and stutter.

The fourth step of the process is the determination of the approximate mixture ratio. This can be important for determining if a major component can be discerned (Schneider et al. 2008; Bille et al. 2019), determining which genotype sets are viable, and assessing the NOC. The mixture ratio can be calculated manually or by computer using tools such as probabilistic genotyping software. In a high-level two-person mixture, the mixture ratio approximation may be fairly straightforward. Typically, this process involves examining loci with the maximum number of expected alleles for the assumed NOC to ensure there is no allele sharing. In this case, loci containing four detected alleles would be evaluated. Ideally, there would be more than one locus and the loci would represent small molecular weight and high molecular weight targets to account for the possible degradation of one or both components. In a 5:1 or 10:1 mixture, there is a clear distinction in the peak heights for the two contributors making the assumed pairings simple. The two alleles with the greatest RFU are summed and divided by the sum of the peak heights of the remaining two (minor) alleles. Another way of calculating the contributions of the two contributors would be to sum the RFU for all four peaks and calculate the percentage of the total RFU for the two contributors. As with the other aspects of mixture interpretation, approximating the mixture ratio can be very complicated and inexact. If the DNA from one or more contributors to the mixture is degraded, the ratio will vary across the profile from low molecular weight loci to high molecular weight loci (left to right on the electropherogram). In some instances, an undegraded minor component in the small molecular weight loci may become the major component at the larger molecular weight loci if the other component is severely degraded. In this case, the ratio may be 5:1 at the low molecular weight loci, 1:3 at the high molecular weight loci, and somewhere in between for the loci in the middle. Calculating the ratio manually for mixtures containing three, four, or more contributors becomes even more difficult as few if any loci will contain the maximum expected alleles and each of the components may have various degrees of degradation.

The final step of the interpretation prior to the comparison(s) to reference samples is the determination of what genotypes are possible contributors. Ideally, it would be possible to completely separate, or deconvolute, the genotypes of each of the contributors across all loci. Unfortunately, this is rarely possible due to potential allele sharing, mixture ratios close to one, artifacts such as stutter, degradation, inhibition, stochastic effects, and other factors. In instances where a locus in a two-person mixture contains two alleles for the major component and two distinct alleles for the minor component, each of the contributors' genotypes can be confidently discerned. However, if only a single distinct allele (A) was detected for the minor component with two major component alleles (B and C), there are three possible genotypes for the minor component (AA, AB, and AC) barring the possibility of allele dropout. Because of these ambiguities, the interpretation typically results in a pool of potential contributing genotype combinations that can explain the observed mixture. Each combination is typically referred to as a genotype set. For example, a two-person mixture containing a locus where three alleles are detected (A, B, and C) would potentially have six genotype sets (see Table 7.1c) when not considering peak height. The listed genotype sets do not take into account contributor order. Meaning, the AA/BC set represents Individual 1 with genotype AA mixed with Individual 2 with genotype BC or Individual 1 with a genotype of BC mixed with Individual 2 with a genotype of AA. This distinction becomes important during statistical calculations.

The genotype sets listed in Table 7.1 do not consider the possibility of allele dropout. Allele dropout is an extreme case of heterozygous imbalance where one allele of a heterozygous pair is not detected. When dropout should be considered as a possibility is another complicating factor for mixture interpretation. Allele dropout was considered

Table 7.1   **Possible Genotype Sets for a Two-Person Mixture at a Locus Containing (a) One Allele, (b) Two Alleles, (c) Three Alleles, and (d) Four Alleles**

(a) Genotype Sets		
AA	&	AA
(b) Genotype Sets		
AA	&	AA
AA	&	BB
AB	&	AB
AB	&	BB
(c) Genotype Sets		
AA	&	BC
AB	&	AC
AB	&	BC
BB	&	AC
BC	&	AC
CC	&	AB
(d) Genotype Sets		
AB	&	CD
AC	&	BD
AD	&	BC

Notes: The genotype sets do not consider contributor order. For example, set D would have three additional combinations.

during interpretation in the earliest days of PCR-based analysis. The "C" and "S" control dots incorporated on the test strips for AmpliType® DQA1 and PM analyses, respectively, were a measure of reliability for the detected alleles (Walsh et al. 1992). Dot intensities weaker than the control dots were not considered reliable and may be associated with allele dropout. This threshold, now typically referred to as the stochastic threshold, is evaluated during the validation.

Technology has advanced both for forensic DNA analysis and for the interpretation of the results. When an individual is not excluded as a possible contributor to a mixture, thus included as a possible contributor, it is necessary to calculate a statistical weight associated with the comparison to convey the meaning of the association. Initially, "binary" statistical approaches were employed, i.e., the genotype is either considered a possible contributor or not. These methods included random match probability (RMP) (Bille et al. 2013), combined probability of inclusion (CPI) (Bieber et al. 2016), and the likelihood ratio (LR) (Weir et al. 1997; Buckleton et al. 2016). Laboratories in the United States primarily used RMP or CPI, while the majority of laboratories outside the United States used LR calculations. All three of these methods do not take advantage of a significant portion of the data such as peak heights contained within the electropherogram (epg) (Buckleton et al. 2008), which may be used to further discriminate between possible contributors.

Probabilistic genotyping methods place a probability or weight between zero and one on possible contributing genotypes. Those genotypes that more reasonably explain the observed profile are given greater weight than those that are theoretically possible but less likely. The statistical approach employed by the laboratory affects the amount of data in the epg utilized and to what extent the possible contributing genotypes are evaluated. Bille et al. (2014) noted the differences in the statistical calculations that can be seen based on the approach used, and the amount of data in the DNA profile is considered. Because probabilistic genotyping software removes some of the subjectivity of the analyst's decision-making during the interpretation, allows for the interpretation of more complex mixtures, and makes better use of the data within the DNA profile, more laboratories are moving toward the use of probabilistic genotyping software.

7.2 Conventional Methods for Mixture Interpretation

As discussed earlier, the statistical approach utilized to calculate the weight of an association of a reference profile to an evidence profile influences the interpretation approach used by a laboratory. The CPI and RMP are both binary approaches, but require different interpretation approaches and considerations. The LR calculation can be applied in either a binary approach or in a probabilistic genotyping approach. In this section, the LR calculation will be discussed as a binary approach.

In the early stages of DNA analysis, most samples contained high levels of template DNA, either from sexual assault investigations or investigations involving blood or other bodily fluid evidence. In the late 1990s, the emergence of "touch" or "trace" DNA began to change the types of samples submitted to the forensic laboratory (Oorschot et al. 1997). These samples from handled objects or even individual fingerprints contain smaller quantities of DNA and routinely produced complex mixed DNA profiles. Coupled with improvements in STR kit chemistries and increased sensitivity with capillary electrophoresis instrumentation, complex mixtures are now much more prevalent. The original statistical analysis approaches and the inherent interpretation required by CPI and RMP became less applicable for these sample types.

The interpretation of a DNA profile, no matter what approach is employed, must be done prior to any comparison. The determination of which, if any, loci are suitable for comparison purposes, whether or not a major component can be discerned, and the assumed number of contributors should not be influenced by the reference samples. In some circumstances, an individual's DNA profile can be considered during the interpretation and statistical analysis. In these instances, the case specifics indicate that it can be reasonably assumed that the individual is a contributor to the observed DNA profile. These circumstances include, but are not limited to, when a sample is taken from the body of an individual (e.g., bite mark swab or vaginal/cervical swab), a piece of clothing removed from an individual, or even the steering wheel of a stolen car.

Mixture interpretation when using the CPI calculation is the simplest of the approaches discussed. The CPI calculation itself is the sum of all the possible genotypes from the detected alleles with no consideration of peak height, mixture ratio, NOC, or other factors. Therefore, the only determination the analyst must make during interpretation is whether it is reasonable to assume that dropout did not occur for each locus. The CPI calculation may not require an assumed NOC, but determining the NOC may be necessary to determine if allele dropout is reasonable. The stochastic threshold (the RFU value that, when exceeded, dropout of a sister peak of a heterozygous pair is not expected to occur) developed during validation for binary methods is only applicable to single-source DNA profiles. Additive effects of allele sharing in a DNA mixture can cause an allelic peak to exceed the stochastic threshold, but still be associated with allele dropout. Any locus where it is reasonable that allele dropout has occurred cannot be used for comparison or statistical purposes. Therefore, applying the CPI approach to low-level samples commonly encountered when analyzing trace DNA evidence often results in the use of few if any loci.

Misuse of the CPI interpretation and statistical approach has resulted in the reanalysis of hundreds of cases and the temporary shutdown of more than one laboratory in the US. In these laboratories, reference DNA profiles were used during the interpretation stage to determine which loci were suitable for comparison and statistical purposes. Until this issue arose, there was no peer-reviewed journal paper describing the proper interpretation and use of the CPI statistic. Subsequently, a paper by Bieber et al. (2016) described the proper interpretation of complex mixtures when utilizing the CPI statistic.

The RMP calculation was originally applied to single-source DNA profiles. Use with mixed DNA profiles gained greater prevalence with the increased analysis of trace DNA samples containing low levels of template DNA. The use of the RMP approach requires the assumption of the number of contributors. While the RMP approach can be applied to any mixed DNA profile, the complexity of the interpretation typically limits its application to two-person mixtures. Similar to the CPI calculation, the RMP calculation is a sum of the possible contributing genotypes. The RMP approach can be applied without consideration of peak height information (unrestricted), but becomes much more powerful when the peak height information is included in the interpretation (restricted). When applying the RMP statistic, the analyst must consider each possible genotype set and determine if the genotype set can reasonably explain the observed profile. The possible genotype sets are listed in Table 7.1 for a two-person mixture. Unlike the CPI calculation, a restricted RMP approach requires the analyst to consider allelic peak heights, stutter, degradation, allele sharing, mixture ratio, and heterozygous balance during this determination. A simple example is a locus with four alleles detected: A (900 RFU), B (700 RFU), C (250 RFU), and D (200 RFU). Based on the expected heterozygous balance, the only two possible contributing genotypes are AB and CD, since both A and B alleles would not be expected to pair with either C or D. This same locus using the CPI approach would consider all combinations of the four alleles as possible contributing genotypes (six possible genotypes). The interpretation becomes more complicated when the contribution levels for each individual become more equal and/or stochastic effects such as imbalance or elevated stutter become more prevalent.

Another complexity that can be incorporated into the RMP approach is inclusion of genotype sets with allele dropout. Similar to the CPI approach, the analyst must decide if allele dropout is reasonable based on a predetermined stochastic threshold and considering the additive effects of possible allele sharing. The genotype sets now evaluated increase from those listed in Table 7.1 to those listed in Table 7.2.

The likelihood ratio, unlike the CPI or RMP, is not a probability. It is the ratio of two conditional probabilities. The analyst evaluates the evidence under two (or more) hypotheses (also called propositions). In the first (numerator) hypothesis (H1), the probability of the evidence is determined if the person of interest (POI) is a contributor to the profile. The second (denominator) hypothesis (H2) is an evaluation of the evidence if the POI is not a contributor, but instead an unrelated and unknown person in the population is the true contributor to the profile. The interpretation when using the LR approach as a binary method is similar to the RMP approach. In fact, in its simplest form, the binary LR is 1/ RMP. As with the RMP approach, the number of contributors must be assumed and the

Table 7.2 **Possible Genotype Sets for a Two-Person Mixture Including Possible Allele Dropout at a Locus Containing (a) One Allele, (b) Two Alleles, (c) Three Alleles, and (d) Four Alleles**

(a) Genotype Sets		
AA	&	AA
AX	&	AA
AX	&	AX
(b) Genotype Sets		
AA	&	AB
AA	&	BB
AB	&	AB
AB	&	BB
AX	&	BB
AX	&	AB
BX	&	AA
BX	&	AB
AX	&	BX
(c) Genotype Sets		
AA	&	BC
AB	&	AC
AB	&	BC
BB	&	AC
BC	&	AC
CC	&	AB
AX	&	BC
BX	&	AC
CX	&	AB
(d) Genotype Sets		
AB	&	CD
AC	&	BD
AD	&	BC

Notes: "X" denotes allele dropout. The genotype sets do not consider contributor order. For example, set d would have three additional combinations.

evaluation can be restricted or unrestricted. Additional information about the LR can be found in Bright and Coble (2019).

7.3 Number of Contributors

The true number of contributors to a DNA mixture obtained from an evidence sample is always unknown. Assigning this number is a crucial part of the interpretation process (SWGDAM 2017). This step is necessary prior to performing allele attribution and statistical calculations (excluding CPI). With most probabilistic genotyping software, the analyst must decide and input the assumed number of contributors and occasionally perform multiple analyses with different assumptions. The simplest approach is to estimate the minimum number of contributors to a mixture based on the maximum allele count (MAC) at each locus (Paoletti 2005). Three or four alleles would indicate at least two contributors, five or six would indicate at least three contributors, and so on. This approach does not take into account peak heights, just the number of detected alleles. As the number of individuals in the mixture increases, the chance of these contributors sharing alleles also increases, and thus the chance of underestimating the number of contributors starts to increase.

Coble et al. (2015) demonstrated that using the 20 CODIS (Combined DNA Index System) loci there is a 99.82% chance that a six-person mixture will appear as a five-person and a 93.3% chance of a five-person appearing as a four-person mixture using African American allele frequencies. The method is more accurate as the number of individuals in the mixture decreases and as the number of loci tested increases. For example, there is a 0.17% chance that a three-person mixture would appear as a two-person with the same set of loci. Another approach is the total allele count (TAC), which consists of counting the total number of alleles detected throughout the entire profile (Clayton et al. 1998; Young et al. 2019).

A mixture simulation study conducted by Young et al. (2019) examined the distribution of the total number of alleles detected in two- to five-person mixtures. The authors noted that the overlap between the distribution of total number of alleles detected in two- and three-person mixtures is minimal but the overlap increases between three- and four-person mixtures and even more so between four- and five-person mixtures. This study also highlighted the prediction accuracy potentially provided by massively parallel sequencing (MPS), which allows one to differentiate by sequence alleles that are indistinguishable by size (Young et al. 2019). Haned et al. (2011) proposed an alternative method based on the maximum likelihood principle, which showed prediction values above 94% for mixtures of two to three contributors and 69% for four-person mixtures.

The methods discussed so far only consider the presence or absence of alleles and their frequency, and do not take into account peak height or the possibility of alleles not being detected due to DNA degradation or PCR inhibition. Complex mixtures often contain degraded or low template minor contributors, and depending on the source, may also contain inhibitors. Thus, computer-based methods that take into account the physical characteristics of the entire profile are likely to increase the accuracy of the prediction, as they access and utilize more information. An example of such an approach is NOCit (Swaminathan et al. 2015). This application calculates the "a posteriori" probability on the number of contributors and is based on calibration data from single-source known samples. For its calculations, it utilizes peak heights, population alleles' frequencies, and the possibility of allele dropout and stutter. Studies have shown that NOCit can increase the accuracy of the prediction by 12% compared to previous methods (Swaminathan et al. 2015).

Another example of non-traditional estimation of the number of contributors in a mixture is PACE (Probabilistic Assessment of Contributor Estimation), which is a machine learning-based method that is trained on a set of known mixtures and tested on another separate set of known mixtures. The training process enables the application to develop a prediction model, the output of which is a probability of the mixture being from x number of individuals. Results showed an accuracy of over 98% in identifying the correct number of contributors in mixtures of up to four individuals

7.4 Probabilistic Genotyping

Stochastic thresholds are necessary for mixture interpretation, as they provide the analyst an indication of allele dropout in the evidence profile. Traditional approaches to provide statistical support for inclusion, including CPI, RMP, and the binary LR, require a stochastic threshold to determine if alleles are "in" or "out" of the calculation. If the laboratory uses CPI, minor alleles above the stochastic threshold are included in the statistical calculation, while alleles below the stochastic threshold are thrown out and thus prevent the use of that locus in the statistical calculation. Of course, this process also depends on the assumptions made during interpretation. For example, if the analyst has assumed a two-person mixture and a locus has two alleles above the stochastic threshold (aligned, for example, with the complainant) and two alleles are below the stochastic threshold, then this locus could be used in the CPI calculation since there are four alleles at a locus in a two-person mixture, despite the fact that the two minor alleles are below the stochastic threshold.

Rather than discard a locus with a single allele below the stochastic threshold when using the RMP, the analyst can apply the "2p" rule to account for the possibility of dropout in the profile (NRC II 1996). Buckleton and Triggs (2006) note that the 2p rule may not always be conservative. For example, if the DNA evidence stain is clearly homozygous at a locus (e.g., 15,15) and the allele is well above the stochastic threshold and the POI has a heterozygous genotype (12,15), then the POI would be excluded as a contributor to the evidence. However, if that single 15 allele is below the stochastic threshold, now the POI can be included (invoking dropout of his 12 allele) and a statistic of 2p can be assigned using the frequency of the 15 allele. In this example, we go from an exclusion to an inclusion with the weight assigned to the locus.

As noted previously, recent improvements in STR kit chemistry and capillary electrophoresis instrumentation have increased the sensitivity of detecting low-level profiles in evidence. Additionally, the type of samples submitted for DNA testing has moved from high-quality, high-quantity sources such as blood stains and semen to lower quality and quantity samples such as swabs from guns, cell phones, and steering wheels. This shift in the quality and quantity of samples submitted and the increased sensitivity in detecting STR profiles has produced more and more challenging DNA mixtures for the analyst to interpret. With the limitations of CPI and RMP for reporting the statistical association of a DNA match, there was a need to handle DNA profiles with low-level contributors below the stochastic threshold.

Probabilistic genotyping, according to the SWGDAM Guidelines for the Validation of Probabilistic Genotyping Systems (SWGDAM 2015), is "the use of biological modeling, statistical theory, computer algorithms, and probability distributions to calculate likelihood ratios (LRs) and/or infer genotypes for the DNA typing results of forensic samples." Rather than using a stochastic threshold to include or exclude alleles, probabilistic genotyping software treats alleles probabilistically. This uncertainty in the probability of allele dropout in a low-level DNA profile can be incorporated into the LR. In a nutshell, probabilistic genotyping software attempts to determine optimal genotypes in a mixture and provide weight to certain genotypes that best explain the evidence.

There are two classes of probabilistic genotyping software: semicontinuous modeling and fully continuous modeling (reviewed in Coble and Bright 2019). The two classes are determined by how they treat alleles. Semicontinuous models of interpretation do not consider the peak heights of the alleles in the mixture, only the alleles present. Advocates of this modeling contend that at very low-level DNA concentrations of a minor component in a mixture (e.g., at 50 pg input), biological models of interpretation such as peak height ratios and stutter frequencies do not behave the same way as when there is a greater amount of DNA for that minor contributor (e.g. at 500 pg input). Therefore, the "information content" in the mixture is the alleles that one observes. If an allele is missing (dropout), then a probability of dropout, Pr(DO), can be incorporated in the LR. If extraneous alleles are present, then a probability of drop-in, Pr(DI), can be used in the LR.

7.4.1 Semicontinuous Modeling

Semicontinuous models were the first models described in the literature (Gill et al. 2000; Gill et al. 2007). As an example, we use the method described by Balding and Buckleton (2009) called the "drop model." For this model, the only information required is the alleles present in the evidence, the probability of dropout, and the probability of drop-in. Consider the following locus (D16S539) from a two-person mixture of an alleged sexual assault (Figure 7.1). An interpretation of the mixture has determined that this is a mixture of at least two individuals, and assuming the presence of the complainant (the major contributor), the POI cannot be excluded as the minor contributor profile in the mixture, although there is evidence of allele and locus dropout in the profile. For example, at D16S539, the complainant is homozygous 13,13 and the POI has the 9,11 genotype. The analytical threshold for this laboratory is 75 RFU and the stochastic threshold is 200 RFU.

Since the POI cannot be excluded (he matches the 9 allele in the evidence profile), a statistic is required to be reported for this evidence. If the laboratory uses CPI for this locus, there would be no statistic provided since that 9 allele is below the stochastic threshold. If the laboratory uses RMP for its statistical reporting, then the "2p" rule would be used. If the frequency of the 9 allele is found in 10% of the population, then the RMP = 2 × (0.10) = 0.20, or approximately 20% of the population would be expected to have at least a 9 allele, and would be expected to be a possible contributor to this evidence. Using the LR, the first (inclusionary) hypothesis (H1) would state the DNA mixture contains the complainant and the POI. The second (exclusionary) hypothesis (H2) would state that the DNA mixture contains the complainant and an unknown individual, unrelated to the POI. Since the complainant is conditioned in both hypotheses, the LR reduces to 1/RMP (1/0.20 = 5). Using the aforementioned allele frequency data, the LR report would read (for this locus): The evidence is 5 times more likely to be observed if the complainant and the POI are in this mixture rather than if the complainant and an unknown individual, unrelated to the POI, are in this mixture.

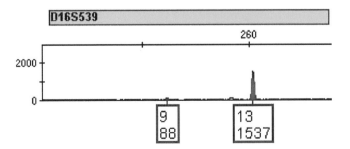

Figure 7.1 An example of a two-person mixture at the D16S539 locus.

Again, the issue with using a stochastic threshold when there are low-level DNA contributors and the potential of dropout is that the probability of the evidence in H1 is assumed to be 1 (certainty) when in fact only half of the POI's genotype matches (the 11 is missing in the evidence). Rather than using a probability of 1, the probability should be something between 0 and 1 to reflect this uncertainty. Using a semicontinuous model of interpretation, we would determine the probability that the POI and the complainant are contributors to the mixture when we only observe the 9 allele from the POI and the 11 allele has dropped out. Suppose we have determined from a sensitivity study that observing low-level alleles (at 88 RFU) has a probability of a missing sister allele of 0.80. We have also determined from negative controls that the probability of observing a spurious contaminating allele is approximately 0.01. With this information, we can assign the H1 probability of having no dropout of the 9 allele from the POI and a dropout of the 11 allele from the POI, and no other alleles have dropped in to the profile (the 13 allele is from the complainant):

H1: Pr(No dropout of the 9 allele) × Pr(Dropout of the 11 allele) × Pr(No drop-in alleles)
H1: (1 − 0.80 = 0.20) × (0.80) × (1 − 0.01 = 0.99)
H1: (0.20) × (0.80) × (0.99)
H1: 0.1584

Note that this probability is not 1, as we used in the binary LR for the 2p rule. It reflects the uncertainty in the evidence from the possibility of dropout.

For the H2 probability, we now ask the question, Who else in the population, unrelated to the POI, could have left this evidence?

We first propose someone with a 10,11 genotype. We now use our dropout and drop-in information to explain how a person with a 10,11 genotype left the evidence in Figure 7.4. That is, the 10 allele has dropped out, the 11 allele has also dropped out, and the 9 allele is a contaminating drop-in allele. The probability of this event would be

H2(10,11): Pr(Dropout of the 10 allele) × Pr(Dropout of the 11 allele) × Pr(Drop-in of the 9 allele)
H2(10,11): (0.80) × (0.80) × (0.01)
H2(10,11): 0.0064

This represents a very low probability of having two alleles from the true contributor dropping out and a contaminating allele that dropped in to the profile.

Since we are actually talking about a random person in the population, we must multiply this probability times the RMP of a person in the population having the 10,11 genotype.

Using the product rule for a heterozygous genotype (and allele frequencies of 0.06 for the 10 allele and 0.27 for the 11 allele):

HWE (10,11) = 2pq
HWE (10,11) = 2 × (0.06) × (0.27)
HWE (10,11) = 0.0324

Multiplying the probability times the genotype frequency for the 10,11 genotype gives

0.0064 × 0.0324 = 0.00020736

which represents a "weighted" probability of someone in the population with a 10,11 geno-type who contributed to the evidence when two alleles (10,11) have dropped out and one allele (9) has dropped in to the evidence.

We continue to the next possible random man in the population (8,9 for example). In fact, we will consider *all* possible heterozygous genotypes and homozygous genotypes using the alleles in the allelic ladder to determine *anyone* who could have left the evidence, all the while using dropout and drop-in to explain the absence of alleles in the profile. We will even include alleles not observed in the ladder by using a "Q" designation. The frequency of this Q allele will be

Q = 1 – Sum of the allele frequencies for all alleles observed in the ladder

Finally, we add all of the weighted genotype probabilities for the H2 combinations to obtain the denominator probability for the LR.

Of course, the model presented here is very simplistic for this demonstration of how a semicontinuous approach would work. We have not considered the possibility of having different dropout probabilities for heterozygous alleles versus homozygous alleles. Nor have we incorporated a theta correction such as NRC II Recommendation 4.2 (NRC II 1996). However, there are several attractive features of this probabilistic approach. First, *all* of the data is used. There is no throwing away of loci because alleles are below a stochastic threshold. Next, since we consider all possible genotype combinations in H2, the math associated with the determination of the denominator can be easily programmed into a software program and once a value for allele frequencies, probability of dropout, and probability of drop-in are entered into the program, the computer can rather quickly calculate the LR. Of course, more complex mixtures with, for example, four contributors will take longer to calculate than a two-contributor mixture. However, it should be noted that if two separate analysts agree on the same alleles in the input file, and use the same probabilities of dropout and drop-in, allele frequencies, and theta, they will obtain the exact same answer every single time they initiate the program.

There are also limitations of semicontinuous methods. Since there is no deconvolution of the mixture, then there is no ability to search a profile in a DNA database. The analyst is still tasked with an interpretation of the evidence and must make decisions such as the number of contributors and if artifacts, especially stutter, should be removed or considered to be allelic. This can have a huge effect on the consistency within a laboratory as was the case in Barrio et al. (2018) where the Spanish and Portuguese Speaking Working Group of the ISFG conducted an interlaboratory study using the semicontinuous software LRmix Studio. All laboratories were provided an epg and given the probabilities of dropout, drop-in, theta, and allele frequencies. None of the labs achieved the same LR result with values that ranged from 172 trillion to 320 trillion. This was because each lab made decisions about the inclusion (or exclusion) of stutter peaks causing them to each have different input files, resulting in a wide range of answers.

For more information on semicontinuous methods, refer to chapter 6 of Bright and Coble (2019). Although semicontinuous models are an improvement over disregarding loci, thus not using available information based on a stochastic threshold, they are not fully efficient since they ignore peak height information. The forensic DNA community has moved solidly in the direction of fully continuous methods of interpretation.

7.4.2 Continuous Modeling

Continuous models of interpretation attempt to deconvolve mixtures into their individual components. This requires the software to model the biological processes that occur during PCR amplification, including peak heights and their variability, stutter ratios, degradation, etc. Some of these programs also have advanced modeling parameters such as locus-specific effects and allowances for PCR replicates. For these continuous methods, dropout and drop-in probabilities are not values that are entered into the program. Instead, the software will model dropout, and potentially invoke drop-in, depending on the casework profile.

Continuous models can be classified into at least two groups: those that attempt to model peak heights by using simulation (the Markov Chain Monte Carlo [MCMC] process), and those that use a "brute force" method to address the issue mathematically. Programs that use the MCMC simulation process include STRmix™ (Taylor et al. 2013) from the Institute of Environmental Science and Research Limited (ESR) and TrueAllele® (Perlin et al. 2009). These programs attempt to find the genotype combinations that best describe the profile and assign a probability or "weight" to these combinations. Combinations that best explain the mixture are given a higher weight than those that poorly describe the mixture. Other programs that use the "brute force" approach include DNA-VIEW® Mixture Solution (Charles Brenner, DNA-View website) and Kongoh (Manabe et al. 2017). For this limited introduction, we focus on MCMC methods, particularly STRmix as this program is currently the most widely used probabilistic genotyping software in the United States.

The software STRmix uses a model for calculating *expected* peak heights by proposing values for a set of parameters (including genotype sets, template quantity, degradation, stutter ratios, etc.). The software then compares the expected heights to the *observed* peak heights in the evidence profile. In simplistic terms, the software draws a set of hypothetical peaks in a hypothetical profile based on a set of parameters, and then compares this hypothetical profile to the actual data from the electropherogram. If the model is a reasonable explanation of the data, then the model is accepted and the genotype sets are recorded as possible explanations of the data. If the model poorly explains the profile, then it is rejected and a new model is proposed.

At each MCMC iteration, a randomly varied genotype set at one locus and other parameter values are proposed, expected peak heights are calculated based on the models, the expected profile is compared to the observed profile, and a decision is made on whether to accept or reject the newly proposed genotype set and parameter values depending on how similar the expected peak heights are to the actual peak heights in the epg (this is the Metropolis–Hastings algorithm). The software will undertake thousands or millions of iterations within the MCMC in a single run.

For every accepted iteration, the genotype combinations are recorded. At the end of the deconvolution, the count of the accepted genotypes is normalized to obtain the weights. These "weights" are then used in the LR assignment. Consider the low-level mixture in Figure 7.1. After deconvolution from STRmix, the weights in Table 7.3 were obtained assuming two contributors were present in the mixture.

In casework, a review of these weights would be undertaken by the analyst to check if they are intuitive. Note that there are four genotype combinations accepted by the software. In all of these combinations, Contributor 1 has the 13,13 genotype, which aligns with the complainant, of whom we assumed was a contributor to the mixture. For Contributor 2, there are four possible genotypes, all having the 9 allele. The first combination (9,13) was given the highest weight by the software during the MCMC process (in roughly 39% of the simulations). The second

Table 7.3 Output of Weights from Proposed Genotype Combinations in Figure 7.4

	Contributors		Weight
Locus	1	2	
D16S539	13,13	9,13	0.3941
	13,13	9,9	0.3552
	13,13	9,Q	0.2151
	13,13	9,12	0.0356

highest weight of the minor contributor was the homozygous 9,9 genotype (0.3552). The 9,Q genotype with a weight of 0.2151 considers the possibility that a sister allele of the 9 has dropped out at this locus. Finally, the lowest weighted genotype from the minor contributor was 9,12. Here, the software considered the 12 allele (not labeled in Figure 7.1), which falls in the stutter position of the major contributor's 13 allele, to be allelic and not a stutter artifact.

An assignment of the LR for this locus would again follow the logic outlined in the semicontinuous section above. For the H1 proposition, we use the weight that aligns with the genotype of the POI (9,11); this would be 9,Q in Table 7.3 since the Q allele would represent the missing 11 allele.

Pr(E|H1): 0.2151

Note that this probability is not 1, as we used in the binary LR using the 2p rule. It reflects the uncertainty in the evidence from the possibility of dropout.

For the H2 probability, we again ask the question, Who else in the population, unrelated to the POI, could have left this evidence in addition to the complainant?

Using the list of genotypes in Table 7.4, we multiply the weight determined from the software times the genotype frequency in the population for each of the four genotypes, and then sum all of these together:

Table 7.4 List of Possible Contributing Genotypes for the Four-Allele Locus and Three-Allele Locus from the 1:1 Mixture Using the Three Interpretation Approaches

Analysis Approach	1:1					
	Four Alleles			Three Alleles		
CPI	11,13; 11,14; 11,16; 13,14; 13,16; 14,16; 11,11; 13,13; 14,14; 16,16			21,23; 21,27; 23,27; 21,21; 23,23; 27,27		
RMP	11,13; 11,14; 11,16; 13,14; 13,16; 14, 16			21,21; 21,23; 21,27; 23,23		
STRmix™	11, 14	13, 16	2.9273E-1	21, 23	21, 27	2.8663E-1
	11, 13	14, 16	2.4872E-1	21, 27	21, 23	2.7444E-1
	13, 14	11, 16	1.7355E-1	23, 27	21, 21	2.5123E-1
	11, 16	13, 14	1.1464E-1	21, 21	23, 27	1.8280E-1
	14, 16	11, 13	8.9607E-2	21, 23	27, 27	2.1949E-3
	13, 16	11, 14	8.0743E-2	21, 27	23, 23	1.9690E-3
				23, 23	21, 27	2.7972E-4
				27, 27	21, 23	2.6301E-4
				21, 27	23, 27	1.1044E-4
				21, 23	23, 27	4.7953E-5
				23, 27	21, 23	2.6156E-5
				23, 27	21, 27	1.3078E-5

Note: The STRmix™ approach also lists the associated weight for each genotype set.

Pr(E|H2$_{(9,13)}$): 0.3941 × HWE (9,13) +
H2(9,9): 0.3552 × HWE (9,9) +
H2(9,Q): 0.2151 × HWE (9,Q) +
H2(9,12): 0.0356 × HWE (9,12)

Recall, the Q allele would be the frequency of all alleles at D16S539 that are not 9, 12, or 13.

The advantages of fully continuous models of interpretation include the ability to deconvolute the mixture and potentially search a DNA database based on the results. Also, as can be seen with this example, the strength of the evidence is down weighted when the POI is not a good fit (e.g., the numerator probability is not simply 1, but is much less than that [0.2151]). There is no longer the need for a stochastic threshold, although some laboratories may continue to use the stochastic threshold to assist with assigning the number of contributors. Finally, artifacts such as stutter, which need to be removed by the analyst using semicontinuous programs are modeled (and hence, included) in fully continuous software. Overall, this allows a laboratory to now interpret data that previously would have been thrown away.

The following examples demonstrate the different interpretations of loci produced by mixtures of two individuals at three different ratios. The possible contributing genotypes are listed for each using the CPI, RMP, and STRmix approaches (Figures 7.2–7.4 and Tables 7.4–7.6).

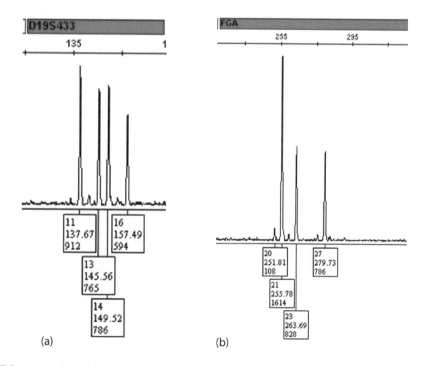

(a)   (b)

Figure 7.2 Examples of loci produced from a 1:1 mixture. (a) D19S433 containing four alleles and (b) FGA containing three alleles. The "20" peak in (b) is assumed to be stutter.

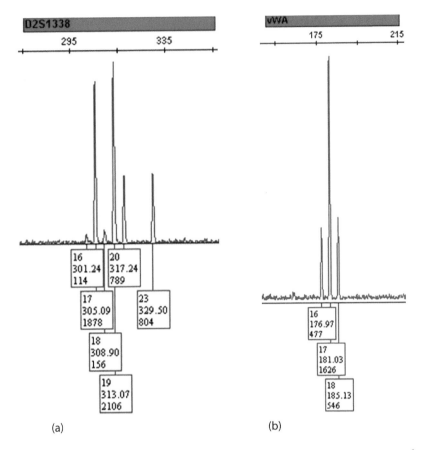

(a)   (b)

Figure 7.3 Examples of loci produced from a 2:1 mixture. (a) D2S1338 containing four alleles and (b) vWA containing three alleles. The "16" and "18" peaks in (a) are assumed to be stutter. While the "16" peak in (b) falls in the stutter position, it significantly exceeds the expected stutter percentage and can be assumed to be an allelic peak, although the peak height may be affected by stutter contribution.

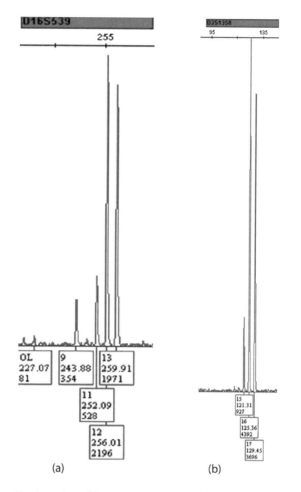

(a)   (b)

Figure 7.4 Examples of loci produced from a 9:1 mixture. (a) D16S539 containing four alleles and (b) D3S1358 containing three alleles. While the "11" peak in (a) and "15" peak in (b) both fall in the stutter position, both significantly exceed the expected stutter percentage and can be assumed to be allelic peaks, although the peak heights of both may be affected by stutter contributions.

Table 7.5   List of Possible Contributing Genotypes for the Four-Allele Locus and Three-Allele Locus from the 2:1 Mixture Using the Three Interpretation Approaches

Analysis Approach	2:1					
	Four Alleles			Three Alleles		
CPI	17,19; 17,20; 17,23; 19,20; 19,23; 20,23; 17,17; 19,19; 20,20; 23,23			16,17; 16,18; 17,18; 16,16; 17,17; 18,18		
RMP	17,19; 20,23			16,18; 17,17; 16,17; 17,18		
STRmix™	17, 19	20, 23	9.3953E-1	17, 17	16, 18	9.0395E-1
	19, 23	17, 20	1.9485E-2	17, 18	16, 17	7.5616E-2
	17, 23	19, 20	1.6781E-2	16, 17	17, 18	1.1208E-2
	19, 20	17, 23	1.0811E-2	17, 17	17, 18	4.3435E-3
	17, 20	19, 23	8.7925E-3	17, 18	17, 17	2.9127E-3
	20, 23	17, 19	4.6043E-3	16, 18	17, 17	9.4999E-4
				17, 17	18, 18	9.4610E-4
				16, 17	18, 18	7.2644E-5

Note: The STRmix™ approach also lists the associated weight for each genotype set.

Table 7.6 List of Possible Contributing Genotypes for the Four-Allele Locus and Three-Allele Locus from the 9:1 Mixture Using the Three Interpretation Approaches

Analysis Approach	9:1					
	Four Alleles			Three Alleles		
CPI	9,11; 9,12; 9,13; 11,12; 11,13; 12,13; 9,9; 11,11; 12,12; 13,13			15,16; 15,17; 16,17; 15,15; 16,16; 17,17		
RMP	9,11; 12,13			15,15; 15,16; 15,17; 16,17		
STRmix™	12, 13	9, 11	8.4209E-1	16, 17	15, 16	3.7034E-1
	12, 13	9, 12	6.1363E-2	16, 17	15, 15	2.2006E-1
	12, 13	9, 9	5.2753E-2	16, 17	15, 17	1.9241E-1
	12, 13	9, 13	4.3791E-2	16, 17	16, 17	1.0097E-1
				16, 17	16, 16	8.7458E-2
				16, 17	17, 17	2.8763E-2

Note: The STRmix™ approach also lists the associated weight for each genotype set.

7.5 Validation of Probabilistic Genotyping

Prior to any method, instrumentation, or software being implemented for use on forensic evidence, it must be properly validated. These validation studies are used to define the limitations of the method and the range of samples where the method can be reliably applied. This includes probabilistic genotyping software. Validation has been defined by the United Kingdom Forensic Science Regulator as "the process of providing objective evidence that a method, process or device is fit for the specific purpose intended" and by the Quality Assurance Standards for Forensic DNA Testing Laboratories (QAS) as "a process by which a method is evaluated to determine its efficacy and reliability for forensic casework analysis." While there are several documents that have been published describing the validation of probabilistic genotyping software (Coble et al. 2016; ANSI-ASB 2020), the following discussion focuses on the validation requirements and recommendations described by the QAS and SWGDAM guidelines (FBI QAS 2020; SWGDAM 2015). The QAS divides validation into two distinct sections: developmental validation, usually conducted by the manufacturer or provider; and internal validation, usually conducted by the end user. In general, the main objectives of the validation studies remain the same as those previously applied to PCR amplification kits or other DNA analysis methods, but the application differs. The following is a summary of the QAS requirements for the validation of software:

Developmental Validation
1. The underlying scientific principles must be publicly available.
2. Functional and reliability testing.
3. As applicable, accuracy, precision, sensitivity, and specificity.
Internal Validation
1. Functional and reliability testing.
2. As applicable, accuracy, precision, sensitivity, and specificity.

For conventional mixture interpretation approaches, several parameters that are used during the steps of interpretation are addressed. The three main parameters are the analytical threshold, stochastic threshold, and stutter thresholds. Each of these leads to binary decisions, e.g., a peak exceeding the analytical threshold can be considered allelic, a single peak above the stochastic threshold would not be expected to be associated with allele dropout, and a peak above the stutter threshold would not be considered stutter. Other factors such as heterozygous balance, variation in mixture ratios across the profile, saturation point, effects of degradation and inhibition, and locus to locus/dye channel to dye channel variation are also evaluated. However, for the most part, these factors are used more subjectively by the analyst during interpretation.

For probabilistic genotyping, most thresholds are now replaced with some form of modeling based on the laboratory's generated data. Variations of the aforementioned factors can be "learned" from analyzing profiles generated by the laboratory and the probabilities applied to the deconvolution of the profile. For example, if the stochastic threshold for the conventional approach is 200 RFU, the probability of dropout associated with a single peak at 50 RFU is the same as a single peak at 199 RFU which is 1. The probabilistic genotyping approach will evaluate a set of data and develop a probability gradient where the probability of dropout decreases as the allele's peak height approaches the stochastic threshold. The number of factors modeled depends on the complexity of the software. As the complexity increases, the computing power required also increases.

For the internal validation, the requirements must be assessed by the laboratory over the expected range of samples typically encountered during casework. Since probabilistic genotyping software commonly has two functions—deconvolution and statistical analysis—the validation and assessment of both functions need to be addressed. This includes the following: single-source samples; degraded samples; samples amplified in the presence of inhibitors; casework-like samples (either adjudicated case samples or simulated); and DNA mixtures covering a range of the number of contributors, ratios of contributions, and levels of template DNA.

The accuracy of the probabilistic software can be difficult to evaluate. For the deconvolution, the output can be compared to the analyst's logical expectations, however, this becomes more difficult as the complexity of the DNA profile increases. This is true for statistical calculations as well. One method of evaluation is to hand-calculate the LR and compare that to the probabilistic genotyping software-generated LR. For complete single-source DNA profiles, this may be possible, however, the calculations become much more complex when applied to partial profiles and/or mixtures. Another suggested way of evaluating the accuracy of the results is comparing the results to another probabilistic software, however, this would require obtaining the software and proper training on its use.

Prior to the introduction of probabilistic genotyping software, precision studies commonly referred to the evaluation of a system's ability to consistently size DNA fragments. The precision of the software is assessed through repeated analyses of the same dataset. As noted earlier, the datasets used should cover the range of samples expected to be encountered during casework analyses (template amounts, mixture ratios, degradation, etc.). Software utilizing iterative processes such as MCMC are not expected to always produce the same result when analyzing the same dataset. Therefore, the range of results expected for repeated analyses should be evaluated. For example, the LR for a complete single-source sample may demonstrate no variation, but the LR for a low-level three-person mixture may demonstrate a range of an order of magnitude or more.

The specificity and sensitivity of a probabilistic genotyping system are evaluated by examining the system's reliability across various template amounts, mixture ratios, numbers of contributors, degrees of degradation, and levels of inhibition. Type I and type II errors (false inclusion of noncontributors and false exclusion of true contributors, respectively) are produced as the limits of the software are reached. Care should be taken when referring to "false inclusion." A false inclusion or false positive in this sense is an adventitious match. For example, if one were to perform ABO typing on a dried blood stain recovered from a crime scene and concluded the blood was from an ABO-type A individual, approximately 40% of the population is ABO-type A, but only one person is most likely the contributor of the blood. Other individuals that happen to be ABO-type A are correctly included as potential contributors. If more markers were analyzed and the amount of information increased, the number of adventitious matches would decrease. This is a trend that should be observed during the specificity study. As the amount of genetic information decreases and the complexity of the mixture increases, the specificity should decrease concordantly. Due to the complexity of the calculations used by the software, the laboratory may not be able to duplicate the process by hand, but the results of the study should be intuitive. The general trend reflected in the LR is that with more information in the profile, LRs of true contributors yield values much greater than 1 and are capped at 1/RMP. With less information in the profile (low-level contributors, for example) the LR trends downward to 1 (uninformative).

When an LR is the product of the statistical analysis, the effects of different sets of hypotheses should be evaluated. Examples of different hypothesis sets for profiles are described next.

1. A two-person mixture recovered from a vaginal swab. The analysis can be performed with and without assuming the female complainant is a contributor to the profile.

   H1: The person of interest (POI) and one unknown individual are contributors to the mixture.
   H2: Two unknown individuals are contributors to the mixture.
   or
   H1: Person A and the complainant are contributors to the mixture.
   H2: The complainant and one unknown individual are contributors to the mixture.

2. A mixture containing two contributors can be evaluated for each person individually or both together.

   H1: Person A and one unknown individual are contributors to the mixture.
   H2: Two unknown individuals are contributors to the mixture.
   or
   H1: Person A and Person B are contributors to the mixture.
   H2: Two unknown individuals are contributors to the mixture.

The effects of other factors should also be evaluated when using the software. If the number of contributors is input by the analyst, then the effects of the correct and incorrect number should be evaluated. The presence of artifacts such as electrophoretic spikes, spectral pull-up, or even nonhuman products can adversely affect the analysis and should be evaluated.

If the laboratory has more than one method for generating a DNA profile (e.g., different injection times, PCR cycle numbers, etc.), the performance of the software should be evaluated under each of the methods. This holds true too if the laboratory modifies a method that may affect sensitivity or transitions to a new capillary electrophoresis instrument (3130 to 3500 Genetic Analyzer, Thermo Fisher Scientific).

The results of the validation studies are then used to inform the methods used by the laboratory. It is critical that the methods developed and implemented by the laboratory are based on its validation.

7.6 Challenges to the Process

One complicating factor that affects the process that is discussed less extensively in the literature is the presence of nonhuman peaks. Nonhuman DNA can produce peaks that may or may not be reproducible and may or may not be sized within allelic bins. There may be a single peak or multiple peaks detected across the profile in one or more dye channels. Part of the developmental validation of DNA typing kits outlined by SWGDAM and others is the evaluation of nonhuman DNA. Due to the sheer number of possible scenarios, not all species can be tested. For example, raccoon blood may be found on the bumper of a car suspected to be involved in a hit-and-run accident or fungal and bacterial DNA may be found in the soil adhering to the fragment of a pipe bomb recovered at a blast scene. While no cross-reactivity may be observed at lower template quantities (1–10 ng), it may occur at significantly higher levels. If the quantitation methods used are human-specific, the DNA analyst will have no idea how much nonhuman DNA is being added to the PCR reaction. Currently, there is no method to identify nonhuman amplified product in an epg, but the awareness of the possibility in certain sample types and evaluating the profile as a whole can mitigate the impact of this complicating factor.

Mixture interpretation is a complex process and can be influenced by an analyst's training, background, confidence, and potentially by background information of the investigation. The NIST 2005 and 2013 interlaboratory studies demonstrated the wide range of differences in interpretation in the forensic DNA community (Butler et al. 2018). While there are thresholds for some decisions the analyst must make, depending on the interpretation approach utilized, many decisions come down to the analyst and may be more subjective than objective. Cognitive bias, or specifically confirmation bias, is when an analyst's decisions are influenced by information other than that directly related to the data being evaluated (Dror et al. 2011). For example, a DNA analyst may be influenced if he/she knows the POI was caught at the scene or compares the POI's DNA profile to the evidence profile during interpretation. Methods have been proposed to reduce the effects of confirmation bias. A more complicated method suggests sequential unmasking to limit the information given to the DNA analyst (Krane et al. 2008). Another method is to only analyze reference samples after the interpretation of the evidence samples has been performed and documented. Laboratories and individual analysts need to be aware of confirmation bias and take steps to avoid it during the interpretation of DNA profiles.

7.7 The Future of Mixtures and Probabilistic Genotyping

It is likely that in the very near future, computational methods for determining the number of contributors will replace the manual performance of this step in most laboratories.

The expectation is that it will increase objectivity, repeatability, consistency between analysts, and throughput. Furthermore, MPS, now the mainstay in clinical genetics, has the potential to increase both the accuracy of the prediction of the number of contributors and the overall interpretation of complex mixtures. This technology not only allows the determination of the actual STR allele sequence, which increases the number of detectable alleles per locus, but also provides access to a whole new set of mixture-informative genetic markers such as microhaplotypes (Oldoni et al. 2019).

The concepts of probabilistic genotyping for mixture interpretation are not new; we are now into the second decade of dealing with allelic dropout in low-level, challenging mixtures. Additional Bayesian statistical methods have been proposed to address the issue surrounding DNA transfer and activity-level propositions (Gill et al. 2018; Gill et al. 2020). Software to probabilistically interpret mixtures is here to stay even when the next suite of DNA markers replaces STRs. This is necessary since there will always be uncertainty in interpretation. Just as the famous Bayesian statistician Professor Dennis Lindley stated, "Whatever way uncertainty is approached, probability is the only sound way to think about it" (Lindley 2013).

References

ANSI/ASB. 2020. Standard 018, First Edition. Standard for Validation of Probabilistic Genotyping Systems. http://www.asbstandardsboard.org/wp-content/uploads/2020/07/018_Std_e1.pdf.

Balding, D.J., and Buckleton, J. 2009. Interpreting Low Template DNA Profiles. *Forensic Sci Int Genet* 4(1):1–10. doi:10.1016/j.fsigen.2009.03.003.

Barrio, P.A., et al. 2018. GHEP-ISFG Collaborative Exercise on Mixture Profiles (GHEP-MIX06). Reporting Conclusions: Results and Evaluation. *Forensic Sci Int Genet* 35:156–163. doi:10.1016/j.fsigen.2018.05.005.

Bieber, F.R., et al. 2016. Evaluation of Forensic DNA Mixture Evidence: Protocol for Evaluation, Interpretation, and Statistical Calculations Using the Combined Probability of Inclusion. *BMC Genet* 17(1):125. doi:10.1186/s12863-016-0429-7.

Bille, T., S. Weitz, J.S. Buckleton, and J.A. Bright. 2019. Interpreting a major component from a mixed DNA profile with an unknown number of minor contributors. *Forensic Sci Int Genet* 40:150–159.

Bille, T., et al. 2013. Application of Random Match Probability Calculations to Mixed STR Profiles. *J Forensic Sci* 58(2):474–485. doi:10.1111/1556-4029.12067.

Bille, T., et al. 2014. Comparison of the Performance of Different Models for the Interpretation of Low Level Mixed DNA Profiles. *Electrophoresis* 35(21–22):3125–3133. doi:10.1002/elps.201400110.

Bright, J.A., and Coble, M. 2019. *Forensic DNA Profiling: A Practical Guide to Assigning Likelihood Ratios*. CRC Press, Taylor & Francis Group.

Buckleton, J., and Curran, J. 2008. A Discussion of the Merits of Random Man Not Excluded and Likelihood Ratios. *Forensic Sci Int Genet* 2(4):343–348. doi:10.1016/j.fsigen.2008.05.005.

Buckleton, J., and Triggs, C. 2006. Is the 2p Rule Always Conservative? *Forensic Sci Int* 159(2–3):206–209. doi:10.1016/j.forsciint.2005.08.004.

Buckleton, J., et al. 2016. *Forensic DNA Evidence Interpretation*. CRC Press, Taylor & Francis Group.

Budowle, B., et al. 2009. Mixture Interpretation: Defining the Relevant Features for Guidelines for the Assessment of Mixed DNA Profiles in Forensic Casework. *J Forensic Sci* 54(4):810–821.

Butler, J.M., et al. 2018. NIST Interlaboratory Studies Involving DNA Mixtures (MIX05 and MIX13): Variation Observed and Lessons Learned. *Forensic Sci Int Genet* 37:81–94. doi:10.1016/j.fsigen.2018.07.024.

Clayton, T.M., et al. 1998. Analysis and Interpretation of Mixed Forensic Stains Using DNA STR Profiling. *Forensic Sci Int* 91(1):55–70. doi:10.1016/s0379-0738(97)00175-8.

Coble, M.D., and Bright, J.A. 2019. Probabilistic Genotyping Software: An Overview. *Forensic Sci Int Genet* 38:219–224.

Coble, M.D., et al. 2015. Uncertainty in the Number of Contributors in the Proposed New CODIS Set. *Forensic Sci Int Genet* 19:207–211. doi:10.1016/j.fsigen.2015.07.005.

Coble, M.D., et al. 2016. DNA Commission of the International Society for Forensic Genetics: Recommendations on the Validation of Software Programs Performing Biostatistical Calculations for Forensic Genetics Applications. *Forensic Sci Int Genet* 25:191–197. doi:10.1016/j.fsigen.2016.09.002.

Dror, I.E., and Hampikian, G. 2011. Subjectivity and Bias in Forensic DNA Mixture Interpretation. *Sci Justice* 51(4):204–208. doi:10.1016/j.scijus.2011.08.004.

Federal Bureau of Investigation Quality Assurance Standards for Forensic DNA Testing Laboratories. 2020. https://www.fbi.gov/file-repository/quality-assurance-standards-for -forensic-dna-testing-laboratories.pdf/view.

Forensic Science Regulator Guidance - DNA Mixture Interpretation. FSR-G-222 Consultation. 2018. https://assets.publishing.service.gov.uk/government/uploads/system/uploads/attachment_ data/file/752164/G222_DNA_Mixture_Interpretation__Issue2.pdf.

Gill, P., C. Brenner, J. Buckleton, et al. 2006. DNA commission of the International Society of Forensic Genetics: Recommendations on the interpretation of mixtures. *Forensic Sci Int.* 160:90–101.

Gill, P., et al. 2000. An Investigation of the Rigor of Interpretation Rules for STRs Derived from Less than 100 Pg of DNA. *Forensic Sci Int* 112(1):17–40. doi:10.1016/s0379-0738(00) 00158-4.

Gill, P., et al. 2007. LoComatioN: A Software Tool for the Analysis of Low Copy Number DNA Profiles. *Forensic Sci Int* 166(2–3):128–138. doi:10.1016/j.forsciint.2006.04.016.

Gill, P., et al. 2018. DNA Commission of the International Society for Forensic Genetics: Assessing the Value of Forensic Biological Evidence - Guidelines Highlighting the Importance of Propositions. *Forensic Sci Int Genet* 36:189–202. doi:10.1016/j.fsigen.2018.07.003.

Gill, P., et al. 2020. DNA Commission of the International Society for Forensic Genetics: Assessing the Value of Forensic Biological Evidence - Guidelines Highlighting the Importance of Propositions. Part II: Evaluation of Biological Traces Considering Activity Level Propositions. *Forensic Sci Int Genet* 44:102186. doi:10.1016/j.fsigen.2019.102186.

Haned, H., et al. 2011. The Predictive Value of the Maximum Likelihood Estimator of the Number of Contributors to a DNA Mixture. *Forensic Sci Int Genet* 5(4):281–284. doi:10.1016/j. fsigen.2010.04.005.

Kelly, H., et al. 2012. Modelling Heterozygote Balance in Forensic DNA Profiles. *Forensic Sci Int Genet* 6(6):729–734. doi:10.1016/j.fsigen.2012.08.002.

Krane, D.E., et al. 2008. Sequential Unmasking: A Means of Minimizing Observer Effects in Forensic DNA Interpretation. *J Forensic Sci* 53(4):1006–1007.

Lindley, D.V. 2013. *Understanding Uncertainty*. John Wiley & Sons.

Manabe, S., et al. 2017. Development and Validation of Open-Source Software for DNA Mixture Interpretation Based on a Quantitative Continuous Model. *PLOS ONE* 12(11):e0188183. doi:10.1371/journal.pone.0188183.

National Academy of Science. 2009. Strengthening Forensic Science in the United States: A Path Forward. https://www.ncjrs.gov/pdffiles1/nij/grants/228091.pdf.

National Research Council, and Commission on DNA Forensic Science. 1996. *The Evaluation of Forensic DNA Evidence*. The National Academies Press. doi:10.17226/5141.

Oldoni, F., et al. 2019. Microhaplotypes in Forensic Genetics. *Forensic Sci Int Genet* 38:54–69. doi:10.1016/j.fsigen.2018.09.009.

Paoletti, D.R., et al. 2005. Empirical Analysis of the STR Profiles Resulting from Conceptual Mixtures. *J Forensic Sci* 50(6):1–6. doi:10.1520/jfs2004475.

Perlin, M.W., and Sinelnikov, A. 2009. An Information Gap in DNA Evidence Interpretation. *PLOS ONE* 4(12):e8327. doi:10.1371/journal.pone.0008327.

President's Council of Advisors on Science and Technology, 2016. *Forensic Science in Criminal Courts: Ensuring Scientific Validity of Feature-Comparison Methods*. https://obamawhitehouse .archives.gov/sites/default/files/microsites/ostp/PCAST/pcast_forensic_science_report_final .pdf

Schneider, P.M., et al. 2008. The German Stain Commission: Recommendations for the Interpretation of Mixed Stains. *Int J Legal Med* 123(1):1–5. doi:10.1007/s00414-008-0244-4.

Scientific Working Group on DNA Analysis Methods. 2015. Guidelines for the Validation of Probabilistic Genotyping Systems. https://1ecb9588-ea6f-4feb-971a-73265dbf079c.filesusr .com/ugd/4344b0_22776006b67c4a32a5ffc04fe3b56515.pdf.

Scientific Working Group on DNA Analysis Methods, Interpretation Guidelines for Autosomal STR Typing by Forensic DNA Testing Laboratories. 2017. https://1ecb9588-ea6f-4feb-971a -73265dbf079c.filesusr.com/ugd/4344b0_50e2749756a242528e6285a5bb478f4c.pdf.

Swaminathan, H., et al. 2015. NOC It: A Computational Method to Infer the Number of Contributors to DNA Samples Analyzed by STR Genotyping. *Forensic Sci Int Genet* 16:172–180. doi:10.1016/j.fsigen.2014.11.010.

Taylor, D., et al. 2013. The Interpretation of Single Source and Mixed DNA Profiles. *Forensic Sci Int Genet* 7(5):516–528. doi:10.1016/j.fsigen.2013.05.011.

Van Oorschot, R.A.H., and Jones, M.K. 1997. DNA Fingerprints from Fingerprints. *Nature* 387(6635):767–767. doi:10.1038/42838.

Walsh, P.S., et al. 1992. Preferential PCR Amplification of Alleles: Mechanisms and Solutions. *Genome Res* 1(4):241–250. doi:10.1101/gr.1.4.241.

Weir, B.S., et al. 1997. Interpreting DNA Mixtures. *J Forensic Sci* 42(2):213–222. doi:10.1520/ jfs14100j.

Young, B.A., et al. 2019. Estimating Number of Contributors in Massively Parallel Sequencing Data of STR Loci. *Forensic Sci Int Genet* 38:15–22. doi:10.1016/j.fsigen.2018.09.007.

Rapid DNA

8

CHRISTOPHER MILES

Contents

DOI: 10.4324/9780429019944-9

8.1 Introduction

8.1.1 Overview

An investment in science today yields untold rewards in the future. An investment, more specifically, in DNA science changes society and assures equal access to programs and justice for every person. In 2015, the Federal Bureau of Investigation (FBI) Director James B. Comey testified before Congress that Rapid DNA technology "would help us change the world in a very, very exciting way" (Jackman, 2018). This chapter discusses the development of Rapid DNA told from the perspective of the United States (US) Department of Homeland Security (DHS) Science and Technology Directorate (S&T). While Rapid DNA technology has made significant progress, the technology has not yet realized its full potential.

DNA is the most informative and robust biometric available in the world today. DNA is the only biometric that allows for verification of biological relationship/kinship claims. This unique ability has vital security, social, and cost-saving applications; and may reduce immigration fraud and human trafficking/smuggling and ensure the accurate reunification of families. Forensic DNA testing is responsible for the accurate conviction of criminals and is also responsible for the exoneration of those who are wrongly accused. DNA reveals an individual's true identity without the distracting lens of social or confirmation bias and allows every person to be objectively considered.

Rapid DNA technology is an innovative technology that expedites the human DNA analysis process from three to six months to 90 minutes. The Rapid DNA instrument is a printer-sized device that allows a nonscientific individual, who receives a brief orientation, to operate the instrument at the collection site and receive immediate, accurate, and valid scientific results. This technology removes the barriers of traditional DNA processing and allows end users to obtain scientific support for immediate intelligence/investigation and law enforcement decisions.

The Rapid DNA instrument has been tested and proven in extreme field conditions, can be easily shipped to requested locations, and runs on generator power. Rapid DNA removes the operational, financial, and infrastructure barriers inherent with traditional DNA testing and analysis, effectively providing accurate and reliable results for criminal investigations, biological relationship/kinship verifications, intelligence-based operations, and humanitarian efforts such as mass fatalities and missing persons.

DNA testing capability can quickly and accurately focus an investigation that may save time and money and, very importantly, ensure community safety and justice are served. Allowing an arrestee's DNA to be tested and matched against existing criminal profiles while in custody may enhance community safety.

8.1.2 How Rapid DNA May Expand Use of DNA Testing

The Rapid DNA instrument quickens the DNA testing process from days and weeks to an astonishing 90 minutes. The instruments are typically a desktop device that allows a nonscientific, briefly trained individual to operate the instrument at or near the point of collection and receive immediate, accurate, and valid DNA profile results. This technology removes the operational delays of traditional DNA processing. It enables law enforcement in the field to quickly receive the benefits of a complex science and receive DNA matching reports in near real-time. This ability quickly contributes objective and quantitative data to a criminal investigation, and in some cases, this investigative tool is used while the arrestee is in custody.

Given the accelerated pace of DNA testing, it is now possible for law enforcement to quickly expand collections for specific circumstances. For example, as migrants flow across the US border, DNA may be quickly tested so that identity may be verified, or family relationships may be confirmed, thus enhancing national security.

8.1.3 Rapid DNA Instrument: How Rugged?

The Rapid DNA instrument has been tested and proven effective in extreme and rugged field conditions, showing that it can be easily shipped to remote locations and operated using generator power. The instrument calibrates itself between each run and uses disposable components to remove any possibility for contamination between tests. Additionally, the test kits needed to be equally resilient and able to withstand extreme temperatures.

8.1.4 Rapid DNA: Especially Important to DHS

Before the recent introduction of Rapid DNA, there were significant barriers for DHS to perform in-house human DNA testing. The primary inhibitor of using DNA to its full capacity at DHS hinged on the inability of DHS components to obtain immediate and actionable test results at its 328 ports of entry (Department of Homeland Security, 2018). Historically, DNA samples were shipped to specialized (traditional) forensic DNA laboratories. DNA testing typically took weeks to months to perform and, therefore, did not offer the necessary immediate actionable information that would be most helpful. DNA testing was unable to be used to its full capacity.

Rapid DNA gives DHS the ability to quickly expand or contract human DNA collection and testing efforts across DHS. The benefit of expanding human DNA testing capabilities to national security is significant as it provides the most accurate and reliable biometric for human identification purposes and is the only biometric that can verify biological relationships/kinship.

The innovative Rapid DNA technology:

- Provides DHS decision-makers with a new, actionable biometric that rapidly verifies family biological relationships to support border enforcement, disaster recovery, and immigration
- Offers a field ready, fully integrated, and automated DNA tool to conduct faster, more comprehensive identifications of persons or objects in support of critical incidents or criminal investigations
- Allows for the accurate and rapid determination of family biological relationships to expedite valid immigration cases and deter immigration fraud and human trafficking
- Expands the ability to screen persons (e.g., non-US citizens, undocumented individuals, confidential informants, employees) to ensure that individuals are not on known or suspected criminal/terrorist DNA watch lists

8.2 Rapid DNA Development

8.2.1 Overview

Rapid DNA is an ideal example of how US federal funding and interagency cooperation, coupled with university involvement, private sector development, and the scientific community support can result in a much-needed technology. The primary federal agencies who made a significant contribution include the

- Department of Defense (DOD)
- Department of Justice (DOJ)
- National Institute of Justice (NIJ)
- National Institute of Standards and Technology (NIST)
- Department of Homeland Security (DHS)

It is difficult to describe the complex array of actions, interactions, and relationships that had to take place in order for Rapid DNA to enter the marketplace.

The first high-priority requirement for Rapid DNA testing at DHS was presented in 2005, and the final envisioned product was released to the marketplace in 2014. The US government contributed funding to stimulate the private sector to develop a high-risk, innovative, scientifically valid, and legally sound new DNA testing instrument that would better serve US national security interests. The new instrument was envisioned as a solution and tool for law enforcement to quickly verify human identification, mitigate immigration fraud, and accurately identify claimed family relationships.

Along with the technology development, additional efforts were made to ensure privacy, regulatory and legal compliance, and the development of additional support to augment the needs of a field-based law enforcement tool previously only available in a fully accredited laboratory managed by DNA specialists. Rapid DNA timeline is shown in the Figure 8.1.

8.2.2 Legislative and Regulatory Overview

In 2017 Congress introduced the Rapid DNA Act, which became law in August 2017. Sec. 2 of the bill amends the DNA Identification Act of 1994 to require the Federal Bureau of Investigation (FBI) to issue standards and procedures for using Rapid DNA instruments to analyze DNA samples of criminal offenders. It was written "to establish a system for integration of Rapid DNA instruments for use by law enforcement to reduce violent crime and reduce the current DNA analysis backlog" (Rapid DNA Act, 2017).

The DOJ continues to effectively integrate Rapid DNA into its operations to identify persons and search unsolved crimes while suspects are in custody during the booking process. This use of DNA evidence has been legally reviewed up to the Supreme Court (*Maryland v. King*, 2013), which found that DNA is a unique identifier and that collection and processing to determine association with a crime scene or victim are constitutionally reasonable and a "negligible" physical intrusion. Congress also reviewed the authority to use Rapid DNA in law enforcement booking stations (Rapid DNA Act, 2017).

1996	2005 - 2007	2008	2009	2011 - 2012	2014	2017	2019
The first concept: the National Institute of Justice (NIJ) sponsored book, Automated DNA Typing: The Method of the Future? discussed the transition from the primary method of southern blotting analysis of restriction fragment length polymorphism–variable number of tandem repeat (RFLP-VNTR) to a new molecular biological technique based on the polymerase chain reaction (PCR).	The first high priority requirement: The U.S. Citizenship and Immigration Services (USCIS), a member of the PS-IPT, presented a high-priority need for an objective, technology-based, way to determine kinship in support of the validation of family relationships for immigration purposes and foreign adoptions.	The first product: NIJ funded,development of a sample collection system for law enforcement agents to use in collecting, protecting, and documenting biological evidence. The Homeland Security Institute (HSI) was asked to perform an exploratory gap assessment and refinement of the capability gap. The study suggested the gap be refined to read "Rapid (ultimately within 45 minutes) and cost effective (under $100 per test) DNA testing."	DHS S&T released a Small Business Innovation Research (SBIR) solicitation seeking proposals for a "Low-Cost and Rapid DNA-based Biometric Device."	In 2011, a Rapid DNA manufacturer partnered with the Palm Bay Police Department to field test its RapidHIT200 in the world's first law enforcement use of the technology.The Palm Bay, Florida, Police Department received the first commercially produced Rapid DNA unit placed in a police agency.	In January 2014, after a year of testing and validation, the Palm Bay Police began generating leads on real-world criminal cases, including first ever application in a criminal investigation producing a DNA profile in 90 minutes that linked a suspect to a burglary of a U.S. soldier's home that occurred while he was in Afghanistan.	Rapid DNA Act of 2017- (Sec. 2) This bill amends the DNA Identification Act of 1994 to require the Federal Bureau of Investigation (FBI) to issue standards and procedures for using Rapid DNA instruments to analyze DNA samples of criminal offenders. Rapid DNA instruments carry out a fully automated process to create a DNA analysis from a DNA sample. DNA samples prepared by criminal justice agencies using Rapid DNA instruments in compliance the FBI-issued standards and procedures may be included in the Combined DNA Index System (CODIS). (Sec. 3) The bill amends the DNA Analysis Backlog Elimination Act of 2000 to allow the FBI to waive certain existing requirements if a DNA sample is analyzed using Rapid DNA instruments and the results are included in CODIS.	In May 2019, U.S. Immigration and Customs Enforcement and U.S. Customs and Border Protection piloted a program with Massachusetts based ANDE. The program's intention is to expose "family unit fraud," the idea that asylum seeking adults and children traveling together are illicitly posing as biologically related. In June 2019, a contract to expand the program was awarded to Bode Cellmark Forensics, Inc.

Figure 8.1 Rapid DNA timeline.

8.2.3 Initial Concepts and Efforts

The Department of Justice's (DOJ) National Institute of Justice (NIJ) mission is developing programs and providing grants that assist and support law enforcement by accelerating forensic technology and science innovations. Each year NIJ researchers (physical scientists, technologists, and social scientists) solicit research ideas and review thousands of grant requests with the sole mission of identifying the highest value topics for financial support. The peer-reviewed grant recipient identification process guarantees that even the most novel proposals will receive consideration. As a research institute, NIJ is committed to supporting promising approaches that may fail or never be initiated without federal support or assistance. For many years NIJ has been a visionary leader in the drive to automate and integrate RAPID processing of DNA. In 1997, the NIJ-funded research paper *Automated DNA Typing: The Method of the Future?* discussed the transition from the primary method of southern blotting analysis of restriction fragment length polymorphism–variable number of tandem repeat (RFLP-VNTR) to a new molecular biological technique based on the polymerase chain reaction (PCR). The PCR method made copies of a section of DNA extracted from evidence or a known sample and amplified known portions of the DNA. The paper discussed the Baylor College of Medicine–Houston studies to computerize the PCR technique for 13 so-called short tandem repeat (STR) loci. Acceptance of the new technique PCR was projected to be "several years away."

In 2001, "Validation of STR Typing by Capillary Electrophoresis" was published in the *Journal of Forensic Science* (Moretti, Baumstark, Defenbaugh, Keys, Brown, and Budowle, 2001). This NIJ-funded study described the validation of STR typing by capillary electrophoresis (CE), allowing high-resolution electrophoretic separation of STR loci to be achieved in a semiautomated fashion. The study concluded that CE provided efficient separation, resolution, sensitivity, and precision; the analytical software provided reliable genotyping of STR loci; and the analytical conditions of the study were suitable for typing samples such as reference and evidentiary samples from forensic casework.

8.2.4 Department of Homeland Security (DHS)/Small Business Innovation Research (SBIR) Program

8.2.4.1 *People Screening (PS)–Integrated Product Team (IPT)*

The DHS was created in 2002. In the early years, the DHS, Science and Technology Directorate (S&T), managed people screening (PS) operational requirements with an Integrated Product Team (IPT). The US Citizenship and Immigration Services (USCIS), a member of the PS-IPT, presented a high-priority need for an objective, technology-based way to determine kinship in support of the validation of family relationships for immigration purposes and foreign adoptions. Family relationship determination at the time was more of an art than a science, and based on in-person interviews and potentially fraudulent paperwork.

8.2.4.2 *PS-IPT/Homeland Security Institute (HSI) Partnership*

The PS-IPT contracted with the Homeland Security Institute (HSI) to work with DHS and other federal agencies to establish the need and parameters for an instrument that could provide a rapid (ultimately within minutes) DNA test result. At that time, the cost of private laboratory DNA testing was $600–$1500/case (Lewis, 2010). The Federal Emergency

Management Agency (FEMA) also expressed a PS-IPT need for a more rapid technical aid for mass casualty situations and potentially in mass evacuations. This USCIS high-priority requirement persisted from 2005 to 2007 without S&T finding a solution.

8.2.5 Federal Funding

Building an innovative, high-risk technology would require the federal government's assistance, and the funding, in large part, was contributed by the DOD, DHS, and DOJ. In many ways, the Rapid DNA instrument is an ideal example of how the federal government may help expedite and stimulate the development of a specific technology. Without federal assistance, the Rapid DNA instrument may not have reached commercialization so quickly. Furthermore, it is only through the collaboration and coordinated efforts of many federal partners that Rapid DNA became a reality.

In 2008, DOD, DOJ, and NIJ funded the development of a sample collection system for law enforcement agents to collect, protect, and document biological evidence (Tan, 2011). The sample collection system consisted of an evidence-collection device (a collecting swab and accompanying tube for swab insertion) and a sample processing cartridge. This "Smart Cartridge" lyses cells, makes the cells soluble and concentrates the DNA, and transfers the DNA solution to a microfluidic biochip. The microfluidic biochip would process the DNA solution to generate purified DNA compatible with STR analysis. The sample collection system performed initial processing steps in a format compatible with rapid sample-in to results-out microfluidic DNA analysis. Eugene Tan at NetBio conducted this research.

8.2.6 HSI Gap Assessment

In 2008, the DHS Homeland Security Institute was asked to perform an exploratory gap assessment and refinement of the capability gap. To this end, HSI met with several agencies, including but not limited to USCIS, FEMA, Customs and Border Protection (CBP), Immigration and Customs Enforcement (ICE), and the State Department. The gap refinement study suggested the gap be refined to read "Rapid (ultimately within 45 minutes) and cost-effective (under $100 per test) DNA testing." HSI also conducted an analysis of alternatives (AoA) that suggested the gap could be attained in approximately two to three years with an infusion of $8 million to $10 million in development resources.

S&T, in response, released a Small Business Innovation Research (SBIR) solicitation in 2009 seeking proposals for a "Low-Cost and Rapid DNA-based Biometric Device."[1] The SBIR program is broken into three phases. In Phase I, six-month awards were up to $100,000 (at that time) for a feasibility assessment of the proposed idea. Phase II 24-month awards were up to $750,000 in cost for the principal research and development effort to produce a well-defined deliverable prototype. One Phase II award is usually the result of a down select from the Phase I awards. Phase III is a follow-up with external funding from the private sector and/or government funding outside the SBIR program.

8.2.6.1 Phase I SBIR technical objectives
- To conduct a needs and requirements assessment for all DHS components
- To establish specific performance metrics for the device
- To study candidate enabling technologies and risks
- To define a viable technology architecture for the proposed device

8.2.6.1.1 Conducting a Needs and Requirements Assessment for All DHS Components

The assessment identified the requirements for a fully integrated STR DNA typing instrument for routine use in kinship verification of persons seeking to immigrate to the US and children being adopted overseas. Requirements include instrument throughput, instrumentation and per-sample cost, site and service requirements, data security requirements (including sample archiving), concept of operations, and user qualifications.

Based on a series of discussions with DHS stakeholders, it became clear that there are many applications for DHS Rapid DNA Analysis technology, and six potential DHS applications were initially identified (this list later expanded):

- Citizen and Immigration Services (CIS)
  - Applications support centers
  - Refugee/asylee processing
  - International adoptions
- Customs and Border Protection (CBP)
  - Border security at ports of entry
  - Border security between ports of entry (mobile applications)
- FEMA: Disaster Mortuary Operational Response Team
  - Border security at ports of entry

8.2.6.1.2 Establishing Specific Performance Metrics for the Device

The functionalities and specifications required for DHS use in the immigration setting include time to answer and accuracy of the STR profiles.

Based on the needs and requirement assessment conducted, it was determined that the Rapid DNA system should have the following performance metrics and features:

Science/Throughput
- Accepts a minimum of five buccal swab samples with a goal of 16.
- Performs DNA extraction and purification, PCR amplification, and separation and detection for 5–16 samples in a minimum of 90 minutes with a goal of 45–60 minutes.
- Generates full STR profiles using a primer set with 25 or more loci (which also contains the 13-core Combined DNA Index System CODIS loci) and an integrated expert system to analyze the electropherograms.
- Provides identity or family relationship verification results in less than one hour.
- Provides STR profiles that adhere to standard interpretation guidelines.
- Yields unambiguous analysis results: match versus no-match.

Technology
- Transfers STR results in standard message/database formats for compatibility with existing infrastructure.
- Communicates STR profiles and electropherogram data externally through both wired and wireless interfaces.
- Contains all necessary reagents onboard the consumables.
- Operates using consumables with minimum three-month shelf life.

Physical Features/Security
- Weighs less than 50 kg and occupies a volume of less than six cubic feet.
- Ruggedized for transport (MIL-STD-810F) and field operation at 20°C–30°C with no routine alignment or manual recalibration required after transport or during operation.
- Can be set up in less than 15 minutes (uncrating to readiness for analysis).
- Fully automated, requiring no human intervention after sample input.
- Supports use in an office environment by minimally trained office personnel with no formal training in biology.
- Incorporates physical and data security to prevent tampering with the device, prevent unauthorized access to data or results, and protect an individual's privacy.

8.2.6.1.3 Studying Candidate Enabling Technologies and Risks

The major industrial and academic groups developing human ID technologies were surveyed. An overview of the basic technical approaches and risks of each approach was developed, with a focus on the functionality of each module regarding required performance metrics (DNA purification, STR amplification, and separation and detection), technology readiness level of each technology/component, ability to integrate the modules, manufacturability, cost of the instrument and disposables, and likelihood of system completion within the next two years. The results of this study are corporate proprietary and are not included here.

8.2.6.1.4 Defineing a Viable Technology Architecture for the Proposed Device

This work encompassed three activities:

1. The technology architecture elements currently developed (including DNA purification, STR amplification, separation and detection, and STR allele calling) were reviewed considering the needs and requirements assessment and the performance metrics.
2. The technology architecture was customized to address DHS needs and requirements.
3. A detailed risk assessment (technology maturity, readiness, suitability) of each technology component was performed.

8.2.6.2 The Objective of the 2009 DHS S&T SBIR Phase II and III

The objective was to demonstrate a desktop-sized prototype device that could be used in an office environment by minimally trained office staff to conduct verification of identity or a family relationship in less than one hour at the cost of less than $50 per verification. Verification of identity would be against a previously collected DNA sample, and verification of a family relationship would be by comparing the DNA of two individuals. The device results were intended for initial screening purposes only; any questionable results or legally actionable cases would require DNA to be sent to a certified forensic laboratory. Device security, access to data or results, and protection of an individual's privacy were required to be an integral part of the design and prototype device construction. The Phase II SBIR effort would develop, test, and deliver at least one prototype DNA-based biometric device. Phase III would focus on completing a commercial Rapid DNA-based biometric device with applications in mass casualty situations, reunification of family members following mass evacuations, identification of missing persons, rapid processing of crime scene and suspect DNA, and various scientific and educational uses.

Although three contractors were selected for the Phase I awards, the Phase II and III down select was made only to NetBio, as it completed the integration and automation of their PCR and STR typing by CE work creating a standalone Rapid DNA biometric device. A prototype was delivered in September of 2014 that processed the traditional 15 STR loci. A finalized commercial device was delivered in October 2016, which had been advanced to process 27 STR loci to support the DHS required verification of grandparent–grandchild and sibling relationships.

The 27plex Rapid DNA multiplexed assay (Grover, Jiang, Turingan, French, Tan, and Selden, 2017) is modeled after the Fusion 6C kit with two additional Y-STR loci (DYS570 and DYS576) and the substitution of PentaD with D6S1043, an important STR marker used in law enforcement in China. The 19.5 minute assay (Schumm, Gutierrez-Mateo, Tan, and Selden, 2013) incorporated newly developed six-dye chemistry analyzed using the novel Rapid DNA microfluidic electrophoresis instrument capable of simultaneous detection and discrimination of eight or more fluorescent dyes. The 27-locus multiplex system was developed comprising all 15 STR loci of the European standard set, the current 13 STR loci of the CODIS core, the proposed 20 STR loci of the expanded CODIS core, 4 additional commonly used STR loci, and the amelogenin locus. This multiplex has three sets of linked loci (vWA and D12S391, CSF1PO and D5S818, and SE33 and D6S1043) (Graham, 2018). This expanded set allows laboratories in any jurisdiction to use a single reaction to determine loci profiles they typically generate and create an expanded common STR profiling set of interest to the global community. The linked loci just need to be taken into account when kinship or other DNA comparisons are calculated.

8.2.6.3 Massachusetts Institute of Technology/ Lincoln Laboratory Solicitation

Very shortly after DHS S&T published its SBIR solicitation, the DOD and DOJ both stated that they were also interested in supporting an effort to develop and deliver Rapid DNA systems. The DOD needed a tool that could be deployed into the battlefield to quickly identify the makers of improvised explosive devices (IEDs) from DNA left behind in these devices. The DOJ needed a DNA collection instrument that police officers could use in booking stations to collect and rapidly process DNA samples for comparison to the FBI DNA database. As previously stated, DHS needed a tool to verify family relationship claims to reduce fraud and expedite legal immigration.

Out of these joint federal needs came the requirement to reduce the multimillion-dollar forensic laboratory down to a fully automated, rugged, and portable device—a device that could produce DNA profiles from at least five persons within 90 minutes, with the accuracy of a full DNA forensic laboratory. It needed to be no larger than a laser printer and simple to operate by field personnel in an office setting, mobile laboratory, or the back of a van equipped with a generator.

Additionally, the DOD wanted the tool to be ruggedized to military standards and be able to process forensic samples. DHS paid to add analysis software to verify claimed family relationships at greater than 99.5% probability for parent-to-child relationships and 90% for sibling-to-sibling and grandparents-to-grandchild relationships. And the FBI paid to validate that it met all of the quality assurance standards required for results to be automatically loaded and searched against its DNA database.

The Massachusetts Institute of Technology–Lincoln Laboratory (MIT-LL) conducted a solicitation on behalf of all three federal agencies for the Field-Deployable Accelerated Nuclear DNA Equipment (ANDE) Program. ANDE was to be an 18-month technology development program, described in a statement of work, to deliver five production-ready prototype systems capable of rapidly acquiring and processing human DNA profiles. The Departments of Defense, Justice and Homeland Security along with MIT-LL developed the system requirements over a period of six months. Some requirements contained both a minimum threshold as well as a goal to attempt to achieve, as shown in Table 8.1.

The ANDE program was awarded to NetBio Systems by MIT-LL. Because of several technical challenges, the 18-month planned program became a 24-month development program. Nevertheless, five prototype systems were delivered by October 2014, and fully commercialized systems were delivered in September 2016.

8.2.6.4 Applications and Field Studies

With such revolutionary technology available, Rapid DNA found its way into use in sheriffs' offices, police stations, and forensic laboratories, and has been evaluated for use in other applications. Many law enforcement agencies have used Rapid DNA to test forensic DNA samples and cheek swabs collected directly from persons. Rapid DNA was used for the first time in July 2014 by the Richland County (South Carolina) Sheriff's Office (Jung, 2015, *The State*, 2015).

In 2015, the Arizona Pima County Sheriff's Office and Tucson Police Department were not only using Rapid DNA to solve crimes. They used it to certify sheriff and police detectives to operate Rapid DNA instruments themselves when the crime lab is closed (Jung, 2015). The many cases solved with Rapid DNA include one that linked blood left behind by a car thief to a suspect in the Arizona DNA database in just a few hours. Other sheriffs' offices already using Rapid DNA include Alameda County in California (Alameda County, 2017).

Having been proven in the laboratory and validated for use with closely trained and managed users, Rapid DNA moved out of the lab and into office and field locations. Rugged enough to handle transportation, it calibrates itself between each run, and uses disposable components to ensure there isn't any contamination between tests.

With DNA being the only biometric that accurately verifies biological family relationships, DHS has been evaluating the use of Rapid DNA to fight human trafficking along US borders and reunify families following critical or mass casualty incidents. Rapid DNA instruments and their supplies are already on the FEMA authorized equipment list[2] and can be purchased by state and local agencies using state FEMA preparedness grant funds.

DHS S&T participated in a large-scale disaster exercise with FEMA's Task Force One Urban Search & Rescue (US&R) team in July 2016. Sponsored by NORTHCOM (US Northern Command) and the National Guard, the Vigilant Guard (VG16) exercise showed that Rapid DNA could identify critical or mass casualty victims before body recovery, even if the remains were contaminated and couldn't be transferred immediately to a morgue.

FEMA's US&R team conducted canine searches for four samples of human remains (bone, muscle, and two other tissues) hidden at four different mock disaster locations. Once the human remains were found, a medical examiner documented and collected the human remains samples. While maintaining chain of custody, the samples were

Table 8.1   System Requirements for the MIT-LL ANDE Solicitation

Physical Features/Throughput/Maintenance/Training

Power: Must operate on 120 V (60 Hz) line power with <15 amp peak load.

Weight: <50 kg.

Size: ≤6 cubic feet with no dimension greater than 30 inches. Goal: <4.5 cubic feet with no dimension greater than 30 inches.

Setup time: ≤15 minutes to unpack and be ready to process samples.

Analysis time: <1-hour processing of buccal swab to STR profile. Goal: ≤45 minutes.

Instrument throughput: Able to simultaneously produce multiplexed STR profiles from a minimum of eight samples (may include ± controls and allelic ladders). Goal: ≥16 samples. Also, two basic plans to customize prototype systems that accept four or eight samples.

Recalibration: No routine alignment or recalibration after transport or during operations.

User manipulation: No manual user manipulations other than a simple replacement of preloaded consumables, insertion of swabs, and sensor run initiation.

Portability and ruggedization: Two-person lift with no lifting equipment required. Operates in an enclosed deployable lab or climate-controlled buildings. The prototype must conform to relevant sections of MIL-STD-810F (e.g., transportation vibration).

Hardware reliability: Must be specified by the manufacturer. Goal: Must include the threshold plus a confidence level.

Maintenance: Routine maintenance intervals of ≥1 month. Description of all expected routine maintenance, intervals, and personnel qualified to do maintenance. System self-test with a display of subsystem name/code in need of maintenance. Goal: Routine maintenance intervals of ≥1 year.

Operator training: Untrained personnel (i.e., no lab experience) with 1-hour training able to operate and reload the prototype sensor for processing of buccal swabs.

Costs/Consumables/Maintenance/Service Levels

Cost per sample: ≤$100 for initial low-volume consumable production. Goal: ≤$20.

Cost per instrument: System ≤$275,000 in initial low-volume production. Goal: System cost <$75,000.

Reagent storage: Pre-packaged into consumables. Stable for a minimum of 3 months for temperatures between 20°C and 30°C. Goal: 6-month reagent/consumable stability for 10°C to 50°C.

Consumables: Disposable, shipped with preloaded reagents (or automatically loaded by the system). Sealed to prevent cross-contamination. Human DNA-free and DNAase-free.

Delivered consumables: 100 sets delivered per prototype system. Pricing and lead time for orders of 30 additional sets.

Delivered prototype systems: Five delivered at the completion of the contract. Scaled up manufacturing within the first 6 months. Optional pricing for additional systems in quantities of 1, 10, and 100.

Component spares and service: Spare parts and service for 12 months following contract. Optional pricing for additional parts and services is also provided.

Period of performance: One prototype delivered 16 months and four more delivered within 18 months after contract start.

Other

Technology readiness level: Prototypes delivered must be TRL level 6 and suitable for field testing in a relevant environment. Goal: Prototypes delivered in final form, ready for field testing in an operational environment (≥TRL 7).

Extraction efficiency, DNA purity: Capable of extracting and purifying DNA from buccal swabs with efficiency and purity comparable to common bench methods. The process must include means for limiting DNA extraction yield, so it is in the optimum range for multiplexed STR amplification.

Separation/detection: For 16 or more loci and provide single base-pair resolution of DNA amplification products with sizes up to 500 base pairs. Color-separated electropherograms and description of hardware used to generate them delivered with final system.

Sample types: Process fresh or dried buccal swabs and purified DNA samples prepared in separate manual processes from forensic evidence. Goal: Threshold plus automatic processing of dried blood reference samples and/or swab samples collected from forensic evidence. Data on additional sample types and anticipated performance included.

Trace/low copy number (LCN) performance. Goal: Additional abilities to process DNA samples from forensic trace and LCN evidence. Consistent, full-profile results from as little as 100 pg of human genomic DNA and partial profiles from as little as 50 pg.

transported to the Rapid DNA instrument in a DHS Customs and Border Protection mobile laboratory van for processing. The van offered generator power and a clean sample preparation space while transporting the Rapid DNA system to three different working sites in two days.

The Massachusetts Office of Chief Medical Examiner (MAOCME) purchased a Rapid DNA system, validating it to identify human remains and reunite those remains with family members. The office worked with DHS S&T to develop guidelines and best practices for regional medical examiners' offices to use Rapid DNA for daily morgue operations and to be ready to respond rapidly to regional mass casualty events when they occur.

The DOD also expanded its original plans for the use of Rapid DNA. The DOD Special Operations Command began evaluating Rapid DNA for sensitive site exploitation and verifying target identity before and after raids. Without Rapid DNA, samples had to be sent back to the United States for testing, and the DOD would not get results for two to three weeks.

8.3 Advantages and Limitations of Rapid DNA Technology

DNA is the single most informative and robust biometric available in the world today and the only biometric that cannot be altered and allows for verification of biological relationship/kinship claims. DNA is the only biometric that remains relatively unchanged from birth and remains viable for identification purposes beyond death. Its unique kinship verification ability has vital security, social, and cost-saving applications. It can effectively reduce immigration fraud, human trafficking/smuggling, fight terrorism, and ensure the accurate and timely reunification of families. As a biometric, DNA has a significant advantage over other biometrics as it can verify human identification and confirm claimed kinship/biological relationships.

8.3.1 Technology Strengths

- **Provides immediate human identification results**. The Rapid DNA instrument generates a DNA profile in 90 minutes and offers a field ready, fully integrated and automated DNA tool to conduct faster, more comprehensive identifications of persons or objects in support of critical incidents or criminal investigations.
- **Is proven ruggedized and resilient.** Tested in extreme conditions and shown to perform in a hot desert environment. Some of the instruments are stackable.
- **Offers ease of use.** Minimally trained staff may successfully operate the instrument and generate DNA profiles in a field or laboratory setting.
- **Allows for quick expansion of DNA testing capacity.** The ability to quickly expand the screening of individuals (e.g., aliens, confidential informants, employees) to ensure that individuals are not on known or suspected criminal/terrorist DNA watch lists.
- **Generates kinship/biological family analysis.** Offers law enforcement decision-makers with a new, actionable tool that can quickly verify claimed family biological relationships to support border enforcement, disaster recovery, and immigration applications. It may be used to expedite valid immigration cases and deter immigration fraud and human trafficking.

8.3.2 Technology Limitations

- **Does not quantitate the DNA.** Because it does not quantitate the DNA, the FBI guidelines are not followed, and therefore unknown samples cannot be submitted to CODIS.
- **Sample consumption.** The sample can be consumed during the testing process. Therefore, end users need to be careful which samples are put through the Rapid DNA instrument.
- **Need for scientific reachback support.** Some human analysis will be required to support the instrument. Additional analysis is required for interpretation of problematic samples where the DNA may be degraded and pristine samples have genetic anomalies (tri-alleles).

8.3.3 Technology Applications

The instrument is currently used in sheriffs' offices, police stations, and forensic laboratories, and has been evaluated for use in other applications. The applications include the following.

Booking stations. Currently, an arrestee may only be connected to a crime after being released into the community, even though approximately 3% of arrestees in California are associated with unsolved crime. This leaves 300 potentially dangerous criminals on the street per month while their DNA is being processed (Babak, 2016). This slow process allows potential criminals to commit additional crimes or flee the state by the time the DNA results are received.

Identifying human trafficking victims. With DNA being the only biometric that accurately verifies biological family relationships, Rapid DNA can be used to verify claimed family relationships when purported families are crossing the border. Often, trafficked individuals cannot speak up.

Expediting testing of evidence. Rapid DNA can be used to detect DNA profiles associated with property crimes. Typically, this evidence is given a lower priority at larger public crime laboratories.

Mass fatality human identifications. Following a mass fatality, the local medicolegal authority can use DNA profiling to identify human remains. DNA profiling may be the only viable identification method in circumstances where there is extensive fragmentation or decomposition of bodies. Historically, DNA technology has been used as a last resort in identifying disaster victims, as traditional DNA testing is a lengthy, time-consuming, and costly process, as samples are sent to third-party laboratories. Rapid DNA allows DNA to be used at the family assistance center and temporary morgue.

Identifying persons of interest. The need for a field-based DNA testing instrument is significant for the military, as theater operations are often conducted in countries where individuals do not have unique identifiers or are paperless due to unrest or migration.

8.4 Transition of Rapid DNA into Operations

The criminal justice community is generally cautious and slow to adopt new technologies related to the testing of forensic evidence, as the criminal justice community needs to know that the new technology is reliable, scientifically valid, and able to withstand the scrutiny of a legal proceeding. However, adoption of Rapid DNA has been slower than anticipated for several reasons, including:

- **Costs.** Law enforcement agencies could conduct DNA testing, although it was slower.
- **Initial issues.** There were early manufacturing quality issues with the instruments leading to some inconsistent results, which created a general concern about the reliability of the technology as a whole.
- **Hesitation about new technologies.** There were concerns from the crime laboratory community that the technology would be used improperly, resulting in legal challenges to the results and loss, or restricted use, of DNA usage in court.

With increased adoption and criminal justice success, it is expected that there will be more widespread adoption of Rapid DNA as crime laboratories, booking stations, and local jurisdictions realize the overall cost savings from using DNA evidence early in an investigation.

8.4.1 Examples of Successful Rapid DNA Applications

8.4.1.1 Law Enforcement Offices

In July 2014, the Richland County (South Carolina) Sheriff's Office, for the first time in history, used Rapid DNA to verify that DNA on a subject's clothing matched the DNA of an attempted murder victim. "This is an important new crime-fighting tool when it comes to saving time, money, and manpower, and it is the future of DNA analysis," Sheriff Leon Lott said at the time. "It gets the criminal off the streets quickly, which prohibits them from committing additional crimes" (Jung, 2015, *The State*, 2015).

8.4.1.2 Rapid Processing of Casework in the Crime Laboratory

In 2019 Kentucky became the first state in the US to use the Rapid DNA instrument to test DNA from sexual assault evidence kits (SAEKs). According to Public Safety Cabinet Secretary John Tilley:

> Rapid DNA allows us to use DNA immediately for investigating sexual assault cases. It gives us the most powerful witness of all: scientifically proven identity. Kentucky is proud to pioneer this use of Rapid DNA technology. Our goal is to identify suspects and exonerate the innocent to provide justice as soon as possible. Testing sexual assault cases quickly is critical as many rapists are serial offenders—and this can prevent them from committing more crimes.[3]

Sexual assault kits are associated with a significant backlog issue and can take months to years to process. The initiative has been successful, with some DNA identifications being made hours after submission to the Kentucky State Police Forensics Laboratories.

8.4.1.3 Rapidly Identifying Human Remains in Medical Examiner Offices

The Massachusetts Office of Chief Medical Examiner (MAOCME) was the first OCME office to purchase a Rapid DNA instrument to identify human remains and reunite those remains with family members. Rapid DNA provides a faster and much higher confidence identification than the traditional visual identification of human remains or dental record checks. MAOCME validated the ANDE instrument and worked with the American Association of Blood Banks (AABB)[4] to become the first accredited relationship testing facility to use Rapid DNA.

8.4.1.4 Mass Fatality Planning and Response

FEMA allows for Rapid DNA instruments and associated supplies to be on the authorized equipment list,[5] and can be purchased by state and local agencies using state FEMA preparedness grant funds. If Rapid DNA can become an integrated part of all mass fatality responses, this may expedite the time required to identify human remains and/or locate missing persons.

In 2015 DHS S&T started exploring the potential use of Rapid DNA technology in mass fatality response operations. DHS participated in a series of stepwise mass fatality exercises to assess the feasibility of conducting Rapid DNA testing in mass fatality DNA response operations. In the first exercise, DHS deployed the Rapid DNA instruments to a family assistance operation. DHS S&T tested whether the instruments could be shipped on short notice. The exercise was successful and informative. Over the next year, the drills became more complex. As a result, DHS S&T was able to identify and mitigate potential logistical challenges for more complex deployments. The technology proved to be robust, effective, and easy to use. Rapid DNA integrated well into the overall response operations. Just-in-time training was productive, and the instruments produced real-time results that contributed to the identification effort. DHS interviewed the actors participating in the exercises as family members and polled exercise participants on how supportive they were of the government's use of Rapid DNA to identify missing persons in disasters. Results showed that 100% of respondents were supportive.

Mass fatality responses have significant challenges, and Rapid DNA offers many solutions. However, due to the slower nature of traditional DNA testing, the US has not developed a coordinated response for DNA operations like the Disaster Mortuary Operations Response Team (DMORT) morgue operations. In April 2017, the American Society of Crime Laboratory Directors (ASCLD) identified Rapid DNA as an important tool in forensic testing and started a Rapid DNA Task Force to address the responsible implementation of Rapid DNA. Four months later, the ASCLD Disaster Victim Identification (DVI) subcommittee was formed to determine best practices, develop models and policies, and be an important resource during times of need.

National, state, and local crime laboratories have expertise in DNA identification and are well situated across the nation to be a willing partner in a national DVI response. Leveraging the expertise, location, and responsibility of crime labs represents a tremendous opportunity to build a national network.

In November 2018, the Camp Fire in Butte County, California, took the lives of 84 people. Due to the heat and intensity of the fire, traditional methods of identification were not feasible. The Butte County Sheriff's Office used Rapid DNA to scientifically identify the victims. ANDE provided instrumentation and staff in support of the Rapid DNA testing. The resulting quick and accurate victim identifications demonstrated that Rapid DNA is a viable option for DVI. Rapid DNA can save time and money and bring information to the families of the victims much faster than traditional DNA testing methods. This successful DVI response demonstrated how Rapid DNA technology can transform mass fatality DNA response operations moving forward.

A joint initiative by the ASCLD DVI Rapid DNA Subcommittee, DHS S&T, NIJ Forensic Technology Center of Excellence, and SNA International resulted in an interactive Rapid DNA DVI workshop in May 2019 at the ASCLD Annual meeting. The daylong workshop provided participants with hands-on and realistic scenarios that may occur during a DVI response, relying on the use of Rapid DNA. This exercise provided participants with a genuine representation of the complexity of a DVI response. The workshop helped identify the next steps needed to formalize ASCLD's DVI Rapid DNA Subcommittee deliverables and the Rapid DNA Cooperative concept. In 2019, ASCLD continued to strategically plan the Rapid DNA Cooperative concept conducting a tabletop exercise that involved the deployment of Rapid DNA instruments at the Human Identity Trade Association workshop held at the International Symposium on Human Identification.

On September 2, 2019, a boat fire claimed the lives of 34 passengers on a scuba diving trip off the coast of Santa Cruz Island, California. Five crew members escaped with injuries, but the intense fire aboard the *Conception* dive boat left the other victims unrecognizable. The Sacramento County Coroner, Kimberly Gin, and one of her deputies brought their ANDE instrument to help the Santa Barbara Sheriff-Coroner identify all 34 victims.

The day after Christmas in 2019, an Airbus AS350 B2 tour helicopter crashed into the tropical mountainous terrain on the island of Kaua'i, killing all seven people aboard. The Kaua'i Police Department's Crime Scene and Laboratory Unit responded to the helicopter tour crash using the Thermo Fisher RapidHIT ID and positively identified the pilot and six passengers.

8.4.1.5 Combating Family Unit Fraud

In 2018, the US Immigration and Customs Enforcement (ICE) became aware of an increase in family unit fraud encountered at the US border. This fraud scheme generally involved adult non-US persons and unrelated children posing as biological parent and child to DHS authorities to receive special treatment in accordance with law and policy. Family unit fraud is believed to be associated with other crimes, including immigration violations, identity and benefit fraud, alien smuggling, human trafficking, foreign government corruption, and child exploitation. In 2019 ICE conducted a pilot study. If ICE suspected family unit fraud, it administered a voluntary Rapid DNA kinship test to validate the claimed relationship. All DNA profiles were destroyed after the test. The pilot lasted three days, and 84 family units that presented indications of fraud were tested. Of those tested, 16 family units were identified as fraudulent (Department of Homeland Security, 2019). ICE then continued with the testing to protect children from being trafficked, ensuring they are with their parents and not being used to exploit immigration loopholes.

8.5 Lights out Rapid DNA for Biometric Uses

To successfully adopt Rapid DNA technology, which can significantly expand the use of DNA to fight terrorism and human trafficking, and identify immigration fraud, the DHS Office of Biometric Identity Management developed a software prototype platform for DNA. This software platform integrates with Rapid DNA instruments and offers a secure location for DHS DNA collection results that support direct and familial matching algorithms, ensures standards-based real-time data sharing with law enforcement and intelligence community stakeholders, and is compliant with civil liberties and federal records retention requirements.

8.5.1 Expanding Capabilities: Software Support

The DNA SMS Prototype Software DNA identity management software adheres to privacy and civil liberties requirements while providing lights out DNA identity functions, including:

- Verifying family relationship claims.
- Locating direct DNA matches of one person's DNA profile against a database of DNA profiles.
- Creating profiles in a format that will allow export to databases in the DOD and DOJ, and accept results returned from those databases.

The DNA SMS Prototype Software is built on a service-oriented architecture compatible with the Office of Biometric Identity Management OBIM existing biometric system (IDENT) and the future Homeland Advanced Recognition Technology (HART) system. It can also be deployed and operated as a standalone web application without OBIM's biometric system. The Prototype Software uses standards established by the National Information Exchange Model (NIEM) and American National Standards Institute/National Institute of Standards and Technology (ANSI/NIST) to ensure interoperability with other US government agencies and foreign partners.

The Prototype Software currently runs in a commercial cloud powered by Amazon Web Services (AWS) in a standalone mode. Artificially generated data and data acquired from commercial laboratories were used to provide examples to demonstrate the matching algorithms, kinship analysis, and identity management functions.

The Prototype Software supports five core use cases:

- Kinship analysis when all family members are in one location.
- Kinship analysis when family members are separated geographically.
- Finding direct DNA profile matches.
- Finding missing relatives in a database of DNA profiles from unidentified human remains.
- Searching against external DNA databases (e.g., searching missing children in INTERPOL).

The software also supports DNA sample collection, sample transporting, and testing and reporting processes.

8.5.2 Expanding Capabilities: Reachback Support

DNA laboratories have an existing structure that allows for additional scientific support as needed. However, once the DNA analysis capability is removed from the laboratory and placed in the field, laboratory scientific support ceases to exist, leaving non-scientists alone with a complex scientific process and instrument. If a sample fails, they will not know if it is the sample, the test kit, or the machine.

For the Rapid DNA instrument's potential to be fully realized, there must be some reachback support offered for each instrument. DHS is presently incorporating Rapid DNA technology into field operations, which will reduce some dependency on a full DNA laboratory but still require the support of forensic scientists. This prototype service being developed at DHS is referred to as Reachback Support Solution (RBSS). The RBSS ensures that DNA profiles generated by Rapid DNA instruments in the field are technically valid, kinship matches are precise, and there is adherence to established federal privacy and physical security policy requirements and guidelines supporting DNA science. The review of DNA profiles by RBSS analysts will allow for the maximum rate of usable Rapid DNA samples. The DNA RBSS Software has the following features:

- Data transfer—automatic transfer of flagged data from Rapid DNA instruments to the DNA forensic DNA analysts.
- Tracking—track and manage samples, profiles, and cases while being reviewed.
- Field-based communication—function as a conduit for field-based agents and officers to communicate with the DNA RBSS.
- Metrics—function as a metrics reporting tool to enhance quality and improve efficiency in the field.
- Guided workflow—user alerts facilitate a proactive and guided workflow.

DNA testing in the field is a relatively new capability, and the DNA RBSS provides a unique single point-of-contact for Rapid DNA user support. This augmented support may help with the adoption of Rapid DNA, as it provides a strategic methodology to transition from vendor laboratory dependency and places the ownership of the DNA testing process to local operations.

8.6 Rapid DNA Over the Next 20 Years

As law enforcement agencies are becoming more interested in the use of Rapid DNA at crime scenes, the DNA Working Group of the European Network of Forensic Science Institutes (ENFSI) along with the Scientific Working Group of DNA Analysis Methods (SWGDAM) and the FBI's Rapid DNA Crime Scene Technology Advancement Task Group issued a joint position statement on the use of Rapid DNA for crime scene use. This position paper identified five major areas requiring enhancement before Rapid DNA profiles from a crime scene can be entered into state and national databases. The Rapid DNA instruments must:

1. Have an integrated method of human-specific internal positive controls to identify low quantity, degradation, and inhibition.
2. Be able to export analyzable raw (optical preprocessed) data.
3. Have an onboard, fully automated expert system programmed with rules to accurately flag allele calls in single-source and mixture data requiring analyst evaluation.
4. Must achieve improved peak height ratio balance (per locus and across loci) for low quantity and mixture samples (high and low quantity) commonly found in forensic evidence.

5. Undergo well-defined publicly available developmental validation on a wide variety of forensic evidence-type samples commonly encountered in the forensic DNA laboratory.

While some of these will be challenges, they will most likely result in other tangible outcomes, including:

- An increase in instrument adoption.
- A lower cost of Rapid DNA instrumentation and DNA test kits. Prices may continue to drop as widespread adoption occurs, leading to economies of scale.
- The ability for all US law enforcement to quickly verify arrestees in custody against NDIS.

In 1996 Rapid DNA technology was a seemingly impossible concept, but scientists and technologists, who refused to be limited by the current restrictions, dared to dream of a better way to use the power of DNA. By 2014 that vision, through federal funding, became a reality. The development of Rapid DNA technology proves that scientific vision can become a reality in the US with appropriate funding, collaboration, vision, persistence, scientific competence, and dedication.

Notes

1. The full SBIR solicitation was issued on November 12, 2008, with a response date of January 5, 2009, and can be found at https://www.sbir.gov/sbirsearch/detail/385974.
2. Rapid DNA equipment and supplies can be purchased under DHS/FEMA Authorized Equipment List numbers 07BD-02-DNRN and 09MY-01-DNAK: www.fema.gov/authorized-equipment-list.
3. https://kentuckystatepolice.org/hq-4-10-19/.
4. The AABB (www.aabb.org) develops standards that combine internationally accepted quality management with technical requirements appropriate for relationship testing. These standards form the basis for AABB's accreditation of laboratories that conduct relationship testing in the United States and in multiple foreign countries.
5. Rapid DNA equipment and supplies can be purchased under DHS/FEMA Authorized Equipment List numbers 07BD-02-DNRN and 09MY-01-DNAK: www.fema.gov/authorized-equipment-list.

References

Alameda County, Office of the District Attorney. (2017, August 22). Rapid DNA Act Signed into Law [Press Release]. https://www.alcoda.org/newsroom/2017/aug/rapid_dna_act.

Babak, Kevin. (2016, November 11). The Evolution of Rapid DNA in the United States. https://www.ishinews.com/the-evolution-of-rapid-dna-in-the-united-states/.

Department of Homeland Security, Immigration and Customs Enforcement. (2019, June). ICE Awards New Contract for Rapid DNA Testing at Southwest Border, Expands Pilot Program. https://www.ice.gov/news/releases/ice-awards-new-contract-rapid-dna-testing-southwest-border-expands-pilot-program.

Graham, Timothy. (2018). Genes of a Feather Stick Together: Evaluating Linked Markers [Webinar]. George Washington University. https://nij.ojp.gov/events/genes-feather-stick-together-evaluating-linked-markers-part-dna-kinship-webinar-series#-1.

Grover, R., H. Jiang, J. Turingan, E. Tan French, and R. Selden. (2017). FlexPlex27-Highly Multiplexed Rapid DNA Identification for Law Enforcement, Kinship, and Military Applications. *Int J Legal Med* 131(6):1489–1501.

Jackman, Tom. (2018, December 13). FBI Plans Rapid DNA Network for Quick Database Checks Against Arrestees. *The Washington Post, True Crime*.

Jung, Yoohyun. (2015, September 29). TPD, Sheriff's Department Using 'Rapid DNA' Technology, Arizona. Daily Star.

Lewis, Kristen. (2010, April 14). The Role of DNA in Kinship Analysis. National Institute of Standards and Technology Biochemical Science Division Applied Genetics Group – DNA Biometrics Project. https://strbase.nist.gov//pub_pres/Lewis-LM-kinship-Apr2010.pdf.

Maryland v. King, No. 12–207 (Supreme Court of the United States 2013).

Moretti, T. R., A. L. Baumstark, D. A. Defenbaugh, K. Keys, A. Brown, and B. Budowle. (2001). Validation of STR Typing by Capillary Electrophoresis. *J Forensic Sci* 46(3):661–676.

Rapid DNA Act of 2017, S.139/H.R. 510, 115th Congress, Introduced January 12, 2017.

Schumm, J. W., C. Gutierrez-Mateo, E. Tan, and R. Selden. (2013). A 27-Locus STR Assay to Meet All United States and European Law Enforcement Agency Standards. *J Forensic Sci* 58(6):1584–1592.

Tan, E. (2011). *Sample Collection System for DNA Analysis of Forensic Evidence: Towards Practical, Fully-Integrated STR Analysis*. U.S. Department of Justice: Document No.: 236826.

The State. (2015, December 9). Sheriff: RCSD First in Nation to Use RapidHit DNA to Put Criminal behind Bars. [Staff Report]. https://www.thestate.com/news/local/article48816510.html.

Uses and Applications

II

Collection and Preservation of Physical Evidence

9

HENRY C. LEE, TIMOTHY M. PALMBACH,
ANGIE AMBERS, DRAGAN PRIMORAC,
AND ŠIMUN ANĐELINOVIĆ

Contents

9.1 Introduction

Despite significant advances in the ability to analyze blood and other forms of biological evidence, if the sample is not properly collected and preserved, no level of sophistication will correct for mishandling a biological sample (Marjanović and Primorac, 2013). In fact, most legal challenges regarding physical evidence such as blood evidence focus on the recognition, collection, and preservation matters rather than the scientific methodologies that were used to analyze the sample (Fisher and Fisher, 2012). The National Institute of Standards and Technology (NIST) Organization of Scientific Area Committees

DOI: 10.4324/9780429019944-11

for Forensic Science (OSAC) is responsible for developing and disseminating best practices for the collection of blood found at crime scenes. One scientific area committee, the Bloodstain Pattern Analysis Subcommittee, has published ANSI/ASB Standards 030, 031, 032, 033, and 072. The OSAC Crime Scene Investigation & Reconstruction Subcommittee is currently working on relevant standards for this topic.

Not only does the collection and preservation have to be conducted in a manner that will preserve the evidentiary nature of the stain, but it has to be done in a process in which there is thorough documentation. The collection, packaging, and preservation of blood evidence at the crime scene should never take place until the crime scene analyst or forensic scientist has taken extra care to make sure that the bloodstain patterns have been extensively documented. This documentation may include written or audio notes, sketches, photographs, and/or video documentation. Generally, proper documentation will require several of these documentation methods. The particular method may have to be tailored to address a specific relevant fact. For example, the condition of a bloodstain—is it wet, dry, or coagulated—is an important factor for estimating the time frame since the blood was deposited (Lee et al., 2010). Recording this information will require notes as well as demonstrative methods such as photography. Furthermore, the general characteristics of the biological stain, its pattern, and the location of the pattern will be of great value in the reconstruction of events, as well as a means for laboratory analysts to best determine which samples are most relevant (Lee and Harris, 2000).

Once a potential blood or biological stain is located, thorough documentation must be completed. Photographs must include overall views showing the object or pattern in relationship to the overall scene. Then there should be an intermediate photograph that shows the observed or potential pattern and its orientation to the object it is located on. Finally, there needs to be close-up photographs. Close-up photographs must be taken with the camera perpendicular to the stain or pattern area to prevent distortion of the pattern. In addition, the photograph should be taken in a manner to ensure that the image is clear, properly exposed, and of high contrast. Finally, the close-up photograph must be taken in a 1:1 format, which is rarely possible, or with a scale included. It is best to use a ruler with a circle and diagonal lines, as this is a method of establishing that the photograph was taken properly. In addition to quality photos, it is very helpful to sketch the stain. Overall stain dimensions should be obtained. In addition, it is important to measure and record the precise location of that stain or pattern area on the object or within the crime scene (Figure 9.1).

9.1.1 Sample Collection from Victim or Suspect

The real value of unknown biological samples from the crime scene or evidence is to provide a link or association with the victim and/or suspect (Lee et al., 2001; Primorac and Marjanović, 2008). Proper known standards must be obtained.

9.1.1.1 Known Oral Swab Standards

Obtain oral swabs of known origin from rape victims and suspects to determine their DNA type. There are numerous swab units that can be obtained from commercial suppliers of forensic science laboratory and crime scene supplies. Be sure to properly dry and package the swabs so that there is no degradation of the DNA sample. With the advent of Rapid DNA technology, it is possible for known DNA samples to be obtained by collection

Figure 9.1 Examination quality, close-up view with scale of bloodstains on clothing.

of a buccal sample and processed in a Rapid DNA instrument, generally in under 1.5 hours (Carney et al., 2019).

9.1.1.2 Liquid Urine and/or Fecal Material
These samples may be collected in a clean specimen jar. Label accordingly and refrigerate until submission to the laboratory for DNA and toxicological analysis.

9.1.1.3 Vaginal Materials
Vaginal secretions are usually encountered in connection with a sexual assault case. Most often, these samples are included in the sexual assault evidence kit and should be collected according to directions. Usually, these samples are obtained with a swab and may include a glass slide prepared for microscopic examination.

9.1.1.4 Nasal Mucous
Nasal mucous is occasionally found at crime scenes on clothing, a handkerchief or tissue, or on a body. These materials should be air-dried and packaged in a paper druggist fold. These samples can yield good DNA results.

9.1.1.5 Bite Mark Evidence
If evidence of a bite mark is found, collect a swabbing of the bite mark of the areas adjacent to and inside the observed bite mark. It is important to obtain examination-quality photographs of the bite mark. Often the bite mark will visibly improve the following day as a result of the bruising and healing process. It may be beneficial to consult with a forensic odontologist during the photographic and documentation process.

9.1.1.6 Skin Tissue
Skin and tissue may be present under the fingernails in cases of sexual assault or violent confrontation. Package fingernail scrapings, which should be taken with a new orange

stick, file, or any other device, in paper and then properly label. Alternatively, collect the fingernails individually by clipping and packaging them separately.

9.1.1.7 Clothing or Personal Items

All evidence potentially containing body fluid evidence should be handled with gloved hands to protect the collector and to reduce the possibility of contamination of the DNA evidence. In addition, care should be exercised to prevent cross-contamination between subsequently collected items. Change gloves often, and with instances involving small items the use of disposable tweezers is recommended. The proper packaging and storage of clothing are vital to the viability of any potential DNA evidence.

1. It is of primary importance that clothing that is wet be allowed to air-dry at room temperature before packaging. The item is then packaged in paper bags or wrapped flat in paper. Portable drying units are available and can be of great assistance in the drying process.
2. Each item of clothing must be packaged separately.
3. Under no circumstances should this evidence be packaged in plastic bags or other airtight containers.
4. In order to preserve the trace evidence that may be present on the item, avoid excessive handling of clothing, and wrap in butcher paper before placing it in the outer container, usually a paper bag.
5. All personnel who handle objects with suspected biological evidence should wear face masks and gloves at a minimum. In most cases, disposable gowns, head covering, and boot covers are recommended as well.

9.2 Recognition and Identification of Blood Evidence

First, a particular stain or sample must be recognized as potential blood (James et al., 2005). Essential to this process is a preliminary evaluation of the scene or the object upon which the suspected stain is located. An understanding of basic bloodstain pattern analysis is of great value at this point in the process. It is critical to understand the basic principles in drop and pattern formation, how various forces or mechanisms that interact with the blood sample will affect the resulting pattern, and what—if any—interferences exist that may alter the resulting stain or pattern. Furthermore, an initial reconstruction of the crime scene, utilizing all available information—investigative, medical or injury patterns, physical evidence, and various forms of pattern evidence—can assist in searching for potential blood samples in appropriate locations.

Following criticism and concern in reports such as "Strengthening Forensic Science in the United States: A Path Forward" (http://www.nap.edu/catalog/12589.html), there was a call for adoption of standards, terminology, and general guidance. The OSAC Crime Scene Investigation & Reconstruction Subcommittee is currently working on relevant standards.

There are a variety of presumptive blood tests available that are very sensitive and can be used in laboratory, field, and crime scene applications. Despite the ability to determine the potential for finding blood based on a careful analysis of an object or scene, it should not be merely presumed that a stain in question is indeed blood. That can only be ascertained after use of an acceptable confirmatory test.

9.2.1 Presumptive Blood Tests

Presumptive blood tests are designed to detect trace amounts of blood. These tests cannot distinguish blood from human versus non-human sources. They are very sensitive and detect minute traces of blood. They are, however, not specific and thus subject to the potential of false positives. False positives can occur when the sample tested contains chemical oxidants or substances with peroxidase activity. Since many common cleaning agents, such as bleach compounds, are strong chemical oxidants, it is likely that one will encounter this situation at crime scenes or on clothing that has been laundered. Moreover, these tests may yield false negatives because of chemical interferences. If one correctly understands that these are presumptive tests only and that additional confirmatory testing is required, the concerns related to false positives or negatives are properly addressed. As valuable as these tests and reagents may be to investigators and forensic scientists, there are certain precautions that must be heeded. Field tests are designed for screening purposes and should not be used in lieu of laboratory analysis and confirmation testing. As a general rule, if the amount of available sample for testing is so minute that there may not be sufficient material for a full array of testing, then, if at all possible, avoid field tests so as to preserve the sample for laboratory analysis.

Following a simple sequential procedure with prepared reagents is all that is required with most of these presumptive tests. The interpretation of testing results is often straightforward and simple, such as observing a color shift or color formation. Despite the simplicity of the basic procedure, it is helpful to have an understanding of the underlying chemical reaction or mechanism.

The tests are based on a chemical oxidation–reduction reaction in which blood, more specifically the heme portion of blood, acts as the catalyst. Heme, a principle component of hemoglobin, is a ferrous-bearing molecule located within a red blood cell. This hemoglobin molecule with its ferrous content is responsible for carrying oxygen throughout the body. The ferrous element is generally in a reduced state. The presumptive blood tests discussed hereafter are designed to detect the presence of the reduced ferrous molecule through an oxidation–reduction reaction that will convert a colorless reagent to a colored by-product (Figure 9.2).

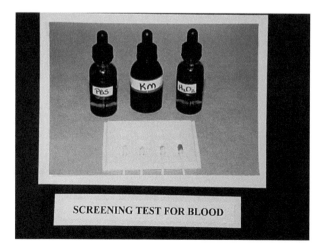

Figure 9.2 Phenolphthalein presumptive test for blood.

Common presumptive blood test reagents include phenolphthalein (Kastle–Meyer reagent), Leucocrystal Violet, and tetramethylbenzidine. Since these different reagents have different sensitivities and may react slightly differently in a more alkaline or acidic environment, the use of more than one reagent is recommended if questionable results are expected or obtained. These same reagents can be used as bloody print enhancement reagents. Enhancement reagents are used to visualize and increase the color contrast of transfer patterns such as bloodstains, fingerprints, footprints, shoe imprints, and other physical patterns. In some instances, enhancement reagents can be used for dual purposes: presumptive tests for biological substances and the development of physical patterns. To use reagents in this capacity, they must be mixed fresh and applied with a spray nebulizer bottle. If the reagents are to be applied to a vertical surface, great care must be exercised to avoid runs or drips through the pattern. Alternatively, these reagents (commonly referred to Blood Enhancement Reagents) can be mixed in a colloidal suspension with chloroform and ether so that there will not be any dripping on vertical surfaces (Figure 9.3).

9.2.2 Confirmatory Blood Tests

Additional samples must be taken and submitted to a forensic laboratory for confirmatory testing. There are a variety of confirmatory tests available in the laboratory. There is an immunoassay procedure, such as an ABA Hema Trace card, which is a simple and fast confirmatory test for human hemoglobin that is capable of being performed at the scene. This is an immunological (antigen/antibody) test that requires only the stain to be tested, a

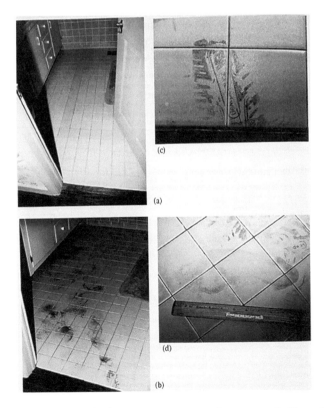

Figure 9.3 Enhancement of bloody foot impressions with tetramethylbenzidine (TMB).

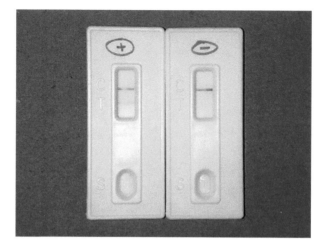

Figure 9.4 Hema trace card.

little pure water, and the commercial test kit. It has quickly become widely accepted and a good addition for modern crime scene processing kits (Figure 9.4).

If there is an insufficient sample available to collect for subsequent confirmatory testing, then do not conduct any field testing—rather, send the entire limited sample into the laboratory for analysis. There are minor differences in sensitivities of various reagents known; thus, controls should be conducted to understand the performance of a particular reagent.

LUMINOL, BLUESTAR, OR FLUORESCEIN

The reagents Luminol, Bluestar, and fluorescein have fluorescent properties. These reagents are commonly utilized as bloody print enhancement reagents. They have great value on dark substrates, but there are added difficulties in photographing the enhanced patterns because of the need to work in low-light conditions. With Luminol and Bluestar, the reagent is simply sprayed onto the area to be checked for the presence of suspected bloodstains, but it must be viewed in total darkness. Bloodstains will yield luminescence or fluorescence within a few seconds. Over time the reaction will fade, but it can be refreshed with additional reagents. Luminol and Bluestar are very sensitive reagents, and work well with very small amounts of blood or older stains (Figure 9.5). However, that same level of sensitivity also makes them very susceptible to false positives. With fluorescein, after the reagent is applied the sample must be viewed with ultraviolet light.

9.3 Collection Methods for Blood

After recognition and documentation of the blood evidence at the crime scene, the collection can begin. Following relevant and peer-reviewed guidelines is important. One such

Figure 9.5 Luminol used to enhance blood stains covered over by a coat of latex paint.

document is the OSAC's "Standard for On-Scene Collection and Preservation of Physical Evidence" (https://www.nist.gov/system/files/documents/2021/04/07/2021-N-0018_Standard%20for%20On-Scene%20Collection%20and%20Preservation%20of%20Physical%20Evidence_DRAFT%20OSAC%20PROPOSEDv2.pdf). Generally, blood is most susceptible to being destroyed or damaged. These are bloodstains that may be located in high-volume traffic areas of the scene; bloodstains found in doorways, hallways, or roadsides; or blood that may be present at outdoor crime scenes. Bloodstains found on movable objects can be protected by temporarily moving the object to a safer location until the stain or object may be collected.

Blood at crime scenes will be found either as a dried stain or in a liquid state. If blood at the crime scene is liquid and not large enough to collect a liquid sample, then let it air-dry. If the object with the bloodstain is movable, then collect the entire object. To package blood and other biological fluids, there are several basic guidelines and precautions. Blood and bloodstained evidence should never be placed in airtight containers.

Bloodstained evidence frequently will have other trace evidence present. If the blood loosely holds the trace evidence, then it is proper to collect that trace evidence and place it in a druggist fold and then a sealed envelope. However, if a dried bloodstain holds the trace evidence tightly, do not attempt to collect the trace evidence. Carefully cover or protect the trace evidence while packaging the bloodstain. Bloodstained items should always be packaged separately to prevent cross-contamination. The packaging should be sealed at the scene before transportation. Do not allow bloodstained evidence to be exposed to excessive heat or humidity. If possible, bloodstained evidence should be refrigerated, but the dried bloodstained evidence can be stored safely at room temperature and then submitted to forensic laboratories as soon as possible.

9.3.1 Dried Blood Stains

If possible, generally so with a moveable object, leave the blood on the object and package the entire object in a breathable container such as a paper bag. In most cases, wrapping the object in butcher paper first and then sealing it in a paper bag is recommended. If the bloodstain is located on a surface that can easily be removed, then cut an area around the bloodstain including the stain. If the surface material is fragile or easily broken, care should be taken to secure the cutout section to prevent breakage of the sample. Wrapping in a druggist's fold and then securing the object in a box works well. If the object cannot be moved, then the sample must be obtained by one of three methods: swabbing, scraping, or tape lifting. In the vast majority of cases, swabbing is the preferred method. With scraping, use care to avoid contamination with the substrate from which you are scraping the sample. Tape lifting is rarely a good choice, as subsequent extraction of the blood from the adhesive surface is complicated. The only rationale for that method is for preservation of the overall pattern. However, good documentation methods are generally a more effective means of dealing with the pattern component. Collection begins with a swab dampened with distilled water. Carefully swab the bloodstain with the swab. Absorb all the stain with a minimum of area consumed. Insert the swab into the swab-drying box. Allow the swab to dry once the stain has been collected. Seal the box and label it. Swab the bloodstain trying to avoid debris or other contaminants. Let the swab dry before placing in a swab box or container.

9.3.2 Liquid Blood Samples

If the wet blood is small, then it should be collected with sterilized cotton swabs and then allowed to air-dry. The stain can be collected by the procedure discussed for dried bloodstains. If the liquid bloodstain is large, then it may be collected by the following choices of collection procedures:

1. Absorb the liquid blood sample onto sterile cotton swabs. Allow the swatch to dry. The cotton swab must immediately be inserted into a swab box and sealed with evidence tape.
2. If the large wet bloodstain is located on a movable object (e.g., clothes or bed sheets), then wait until the bloodstain dries. Next, place paper in between layers of the clothing and collect the item carefully to avoid transfer or alteration of the bloodstain pattern. At the secure location, the object should be unwrapped and laid out for continued air-drying. If new packaging is required because the blood soaked the original packaging, the original packaging should be maintained as trace evidence and may be located on or within that original packaging. Sometimes the object, such as bloodstained clothing, must be cut to remove it from the scene; *do not* cut through the bloodstains as the pattern is often very important.

9.3.3 Seminal Stains

Semen stains found at the crime scene or on objects such as clothing are collected, packaged, and preserved in a similar way to bloodstains.

In sexual assault investigations, the crime scene investigation will also include the collection of evidence from the victim or suspect at the hospital. The victim of a sexual assault should be taken to the hospital for examination as soon as possible. A doctor assisted by a forensic nurse will usually conduct the examination and collect evidence. Forensic training for nurses is available nationwide and should be encouraged for hospital staff. Common training includes sexual assault nurse examiners (SANEs) training and certification. Commercially available or forensic-laboratory-prepared sexual assault collection kits are used for the collection of hospital specimens. These kits will include swabs, microscope slides, and various containers for the collection of a variety of evidence from the victim.

Recognition of a semen stain at a crime scene or on objects will often require examination with an alternate light source (ALS units) or the use of a presumptive screening test. ALS units have proven to be an effective method for the detection of dried semen stains. ALS does not work well on moist or wet semen stains. The most common ALS applications involve the use of a spectrum in the blue range, used in conjunction with an orange barrier filter. Alternatively, the use of an ALS unit with an ultraviolet wavelength, used with a yellow barrier filter, will also work to detect semen stains. Once the suspected stains are located, they should be thoroughly documented. With the use of ALS, specialized photographic methods are required. Once photographed, the entire object containing the stain should be seized and packed in a non-airtight container. If the stain is on a larger object that can be cut, such as a sofa cushion, then cut out the entire stain, making sure not to cut through any part of the pattern. As a last resort, as with dried bloodstains, absorb samples onto sterile swabs. However, because of the nature of seminal stains, minimize handling of suspected stains. The collected evidence should be placed in a primary container—swab box or druggist's fold—followed by an outer, secondary container that is not airtight. The container is sealed with evidence tape, marked appropriately, and preserved by refrigeration if possible. As with bloodstains, seminal stains may be stored at room temperature for a limited period until submission to the laboratory as soon as possible (Figure 9.6).

The use of an acid phosphatase presumptive test is fairly easy and applicable for both laboratory and crime scene purposes. The suspected stain is swabbed with a moistened swab, and a two-reagent process—alpha-naphthyl and Fast Blue B—is used. A positive presumptive test for semen yields a garnet-red color formation. In addition, there is an ABA card immunoassay that will detect P30, a male-specific seminal enzyme.

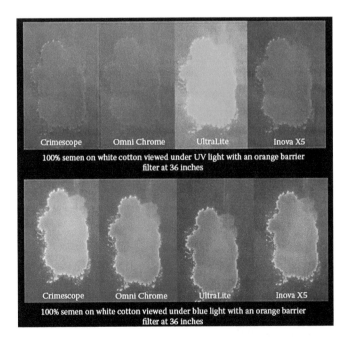

Figure 9.6 Semen stains viewed with various alternate light source (ALS) units.

9.3.4 Stains from Other Physiological Fluids

Other physiological fluid stains are collected, packaged, and preserved in a similar way to the methods and techniques described for blood and seminal stains. These stains may be composed of saliva, urine, perspiration, vaginal excretions, or fecal material. Liquid samples should be collected with the sterile cotton swab method. Dried stains can be swabbed with a moist sterile swab. Saliva stains or bite marks can be collected with a moistened swab. Swab the area with a single swab and concentrate it in a limited area. The saliva standard is collected by use of sterile swabs, air-drying, and the use of swab boxes. The saliva standard is best stored refrigerated but can be stored at room temperature for a limited period before submission to the forensic laboratory.

In addition to traditional methods listed next, messenger RNA (mRNA) is a powerful new tool for body fluid identification as well as an effective method for wound and bruising age estimation, post-mortem interval determination, and stain age estimation.

1. Saliva

   The presence of a starch-digesting enzyme called amylase is the basis of tests for saliva. Amylase is found in high concentrations in saliva, and the detection of this enzyme indicates the presence of saliva. An immunological test in the form of an ABA card test similar to the blood test is now available.

2. Urine

   The identification of urine is based on its characteristic color and odor, as well as the presence of characteristic chemical components, such as creatinine or urea.

3. Fecal material

   Fecal material usually has a characteristic color and odor. Chemical tests for the presence of urobilinogen are also conducted to identify feces.

4. Gastric fluid

   The identification of gastric fluid or stomach contents (vomit) occurs through chemical and microscopic analysis in addition to the detection of digestive enzyme activity.

5. Perspiration

   Perspiration is often present on clothing or other items that are submitted to the laboratory for examination. Its presence must be considered when analyzing evidentiary material for the presence of other body fluids. When testing articles of clothing, test a control sample to ensure that contamination from perspiration is not a factor. This is especially important if DNA analysis will be conducted.

6. Epithelial cells and tissues

   Microscopic examination is conducted of samples containing cellular materials. The morphological characteristics of epithelium and other tissues are used for their identification. Any nucleated cell has the potential for yielding DNA useful for analysis.

9.4 Blood Stain Pattern Analysis

Bloodstain patterns can reveal many facts about crimes—where, how, and when blood has been shed. Forensic scientists analyze the patterns of blood left behind at a crime scene in order to reconstruct the sequence of events; and the patterns of bloodstains can sometimes be as useful as the DNA results. Whereas DNA can tell us about the who of a crime, blood-stain patterns can tell us the where, what, how, and when.

By carefully documenting and examining bloodstain patterns at a crime scene, investigators are able to reconstruct what happened during the crime. This must be based on detailed measurements and analysis of blood patterns. As with all pattern evidence, blood patterns must be interpreted with extreme care. Blood pattern examiners must also conduct controlled laboratory experiments to verify the interpretations they have made of the blood patterns. The following are some examples of what blood patterns at a crime scene can inform us:

1. The approximate distance between the blood source and the surface upon which the blood landed (also known as the target surface).
2. The approximate energy needed to create the resulting blood droplets at the impact site.
3. The approximate direction the blood droplets were traveling in when they impacted the target surface.
4. The position of the victim(s)/witness.
5. The position of the suspect(s).
6. The direction in which a weapon or hand may have been swinging.
7. The direction in which blood may have trailed after impact.
8. Whether blood was then wiped or smeared by a person or object.
9. The estimated amount of time that has elapsed since the blood was deposited on the target surface.
10. The movement of the blood between focal points.
11. The type of injury that caused the blood deposits.
12. The approximate time lapse after the blood was deposited.
13. The sequence of events during a crime such as shooting, dragging, bleeding.
14. The sequence after the crime (such as cleaning, staging, alteration).

9.5 Crime Scene Reconstruction

Crime scene reconstruction is the process of determining the sequence of events, criminal activities, and logical predictions about what occurred during and after the crime. As such, reconstruction is a scientific fact-finding process. These five stages in the reconstruction process parallel the steps in the scientific method. Like any scientific method, reconstruction must concentrate on the "testing stage." Only after exhaustive testing can one have confidence in the reconstruction. Each step in the process should follow a logical analysis model. It involves the scientific analysis of a crime scene, the interpretation of crime scene pattern evidence, the laboratory examination of forensic evidence, and also a systematic study of related information and the formulation of a logical hypothesis.

Reconstruction is a combination of the inductive and deductive aspects of logic and the combination of science and art. The steps and stages of reconstruction closely follow a basic scientific method approach. It involves consideration and incorporation of all investigative information along with results of examination of physical evidence and interpretation into a reasonable explanation of the crime and related events. Logic, careful observation, and considerable experience, both in crime scene investigation and forensic examination, are necessary for correct interpretation, objective analysis, and, ultimately, reconstruction. The following are the five separate steps commonly used in the process of reconstruction:

1. Data collection

    This step requires the accumulation of all information obtained at the scene, from physical evidence, and from the statements of witness/victim. This includes the condition of the evidence, patterns and impressions, injury of the victim, and the relative positions of victim and evidence. Investigators should review and organize all of these pieces of information and put the puzzle of the crime together.

2. Conjecture

    Before making any detailed analysis of the evidence, investigators may infer a possible explanation of the events involved in a crime. However, it is important to note at this stage that this possible explanation does not become the only explanation being considered at this time. In many cases, there may be several possible explanations.

3. Hypothesis formation

    Further accumulation of data is based on the detailed examination of the physical evidence, the continuing investigation, and additional reports. Scene examination includes interpretation of blood and impression patterns, gunshot patterns, fingerprint evidence, and analysis of trace evidence. As these findings become clearer and their interrelationships emerge, it will lead to the formulation of an educated guess as to the probable course of events, that is, a hypothesis.

4. Testing

    Once a hypothesis is formulated, further testing must be done to confirm or disprove the overall interpretation or specific aspects of it. This stage includes comparisons of samples collected at the scene with known standards and alibi samples, as well as chemical, microscopic, and other analyses, and additional testing, as necessary. Some of this "testing" is the mental exercise of careful reexamination and evaluation of the evidence in terms of the hypothesis. At times, testing will require an experimental design to address specific information associated with the case. It is important to understand and account for the potential variables. An example of this would be to construct and experiment to determine how long it would take for a given volume of blood to dry on a specific substrate.

5. *Theory formation*

    Investigators may acquire additional information during the investigation about the condition of the victim or suspect, the activities of the individuals involved, the accuracy of witness accounts, and other information about the circumstances surrounding the events. Testing and confirming the hypothesis involves integrating all the verifiable investigative information, physical evidence analysis and interpretation, and the results of experiments. When it has been thoroughly tested and verified by analysis, the hypothesis can be considered a plausible theory. Complete reconstructions are often not possible; however, partial reconstruction, or reconstructing certain aspects of the events without necessarily being able to reconstruct all of them, can be extremely valuable. Information developed through reconstruction can often lead to the successful resolution of a case.

It is important to understand that crime scene reconstruction is very different from "reenactment," "re-creation," or "criminal profiling." Reenactment generally refers to having the victim, suspect, witness, or other individual reenact events, based on their knowledge and recollection of the crime. Re-creation, on the other hand, is to replace the necessary items or actions back at a crime scene through original scene documentation. And criminal profiling is an analysis based on statistical and psychological factors of the criminal. A core component of reconstructions is strict adherence to the scientific method.

9.6 Case Examples

9.6.1 Murder in Texas: Artificial Intelligence (AI) Resolves DNA Evidence Recovered from the Crime Scene

Shortly before midnight on December 10, 2010, patrons of a nightclub in Houston, Texas, watched in horror as a young man was beaten and stabbed to death in the parking lot. None of the witnesses knew the victim or the perpetrator, nor were they privy to the events that led up to the brutal murder. The Houston Police were called and, throughout the night and the next day, eyewitnesses provided video-recorded statements to investigators. Accounts from eyewitnesses varied in some ways, but were consistent in other respects. The victim (a Caucasian male in his twenties) had run into the parking lot and up the steps leading to the front door of the club. He begged to be permitted to enter and to be protected from someone who was trying to kill him. Employees of the establishment failed to understand the seriousness of the situation and refused to let him inside for sanctuary. The perpetrator appeared moments later. He was an African American male described as being 6′0″–6′6″ tall with an athletic build, weighing approximately 200–260 pounds, in his late 20s or early 30s, and wearing an orange shirt. The perpetrator and the victim scuffled in the parking lot; at one point during the struggle, the victim broke loose, but was ultimately caught and murdered. The perpetrator walked away into the night.

The following evening, Lydell Grant was parking his car in a parking lot adjacent to the club where the murder occurred, as he prepared to meet friends at a neighboring bar. One of the employees who had witnessed the previous night's murder saw him and believed he looked similar to the perpetrator. Indeed, Lydell Grant was an African American male who otherwise fit within the broad and general descriptions provided by eyewitnesses. The bar employee wrote down the vehicle identification number (VIN) of Grant's car, called Crime Stoppers to report the sighting, and eventually collected a cash reward for the information. Based on the Crime Stoppers tip, police investigators assembled a photo lineup of six African American males, including a photo of Grant. Over the next day and a half, all six eyewitnesses selected Grant's photo as that of the perpetrator. The lead detective was allowed to administer the identification process in a non-double-blind manner. That is, he was aware of which photo was the "suspect" as he displayed the spread, and he observed each eyewitness carefully as they were deciding on a choice. This approach permitted the detective to influence or cue the witness during deliberation and to reinforce their decision when the "correct" suspect was chosen. Six months later, the Texas legislature passed a law discouraging such an approach for police lineups, as these procedures have led to hundreds of wrongful convictions of innocent people.

A week after the photo lineup, Lydell Grant was arrested on a murder warrant. He professed his innocence and had a solid alibi for the night of the murder. During trial, a forensic DNA analyst who worked for the Houston Police Department Crime Laboratory testified that she had tested the victim's fingernails collected during autopsy in an effort to obtain the DNA profile of the perpetrator (i.e., from skin cells that may have transferred during the struggle in the parking lot). She testified that there was at least one unknown male donor to the DNA mixture (i.e., there was a DNA profile present in the mixture that was not the victim's profile). She first testified that it was "inconclusive" whether Grant could be excluded as the "unknown male donor" to the mixture. Upon further prodding by the prosecutor, she testified that this meant that Grant "could not be excluded" as the unknown male donor to the mixture. This testimony was misleading and, in conjunction with the eyewitness identification, was obviously damaging to Grant's case. In December 2012, he was convicted of capital murder based primarily on the testimony of six eyewitnesses who had identified him in the police photo spread. The jury completely disregarded Grant's alibi and he was sentenced to life imprisonment.

Post-conviction, Grant lost all of his appeals and eventually contacted the Innocence Project of Texas (IPTX) for help in pursuing his claim of actual innocence. In 2019, IPTX and the Henry C. Lee Institute of Forensic Science began reviewing the case. The team quickly noticed something was off with the DNA evidence. After studying the DNA mixture for a short period of time, it became obvious that a foreign male DNA profile was present underneath the murder victim's fingernails. That is, an autosomal Short Tandem Repeat (STR) profile was present in the DNA mixture that did not match the victim or Lydell Grant.

The original DNA testing conducted in 2011 involved typing of 15 autosomal STR loci (including the 13 core Combined DNA Index System (CODIS) loci) plus amelogenin, for a total of 16 loci tested. The official case report issued by the crime laboratory stated that "no conclusions will be made regarding Lydell Grant as a possible contributor to this mixture." Although inconclusive statements in DNA laboratory reports are often viewed as a conservative approach to avoid making erroneous inclusions or exclusions (which could lead to wrongful convictions or false acquittals, respectively), this case illustrates the need to further consider the implications of "inconclusive" statements regarding DNA mixtures. In this case, it was not difficult for a trained eye to recognize the presence of a foreign male contributor. Given that, as a matter of practice, DNA case reports are shuffled through several internal technical reviews prior to finalization and release, it's concerning that none of these review steps elicited discussion about the apparent exclusion of Grant from the mixture.

For simple presentation purposes (and not intended as a formal analysis), Figure 9.7 displays the DNA mixture data obtained from the murder victim's fingernails and identifies the foreign alleles present in the mixture that do not belong to the victim or to Lydell Grant. In total, there were 26 alleles present in the DNA mixture that are not consistent with the STR profiles of the victim or Grant.

Grant was eventually exonerated, on May 27, 2021, the prosecution dismissed the case.

9.6.2 Concetta "Penney" Serra Homicide

On Friday, July 16, 1973, in New Haven, Connecticut, 21-year-old Penney Serra drove her father's blue Buick into a parking garage on Temple Street at 12:42 P.M. and parked the car on the ninth level. As Serra made her way through the garage to get down to Chapel Street, witnesses later said that Serra was chased on foot by a tall, skinny man with long dark hair. At 1:00 P.M., an employee spotted the body of someone lying in the fetal position at the base of stairwell, on the tenth level.

STR locus	Lydell Grant	Victim	Alleles present in DNA mixture recovered from victim's				
			Allele 1	Allele 2	Allele 3	Allele 4	Allele 5
AMEL	XY	XY	X	Y			
CSF1PO	9,11	12,13	10	11	12	13	
D13S317	11,12	8,11	8	11	12	14	
D16S539	9,11	10,12	8	9	10	11	12
D18S51	14,17	11,16	11	15	16	18	
D19S433	14,15	14,15	12	13	14	15	
D21S11	28,32.2	30,32.2	28	29	30	31.2	
D2S1338	19,25	17,24	16	17	25		
D3S1358	15,15	15,18	14	15	16	17	18
D5S818	12,13	12,12	9	11	12	13	
D7S820	10,11	11,12	10	11			
D8S1179	12,15	13,14	12	13	14	15	
FGA	22,25	22,22	20	21	22	23	24
TH01	6,9	7,9.3	5	6	7	8	9.3
TPOX	6,11	10,11	8	9	10	11	
vWA	16,16	16,19	15	16	17	18	

Figure 9.7 Mixed DNA results from autosomal STR typing of fingernail scrapings from the murder victim. Twenty-six foreign alleles are present in the DNA mixture. Alleles highlighted in yellow and in red font do not match the profile of the victim or the defendant in this case. Probabilistic genotyping software, a form of artificial intelligence (AI), helped resolve this DNA mixture, excluding the wrongfully convicted man and identifying the true perpetrator through a search of the FBI's DNA database CODIS.

When the police arrived, they found Penney Serra dead. Her blue dress was covered in blood. Cuts and scrapes were found on Serra's wrist, finger, knee, and face, and an autopsy later revealed that Serra died of a small deep wound through her fifth and sixth ribs that penetrated the right ventricle of her heart. Serra's chest was full of blood. The medical examiner told police that it was impossible to determine the weapon, only that it was 3 inches in length. The medical examiner said that it took only a minute for Penney Serra to bleed to death. Police cornered off the crime scene and found her unlocked Buick parked at an erratic angle. Serra's car had blood on the outside door handle and door surfaces, on the steering shaft, and driver's side floor. Police also found Serra's purse, wallet with $14.75, her shoes, a parking stamp with an entry time of 12:42 P.M., and unopened envelopes containing invoices for a dental patient. Behind the driver's seat, police found more blood stains on the floor and on a Rite Aid tissue box. Police found a trail of blood drops and splatters leading away from the Buick toward the stairwell up to the ninth and tenth levels. On the seventh level, police found a set of keys. Near the keys, was a man's handkerchief covered in blood. Bloodstains were found on levels 7, 6, and 5. Police were able to determine that the assailant's blood type was type O.

From the car's license plates, police traced the car to Penney Serra's father, John Serra, and contacted him about his daughter. Police learned that Penney Serra had an on-off relationship with a man named Phil DiLieto. After DiLieto's alibi checked out, he was eliminated as a suspect. Police were then presented with another suspect. Anthony Golino's wife said that during one of their vicious arguments he wanted "to do her like Penney Serra." Golino's blood type was type O; however, after some speculation about whether Golino and Serra had a relationship, Golino was eliminated as a suspect. Police also interviewed Martin Cooratal. His dental bill was found on the dashboard of the Buick belonging to the Serras. Cooratal was spotted in the parking garage and matched the description of the suspect. However, like Phil DiLieto, Cooratal's alibi was solid. For nearly two decades, Penney Serra's murder had gone unsolved. John Serra was very critical about how the investigation was handled and put a lot of pressure on the state's attorney to find his daughter's killer, so prosecutors looked to Dr. Henry Lee to investigate.

Lee and his staff began to reexamine all of the physical evidence left at the crime scene. On September 10, 1989, Lee reconstructed the crime scene at the Temple Street garage. With the help of his staff, Lee reviewed documents, photographs, and diagrams. According to a witness, the attacker ran back to the ninth floor and started Penney Serra's Buick and drove down to the eighth floor and parked it at an extreme angle. The attacker then went down to the seventh floor, got into his own vehicle, and drove it out of the garage, leaving a blood-smeared ticket with the garage attendant at 1:01 P.M. Lee also examined latent fingerprints—prints not visible to the naked eye. Lee used chemicals to make these latent fingerprints visible. Still, it would take another five years to get a break in the case.

In 1994, Megan's Law in Connecticut was adopted. Once a person has committed a crime, it is likely they will repeat the same crime. Edward Grant of Waterbury beat his then fiancée so badly she was hospitalized. Grant was then booked and fingerprinted. Grant, who had been badly wounded in Vietnam and had a metal plate in his head, had been subject to violent mood swings. Grant's fingerprints were matched against latent fingerprints left on the bloody tissue box. Grant, who denied his involvement, offered no explanation as to how his prints were found at the crime scene. So, detectives offered him a chance to exonerate himself by asking for a sample of his blood. The sample was sent to Lee's laboratory. Not only was Grant's blood type O, but it was a 300 million-to-1 chance the blood belonged to Edward Grant. Sadly, by the time Edward Grant was arrested, John Serra had passed away.

At the trial, Lee testified about the crime scene reconstruction, latent fingerprints, and DNA blood evidence. Lee explained to the jury that the killer chased Penney Serra, who was running barefoot, up several levels of the garage into the stairwell where she ran into a dead end at the tenth floor. The killer stabbed Serra in the heart and she died on the steps. Lee explained the concept of a primary versus secondary crime scene. The murderer then ran down several levels, got into Serra's car, and left blood all over the car. The killer tried to stop some bleeding on his hand, a defensive wound probably inflicted by Serra, and grabbed the tissues in the back seat to soak up the blood. The killer then parked Serra's car on a sharp angle, got out, ran down another level and got into his own car and fled the scene leaving another bloody print on a parking ticket as he left. The fingerprints and DNA left at the scene matched Edward Grant. After a long trial, the jury found Edward Grant guilty of first-degree murder. Thanks to the advances in forensic science and Dr. Henry Lee, a pioneer in the field, Penney Serra's killer was finally brought to justice (Figures 9.8–9.11).

9.6.3 Brown's Chicken Murders

On January 9, 1993, seven bodies were found in the cooler areas of a Brown's Chicken Restaurant in Palatine, Illinois. On January 28, 1994, Dr. Henry Lee of the Connecticut State Police Forensic Science Laboratory was contacted by Chief Jerry Bratcher of the Palatine Police Department and Attorney Patrick W. O'Brien, special prosecutor in Cook County, Illinois, to review the evidence and to assist in case investigation. From February 10 through 15, 1994, Lee met the investigative team, visited the crime scene, reviewed documents and photographs, and reexamined several pieces of physical evidence. After reviewing documents and photographs, and examining physical evidence, the following observations were made. The conclusions in the case read as follows:

Figure 9.8 Evidence board of Penney Serra case.

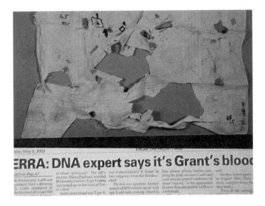

Figure 9.9 Edward Grant's DNA obtained from remnants of handkerchief.

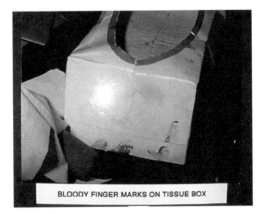

Figure 9.10 View of bloodstain on side of tissue box located in the victim's car.

1. The homicide scene was located at Brown's Chicken Restaurant, 168 W. Northwest Highway, Palatine, Illinois. All seven victims were shot and died as a result of multiple gunshot wounds. In addition, two of the victims also had knife wounds.
2. A large amount of cash receipts for the day were missing from the restaurant. The scene is consistent with a robbery–homicide type of case.

Figure 9.11 View of the bloody stain on the tissue box after enhancement.

3. Firearm, footwear imprint, fingerprint, and a variety of other physical evidence were collected from the restaurant by detectives of Palatine Police Department and forensic scientists from Northern Illinois Crime Laboratory. This evidence was submitted to the Northern Illinois Crime Laboratory for analysis.

4. Police evidence submission lists indicate that approximately 22 bullet/bullet fragments were submitted for examination. These bullet/bullet fragments were examined microscopically. Reports indicate that trace evidence was observed on some of the items.

5. The exact number of individual latent prints and palm prints developed during the examinations was not determined. All the latent prints developed at the scene and on items of evidence seized from the scene were examined by latent print examiners of the Northern Illinois Crime Laboratory.

6. The exact number of individual footwear imprints developed during the examinations was not determined. All the footwear imprints were examined by examiners of the Northern Illinois Crime Laboratory.

7. One of the items of physical evidence was collected from a garbage can in the west side of the restaurant. This item consists of a chicken meal from the restaurant, labeled as "0093-157-9, 01/08/93–12/04/93." This item was stored in a freezer at Northern Illinois Crime Laboratory.

8. This item was taken out of the freezer and examined macroscopically and microscopically by Lee at the Serology Section of the Northern Illinois Crime Laboratory. The following items were found in the evidence:
   a. Thirty-seven strips of French fries
   b. Two intact biscuits
   c. One intact chicken leg
   d. One intact chicken breast
   e. One portion of chicken
   f. One portion of chicken wing
   g. Five pieces of bone remains
   h. Paper napkins, plates, boxes, stirrers, cups, and a variety of other materials

The five pieces of chicken bone appeared to be the remains of a chicken meal eaten by an individual or individuals. It is more likely that the DNA of the person or persons who

Chicago Tribune
Final Edition
Seven massacred in Palatine
Police find bodies of 2 teenagers, 5 adults in restaurant's coolers

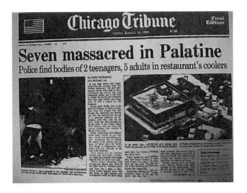

Figure 9.12 News clipping of the Brown's Chicken massacre in Palatine, Illinois.

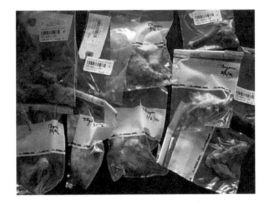

Figure 9.13 Chicken parts and bones located during search.

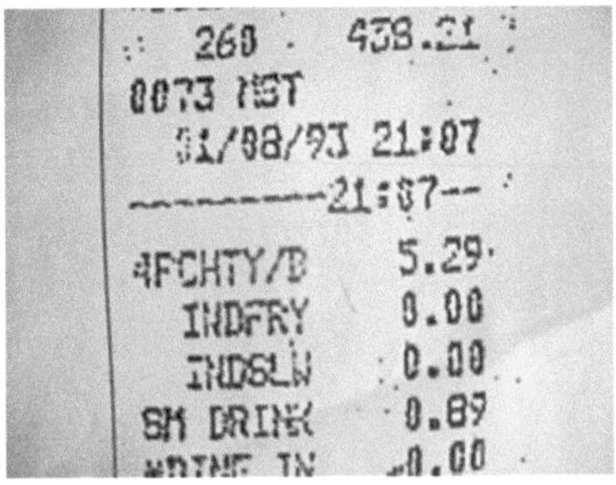
268 · 438.21
0073 MST
01/08/93 21:07
--------21:07--
4FCHTY/D 5.29
INDFRY 0.00
INDSLW 0.00
SM DRINK 0.89

Figure 9.14 Sales receipt showing date and time of purchase of meal consisting of fried chicken.

ate the chicken will transfer and be deposited on those chicken bones. Subsequently, Lee recommended that those pieces of chicken bone be submitted to Cellmark Laboratory for DNA testing.

1. The crime scene documentation, location, and condition of the chicken meal, sales receipts of the last meal, and time that this last chicken meal was sold, all indicate that the physical evidence found on this chicken meal is a very important and leads toward linking a suspect to this case.

Figure 9.15 Chicken bones had DNA mixture.

2. These observations and conclusions are based on a review of the submitted crime scene documentation, and the examination of the previously described items as physical evidence. This conclusion may be subject to change and/or modification based on submission of any new materials or information (Figures 9.12–9.15).The Palatine Police Department took the two suspects, Juan Luna and James Degorski, into custody on May 16, 2002. They were found guilty of all seven counts of murder and were imprisoned.

References

Carney, C., S. Whitney, J. Vaidyanathan, R. Persick, F. Noel, P. Vallone, E. Romos, et al. 2019. Developmental validation of the ANDE rapid DNA system with FlexPlex assay for arrestee and reference buccal swab processing and database searching. *Forensic Sci. Int. Genet.* 40:120–130.

Fisher, A.B., and D.R. Fisher. 2012. *Techniques of Crime Scene Investigation*. Boca Raton, FL: Taylor and Francis.

James, S.H., P.E. Kish, and T.P. Sutton. 2005. *Principles of Bloodstain Pattern Analysis*. Boca Raton, FL: CRC Press.

Lee, H.C., and H.A. Harris. 2000. *Physical Evidence in Forensic Science*. Tucson, AZ: Lawyers & Judges Publishing Company, Inc.

Lee, H.C., E.M. Pagliaro, and K. Ramsland. 2010. *The Real World of Forensic Scientist*. Amherst, NY: Prometheus Books.

Lee, H.C., T.M. Palmbach, and M.T. Miller. 2001. *Henry Lee's Crime Scene Handbook*. San Diego, CA: Elsevier Academic Press.

Marjanović, D., and D. Primorac. 2013. DNA variability and molecular markers in forensic genetics. In: *Forensic Genetics: Theory and Application*, eds. D. Marjanović, D. Primorac, L.L. Bilela, et al., 75–98. Sarajevo: Lelo Publishing (Croatian edition).

Primorac, D., and D. Marjanović. 2008. DNA analysis in forensic medicine and legal system. In: *DNA Analysis in Forensic Medicine and Legal System*, ed. D. Primorac, 1–59. Zagreb: Medicinska Naklada (Croatian edition).

Mass Disaster Victim Identification by DNA

10

MECHTHILD PRINZ, SHEILA ESTACIO
DENNIS, FREDERICK R. BIEBER,
AND ZORAN M. BUDIMLIJA

Contents

10.1 Mass Fatality Incidents

Mass fatality incidents are not rare occurrences; the International Disaster Database globally records several disasters every week. Natural disasters are the most common and claim the most lives; in 2017, a total of 335 natural disasters affected 95.6 million people and killed almost 10,000 people (Center for Research on the Epidemiology of Disasters, n.d., 2018). Other event categories causing mass fatalities are medical outbreaks, and man-made incidents such as industrial or transportation accidents, human rights violations, wars, and terrorist activity. Organizations define a mass fatality incident as any situation where there are more fatalities than can be properly recovered and identified using only local resources, which means the lower limit of the number of fatalities varies based on the resources at a specific site. The presence of biological, chemical, and/or radiological contamination complicates mass disaster response and has to be taken into consideration.

Managing a mass fatality incident (MFI) requires a multidisciplinary scientific and logistic approach, with the ultimate goal of identifying as many victims as possible. The procedures involved are generally known as the disaster victim identification (DVI) process. It has been shown that proper planning and emergency preparedness are essential for mass fatality response and DVI (Butcher and DePaolo, 2011). The US Centers for Disease Control (CDC) has issued capability standards including mass fatality management for preparedness planning from a public health perspective (Centers for Disease Control, 2019). Useful guidance on the DVI command structure and interactions of the different responding agencies and forensic disciplines can be found in the Interpol DVI guide (Interpol, 2018). A clear chain of command, centralized coordination, and a single leadership authority is critical for this complex process. Suggested components for hierarchy and response based on the Interpol standard DVI command structure are shown in Figure 10.1.

The Interpol command structure covers search and recovery, the process of locating and removing human remains, including bodies and body parts as well as the personal

Figure 10.1 Recommended DVI command structure based on Interpol (Interpol, 2018).

effects of the victims under scene processing. It is good policy to treat each and every mass fatality as a crime scene and to treat every recovered item of biological or nonbiological origin under forensic evidence standards regarding documentation, packaging, and chain of custody. The process has to be executed carefully, as it is easy to overlook fragmented remains when rescue and recovery operations occur simultaneously. Too often remains will be removed hurriedly by well-meaning rescue personnel in the search for living victims. Responders are encouraged to leave decedents in place unless they hamper rescue operations and to involve trained specialists such as forensic anthropologists and/or mortuary personnel to assist in the removal of bodies and remains.

10.2 Postmortem or Morgue Operations

Depending on the circumstances of the case, magnitude, and fatality type, the mortuary operation should have as its goal the identification of victims and determination of the cause and manner of death. If possible, a full autopsy should be performed to this effect. The scope of the identification effort and the resulting examination strategy depends on the MFI in question. An "open disaster" is defined as a mass fatality where the actual number of victims is unknown or unverified, such as a terrorist attack at a public gathering. A "closed disaster" is a catastrophic event affecting a fixed identifiable group, such as passengers in a fatal airline crash (Interpol, 2018). Open disasters require investigators to first establish an accurate actual victim list, but it may not be possible to obtain a truly certain number and all fragmented remains will need to be examined. For closed disasters, identification thresholds and sampling strategies may be adjusted based on the progress of identifications. Exploration and recovery at the World Trade Center (WTC) disaster site in New York City is shown in Figure 10.2.

Physical organization of the morgue location should logically translate to the flow of the tasks themselves to maximize the efficiency of movement. Stations should include,

Figure 10.2 Exploration and recovery at the World Trade Center (WTC) disaster site in New York City.

but not be limited to, the following: receiving, triage and cataloging, imaging, personal effects collection, station, fingerprinting, dental imaging, anthropological examinations, pathology procedures, tissue sampling for toxicology, histology and forensic biology analyses, storage, and release. The technology for digital imaging and electronic data sharing now allows for immediate on-site access and remote dental scheme and X-ray evaluation. It is of utmost importance that every single examination is properly documented and all bodies/samples are labeled with unique identifiers. Specimens taken from the same body need to be associated with the unique identification code of the source or primary case. Barcode labels, readers, scanners, and electronic tracking should be used if available. Good practice requires that remains be escorted from station to station, together with the documentation (postmortem charts) necessary for further examination. Postmortem charts and case files will be finalized once all scheduled and/or amended analyses are complete. Morgue operations need to include sufficient space for logistical support (communication, information technology, and supplies) and case reconciliation staff. After the identification process is completed and each single case is technically and administratively reviewed, the remains can be released to the next of kin. It is important to note that the identification process is a process and could be completed in a day, take years, or never be completed.

Although remains recovery and rescue operations should be concurrent, staging of fatality management personnel and equipment should occur in such a way that guarantees full access for life safety operations. Mass fatality incidents may include special conditions, such as chemical, biological, and/or radiological contamination. If the remains are contaminated, they must undergo decontamination by trained and certified personnel prior to further examination. Disaster management must protect the health and safety of assigned personnel and provide personal protection equipment and long-term health monitoring (Centers for Disease Control, 2019).

10.3 Antemortem or Family Assistance Center Operations

The family assistance center (FAC) provides families of possible victims with information that they may need in the days following the incident and facilitates the collection of personal information and biological references to aid in the identification. The FAC should be established quickly after an incident, ensuring easy access, security, and privacy for the families, and should include, among others, grief counselors and interpreters.

It is critical that family member data are recorded correctly from the very first conversation at the FAC, as the opportunity might not present itself again. Antemortem data drive the identification process, and without proper information to compare to recovered remains, the identification practice will fail. Data entry mistakes and confusion about the true nature of biological relationships between family members can interfere with and delay identifications using DNA. Recommendations for the FAC include avoid handwriting, optimize data entry forms, review all collected information while on site with family members present, and involve genetic specialists, if possible (National Institute of Justice, 2006; Donkervoort et al., 2008). Family interviews and data collection may need to be customized based on the incident. For example, dental record keeping varies in different countries and personal effects may be unavailable in natural disasters that destroyed victims' homes. The type of available antemortem data and samples will determine which identification modalities promise to be successful (Wright et al., 2015).

It is reasonable to expect that mass catastrophe incidents will be followed by a level of confusion and chaos. Thus, it is necessary to state that the local authorities are always in charge of an incident happening within their jurisdiction. Professional and general help from outside organizations including multiregional and international agencies may be needed. Family members need to be aware of how to obtain up-to-date information and who is in charge of the identification effort.

10.4 DNA-Driven Victim Identifications: Lessons Learned from the World Trade Center Remains Identification Project

On September 11, 2001, two hijacked airplanes struck the twin towers of the World Trade Center (WTC) in New York City, killing 2753 people. This mass disaster was challenging on many levels to all of the organizations involved; the management of 21,906 recovered human remains fell to the New York City Office of Chief Medical Examiner (NYC OCME). Two major goals in mass fatality investigation are determining the cause of death and identifying the decedents. Identification is necessary for three reasons: to provide closure for the families, to issue a death certificate enabling people to conduct the "business of death" such as life insurance and wills, and to reconstruct the incident for criminal investigation purposes. The sheer magnitude of the disaster on 9/11 is demonstrated in the statistics: as of September 2015, there were 1637 identifications for the 2753 people reported missing. Of these, 1003 were identified by a single means, of which DNA analysis was responsible for 889 of the victims (Office of Chief Medical Examiner, 2015). Since then six more individuals have been identified through DNA, creating a current status of 1643 (or 61%) of victims identified (Annese, 2019). DNA analysis, combined with anthropologic expertise, became the leading identification modality in this disaster where fragmentation was overwhelming

(Bieber, 2002; Hirsch and Shaler, 2002; Biesecker et al., 2005; National Institute of Justice, 2006).

Two of the WTC remains identification project's greatest challenges were the number of fragmented remains and the fact that the incident occurred in a publicly accessible area, without definite knowledge of who was present at the scene at the time of the attacks, meaning investigators dealt with an "open" victim manifest. The collapse of the towers produced 1.7 million tons of debris, including 0.5 million tons of steel (Figure 10.3). The excavation and debris removal strategy included transport of the debris to a vast open field on Staten Island's defunct Fresh Kills landfill where additional search operations and screening for smaller fragments of human remains were performed (Figure 10.4). A total of 16,000 individuals participated in this phase. Nevertheless, more remains were discovered years later during the ongoing demolition and construction of the area in and around the WTC incident (Dwyer, 2006). For each mass fatality, the lead agency must determine if the aim is to identify each victim or every remain. Since WTC was a high fragmentation event with an open manifest, every human remain was tested. This approach gives the best possibility of identifying all of the deceased (Budimlija et al., 2003; Marchi and Chastain, 2002).

The MFI victim identification process has always been a multidisciplinary team effort, one that consists of pathologists, anthropologists, forensic scientists, and law enforcement. Next to DNA, fingerprinting and forensic odontology are considered primary identification methods; medical findings and personal effects are counted as secondary methods (Interpol, 2018). But these traditional forensic identification strategies may not work for some mass fatalities. During the WTC project, environmental conditions, an open manifest, and disintegration of the remains dictated the course of action. When high fragmentation is evident, odontology and fingerprinting will only identify remains containing those particular anatomical parts. This leads to an increased importance of DNA analysis. Other scenarios where DVI becomes a DNA-driven process are efforts with only skeletal remains

Figure 10.3 View from above on WTC site in 2001.

Figure 10.4 Searching for fragmented remains at the Staten Island site in 2002.

like the mass graves in Bosnia, Herzegovina (Huffine et al., 2001), or countries that lack antemortem dental and fingerprint data.

Environmental conditions at the WTC site where fires continuously burned for months, exposed human remains to the elements of sunlight and heat, water, bacterial and insect activity, and mold. Depending on the location, remains like the ones shown in Figure 10.5 appeared more weathered and sun-exposed. All of these factors break down DNA and can interfere with DNA analysis, resulting in negative results for polymorphisms requiring larger DNA targets. Throughout the ongoing efforts of the WTC identification project to date, all of the recovered samples have been tested multiple times using new technologies

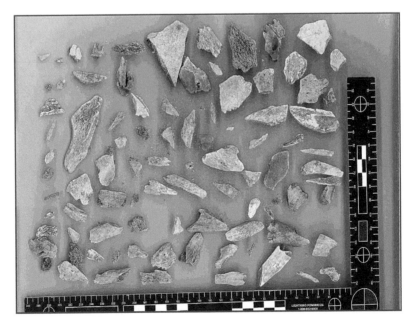

Figure 10.5 Fragments of human remains recovered from the World Trade Center (WTC) complex and buildings in its vicinity.

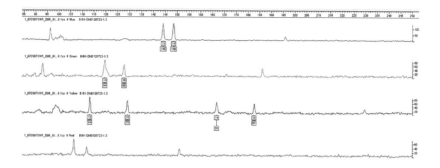

Figure 10.6 Partial profile of the sample using regular short tandem repeats (STRs).

that were developed, validated, and approved for use in human identification. In addition to traditional short tandem repeat (STR) DNA panels, mini-STRs, single-nucleotide polymorphisms, and mitochondrial DNA analysis were used, adding more combined results and making possible additional fragment-to-fragment links or new identifications. DVI projects involving multiple missing relatives and/or DNA degradation will benefit from using different DNA modalities in selected cases (Budowle et al., 2005; Prinz et al., 2007). Figures 10.6 and 10.7 show an example of the information gained by testing mini-STRs or size-reduced STR targets (Wiegand and Kleiber, 2001; Butler et al., 2003).

Research efforts triggered by the WTC victim identification challenge and other humanitarian missing persons efforts have led to improved and higher throughput sample collection and DNA extraction for soft tissue and skeletal remains (Holland et al., 2003; Loreille et al., 2007; Allen Hall et al., 2013; Mundorff et al., 2018). DNA strategies are diverging in two different directions. New methods like Rapid DNA and direct PCR allow for faster STR results. Massive parallel sequencing (MPS) is slower and more expensive but

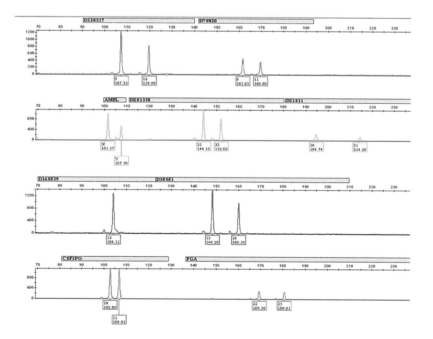

Figure 10.7 Additional results for the same sample using mini-STRs.

offers simultaneous detection of a large number of different types of polymorphisms and phenotypic markers (Watherstone et al., 2018). Target enrichment and capture strategies have been shown to improve MPS results for extremely degraded DNA, making this the method of choice for ancient and severely compromised remains (Templeton et al., 2013; Shih et al., 2018). Antemortem data for DNA identification through direct comparisons are generated by typing personal effects presumably containing the victim's DNA like tooth-brushes, razors, hairbrushes, and clothing, or medical specimens using the same target polymorphisms. It is of the utmost importance that collaborative efforts where more than one agency is involved in DNA typing agree on the target polymorphisms and a central database. Database searching needs to accommodate partial profiles and set appropriate identification thresholds (Leclair et al., 2006).

If the victim cannot be identified via direct match to a personal effect, formal kinship analysis is in an order requiring statistical expertise and computer software designed to perform likelihood ratio analysis. Kinship analysis requires DNA samples from the vic-tim's biological relatives, such as parents, siblings, and children. The closer the biological relationship between the family member and the victim, the more useful the sample will be and appropriate counseling of the family at the FAC is key. More than 17,000 reference samples were collected following the WTC disaster. During the WTC project, some true biological relationships were not determined until after DNA profiling was performed on multiple family members of the same victim.

Statistical calculations were performed on all DNA profiles to make valid compari-sons. The chance of two random people having the same profile is calculated to give assur-ance that the right identification was made. Based on the number of victims, statistics for the threshold of direct matching were finalized at 1 in 200,000,000 for females and 1 in 2,000,000,000 for males (Biesecker et al., 2005). For "closed manifest" incidents, the prior odds can be raised and the identification threshold lowered with each reliably iden-tified victim removed from the pool of the missing (Brenner and Weir, 2003; Budowle et al., 2011). For the WTC effort, the different thresholds for male and female identifica-tions were chosen because there were more male victims. The Office of Chief Medical Examiner team recruited a group of scientific advisers, the Kinship and Data Analysis Panel (KADAP), composed of a group of DNA and statistical experts who made recom-mendations on testing performed and statistical thresholds for identification based on direct or kinship matches (Biesecker et al., 2005; National Institute of Justice, 2006). One remaining open issue was how to combine data from different modalities, where each modality alone would have been insufficient. This complicated issue is still under discus-sion (Wright et al., 2018).

Even with a single modality of victim identification, including DNA, the victim's family is not notified. A formal reconciliation process or administrative review ensures a victim's file does not contain contradictory information, all reference samples have a legiti-mate source, and, if applicable, different identification modalities confirm one another. For fragmented remains, this reconciliation includes the review of all skeletal elements attributed to one set of remains by a forensic anthropologist (De Boer et al., 2018). During the WTC identification effort, the forensic anthropologists not only performed a final anthropological review of each proposed identification, they also proactively reviewed pre-viously examined remains for mismatched skeletal elements and were part of the intake and screening process of all recovered remains (Mundorff et al., 2014). Because of the enormous forces created during the collapse of the 41,514 m high towers (reduced to a 21

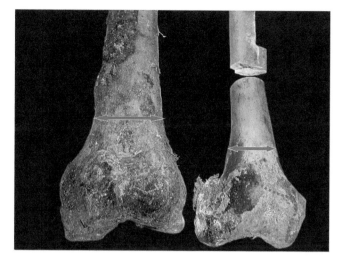

Figure 10.8 Two same-side femurs with the identical DNA profile (conflict).

m high hill dispersed over 6.47 ha [64,700 m^2]), many fragments were crushed together causing commingled or combined remains and anthropological-DNA conflict during the identification process. As a result of the reexamination of remains, 239 "new" cases were created. The final anthropological review prevented misidentifications under the scenario shown in Figures 10.8 and 10.9b. The DNA profile of the soft tissue attached to a bone matched another skeletal element that was either a duplicate or inconsistent with being from the same individual (Budimlija et al., 2003; Mundorff et al., 2014). Another situation where commingling caused conflicting DNA results occurred when both bone and adherent muscle were typed on the same body fragment (Figure 10.9a).

10.5 Reconciliation and Conclusions

Disaster victim identification is a team effort and communication between all stakeholders and forensic experts is critical. Prior to closing an identification case and notifying next of kin, a reconciliation step must ensure that all records associated with the victim in question are complete and information is consistent. Centralized coordination and effective data handling and sharing are essential. Based on WTC identification experiences, the NYC OCME developed a database system to store antemortem and postmortem data, and created a "call center" wherein trained operators take a brief missing persons report from any person believing their loved one was involved in the disaster. This enabled the agency to more quickly develop a missing victim list and request DNA samples and antemortem data from the families of the reported missing. Access to similar systems should be part of any DVI preparedness plan. Multiagency disaster response drills are also essential to make sure communication channels are up to date and should involve all aspects of the mass fatality incident response including the family assistance center, victim identification plans, and morgue operations.

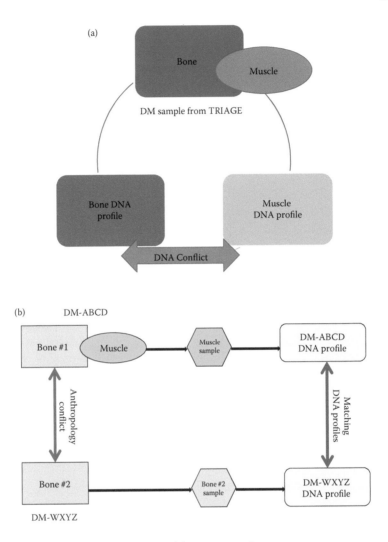

Figure 10.9 Types of DNA conflicts caused by commingling.

References

Allen-Hall, A., and D. McNevin. 2013. Non-cryogenic forensic tissue preservation in the field: A review. *Aust J Forensic Sci* 45(4): 450–460.

Annese, J. 2019. DNA helps identify another World Trade Center victim, nearly 18 years after 9/11 terror attack. Daily News, June 10. https://www.nydailynews.com/new-york/ny-world-trade -center-victim-remains-identified-18-years-after-attacks-20190610-b2rmidygqfhe5ej27p6 6x4xq2a-story.html (accessed June 12, 2019).

Bieber, F.R. 2002. Reckoning with the dead. *Harv Med Alumni Bull* 76(2): 34–37.

Biesecker, L.G., J.E. Bailey-Wilson, J. Ballantyne, et al. 2005. Epidemiology. DNA identifications after the 9/11 World Trade Center attack. *Science* 310(5751): 1122–1123.

Brenner, C.H., and B.S. Weir. 2003. Issues and strategies in the identification of World Trade Center victims. *Theor Popul Biol* 63(3): 173–178.

Budimlija, Z.M., M. Prinz, A. Zelson-Mundorff, et al. 2003. World Trade Center human identification project: Experiences with individual body identification cases. *Croat Med J* 44(3): 259–263.

Budowle, B., F.R. Bieber, and A.J. Eisenberg. 2005. Forensic aspects of mass disasters: Strategic considerations for DNA based human identification. *Leg Med (Tokyo)* 7(4): 230–243.

Budowle, B., J. Ge, R. Chakraborty, and H. Gill-King. 2011. Use of prior odds for missing persons identifications. *Investig Genet* 2(1): 15.

Butcher, B.A., and F. DePaolo. 2011. Mass fatality management. In: *Health Care Emergency Management—Principles and Practice*, eds. M. J. Reilly and D. S. Markenson, 423–445. Sudbury, MA: Jones and Bartlett Learning.

Butler, J.M., Y. Shen, and B.R. McCord. 2003. The development of reduced size STR amplicons as tools for analysis of degraded DNA. *J Forensic Sci* 48(5): 1054–1064.

Centers for Disease Control and Prevention. 2019. Public Health Preparedness and Response Capabilities. Atlanta, GA: US Department of Health and Human Services. Also available online at: https://www.cdc.gov/cpr/readiness/capabilities.htm (accessed June 6, 2019).

Center for Research on the Epidemiology of Disasters. 2018. *J Nat Disasters 2017*. Brussels: CRED. Also available online at: https://cred.be/sites/default/files/adsr_2017.pdf (accessed June 6, 2019).

Center for Research on the Epidemiology of Disasters. n.d. EM-DAT the International Disaster Database. Also available online at: https://www.emdat.be/ (accessed June 6, 2019).

de Boer, H.H., S. Blau, T. Delabarde, and L. Hackman. 2018. The role of forensic anthropology in disaster victim identification (DVI): Recent developments and future prospects. *Forensic Sci Res*. https://doi.org/10.1080/20961790.2018.1480460.

Donkervoort, S., S.M. Dolan, M. Beckwith, T.P. Northrup, and A. Sozer. 2008. Enhancing accurate data collection in mass fatality kinship identifications: Lessons learned from Hurricane Katrina. *Forensic Sci Int Genet* 2(4): 354–362.

Dwyer, J. 2006. Pieces of bone are found on building at 9/11 site. New York Times, April 6. Also available online at: https://www.nytimes.com/2006/04/06/nyregion/pieces-of-bone-are -found-on-building-at-911-site.html (accessed June 12, 2019).

Hirsch, C.S. and R. Shaler. 2002. 9/11 through the eyes of a medical examiner. *J Investig Med* 50(1): 1–3.

Holland, M.M., C.A. Cave, C.A. Holland, and T.W. Bille. 2003. Development of a quality, high throughput DNA analysis procedure for skeletal samples to assist with the identification of victims from the World Trade Center attacks. *Croat Med J* 44(3): 264–272.

Huffine, E., J. Crews, B. Kennedy, K. Bomberger, and A. Zinbo. 2001. Mass identification of persons missing from the break-up of the former Yugoslavia: Structure, function, and role of the International Commission on Missing Persons. *Croat Med J* 42(3): 271–275.

International Police Organization. 2018. *Interpol Disaster Victim Identification Guide*. Lyon, France: Interpol.

Leclair, B., R. Shaler, G.R. Carmody, et al. 2006. Bioinformatics and human identification in mass fatality incidents: The World Trade Center disaster. *J Forensic Sci* 52(4): 806–819.

Loreille, O.M., T.M. Diegoli, J.A. Irwin, M.D. Coble, and T.J. Parsons. 2007. High efficiency DNA extraction from bone by total demineralization. *Forensic Sci Int Genet* 1(2): 191–195.

Marchi, E. and T. Chastain. 2002. The sequence of structural events that challenged the forensic effort of the World Trade Center disaster. *Am Lab* 34: 13–17.

Mundorff, A.Z., S. Amory, R. Huel, A. Bilić, A.L. Scott, and T.J. Parsons. 2018. An economical and efficient method for postmortem DNA sampling in mass fatalities. *Forensic Sci Int Genet* 36: 167–175.

Mundorff, A.Z., R. Shaler, E. Bieschke, and E. Mar-Cash. 2014. Marrying anthropology and DNA: Essential for solving complex commingling problems in cases of extreme fragmentation. In: *Commingled Human Remains: Methods in Recovery and Identification*, eds. B. J. Adams and J. E. Byrd, 285–299. London: Academic Press.

National Institute of Justice. 2006. Lessons Learned from 9/11: DNA Identification in Mass Fatality Incidents. Washington, DC: Office of Justice Programs. Also available online at: https://www .ncjrs.gov/pdffiles1/nij/214781.pdf (accessed June 6, 2019).

Office of Chief Medical Examiner. 2015. World Trade Center Operational Statistics. Also available online at: http://www.nyc.gov/html/ocme/downloads/pdf/public_affairs_ocme_pr_WTC _Operational_Statistics.pdf (accessed 8 June 2019).

Prinz, M., A. Carracedo, W.R. Mayr, et al. 2007. DNA Commission of the International Society for Forensic Genetics (ISFG): Recommendations regarding the role of forensic genetics for disaster victim identification (DVI). *Forensic Sci Int Genet* 1(1): 3–12.

Shih, S.Y., N. Bose, A.B.R. Gonçalves, H.A. Erlich, and C.D. Calloway. 2018. Applications of probe capture enrichment next generation sequencing for whole mitochondrial genome and 426 nuclear SNPs for forensically challenging samples. *Genes* 9(1): e49.

Templeton, J.E.L., P.M. Brotherton, B. Llamas, et al. 2013. DNA capture and next-generation sequencing can recover whole mitochondrial genomes from highly degraded samples for human identification. *Investig Genet* 4(1): 26.

Watherston, J., D. McNevin, M.E. Gahan, D. Bruce, and J. Ward. 2018. Current and emerging tools for the recovery of genetic information from post mortem samples: New directions for disaster victim identification. *Forensic Sci Int Genet* 37: 270–282.

Wiegand, P. and M. Kleiber. 2001. Less is more—Length reduction of STR amplicons using redesigned primers. *Int J Legal Med* 114(4–5): 285–287.

Wright, K., A. Mundorff, J. Chaseling, A. Forrest, C. Maguire, and D.I. Crane. 2015. A new disaster victim identification management strategy targeting "near identification-threshold" cases: Experiences from the Boxing Day tsunami. *Forensic Sci Int* 250: 91–97.

Wright, K., A. Mundorff, J. Chaseling, C. Maguire, and D.I. Crane. 2018. Identifying child victims of the South-East Asia Tsunami in Thailand. *Disaster Prev. Manag* 24(4):447–455.

Bioterrorism and Microbial Forensics

11

ALEMKA MARKOTIĆ, IVAN-CHRISTIAN
KUROLT, LIDIJA CVETKO KRAJINOVIĆ,
JOSIP CRNJAC, AND JAMES W. LE DUC

Contents

11.1 Definitions

Bioterrorism refers to the release of microorganisms or their toxins with the intent to harm, sicken or kill humans, animals, or agricultural crops to further personal, political, or social objectives (CDC 2013).

Microbial forensics is a relatively new scientific discipline used to analyze possible bioterrorism attacks, or inadvertent microorganism or toxin releases (Breeze et al. 2011).

11.2 History of Bioterrorism and Biological Warfare

Throughout history, there are examples of humans using biological organisms as weapons to sicken or kill perceived enemies either as acts of war by organized government states, or acts of terrorism by non-state-affiliated organizations or individuals. As early as in 6th century BC, Assyrians poisoned the wells of their enemies with rye ergot, and Solon of Athens poisoned the water supply with hellebore (skunk cabbage), an herb purgative, during the siege of Krissa. In a naval battle against King Eumenes of Pergamon in 184 BC, Hannibal's forces hurled pots filled with snakes upon the enemy decks and won the battle (Philips 2005). In the 14th century, during the siege of Kaffa, the Tartar army catapulted plague-ridden corpses over the walls of the city and forced the defenders to surrender. During the French and Indian War (1767), the English general Sir Jeffrey Amherst gave blankets contaminated with smallpox to Indians loyal to the French. The resulting epidemic decimated the tribes, and the British successfully attacked Fort Carillon (Breeze et al. 2011; Philips 2005). During WWI the Germans attempted to spread cholera in Italy, plague in St. Petersburg, and dropped bombs with pathogenic microorganisms over

Britain. A German-American, Dr. Anton Dilger, supported by the German government, had grown cultures of *Bacillus anthracis* and *Burkholderia mallei* in his Washington, DC, home. These agents were spread by dockworkers in Baltimore in an attempt to infect 3000 head of horses, mules, and cattle destined for the Allied troops in Europe. As a result, an estimated several hundred troops were affected (Philips 2005).

On June 17, 1925, the Geneva Protocol banned biological weapons. It was the first multilateral agreement that extended the prohibition of chemical agents to biological agents. In spite of the Geneva Protocol, during WWII, the Japanese released plague bacteria at Chuhsien resulting in the deaths of 21 people (1940), and in 1941 the British experimented with anthrax at the Scottish coast (Philips 2005; Kasten 2002). More recently, in 1972, members of the right-wing "Order of the Rising Sun" possessed 30–40 kg of typhoid cultures, which they planned to use to poison the water supply in Chicago, St. Louis, and other Midwest cities in the US (Philips 2005). The same year, the Biological Weapons Convention (aka Convention on the Prohibition of the Development, Production, and Stockpiling of Bacteriological and Toxin Weapons and on their Destruction) was signed by 103 nations. Like the Geneva Protocol, the Biological Weapons Convention was similarly ignored when in 1978 the Bulgarian dissident and writer Georgi Markov, then living in London, was injected with ricin by a specially constructed umbrella and died in several days (Breeze et al. 2011; Philips 2005). In 1995 three briefcases designed to release botulinum were discovered in a Tokyo subway. In Zadar, Croatia in 1999, a perpetrator tried to rob a currency exchange office by using a syringe allegedly containing HIV as a weapon (Philips 2005). In 2001 anthrax letters were sent to several news media and US senators (CDC 2001; Jernigan et al. 2001; Atlas 2002; Greene et al. 2002; Tan et al. 2002; Hoffmaster et al. 2002). There are many other examples and some of them will be described later in the suspicious infectious diseases outbreaks sections.

The threat of a bioterrorist attack has been the topic of international concern. Numerous US National Intelligence Estimates and reports as well as have examined the issues (Breeze et al. 2011).

In December 2008 in accordance with the Implementing Recommendations of the 9/11 Commission Act of 2007 (P.L. 110-53), the Commission on the Prevention of Weapons of Mass Destruction Proliferation and Terrorism submitted its report, "World at Risk" (2009). That report assessed the nation's activities, initiatives, and programs to prevent weapons of mass destruction proliferation and terrorism, and provided concrete recommendations to address these threats. The report provided the following threat assessment that was unanimously expressed:

> Unless the world community acts decisively and with great urgency, it is more likely than not that a weapon of mass destruction (WMD) will be used in a terrorist attack somewhere in the world by the end of 2013. That weapon is more likely to be biological than nuclear.

In the Update on the Recommendations of the Commission on the Prevention of Weapons of Mass Destruction Proliferation and Terrorism, The Aspen Institute Homeland Security Group's WMD Working Group (November 15, 2012), it is underlined that

> rapid detection and diagnosis, adequate supplies of medical countermeasures and the means to rapidly dispense them, and surge medical capacity are among the critical elements required for effective response. While bioattacks cannot be entirely prevented, proper response can

prevent an attack from becoming a catastrophe. The long-range strategy is to develop protective and response capabilities that would minimize the effect of a bioattack and thus remove bioweapons from the category of WMD.

11.3 Classification of Specific Bioterrorism Agents

The US Centers for Disease Control and Prevention (CDC) has classified microbial agents that could be potentially used in bioterrorism attacks in three categories (2013):

A category. Anthrax (*B. anthracis*), botulism (*Clostridium botulinum* toxin), plague (*Yersinia pestis*), smallpox (variola major), tularemia (*Francisella tularensis*), viral hemorrhagic fevers (filoviruses [e.g., Ebola, Marburg] and arenaviruses [e.g., Lassa, Machupo]). Biological agents that can be easily disseminated or transmitted from person to person, result in high mortality rates and have the potential for major public health impact, might cause public panic and social disruption, and require special action for public health preparedness.

B category. Brucellosis (*Brucella* species), epsilon toxin of *Clostridium perfringens*, food safety threats (e.g., *Salmonella* species, *Escherichia coli* O157:H7, *Shigella*), glanders (*Burkholderia mallei*), melioidosis (*Burkholderia pseudomallei*), psittacosis (*Chlamydophila psittaci*), Q fever (*Coxiella burnetii*), ricin toxin from *Ricinus communis* (castor beans), staphylococcal enterotoxin B, typhus fever (*Rickettsia prowazekii*), viral encephalitis (alphaviruses [e.g., Venezuelan equine encephalitis, eastern equine encephalitis, western equine encephalitis]), water safety threats (e.g., *Vibrio cholerae*, *Cryptosporidium parvum*). Biological agents, which are moderately easy to disseminate, result in moderate morbidity rates and low mortality rates, and require specific enhancements of diagnostic capacity and enhanced disease surveillance.

C category. Emerging infectious diseases (such as Nipah virus and hantavirus). Biological agents including emerging pathogens that could be engineered for mass dissemination in the future because of availability, ease of production and dissemination, and potential for high morbidity and mortality rates and major health impact.

Some basic facts about incubation periods, transmission modes, general signs and symptoms, treatment, vaccines for specific bioterrorism agents, and a list of samples that can be utilized for the detection of specific microbial agents are presented in Table 11.1.

11.4 Microbial Forensic Protocols and Practices

Continuing rapid developments in the life sciences offer the promise of providing tools to meet global challenges in health, agriculture, the environment, and social and economic development. However, such advances also bring with them new social, ethical, legal, and security challenges and reflect continuing concerns that the knowledge, tools, and techniques resulting from life sciences research could also enable the development of

Table 11.1 **Basic Facts about Specific Bioterrorism Agents**

A Category

Anthrax (*Bacillus anthracis*)

Incubation	Few hours to 7 days
Transmission mode	Zoonotic infection; cattle or other livestock infected by spores from soil pass infection to humans; person-to-person transmission very rare; articles and soil contaminated with soil, remain infective for decades
Basic signs/ symptoms	Cutaneous: swelling, ulcer formation, black eschar; heals spontaneously in 1–2 weeks (80% cases)
	Pulmonary: edema factor (EF), protective antigen (PA), and lethal factor (LF) cause swelling and necrosis in mediastinum and lungs, septicemia, and death in 2–5 days
	Gastrointestinal: lethal swelling in the intestine
Treatment	Penicillin or ciprofloxacin
Vaccine	Vaccine is protective against invasive disease and recommended only for high-risk populations
Specimens to be collected	Swabs, whole blood, fluids (pleural, bronchial, CSF), fresh tissue, blood clot, serum, citrated plasma, stool

Botulism (*Clostridium botulinum* toxin)

Incubation	12–36 hours
Transmission mode	A spore-forming obligate anaerobic bacillus produces toxin in improperly processed, canned, low-acid, or alkaline foods, and in pasteurized and shortly boiled or cooked food; spores are common in the soil and can contaminate food including honey.
Basic signs/ symptoms	Toxin blocks acetylcholine release and interferes with the transmission of signals for muscle contraction (painful flaccid paralysis; swallowing, vision, and speech are primarily affected, then breathing with case-fatality rate up to 10%)
Treatment	Amoxicillin + polyvalent serum
Vaccine	Pentavalent (ABCDE) toxoid, 3-dose efficacy 100% against 25–250 LD50 in primates
Specimens to be collected	Feces, enema, gastric aspirate or vomitus, serum, tissue or exudates, food specimens, swab samples (environmental or clinical), soil, water

Plague (*Yersinia pestis*)

Incubation	1–7 days
Transmission mode	"Black Death" of the Middle Ages; result of human contact with sylvatic rodents, but also infected domestic pets and their fleas (*Xenopsylla cheopis*); possible person-to-person transmission by *Pulex irritans* fleas; in secondary pneumonic plague respiratory droplets are the source of person-to person transmission; devastating epidemic
Basic signs/ symptoms	Bubonic: inflamed, swollen, and tender lymph nodes (inguinal area 90%); fever; may progress to septicemic plague with bloodstream dissemination with 50%–60% case-fatality in non-treated patients
	Pneumonic: secondary involvement of lungs (due to bloodstream dissemination); pneumonia with mediastinitis or pleural effusion, after interhuman transmission, primary pneumonic or pharyngeal plague
Treatment	Gentamicin, doxycycline
Vaccine	Suspension of killed (formalin-inactivated) *Yersinia pestis*; has yet to be measured precisely in controlled studies; at least 2 vaccinated persons contracted pneumonic plague following *Y. pestis* exposure
Specimens to be collected	Lower respiratory tract (pneumonic): bronchial wash or transtracheal aspirate (≥1 ml); sputum; blood; aspirate of involved tissue (bubonic) or biopsied specimen; liver, spleen, bone marrow, lung.

Smallpox (variola major)

Incubation	7–17 days

(Continued)

Table 11.1 (Continued)   Basic Facts about Specific Bioterrorism Agents

Transmission mode	Person-to-person transmission by infected saliva droplets that expose a susceptible person having face-to-face contact with the ill person; patients with smallpox are most infectious during the first week of illness; some risk of transmission lasts until all scabs have fallen off
Basic signs/ symptoms	Initial symptoms: high fever, fatigue, head and backaches; characteristic rash follows in 2–3 days (most prominent on the face, arms, and legs), which starts with flat red lesions, become pus-filled, and begin to crust early in the second week with scabs developing, and then separating and falling off after about 3–4 weeks. Death occurs in up to 30% of cases
Treatment	Supportive therapy
Vaccine	Two calf-lymph-derived vaccines, Dryvax® (Wyeth Laboratories Inc.; Marietta, Pennsylvania), and Aventis Pasteur vaccine (Swiftwater, Pennsylvania); two newly developed vaccines from Acambis/Baxter Pharmaceuticals (Cambridge, Massachusetts), ACAM1000 (grown in human embryonic lung cell culture [MRC-5]) and ACAM2000 (grown in African green monkey cells [VERO cells)])
Specimens to be collected	Vesicular material, scab specimens, biopsy lesions, serum

Tularemia (*Francisella tularensis*)

Incubation	1–14 days
Transmission mode	Bite of arthropods (ticks, flies, and mosquitoes); handling or ingesting insufficiently cooked meat; drinking contaminated water; by inhalation of dust from contaminated soil, grain or hay.
Basic signs/ symptoms	Skin or mouth, swollen and painful lymph glands, swollen and painful eyes, and a sore throat with dry cough and progressive weakness.
Treatment	Streptomycin, gentamicin, doxycycline
Vaccine	Live attenuated vaccine; 80% protection against 1–10 LD50.
Specimens to be collected	Blood samples, conjunctival swabs, environmental samples (water, mud, etc.)

Viral Hemorrhagic Fevers Filoviruses (e.g., Ebola, Marburg)

Incubation	Ebola virus: 2–21 days Marburg virus: 3–9 days
Transmission mode	Person-to person transmission; direct contact with infected blood, secretions, organs, and semen; frequent nosocomial infections; reservoir not known
Basic signs/ symptoms	Life-threatening multiorgan infection with fever, CNS involvement, bleeding from multiple sites, mostly mucous membranes with mortality of 50%–80%.
Treatment	Supportive therapy, i.v. ribavirin may tentatively reduce mortality if given before day 7 after the onset
Vaccine	Vaccines were 100% effective in protecting a group of monkeys from the disease; based on either a recombinant Vesicular stomatitis virus or a recombinant Adenovirus carrying the Ebola spike protein on its surface; early human vaccine efforts have so far not reported any successes
Specimens to be collected	Antemortem: biopsy material of the lung or bone marrow aspirate or clot; postmortem: spleen, lung, kidney, liver, lymph nodes, heart, pancreas, pituitary, brain, or liver tissue, or heart blood.

Viral Hemorrhagic Fever Arenaviruses (e.g., Lassa, Machupo)

Incubation	6–21 days
Transmission mode	The reservoir/host of Lassa virus—rodent of the genus *Mastomys*; rodents shed the virus in urine and droppings, and the virus can be transmitted through direct contact with these materials, through touching objects or eating food contaminated with these materials, or through cuts or sores; person-to person transmission—direct contact with infected blood, secretions, organs, and semen; frequent nosocomial infections.

(Continued)

Table 11.1 (Continued) Basic Facts about Specific Bioterrorism Agents

Basic signs/ symptoms	Life-threatening multiorgan infection, with fever, retrosternal pain (pain behind the chest wall), sore throat, back pain, cough, abdominal pain, vomiting, diarrhea, conjunctivitis, facial swelling, proteinuria, and mucosal bleeding; neurological problems (hearing loss, tremors, and encephalitis), with mortality of 50%–80%.
Treatment	Supportive therapy, i.v. ribavirin may tentatively reduce mortality if given before day 7 after the onset
Vaccine	A live attenuated vaccine for Lassa fever made by reassortment of Lassa and Mopeia viruses; human vaccine efforts have so far not reported any successes
Specimens to be collected	Antemortem: biopsy material of the lung or bone marrow aspirate or clot. Postmortem: spleen, lung, kidney, liver, lymph nodes, heart, pancreas, pituitary, brain, or liver tissue, or heart blood.
B Category	
Brucellosis (*Brucella* species)	
Incubation	5–60 days
Transmission mode	Contact with tissues, blood, urine, vaginal discharges, aborted fetuses and especially placentas (through breaks in skins); ingestion of raw, contaminated unpasteurized milk or cheese from infected animals; no person-to person transmission.
Basic signs/ symptoms	The symptoms are nonspecific and systemic, with fever, sweats, headache, anorexia, back pain, and weight loss being frequent; the chronic form of the disease can mimic miliary tuberculosis with suppurative lesions in the liver, spleen, and bone; the mortality is up to 5% in untreated individuals.
Treatment	Streptomycin + gentamicin or doxycycline; ciprofloxacin.
Vaccine	No human vaccine available; in 1996, RB51, animal live attenuated strain of *B. abortus* replaced the S19 strain which was also a live attenuated vaccine
Specimens to be collected	Blood, bone marrow, spleen, liver, joint fluid or abscesses, serum
Epsilon Toxin of *Clostridium perfringens*	
Incubation	8–22 hours
Transmission mode	The epsilon toxin could probably be transmitted in contaminated food, water, or by aerosol
Basic signs/ symptoms	Epsilon toxin is one of 12 protein toxins produced by *Clostridium perfringens*, a gram-positive, anaerobic spore-forming rod; toxin is a pore-forming protein; it causes potassium and fluid leakage from cells; little or no information about the effects of epsilon toxin on humans; extrapolation from studies with experimentally infected animals, neurologic signs or pulmonary edema may be possible.
Treatment	Supportive therapy, hyperimmune serum might be helpful if given soon after exposure
Vaccine	No
Specimens to be collected	Stool, environmental samples
Food Safety Threats (e.g., *Salmonella* Species, *Escherichia coli* O157:H7, *Shigella*)	
Incubation	*Salmonella*: 6–72 h *E. coli* O157:H7: 3–8 days *Shigella*: 1–3 days
Transmission mode	Eating contaminated food that may look and smell normal; person-to person transmission if hygiene or handwashing habits are inadequate; this is particularly likely among toddlers who are not toilet trained
Basic signs/ symptoms	With or without fever, abdominal cramps with often bloody diarrhea (dysentery)

(Continued)

Table 11.1 (Continued)   Basic Facts about Specific Bioterrorism Agents

Treatment	Supportive therapy; trimethoprim/sulfamethoxazole; ceftriaxone; no treatment for *E. coli* O157:H7 (risk of enhancement of toxin release and hemolytic uremic syndrome)
Vaccine	DT TAB diphtheria, tetanus, *Salmonella* typhi, Paratyphi A & B AVP (France) only available
Specimens to be collected	Blood, stool, food

Glanders (*Burkholderia mallei*)

Incubation	1–5 days
Transmission mode	Animal exposure, cases of human-to-human transmission have been reported (two suggested cases of sexual transmission and several cases in family members who cared for the patients)
Basic signs/symptoms	Localized, pus-forming cutaneous infections, pulmonary infections, bloodstream infections, and chronic suppurative infections of the skin; generalized symptoms—fever, muscle aches, chest pain, muscle tightness, and headache, also excessive tearing of the eyes, light sensitivity, and diarrhea
Treatment	Tetracyclines, ciprofloxacin, streptomycin, novobiocin, gentamicin, imipenem, ceftazidime, sulfonamides
Vaccine	No
Specimens to be collected	Blood, sputum, urine, or skin lesions

Melioidosis (*Burkholderia pseudomallei*)

Incubation	Two days to many years
Transmission mode	Inhalation of dust, ingestion of contaminated water, contact with contaminated soil especially through skin abrasions, and for military troops, by contamination of war wounds; person-to-person transmission can occur
Basic signs/symptoms	Acute, localized infection: a nodule; can produce fever and general muscle aches Pulmonary infection: mild bronchitis to severe pneumonia; chest pain is common; a nonproductive or productive cough with normal sputum is the hallmark of this form of melioidosis Acute bloodstream infection: underlying illness such as HIV, renal failure, and diabetes; usually results in septic shock; respiratory distress, severe headache, fever, diarrhea, development of pus-filled lesions on the skin, muscle tenderness, and disorientation Chronic suppurative infection: involves the organs of the body (the joints, viscera, lymph nodes, skin, brain, liver, lung, bones, and spleen)
Treatment	Imipenem, penicillin, doxycycline, amoxicillin-clavulanic acid, azlocillin, ceftazidime, ticarcillin-clavulanic acid, ceftriaxone, aztreonam
Vaccine	No
Specimens to be collected	Blood, sputum, urine, or skin lesions

Psittacosis (*Chlamydophila psittaci*)

Incubation	1–4 weeks
Transmission mode	Inhalation of the agent from desiccated droppings, secretions, and dust from feathers of infected birds, but also geese and pigeons may be reservoirs, rare person-to-person infections
Basic signs/symptoms	Fatigue, fever, headache, rash, chills, and sometimes pneumonia
Treatment	Azithromycin, doxycycline
Vaccine	No

(Continued)

Table 11.1 (Continued)   Basic Facts about Specific Bioterrorism Agents

Specimens to be collected	Sputum, bronchoalveolar lavage, throat swab, serum

Q Fever (*Coxiella burnetii*)

Incubation	2–3 weeks
Transmission mode	High numbers of *C. burnetii* are shed in the reproductive fluids and placentas of infected animals; infection of humans—inhalation of agents in small droplets or from inhalation of barnyard dust contaminated with *C. burnetii*; infection via ingestion of contaminated dairy products, including tick bites and human-to-human transmission, are rare
Basic signs/ symptoms	Sudden onset of high fevers, severe headache, confusion, sore throat, nonproductive cough, nausea, vomiting, diarrhea, abdominal pain, and chest pain and pneumonia; weight loss can occur
	Chronic Q fever, characterized by infection that persists for more than 6 months; uncommon but is a much more serious disease
	Patients who have had acute Q fever may develop the chronic form as soon as 1 year or as long as 20 years after initial infection with serious complication; endocarditis and high fatality (65%).
Treatment	Azithromycin, doxycycline
Vaccine	A human vaccine developed and used successfully in Australia; prior to vaccination with whole cell vaccine, persons must have a skin test to verify they have not been previously exposed to *C. burnetii*
Specimens to be collected	Blood samples, serum, heart valve tissue

Staphylococcal Enterotoxin B

Incubation	3–12 hours
Transmission mode	Ingestion of food products containing toxins (contaminated milk and milk products); mostly "food handlers," purulent discharges (infected finger, eye, abscess nasopharyngeal secretions)
Basic signs/ symptoms	A sudden onset of high fever, chills, headache, myalgia, and nonproductive cough, with shortness of breath and retrosternal chest pain; also nausea, vomiting, and diarrhea (if person swallow toxin)
Treatment	Supportive therapy
Vaccine	No
Specimens to be collected	Serum, culture isolates, food, environmental specimens, nasal swab, induced respiratory specimens, urine, stool/gastric aspirates

Typhus Fever (*Rickettsia prowazekii*)

Incubation	1–2 weeks
Transmission mode	By *Pediculus humanus* corporis (body louse), infected by feeding on the blood of patient with acute disease; lice may get infected by feeding on the blood of patients with Brill–Zinsser disease; people infected by rubbing crushed lice or their feces into the bite or abrasions
Basic signs/ symptoms	Headache, chills, prostration, fever, and generalized body aches; a macular eruption becomes petechial or hemorrhagic, then develops into brownish-pigmented areas; changes in mental status are common with delirium or coma; toxemia myocardial and renal failure; lethality 10%–40%
Treatment	Doxycycline
Vaccine	No
Specimens to be collected	Blood samples, tissue samples, including skin biopsies

(Continued)

Table 11.1 (Continued)   Basic Facts about Specific Bioterrorism Agents

Viral Encephalitis (Alphaviruses [e.g., Venezuelan Equine Encephalitis, Eastern Equine Encephalitis, Western Equine Encephalitis])

Incubation	Venezuelan equine encephalitis: 1–6 days
	Eastern equine encephalitis: 5–15 days
Transmission mode	Bite of infected mosquito; viruses are maintained in rodent–mosquito cycle; cycle also involves horses
Basic signs/ symptoms	High fever, vomiting, stiff neck, and drowsiness, generalized, facial, or periorbital edema; paresis; major disturbances of autonomic function (impaired respiratory regulation or excess salivation) with up to 75% lethality; neurological sequelae in survivors; cognitive impairment
Treatment	Supportive therapy
Vaccine	Poorly immunogenic, investigational vaccines
Specimens to be collected	Blood samples, serum, cerebrospinal fluid (CSF), human or animal tissues

Water Safety Threats (e.g., *Vibrio cholerae, Cryptosporidium parvum*)

Incubation	Cholera: 4 hours–5 days
	Cryptosporidiosis: 1–12 days
Transmission mode	Ingestion of fecally contaminated food or water, including water swallowed while swimming; by exposure to fecally contaminated environmental surfaces; and by the fecal–oral route from person to person
Basic signs/ symptoms	Sudden onset of vomiting, abdominal distension, headache, pain, little or no fever, followed by profuse watery diarrhea with a "rice water" appearance; dehydration, hypovolemia, and shock; symptoms may be less prominent in cryptosporidiosis
Treatment	Doxycycline, ciprofloxacin for cholera; paromomycin for cryptosporidiosis
Vaccine	Cholera vaccine exists - not recommended because of its partial efficacy; no vaccine for cryptosporidiosis
Specimens to be collected	Stool, vomitus, food and water samples

C Category

Emerging Infectious Diseases (e.g., Nipah Virus and Hantavirus)

Incubation	Nipah virus: up to 1 month
	Hantaviruses: 9–35 days
Transmission mode	Nipah virus was transmitted to humans, cats, and dogs through close contact with infected pigs; the natural reservoir is still under investigation; preliminary data suggest the bats of the genus *Pteropus*
	Main reservoirs of hantaviruses are persistently infected small rodents (e.g., *Apodemus flavicollis, Apodemus agrarius, Celthrionomys glareolus, Peromyscus maniculatus, Olygoryzomis longicaudatus*, etc.); virus shedding by rodent's secrets and excreta; contaminated dust inhalation; professional exposure; human-to-human transmission possible for Andes virus.
Basic signs/ symptoms	Nipah virus: encephalitis (inflammation of the brain) characterized by fever and drowsiness and more serious central nervous system disease, such as coma, seizures, and inability to maintain breathing
	Two clinical syndromes recognized in infections with hantaviruses: hemorrhagic fever with renal syndrome (HFRS) and hantavirus pulmonary syndrome (HPS); a broad spectrum of clinical conditions have been recognized from unapparent or mild illness to a fulminant hemorrhagic disease with severe renal or cardiopulmonary failure and death
Treatment	Supportive therapy, i.v. ribavirin may tentatively reduce mortality if given before day 7 after the onset

(Continued)

Table 11.1 (Continued)   Basic Facts about Specific Bioterrorism Agents

Vaccine	No vaccine for Nipah virus
	Hantavax: killed vaccine; DNA vaccines in investigation; no firm evidence for efficacy so far
Specimens to be collected	Blood sample, serum, CSF, urine, human and animal tissues

Sources: CDC (2013), Benenson (1995), Begovac (2006), Schönwald et al. (2000), Gilbert (2006), , Meltzer et al. (2001), LeDuc et al. (2001), Jahrling et al. (2004), Engelthaler and Lewis (2004).

bioweapons or facilitate bioterrorism. Certain life sciences research is thus "dual use," that is, although intended to serve beneficial purposes, it could also be misused to cause harm. It is important to build strong common understandings of the essential elements of governance for life sciences research that raises biosecurity and dual-use concerns (NAS 2018). However, thanks to modern technology and a tremendous increase in knowledge accumulated about molecular biology of microbial agents, scientists can more effectively and efficiently identify pathogens from a variety of matrices including those from the environment. Progress in this area has been of interest to the traditional public health community for a very long time. It is also of interest to those communities that will be responsible for attribution following a criminal act or terrorist attack using a bioagent. To ultimately identify the individuals or organizations responsible for such an attack will very likely require rigorous use of modern microbiological analytic tools coupled with traditional forensic disciplines (Breeze et al. 2011; NRC 2014; Budowle 2014). Recently, an expert government panel from the United States acknowledged that the determination of a responsible party for a covertly planned or actual biological attack would be the culmination of a complex investigative process drawing on many different sources of information including technical forensic analysis of material evidence collected during the course of an investigation of a planned attack or material evidence resulting from an attack. Evidence collected as part of a biological attribution investigation will yield unique types of microbiological evidence that may be specific to the nature of the attack. The panel cited examples of such microbiological evidence as viable samples of the microbial agent, protein toxins, nucleic acids, clinical specimens from victims, laboratory equipment, dissemination devices and their contents, environmental samples, contaminated clothing, or trace evidence specific to the process that produced and/or weaponized the biological agent. This group acknowledged that there was a need for research and development efforts to enhance current capabilities in microbial forensics and that this would require effort among many elements of the national government (Breeze et al. 2011). The recommendations found in this national strategy may be relevant to many countries attempting to better prepare for the threat of bioterrorism.

Whether for a public health or a bioterrorism event, the identification, collection, and analysis of appropriate samples are critical prerequisites for the successful detection of microbes. When reporting to a court, investigators must have confidence in their results and understand the power and limitations of the procedures they have used, which can only be gained through accumulated experience in testing and validation. Not only can clinical material from infected patients be utilized for the detection, but very often animal or environmental samples will yield critical information. For the successful identification of suspicious agents, it is important to use both classic microbiological techniques and modern molecular biology techniques. Traditional microbiology and culture are the current gold standard in clinical diagnostics, but it is not always possible or practical to culture microbes in a forensic investigation. Sensitive nucleic acid–based and antibody-based detection technologies also are needed. Nucleic acid–based detection and identification technologies provide the ability to examine the genetic as well as structural information associated with a pathogen. Among the early tools developed to access an organism's genetic information were nucleic acid amplification techniques, such as polymerase chain reaction (PCR). Over time, variations of the basic PCR technique have developed, such as "multiplex PCR," which allows several DNA genomic regions to be amplified in the same reaction set, and quantitative or "real-time" PCR (qPCR), which can measure the quantity of a target sequence in real time. The DNA extraction protocols for some human samples do not necessarily correspond to the successful extraction of environmental samples: recovery may be low or PCR inhibitors may not be properly removed. Different DNA-based methods are necessary for identification of specific bioterrorism agents and, in general, DNA-based analysis for the detection of specific bioterrorism agents includes partial or whole genome sequencing, 16S rRNA sequencing, microarray analysis including pathogenicity array analyses, single-nucleotide polymorphism (SNP) characterization, variable number tandem repeat analyses and also a very important tool—antibiotic resistance gene characterization. Mass spectrometry also is useful in identifying many non-nucleic acid chemical species that may provide clues to microbial identity, origins, and production processes. Proteins, peptides, lipids, carbohydrates, inorganic metals, and organic metabolites may provide information about an organism's source environment, how it was produced, and the level of sophistication of the preparation. Microarray platforms provide similar capabilities but in a more multiplexed format. They have the ability to detect both nucleic acid and protein signatures. Many panels have several million assays on a single array chip. There are custom and standard microarray panels. Microarray panels are available for microbial detection, SNP detection, and genome-wide association studies, gene expression, and protein presence and abundance. For microbial forensics purposes, next-generation sequencing (NGS) is a powerful set of technologies that offers a state-of-the-art approach. Such high throughput detection using NGS generates large amounts of high-quality sequence data for microbial genomics. The value of NGS for microbial forensics is the speed at which evidence can be collected and the power to characterize microbial-related evidence to solve biocrimes and bioterrorist events. However, NGS also poses a new set of issues for implementation such as the possibility that different platforms give somewhat different answers. These issues will have to be resolved or at least understood for the microbial forensics' context. As high throughput technologies continue to improve, they provide increasingly powerful sets of tools to support the entire field of microbial forensics. In general, development of robust, powerful, multiplex molecular diagnostic tools and bioinformatic analysis tools are very important, but also development of rapid and inexpensive point-of-care diagnostic tests (Markotic et al. 1996; Le Duc et al. 2001; FDA 2013; NRC 2014; Budowle 2014; Tadin 2016).

Ideally, laboratories should be capable of identifying bioengineered microorganisms to distinguish a bioterrorist attack from a natural outbreak. For toxin detection, various methods are available today: different immunoassays produced by recombinant technology, biofunctional assays, protein or peptide-based assays, mass spectrometry, and DNA-based assays. Additionally, numerous non-DNA-based methods may be important for identification of microorganisms and to characterize preparation and process-related signatures associated with the microbial forensic samples. Bioengineering horizon scan could also be useful methodology to address potential future bioterrorism agents and event challenges. Horizon scanning is intended to identify the opportunities and threats associated with technological, regulatory, and social change, and to build societal preparedness. Bioengineering is expected to have significant impacts on the society in the near future as applications increase across multiple areas. The use of "platform technologies" to address, for example, emerging disease epidemics and pandemics, has taken on particular significance as a result of the COVID-19 pandemic. Many of the vaccine candidates for COVID-19 were developed from platforms for non-coronavirus candidates such as influenza, SARs, and Ebola (Kemp 2020).

Microbial forensics, as with other traditional forensic disciplines, requires implementation and application of protocols and practices that will ultimately yield results that can be used by decision-makers in law enforcement and national/international security. Identification of appropriate samples, "chain of custody" records (documentation that tracks physical control of samples), the use of valid analytical protocols by trained personnel, adherence to quality assurance measures, and ensuring the secure storage and preservation of samples are all of critical importance for successful attribution investigations. Somewhat unique to this type of evidence is ensuring the safety of anyone involved with the collection and handling of microbial forensic evidence. When dealing with pathogenic microorganisms, there are special handling procedures that must be followed in order to adequately prevent additional harm to personnel and the surrounding community. All personnel involved in response to or handling of these bioagents must undergo specialized training to ensure the safety and security of the individuals involved and the surrounding environment. Manipulations of potentially infectious material should be conducted under the appropriate containment conditions, ideally using a biocontainment laboratory equipped with the necessary safety equipment. Those individuals at risk of direct exposure to infectious material during collection or handling should wear appropriate personal protective equipment (Engelthaler et al. 2004; CDC 2013; Breeze et al. 2011; Hoffmaster et al. 2002; Jahrling et al. 2004).

11.5 Criteria for Considering an Outbreak Unusual

The initial outbreak of an infectious disease may not immediately be recognized as that caused by a bioterrorism event because pathogenic microorganisms occur naturally all over the world. Early determination of an attack versus a naturally occurring outbreak is a goal of organizations responsible for the attribution of bioterrorist events. It is important to identify unusual outbreaks in a timely fashion. In 1998, the World Health Organization (WHO) Ad Hoc Group developed criteria for considering outbreaks unusual (Breeze et al. 2011; Dando et al. 1988). The criteria are:

1. The first appearance of a disease outbreak in a non-endemic region.
2. An outbreak occurring outside of its normal season.
3. An uncommon route of transmission of pathogens.
4. The natural reservoir hosts or insect vectors do not occur in the region when the outbreak occurs.
5. The epidemiology of the diseases suggests an abnormal reduction in the incubation period of disease.
6. Characteristic of the pathogen differs from the known characteristics profile of the agent (strain type, sequence, antibiotic resistance, pattern etc.).

7. An increased virulence of pathogens.
8. The pathogen is capable of establishing new natural reservoirs to facilitate continuous transmission.

These criteria are based on the knowledge that though microbial agents are ubiquitous, it is well known that many of them may survive in some climate conditions or biotopes and not in others, and that some infectious diseases are more common or exclusively appear in some seasons (Benenson 1995; Begovac et al. 2006). For example, detection of human influenza cases during the summer time would be an unusual outbreak worthy of further investigation. Thus, epidemiological investigations to differentiate naturally occurring outbreaks from those that are man-made are necessary. The microbial agents that can be transferred from animals to people (zoonoses) or by insects (vector-borne diseases) are closely linked to their animal reservoirs/vectors and to their specific biotopes. For example, the reservoirs for New World hantaviruses, which cause hantavirus pulmonary syndrome (HPS), are distinct small rodent species that are found only in North or South America. An outbreak of HPS that might appear in other parts of the world would be suspicious as a possible bioterrorism attack and worthy of careful investigations. Also, if an animal reservoir or vector appears to be transmitting a new microbial agent, this should be carefully analyzed as a possible deliberate distribution. It is also important to know the typical disease characteristics caused by potential threat pathogens such as the normal incubation time. If a disease appears with a short incubation period or enhanced or unusual clinical signs and symptoms, especially in a population experienced with that disease, this too could be considered as a suspicious disease outbreak. The transmission route for many pathogens is very well known, as well as consequent pathogenic mechanisms. Significant variance in the route of transmission or pathogenic mechanisms may be the result of an intentional spread of newly engineered microorganisms. This might include the appearance of an unexpectedly high rate of resistance to previously effective antibiotics or enhanced virulence of a microorganism that previously caused only mild or moderate infections (CDC 2013; Breeze et al. 2011; Benenson 1995; Begovac et al. 2006). One very important tool to help differentiate natural epidemic and intentional attack is disease surveillance and global disease reporting system. There are several web-based programs like ProMed (http://www.promedmail.org) or GOARN (WHO's Global Outbreak Alert and Response Network, http://www.who.int/csr/outbreak-network), which are important tools in recognizing outbreaks and possible bioterrorist events. It is important to collect essential information on all significant factors of disease outbreaks that allow us to learn how and why the disease has occurred. For such purposes, routine monitoring of changes in disease patterns, documenting and understanding of factors that have initiated changes, timely dissemination of information to decision-makers, understanding the normal occurrence of diseases, web-based programs for international disease surveillance and for reporting of human and animal diseases can all contribute to a clearer understanding and better recognition of unusual events (Breeze et al. 2011).

11.6 Suspicious Infectious Diseases Outbreaks

As discussed, bioterrorism unfortunately has a rich, long history. However, today with modern microbial forensics methods and a tremendous increase in knowledge about microbial agents, including modern laboratory techniques and global disease reporting systems, it is much easier to distinguish naturally occurring infections from deliberately caused diseases. Several examples follow of outbreaks caused by accident or criminal acts:

- In April 1979, in the city of Sverdlovsk, southeast Ukraine, an explosion occurred at a local military compound. Over the next several days in an area 4 km south and east of the compound, many residents began to develop high fevers and difficulty breathing. Approximately 200 patients died in short period of time. Local doctors diagnosed it as an outbreak of inhalation anthrax; however, the government officially reported the consumption of tainted meat from a cow suffering from anthrax as being responsible for the outbreak. Patients did not, however, display the symptoms of gastric or cutaneous anthrax, and autopsies of human victims indicated severe pulmonary edema and toxemia. Sheep and cattle in six different villages up to 50 km away also died of anthrax. DNA analysis of tissue samples taken from victims indicated they had been exposed to several anthrax strains simultaneously, which is atypical in natural infection in cattle. It was subsequently concluded that the anthrax outbreak was the result of the accidental release from a bioweapons facility (Wade 1980; Walker et al. 1994; Sternbach 2003; Brookmeyer et al. 2001; Wilkening 2006).
- In the 1980s, a number of people became HIV-infected. None of these individuals had lifestyles that would put them at high risk for exposure to HIV. In 1986, when their dentist was found to be HIV positive, it became clear that multiple patients who contracted HIV had visited the same dentist. DNA sequencing of samples from the patients, the dentist, a local control group, and a distinct control was performed. HIV nucleotide sequences from a number of patients were closely related (not exactly the same) to those from the dentist, and distinct from viruses obtained from control patients. The evidence strongly supported but could not conclusively prove HIV transmission from the dentist to patients during invasive dental care (Breeze 2011). This was, however, non-intentional transmission of HIV from the dentist who was not aware that he was an HIV carrier.
- In September 1984, the Rajneeshees, a religious cult, purposely contaminated salad bars in restaurants located in or near Dalles, Oregon, with *Salmonella typhimurium*. Over 750 persons were poisoned and about 40 hospitalized. The purpose of the attack was to influence the outcome of a local election as it was discovered a year later when members of the cult turned informants (Philips 2005; Carus 2001).
- In 1995, on more than ten occasions, Aum Shinrikyo attempted to disperse anthrax, botulinum toxin, Q fever, and Ebola virus against the general population and authority figures in Japan. Fortunately, no reported infections occurred, although they were successful in releasing sarin gas into the Tokyo subway system that year, resulting in 12 deaths and causing an estimated 6000 people to seek medical attention (Philips 2005).
- The 2001 anthrax letter attacks came in two waves. The first letters were sent to several US news media offices in New York (ABC, CBC, NBC, and the *New York Post*) and Florida (American Media Inc.), and three weeks later the second set of letters were sent to two US senators in Washington, DC, Tom Daschle of South Dakota and Patrick Leahy of Vermont. Five people were killed and 17 others were infected between October 4 and November 22. DNA sequencing of the anthrax isolated from the first victim who died from pulmonary anthrax was conducted

and it was showed that the anthrax was identical to the original Ames strain. This strain had only been used for research purposes. Since its original isolation, the Ames strain had been maintained at the US Army Medical Research Institute of Infectious Diseases (USAMRIID) in Fort Detrick, Maryland. USAMIIRD had shared the Ames strain with at least 15 bioresearch laboratories within the US and six locations overseas (CDC 2001; Jernigan et al. 2001; Atlas 2002; Greene et al. 2002; Tan et al. 2002; Hoffmaster et al. 2002; NRC, NAS 2011). The Federal Bureau of Investigation (FBI) was responsible for the investigation of the letter attacks. The FBI assigned squads of special agents who teamed with investigators from the US Postal Service. The investigation, given the major case title of "Amerithrax," lasted nearly nine years and involved almost 600,000 investigator work hours, 10,000 witness interviews, 80 searches, and 26,000 email reviews including analysis of 4 million megabytes of computer memory in six continents. In addition, 29 government, university, and commercial laboratories were involved in scientific analyses, and close collaboration with other federal agencies was established. In June 2002, the FBI scrutinized 20 to 30 scientists who might have had knowledge and opportunity to arrange the "letters attack." On August 6, 2002, a former USAMRIID scientist, Steven J. Hatfill, a biodefense expert, was indicated as a "person of interest" in the investigation. However, on August 8, 2008, the Department of Justice (DOJ) cleared Hatfill of involvement in the anthrax mailings and the federal government awarded Hatfill $5.82 million to settle his violation of privacy lawsuit against the DOJ. On July 29, 2008, another USAMRIID microbiologist, Bruce E. Ivins, committed suicide just as the FBI was about to file criminal charges against him. Following Ivins's suicide, the FBI released information that they believed linked him to the crime. Specifically, he had been responsible for the preparation and oversight of a flask containing a stock supply of *B. anthracis* that was identified as RMR-1029. *B. anthracis* from the letters was believed to have originated from this stockpile. This link as well as other information led the FBI, the DOJ, and the US Postal Inspection Service to officially close the case on February 19, 2010, concluding that Ivins alone mailed the anthrax letters (NRC, NAS 2011).

- Prior to the official closure of the Amerithrax case, the FBI asked the National Research Council (NRC) of the National Academy of Sciences (NAS) to conduct an independent review of the various scientific methods and procedures that had been used throughout the investigation. This was in response to issues that had been raised concerning the FBI's determination of RMR 1029 as the source of the *B. anthracis* used in the letters. For that purpose, a committee of esteemed scientist and experts was appointed. The committee was expected to determine whether microbial forensic techniques used during the FBI investigation that supported the final FBI conclusions met appropriate standards for scientific reliability. The review process focused on novel scientific methods that were developed for purposes of the FBI investigation.

After an extensive analysis of documents and facts provided by the FBI, the committee made the following conclusions:

1. The *B. anthracis* in the letters was the Ames strain and was not genetically engineered.

2. Multiple distinct colony morphological types or morphotypes of *B. anthracis* Ames were present in the letters.
3. The scientific link between the letter material and flask number RMR-1029 indicated by the FBI as an origin source was not conclusive.
4. There was no evidence that silicon found in letter powders was intentionally added.
5. It was not possible to conclude the amount of time needed to prepare the spore material; it was considered that the time might vary from two to three days or to several months.
6. The physicochemical and radiological analyses were of limited forensic value in this investigation.
7. There was no consistent evidence of *B. anthracis* Ames DNA in environmental samples collected from an overseas site.
8. During the time, new powerful scientific tools and methods (e.g., high-through-put next-generation DNA sequencing) were developed, which might give better insight if used during the investigation.

In general, the committee underlined that in future biological attacks, better communication between the public and policymakers should be established before, during and after the investigation (NRC, NAS 2011).

11.7 Does SARS-CoV-2 Have the Potential to be a Bioterrorism Agent?

At the end of 2019, a new severe respiratory infection caused by SARS-CoV-2 spread quickly and resulted in a high morbidity and mortality rate in Wuhan, China (Huang 2020; Čivljak 2020; Rode Đaković 2021). The subsequently developed COVID-19 pandemic is the result of the extremely rapid international spread of SARS-CoV-2, causing millions of infections and deaths, and making it one of the most frightening and comprehensive pandemics in history (Singh 2021). From the very beginning of the pandemic, there has been speculation about the possibility that SARS-CoV-2 was intentionally or unintentionally released into the environment, among humans, and spread around the world. Its potential as a zoonosis, possible initial spread from animals to humans, and transmission potential to humans from different animal species are still being questioned. So far, there is not any firm evidence that the current pandemic resulted from either bioweapons or a laboratory leak (Nie 2020; Knight 2021). Although the threats associated with biological warfare, bioterrorism, and the accidental leakage of deadly viruses from laboratories are possible, the emergence and the spread of these conspiracy theories reflect a series of internationally longstanding and damaging trends, which include deep mistrust, animosities, distracting ideologies, and the true destruction in numerous released propaganda campaigns (Nie 2020). However, natural outbreaks and bioweapons can affect both human and animal populations in similar ways (Knight 2021). Coming back to the SARS-CoV-2 potential to be a bioterrorism agent, it is certainly a microorganism that can be easily disseminated or transmitted from person to person, resulting in the high potential to be spread all over the world in a short period of time and also to quickly make numerous mutations with the emergence of new variants of the virus. Although the mortality rate is not considered high per se, the possibility for causing a pandemic with consequent high morbidity, and thus a large number of deaths worldwide, gives the potential for major public health impact;

might cause public panic, healthcare system burden, and social disruption; and require special action for public health preparedness. It also requires specific enhancements of diagnostic capacity and enhanced disease surveillance. SARS-CoV-2 could be engineered for mass dissemination in the future because of availability, ease of production and dissemination, and potential for high morbidity and mortality rates and major health impact. From this, it is clear that SARS-CoV-2 has a number of elements of microorganisms that can potentially be used in bioterrorism attacks in all three categories (CDC 2013).

The COVID-19 pandemic has firmly demonstrated the power of infectious diseases, and the power of viruses, which have been responsible for many terrifying epidemics throughout history. To better understand how SARS-CoV-2 caused a global pandemic and to be prepared for future health challenges, establishing a better understanding of the important early epidemiological events and of authorities' responses to the spread of some future emerging pathogens is essential. The collected evidence of the global disaster caused by SARS-CoV-2 indicates much room for performance improvement in order to ensure a more efficient and rapid response in the future, particularly for potentially high-impact respiratory pathogens, which are most likely to cause the next pandemic (Nie 2020).

11.8 Biosafety and Biosecurity

The field of forensic microbiology needs to harmonize the efforts of scientists in the field of infectious diseases, epidemiology, and microbiology with people who make political, administrative, and legal decisions in making biodefense policy. This is a prerequisite for building a national capacity to investigate bioterrorism. The research community in infectious diseases, microbiology, epidemiology, and immunology can contribute greatly through their skills and insights to the efforts to build national capacity for biodefense. Basic research, including genomics, vaccines, therapies, and a robust corps of talented and committed scientists who are trained and educated are important to achieve national biosafety and biosecurity. There is also a need to have laboratory and public health experts familiar with law enforcement and national security and vice versa. A program to construct national biocontainment laboratories is essential to the success of biosecurity and biodefense, since it is important to ensure that pathogen collections are safely stored and access to pathogenic strains is appropriately limited. In the US, the select agents' rules are important aspects of each organization's biosecurity program.

Though biosecurity is inherently a government's responsibility, there are professional organizations (institutional, academic, and private sector) and individual experts that can help shape and develop guidelines on equitable and fair access and benefit-sharing of genetic resources, microbial agents, and the ethical implications of biotechnology, international governance of biotechnology and biosafety, vaccination programs, and rulemaking (CDC 2013; Breeze 2011; CDC 2001; Atlas and Reppy 2005; Cook-Deegan et al. 2005; Atlas and Dando 2006; Bork et al. 2007).

References

Atlas, R.M. 2002. Bioterrorism: From threat to reality. *Annu Rev Microbiol* 56:167–85.
Atlas, R.M. and J. Reppy. 2005. Globalizing biosecurity. *Biosecur Bioterror* 3(1):51–60.
Atlas, R.M. and M. Dando. 2006. The dual-use dilemma for the life sciences: Perspectives, conundrums, and global solutions. *Biosecur Bioterror* 4:276–86.

Begovac, J., D. Božinović, M. Lisić, B. Baršić and S. Schönwald, eds. 2006. *[Infectology]*. Zagreb: Profil.

Benenson, A.S., ed. 1995. *Control of Communicable Diseases Manual. An Official Report of the American Public Health Association*. Washington, DC: American Public Health Association.

Bork, K.H., V. Halkjaer-Knudsen, J.E. Hansen and E.D. Heegaard. 2007. Biosecurity in Scandinavia. *Biosecur Bioterror* 5(1):62–71.

Breeze, R.G., B. Budowle, S.E. Schutzer, P. Keim and S. Morse, eds. 2011. *Microbial Forensics* (second edition). Amsterdam: Elsevier Academic Press.

Brookmeyer, R., N. Blades, M. Hugh-Jones and D.A. Henderson. 2001. The statistical analysis of truncated data: Application to the Sverdlovsk anthrax outbreak. *Biostatistics* 2(2):233–47.

Budowle, B., N.D. Connell, A. Bielecka-Oder, et al. 2014. Validation of high throughput sequencing and microbial forensics applications. *Investig Genet* 5:9.

Carus, W.S. 2001. Bioterrorism and Biocrimes. The Illicit Use of Biological Agents since 1900. Washington, DC: Center for Counterproliferation Research National Defense University, 217 pages. http://www.ndu.edu/centercounter/Full_Doc.pdf.

Centers for Disease Control and Prevention (CDC). 2001. Update: Investigation of bioterrorism-related anthrax and adverse events from antimicrobial prophylaxis. *MMWR Morb Mortal Wkly Rep* 50(44):973–6.

Centers for Diseases Control and Prevention (CDC). 2013. *Bioterrorism Agents/Diseases*. http://www.bt.cdc.gov/Agent/agentlist.asp.

Centers for Disease Control and Prevention (CDC), Department of Health and Human Services (HHS). 2001. Requirements for facilities transferring or receiving select agents. Final rule. *Fed Regist* 66(170):45944–5.

Committee on Review of the Scientific Approaches Used during the FBI's Investigation of the 2001 *Bacillus anthracis* Mailings, Board on Life Sciences Division on Earth and Life Studies, Committee on Science, Technology, and Law Policy and Global Affairs Division, National Research Council of the National Academies. 2011. *Review of the Scientific Approaches Used During the FBI's Investigation of the 2001 Anthrax Letters*. Washington, DC: The National Academies Press.

Cook-Deegan, R.M., R. Berkelman, E.M. Davidson, et al. 2005. Issues in biosecurity and biosafety. *Science* 308(5730):1867–8.

Čivljak, R., A. Markotić and I. Kuzman. 2020. The third coronavirus epidemic in the third millennium: What's next? *Croat Med J* 61(1):1–4.

Dando, M., G.S. Pearson, and B. Kriz, eds. 1988. Scientific and technical means of distinguishing between natural and other outbreaks of disease. In *Proceedings of the NATO Advanced Research Workshop*, Prague, Czech Republic, 18–20 October 1988.

Engelthaler, D.M. and K. Lewis, eds. 2004. *Zebra Manual: A Reference Handbook for Bioterrorism Agents*. Phoenix, AZ: Arizona Department of Health Services, Division of Public Health Services.

FDA. 2013. *ORA Lab Manual, Volume IV, Section 2-Microbiology*. http://www.cfsan.fda.gov/~ebam/bam-28.html.

Gilbert, D.N., R.C. Moellering Jr., G.M. Eliopoulos and M.A. Sande, eds. 2006. *The Sanford Guide for Antimicrobial Therapy 2006*. Sperryville, VA: Antimicrobial Therapy Inc.

Greene, C.M., J. Reefhuis, C. Tan, et al. 2002. Epidemiologic investigations of bioterrorism-related anthrax, New Jersey, 2001. *Emerg Infect Dis* 8(10):1048–55.

Hoffmaster, A.R., R.F. Meyer, M.D. Bowen, et al. 2002. Evaluation and validation of a real-time polymerase chain reaction assay for rapid identification of *Bacillus anthracis*. *Emerg Infect Dis* 8(10):1178–82.

Huang, C., Y. Wang, X. Li, et al. 2020. Clinical features of patients infected with 2019 novel coronavirus in Wuhan, China. *Lancet* 395(10223):497–506.

Jahrling, P.B., L.E. Hensley, M.J. Martinez, et al. 2004. Exploring the potential of variola virus infection of cynomolgus macaques as a model for human smallpox. *Proc Natl Acad Sci U S A* 101(42):15196–200.

Jernigan, J.A., D.S. Stephens, D.A. Ashford, et al. 2001. Bioterrorism-related inhalational anthrax: The first 10 cases reported in the United States. *Emerg Infect Dis* 7(6):933–44.

Kasten, F.H. 2002. Biological weapons, war crimes, and WWI. *Science* 296(5571):1235–7.

Kemp, L., L. Adam, C.R. Boehm, et al. 2020. Bioengineering horizon scan 2020. *eLife* 9:e54489. https://doi.org/10.7554/eLife.54489.

Knight, D. 2021. COVID-19 pandemic origins: Bioweapons and the history of laboratory leaks. *South Med J* 114(8):465–7.

LeDuc, J.W., I. Damon, D.A. Relman, J. Huggins and P.B. Jahrling. 2001. Smallpox research activities: U.S. interagency collaboration, 2001. *Emerg Infect Dis* 8(7):743–5.

Markotic, A., J.W. LeDuc, D. Hlaca, et al. 1996. Hantaviruses are likely threat to NATO forces in Bosnia and Herzegovina and Croatia. *Nat Med* 2(3):269–70.

Meltzer, M.I., I. Damon, J.W. LeDuc, and J.D. Millar. 2001. Modeling potential responses to smallpox as a bioterrorist weapon. *Emerg Infect Dis* 7:959–969.

Nie, J.B. 2020. In the shadow of biological warfare: Conspiracy theories on the origins of COVID-19 and enhancing global governance of biosafety as a matter of urgency. *J Bioeth Inq* 17(4):567–74.

National Academies of Sciences, Engineering, and Medicine. 2018. Governance of dual use research in the life sciences. In *Advancing Global Consensus on Research Oversight: Proceedings of a Workshop*. Washington, DC: The National Academies Press. https://doi.org/10.17226/25154.

National Research Council. 2011. *Review of the Scientific Approaches Used During the FBI's Investigation of the 2001 Anthrax Letters*. Washington, DC: The National Academies Press. https://doi.org/10.17226/13098.

National Research Council. 2014. *Science Needs for Microbial Forensics: Developing Initial International Research Priorities*. Washington, DC: The National Academies Press. https://doi.org/10.17226/18737.

Philips, M.B. 2005. Bioterrorism. A brief history. *North East Florida Med J* 56:32–5.

Rode Đaković, O., I.C. Kurolt, I. Puljiz, et al. 2021. Antibody response and the clinical presentation of patients with COVID-19 in Croatia: The importance of a two-step testing approach. *Eur J Clin Microbiol Infect Dis Off Publ Eur Soc Clin Microbiol* 40(2):261–8. https://doi.org/10.1007/s10096-020-04019-y.

Schönwald, S., B. Baršić, A. Beus, et al. 2000. *Manual for the therapy and prevention of infectious diseases*. Zagreb: Birotisak.

Singh, S., C. McNab, R.M. Olson, et al. 2021. How an outbreak became a pandemic: A chronological analysis of crucial junctures and international obligations in the early months of the COVID-19 pandemic. *Lancet* 398(10316):2109–24.

Sternbach, G. 2003. The history of anthrax. *J Emerg Med* 24(4):463–7.

Tadin, A., R. Tokarz, A. Markotić, et al. 2016. Molecular survey of zoonotic agents in rodents and other small mammals in Croatia. *Am J Trop Med Hyg* 94(2):466–73.

Tan, C.G., H.S. Sandhu, D.C. Crawford, et al. 2002. Surveillance for anthrax cases associated with contaminated letters, New Jersey, Delaware, and Pennsylvania, 2001. *Emerg Infect Dis* 8(10):1073–7.

Wade, N. 1980. Death at Sverdlovsk: A critical diagnosis. *Science* 209(4464):1501–2.

Walker, D.H., O. Yampolska and L.M. Grinberg. 1994. Death at Sverdlovsk: What have we learned? *Am J Pathol* 144(6):1135–41.

Wilkening, D.A. 2006. Sverdlovsk revisited: Modeling human inhalation anthrax. *Proc Natl Acad Sci U S A* 103(20):7589–94.

World at Risk Commission on the Prevention of Weapons of Mass Destruction Proliferation and Terrorism. 2009. Available from: www.preventwmd.gov/report/.

Forensic Animal DNA Analysis 12

MARILYN A. MENOTTI-RAYMOND,
VICTOR A. DAVID, SREE KANTHASWAMY,
PETAR PROJIĆ, VEDRANA ŠKARO,
GORDAN LAUC, AND ADRIAN LINACRE

Contents

12.1 Introduction

The application of biological testing to allegations involving animals (other than human) and plants is relatively recent. This is a reflection of a very recent awakening to the scope of illegal trade in endangered species, a previous view that the crimes were of less significance than investigating crimes against people, and that few laboratories were willing and able to assist in such investigations. However, in the past decade there has been enormous increase in interest in the area, leading to a burgeoning number of research papers along with textbooks and recent reviews (Alacs et al. 2010; Linacre 2008, 2009; Linacre and Tobe 2011; Ogden et al. 2009; Tobe and Linacre 2010; Wilson-Wilde 2010a, 2010b).

Since the mid-1990s, DNA testing in the field of forensic science has increased dramatically. Standardized and accredited crime scene processing and laboratory protocols, statistical models, and analytical techniques for human DNA analysis have elevated this forensic science tool to the point of being the gold standard against which other forensic sciences are measured. As more research provides a deeper understanding of DNA analysis, applications have begun to span beyond the realm of human forensic science and into the area of nonhuman forensic DNA analysis.

DOI: 10.4324/9780429019944-14

Nonhuman DNA analysis has grown by leaps and bounds for many reasons both criminal and civil. For decades, laboratories have been testing horses and cattle for pedigree and other purposes. The study of cultivars to look at variation among plant species, the testing of fish and meat sold in the marketplace to verify that it is correctly labeled, the Convention on International Trade in Endangered Species of Wild Flora and Fauna (CITES), as well as the ubiquitous presence of our pets and their fur winding up as trace evidence in crime scene material have all lead to the expansion of the use of DNA technology to identify or characterize nonhuman samples.

Prior attempts at forensic analysis of animal hair relied on morphological criteria (Moore 1988; Peabody et al. 1983); identification could be made at a species or, at most, breed level. Isoelectric focusing of keratins were the first efforts used toward molecular characterization of animal specimens (Carracedo et al. 1987). The potential for genetic individualization of specimens was first demonstrated by Jeffreys et al. (1985a) using highly repetitive minisatellite loci to generate individual-specific "fingerprints" of human DNA, used soon after in a highly publicized immigration test case (Jeffreys et al. 1985b). Jeffreys et al. were also the first to demonstrate the potential of DNA fingerprinting of companion animals—using human minisatellite DNA to generate multilocus DNA fingerprints of dog and cat DNA (Jeffrey and Morton 1987).

Locard's exchange principle (James and Nordby 2003) allows forensic analysts to link original source and target surface, and has a perspective on primary and secondary transfers. Animal biological evidence is not exempted from Locard's rule and has become very important for identification and individualization of trace or transfer biomaterial from crime scenes. In recent years, interest in animal forensic science has piqued, because of the abundance of animal evidence encountered in crime investigations. This is not surprising because more than 50% of households in the United States own at least a cat (86.4 million cats in the United States) or a dog (78.2 million dogs in the United States), leading to high likelihoods of exchange of pet hair between a crime suspect and victim at the scene of the crime (Humane Society of the United States 2012). Because of the vast amounts of hairs shed by dogs and cats, D'Andrea et al. (1998) showed that an intruder couldn't exit a house with a pet without carrying away the animal's hair.

Most DNA analysis of animal biomaterial, however, only provide vital investigative leads and reconstruction of the crime; the analysis of animal DNA cannot contribute to the individualization of any human suspect, and usually this type of evidence needs to be supplemented by other forms of physical evidence. The earliest use of animal DNA analysis in a criminal case was the 1996 case of "Snowball" the cat, performed by Menotti-Raymond et al. (1997a) (see later). Since then, animal evidence has been used in myriad types of litigation including those involving traffic accidents (Schneider et al. 1999), murders (*State of California v. David Westerfeld* 2002), bank robberies (Savolainen and Lunderberg 1999), and dog attack cases where there are human (Pádár et al. 2002) or nonhuman victims (Pádár et al. 2001). Since 1996, canine DNA evidence has contributed to more than 20 criminal cases in Great Britain and the United States alone (Halverson and Basten 2005a).

Animal forensic cases can be divided into three basic categories: (1) when the animal is a passive witness to a crime, such as when dog or cat hair and dander are used to link a suspect or perpetrator to the crime scene or victim; (2) when the animal is the victim, such as abuse, theft, killing (poaching), violation of laws on endangered species, and dog or cock fighting cases; (3) when the animal is a suspect, for example, when a dog or other animal attacks or mauls humans or other animals (Kanthaswamy 2009).

Cases of arson, homicide, and burglary are examples of human-on-human crimes that have been successfully investigated and resolved using DNA tests of material from animal "witnesses" found at the scene or on the suspect and/or victim. One of the earliest animal forensic cases was performed at the Veterinary Genetics Laboratory (VGL), University of California, Davis (UC Davis). The blood trail found leading away from a murdered pub bouncer was identified by the UK Forensic Science Service (FSS) as canine in origin. DNA analysis of the blood samples matched a dog belonging to a man that had earlier been refused admittance into the pub, who subsequently stabbed the bouncer to death, wounding his own dog during the knife attack (Agronis 2002).

One of the most heinous acts against animals was committed by two teenagers in Largo, Florida (Inhumane.org). Samples of clothing, blood-soaked leaves, a metal rod, hockey stick, and llama blood samples sent to the UC Davis animal forensic lab for DNA analysis were determined to match reference samples from three llamas (Dreamin' Demon; Tampa Bay Online 2010).

Animal suspects, victims, or witnesses cannot be interviewed or represent themselves or testify in court, but their biomaterial can independently and objectively link an animal suspect to an animal victim. Animal DNA testing that has standardized, validated, and accredited procedures facilitates greater reliance on objective scientific evidence when eyewitness accounts can be unreliable/biased (Himmelberger et al. 2008; Kanthaswamy et al. 2009; Smalling et al. 2010). Aside from providing important leads and links during an investigation, this testing confers leverage to disprove, refute, or support alibis.

Although there are many parallels between human and animal forensic DNA analysis, there are also some significant differences that impose some of the most formidable challenges. One of the more effective challenges to animal DNA techniques is attacking the integrity of the sample, including collection, preservation, and documentation at the crime scene, and analysis and interpretation of the results and reporting at the forensic laboratory (Scharnhorst and Kanthaswamy 2011). The ability of an analyst to identify probative animal evidence from vast quantities of redundant, irrelevant, or unrelated items is a rare trait. Much of the evidence is not recognized as being probative or important by criminalists/forensic analysts (Scharnhorst and Kanthaswamy 2011).

Compared to human forensic DNA analysis, animal forensic DNA analysis is still in its infancy, and highly likely to experience admissibility issues. In the United States, where the Frye and Daubert standards impart particular rules regarding the admissibility of evidence in the court of law, procedures involving animal evidence must gain general acceptance in the field (Cassidy and Gonzales 2005). Of chief concern has been the lack of an accredited and comprehensive quality assurance system (Budowle et al. 2005; Scharnhorst and Kanthaswamy 2011) for animal forensic DNA testing. Therefore, analysis of animal DNA is frequently not requested and as a result, is seldom performed in the majority of forensic laboratories. However, this situation is changing. At UC Davis, the Veterinary Genetics Laboratory Forensic Unit (VGL Forensics) has become the first accredited crime lab dedicated to animal DNA profiling under the ASCLD/LAB-International program. Additionally, it is now possible to obtain a master of science degree in veterinary forensic sciences (https://www.forensicscience.ufl.edu/veterinary/).

The lack of commercially available forensic DNA kits or comprehensive population genetic databases for most species that incorporate regional and breed-specific information is another indication of how underdeveloped the field of animal forensic science is compared to its human counterpart. Most forensic canine and feline population databases do not contain sufficient sample sizes of outbred animals despite the fact that more than 50% of US dog and approximately 98% of US cat populations, respectively, are outbred animals, which is disconcerting. Comprehensive relational databases are needed for estimating the frequencies of common and rare alleles and haplotypes in a particular population, and to give statistical weight or meaning to matches between genotype/haplotype from the questioned sample and a genotype/haplotype from a known reference sample. Exclusion probabilities, likelihood ratios, random match probabilities, and Bayesian methods are all examples of statistical assessments used to convey the significance of a match (Budowle et al. 2005).

Because the DNA statistical model relies heavily on the Hardy–Weinberg equilibrium, which requires random mating, the high occurrence of inbreeding and intensity of genetic subdivisions within and among domestic animal populations are major concerns especially for estimating animal profile frequencies and random match probability. Issues with inbreeding and genetic subdivision are even greater pitfalls for species that are threatened or endangered, because their numbers are much smaller (Ayres and Overall 1999; Ayres 2000; Kanthaswamy et al. 2009). Excessive inbreeding may require additional panels of markers for parentage and genetic identity analyses for endangered animal populations.

Participation in an interlaboratory proficiency testing program from an outside agency will ensure that the laboratory and its analysts are adhering to a successful quality assurance (QA)/quality control (QC) program that covers standard operating protocols pertaining to sample handling, storage and integrity, casework documentation, and whether the DNA analyses—including genotyping, statistical analysis, and interpretation—are consistent with scientific approaches (Scharnhorst and Kanthaswamy 2011; Budowle et al. 2005).

The sections that follow deal with Felids (cats) (Menotti-Raymond, David), Canids (dogs) (Kanthaswamy), Bovids (cattle) (Projić, Škaro and Lauc), and wildlife forensic DNA testing (Linacre).

12.2 Felid Forensic DNA Testing

Domesticated some 10,000 years ago (Vigne et al. 2004), *Felis silvestris catus* has become one of the world's most popular household pets.

Gilbert et al. (1991) characterized the first cat multilocus minisatellite probes, which they used to assess genetic relatedness of individuals in lion prides and cheetah populations, and ultimately as a metric to infer the social dynamics of lion pride structure.

Short tandem repeat (STR) markers ultimately revolutionized human forensic analysis (Fregeau and Fourney 1993).

With the routine use of STR loci applied to the genetic individualization of human samples came the realization of the potential of samples of nonhuman origin. Genetic linkage maps incorporating STRs with coding loci were rapidly developed in companion and commercial animal species for the mapping and characterization of genes of health-related and commercial interest (Archibald et al. 1995; Barendse et al. 1994; Crawford et al. 1995; Dietrich et al. 1992; Ellegren et al. 1994; Jacob et al. 1995; Mellersh et al. 1997; Menotti-Raymond et al. 1999), thereby opening the potential of STR application to forensic analysis.

The National Cancer Institute's Laboratory of Genomic Diversity (LGD) developed the first genetic linkage and radiation hybrid maps of the domestic cat, incorporating genes and STR loci, as a resource to map and characterize felid models of human hereditary disease (O'Brien et al. 2002; Menotti-Raymond et al. 1999, 2003a, 2009; Murphy et al. 2000,

2007; Pontius et al. 2007). Domestic cat STR loci had additionally been used in conservation genetics applications in exotic felids (Culver et al. 2000, 2001; Eizirik et al. 2001; Johnson et al. 1999).

12.2.1 Case Studies

The first application of STR loci in genetic individualization of felid samples arose from a puma attack on a female jogger in California in 1994 (Culver, unpublished). Early in the morning of December 4, 1994, a woman hiking alone was attacked and killed by a puma in the Cuyamacha Rancho State Park near San Diego. Samples submitted by the California Department of Fish and Game from the mountain lion presumed responsible for the attack provided matching STR profiles for eight felid STR loci from DNA isolated from puma hairs found on the victim (Culver, unpublished). Culver, who was generating a phylogeny of puma subspecies, even had a database of pumas with which to calculate the match probability that another individual could have been the attacker (Culver et al. 2000).

The first application of domestic cat STR loci in genetic individualization of domestic cat samples came in response to an inquiry from the Royal Canadian Mounted Police (RCMP). The case involved a 32-year-old woman from Richmond, Prince Edward Island, Canada, who was reported missing on October 3, 1994. Within a few days, her car was found abandoned in a wooded setting within a few kilometers of her home and soon thereafter, within the same vicinity, a bag containing a man's leather jacket and tennis shoes, stained with blood determined to be that of the missing person. White hairs clinging to the jacket lining were identified by the RCMP's Halifax forensics laboratory (A.E. Evers, RCMP) to be cat hairs. These hairs were potential probative evidence in the case, as the only suspect lived with his white cat, Snowball (Figure 12.1). The RCMP queried whether DNA fingerprint profiles could be generated and compared from DNA extracted from the hairs in the jacket and Snowball.

There had been multiple reports of DNA fingerprint profiles generated from hairs, including profiles from a human hair root (Higuchi et al. 1988) and from single hairs of free-ranging chimpanzees (Morin 1992). Twenty-seven white hairs from the jacket lining and approximately 10 mL of blood drawn from the subpoenaed cat were provided. Microscopic evaluation of the hairs identified one hair with a good root and possibly a

Figure 12.1 Snowball, the suspect's pet cat. (From Menotti-Raymond, M. et al., *Nonhuman DNA Typing, Theory and Casework Applications*, ed. H.M. Coyle, CRC Press, Boca Raton, FL, 2008b. With permission from Taylor & Francis Group.)

small piece of attached tissue. Approximately 17 ng of DNA was extracted from this hair root, but no DNA from three additional hair roots or any of the four hair shafts examined (Menotti-Raymond et al. 1997b).

Ten dinucleotide STR loci had been selected for the analysis (Menotti-Raymond et al. 1997b) based on robustness, absence of linkage with other proposed STRs, and low "stutter" (Hauge and Litt 1993). PCR products generated from DNA extracted from the one hair root (Menotti-Raymond et al. 1997a) and, subsequently, from the blood sample were electrophoresed in the same gel.

A "match window" was established for each STR locus using guidelines developed by the RCMP to empirically determine a size difference threshold, which would define acceptable variation in migration between any two measured alleles to conclude that they match (RCMP Biology Section Methods Guide 1996). The match window provided an empirical determination of precision for each STR locus from a sample of multiple ascertainments of identical alleles (Menotti-Raymond et al. 2005). Using match window criteria, *composite STR genotypes* amplified from hair root and blood DNA were determined to match at all ten loci (seven heterozygous and three homozygous) STR loci (Figure 12.2) (Menotti-Raymond et al. 1997a).

A population genetic database was generated from cats in Prince Edward Island in order to calculate the likelihood of a random match between the hair genotype and a random individual in the population (Menotti-Raymond et al. 1997a). Additionally, a

Comparison of Snowball and Jacket Cat Hair STR Genotypes

STR Locus	DNA Source	Allele	Allele Size (bin)	Size Difference (bp) [Snowball Blood - Hair]		STR Locus Match Window (bp)	Conclusion
FCA 026	Snowball	1	147.83 (F)				
	Jacket cat hair	1	148.11 (F)	0.28	<	0.37	MATCH
	Snowball	2	143.73 (D)				
	Jacket cat hair	2	143.73 (D)	0	<	0.37	MATCH
FCA 043	Snowball	1	126.38 (C)				
	Jacket cat hair	1	126.29 (C)	0.09	<	0.59	MATCH
	Snowball	2	120.52 (B)				
	Jacket cat hair	2	120.50 (B)	0.02	<	0.59	MATCH
FCA 080	Snowball	1	259.27 (F)				
	Jacket cat hair	1	259.20 (F)	0.07	<	0.30	MATCH
	Snowball	2	253.39 (C)				
	Jacket cat hair	2	253.14 (C)	0.25	<	0.30	MATCH
FCA 088 A	Snowball	1	121.91 (F)				
	Jacket cat hair	1	121.91 (F)	0	<	0.42	MATCH
	Snowball	2	110.50 (A)				
	Jacket cat hair	2	110.50 (A)	0	<	0.42	MATCH
FCA 126	Snowball	1	143.41 (D)				
	Jacket cat hair	1	143.41 (D)	0	<	0.53	MATCH
	Snowball	2	141.08 (C)				
	Jacket cat hair	2	141.08 (C)	0	<	0.53	MATCH
FCA 132	Snowball	1	152.73 (G)				
	Jacket cat hair	1	152.73 (G)	0	<	0.27	MATCH
	Snowball	2	150.69 (F)				
	Jacket cat hair	2	150.69 (F)	0	<	0.27	MATCH
FCA 149	Snowball	1	132.02 (E)				
	Jacket cat hair	1	131.80 (E)	0.22	<	0.29	MATCH
	Snowball	2	128.05 (C)				
	Jacket cat hair	2	128.07 (C)	0.02	<	0.29	MATCH
FCA 058	Snowball	1	229.03 (D)				
	Jacket cat hair	1	229.24 (D)	0.21	<	0.36	MATCH
FCA 090	Snowball	1	93.54 (A)				
	Jacket cat hair	1	93.54 (A)	0	<	0.46	MATCH
FCA 096	Snowball	1	210.95 (D)				
	Jacket cat hair	1	211.00 (D)	0.05	<	0.25	MATCH

Figure 12.2 Graphics presented to the Prince Edward Island jury presenting matching composite STR profiles generated from DNA extracted from an evidentiary cat hair and the blood of Snowball. (From Menotti-Raymond, M. et al., *Nonhuman DNA Typing, Theory and Casework Applications*, ed. H.M. Coyle, CRC press, Boca Raton, FL, 2008b. With permission from Taylor & Francis Group.)

second database was generated from outbred cats from the Eastern United States (Menotti-Raymond et al. 1997a). Although the Prince Edward Island database was small, the island sample was adequate (95% confidence interval) to detect any STR allele present at a frequency of 9.5% or higher (Menotti-Raymond et al. 1997a). The two populations showed appreciable allelic variation and remarkable population genetic similarity (as opposed to geographic population substructuring). The incidence of the composite hair genotype for the seven heterozygous loci, estimated using the product rule (Jones 1972) and minimum allele frequency estimates for rare alleles (Budowle et al. 1996), was 2.2×10^{-8} and 6.9×10^{-7} for the Prince Edward Island and the US population databases, respectively (Menotti-Raymond et al. 1997a).

The cat evidence and additional human DNA evidence was presented to the Supreme Court of Prince Edward Island. On July 19, 1996, the jury convicted the defendant of second-degree murder.

12.2.2 Development of a Forensic Typing System for Genetic Individualization of Domestic Cat Samples

This legal precedent introducing animal genetic individualization in a homicide trial stimulated interest in the forensic community for the potential use of animal DNA forensic analysis. Under support from the National Institute of Justice (NIJ), the LGD developed a peer-reviewed forensic typing system for genetic individualization of domestic cat samples.

Ten tetranucleotide STRs were selected for the panel. The loci were unlinked, demonstrated high heterozygosity across multiple cat breeds, and showed an absence of cross-species amplification (Menotti-Raymond et al. 2005). A multiplex amplification protocol was developed so that the loci could be coamplified, including a gender identifying STS on the Y-chromosome amplifying a fragment of the SRY gene (Menotti-Raymond et al. 2005) (Figure 12.3). Validation studies of the multiplex demonstrated that complete product

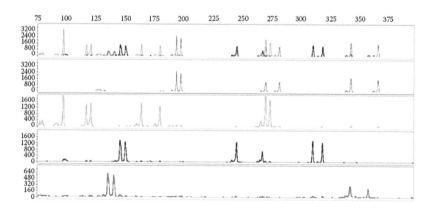

Figure 12.3 The PCR products of the 11 loci were designed in a size range from 100 to 415 bp, labeled with one of four fluorescent tags, with no allele overlap with adjacent loci and the SRY product detectable at 96 bp. Electropherogram of PCR products of 12-member multiplex amplified from 4 ng of male genomic DNA (upper panel). Lower panels demonstrate PCR products labeled with fluorescent tags FAM (3 STR), Vic (3 STR, SRY gene), Ned (3 STR), and PET (2 STR). Note that one locus (FcA736) of the original proposed multiplex has since been dropped due to a high incidence of null alleles observed in cat breeds of Asian ancestry (Menotti-Raymond, M. et al 2008)

profiles could be generated with as little as 125 pg of genomic DNA, with an absence of "allele dropout" (Coomber et al. 2007).

A population genetic database of domestic cat breeds was generated from the multiplex to compute composite match probabilities (Menotti-Raymond et al. 2012) using a sample set of 1043 samples representing 38 cat breeds (Vella et al. 2003). A small sample set of outbred domestic cats ($n = 24$) was additionally genotyped that demonstrated high heterozygosities for the multiplex (Menotti-Raymond et al. 2012).

Heterozygosities for the individual loci using all cats of recognized breed as a single dataset are quite high, ranging from 0.72 to 0.96, and are not significantly different than heterozygosities obtained for the outbred cat sample set (Menotti-Raymond et al. 2012). The multiplex exhibits good potential for genetic individualization of domestic cat samples within cat breeds and in outbred domestic cats, with a probability of match (P) of 6.2 × 10^{-14} using a conservative $\Theta = 0.05$ (Menotti-Raymond et al. 2012).

12.2.3 Validation Studies of Cat Multiplex

Quality assurance standards for DNA analysis advised by the DNA Advisory Board (DAB 1998) and recommendations for animal DNA and forensic testing (Budowle et al. 2005) before analysis of evidentiary samples were carefully followed to ensure the accuracy, precision, and reproducibility of the system. We examined species specificity and reproducibility of the system in DNA extracted from multiple tissue types, in addition to the ability of the typing system to identify DNA mixtures and the mutation rate of the loci (Coomber et al. 2007).

Species specificity of the loci was examined by amplifying DNA extracted from 28 North American mammals as well as two prokaryotes, *Sacchromyces cervesiae* and *Escherichia coli*. Other than two amplification products observed in the brown bear, another member of the Carnivore order, under the amplification conditions used in the multiplex, no other products were observed in any of the samples (Coomber et al. 2007).

Reproducibility of the typing system was examined by amplifying the multiplex in DNA extracted from blood and buccal samples from 13 unrelated domestic cats, producing identical profiles (Coomber et al. 2007).

Sensitivity studies demonstrate that a minimum of 0.125 ng of DNA was required to detect (with a minimum of threshold of 50 relative fluorescent units) both alleles of heterozygous loci.

The most common felid specimens that are likely to be associated with crime scenes are hair samples. Cat hair roots have proven to be a poor source of genomic DNA, with a 30 ng yield of DNA from the very best fresh plucked guard hair roots (Menotti-Raymond et al. 2000). This is on an order of 10–30 times less DNA than is available from a human hair root.

Samples which do not yield sufficient nuclear DNA for felid STR testing are potential candidates for mitochondrial DNA (mtDNA) analysis, a valuable technology utilized for degraded samples or those of insufficient quantity for STR analysis (Budowle et al. 1990; Holland and Parsons 1999). The cat mtDNA genome largely resembles that of other placental mammals (Lopez et al. 1996). However, 50% of the felid mtDNA genome has been transposed into the nuclear genome, extending from the 3′ end of the control region (CR) through 80% of the *COII* gene, exhibiting high sequence similarity with the homologous cytoplasmic mtDNA region, demonstrating 5.1% sequence divergence (Lopez et al. 1994).

This region must be avoided in forensic analysis. Translocated fragments of mtDNA have been reported in more than 64 different animal species (Bensasson et al. 2001).

Polymorphisms within the CR of human mtDNA have generally been used as a source of variation for forensic analysis (Gill et al. 1994; Holland et al. 1993; Holland and Parsons 1999). Fridez first characterized the discriminating power of felid mitochondrial CR using 21 polymorphic sites to identify 14 haplotypes in a survey of 50 outbred cats (Fridez et al. 1999). A felid mtDNA typing system has been proposed (Tarditi et al. 2011) following examination of a 402 bp region of the CR in 174 random-bred cats representing four geographic regions in the United States. Thirty- two mtDNA haplotypes were observed, ranging in frequency from 0.6% to 27% (Tarditi et al. 2011). Grahn et al. (2011) have since reported on a population genetic database generated from 1394 cats, which includes individuals from 25 worldwide geographic locations and 26 cat breeds, for the 402 bp region. Twelve major haplotypes were observed and a random match probability of 11.8%. Additionally, haplotype frequencies have been established for a Canadian population (Arcieri et al. 2016.)

The murder trial of the *State of Missouri v. Henry L. Polk, Jr.* represents the first legal proceeding where cat mitochondrial DNA analysis was introduced into evidence (Lyons et al. 2014). The mtDNA evidence was initially considered inadmissible due to concerns about the size of databases required for animal DNA profiling and the documentation of DNA markers. At a subsequent date, three laboratories provided additional DNA testimony at trial, expanding the size of the database, demonstrating, through expansion and saturation analysis of the cat mtDNA CR, support of the initial interpretation of the evidence.

Thus, the value of cat mtDNA for forensic analysis has been found acceptable in court in the US (Lyons et al. 2014). Additionally, saturation studies support the sufficiency of the US cat mtDNA CR dataset for forensic interpretations (Grahn et al. 2011). The consensus "Sylvester" reference sequence is published for standardization of future cat mtDNA CR studies (Grahn et al. 2011).

A growing area of interest in human forensic analysis is the potential of forensic DNA phenotyping, which focuses on the prediction of the physical appearance of an individual from a DNA sample (Kayser and de Knij 2011). Such phenotypes will be inferred largely by identifying SNPs/deletions/insertions that have been characterized as causative of distinctive morphological characteristics. Mutations have been characterized in the domestic cat responsible for a range of phenotypes, including melanism (black) (Eizirik et al. 2003), variants of the brown locus (brown and cinnamon) (Schmidt- Küntzel et al. 2005), dilute (Ishida et al. 2006), and white (David et al. 2014), as well as hair length (Drogemüller et al. 2007; Kehler et al. 2007), tabby pattern (Eizirik et al. 2010; Kaelin et al. 2012), hair consistency (Gandol et al. 2010), and three mutations associated with albino (Imes et al. 2006), Siamese, and Burmese phenotypes (Lyons et al. 2005; Schmidt-Küntzel et al. 2005). Thus, it would be possible from a non-hair sample to accurately predict the appearance of a cat. A SNP-based miniplex protocol has been developed with high discriminating power focusing on individual and phenotypic identification (Brooks et al. 2016). These SNPplexes can distinguish individual cats and their phenotypic traits, which could provide insight into crime reconstructions.

Animal forensic analysis is still in its infancy relative to human genetics. However, the genomic tools for STR profiling or mtDNA analysis have now been developed and peer reviewed, and population genetic databases have been generated with which to generate match probabilities. Accredited laboratories for animal profiling and advanced degree programs in animal forensics have emerged. Biological evidence from animal specimens has the potential to play an important role in future investigations. What will facilitate the application of these genomic profiling systems is the development of commercial kits with the necessary QC-produced reagents, validation studies, and allelic ladders for the STR kit. An ad hoc survey of detectives across the country by the NIJ's Office of Science and Technology revealed a strong interest in using such evidence if it were available.

12.3 Canine Forensic DNA Testing

Unlike cats, which can be aggressive but rarely inflict significant injury, dogs can be perpetrators that produce significant injury. The occurrence of dog bites at the rate of 3.5 million and 4.7 million dog bite injuries per year result in 238 mauling deaths annually and hundreds of millions of dollars in insurance claims (USA Today.com). Among the most vicious dog breeds, rottweilers, American pit bulls, and their hybrids have been characterized as "dangerous" breeds by the US Centers for Disease Control because they lead the statistics in dog bite–related fatalities (Kanthaswamy et al. 2009). The passage of "Cody's Law" in California to allow prosecutors to charge dog owners with a misdemeanor or a felony if their dogs attack people was prompted by a vicious dog attack (Dog Bite Law). In this incident, a pack of 21 pit bull–like dogs attacked an 11-year-old California boy named Cody Fox. Analysis of canine hairs from the victim's clothing helped with the unambiguous identification of two of the biting dogs. Both dogs were euthanized but since the case preceded Cody's Law (Governor Gray Davis of California signed the legislation in August 1999), the dogs' owner was only found guilty of misdemeanor for training dogs to attack and causing injury despite the intensity of the attack. Besides attacking people, usually children, animals tend to attack other animals. In one case investigation in which the author (Kanthaswamy) was involved, the killing of a cat by neighbor's dog resulted in the victim's owner paying about $500 to prove that the DNA from canine hairs found in the cat's mouth came from the suspected dog. As a result, the owners of the dog had to spend $1000 on a dog fence to prevent further cat kills (Tresniowski et al. 2006).

Animal forensic science is still a much younger science relative to human forensics. Budowle et al. (2005) proposed a set of recommendations for improving the quality standards of this field. In accordance to these guidelines to ease the exchange of information between laboratories, Kanthaswamy et al. (2009) and Tom et al. (2010) developed a DNA identification reagent kit consisting of standardized genetic markers, a common nomenclature, and a publicly accessible database for canine forensic DNA analysis. The reagent kit Canine Genotypes™ Panel 2.1 Kit (catalog number: F864S), which was developed by UC Davis and Finnzymes Oy (Espoo, Finland) specifically for forensic analysis of dog material, including identity and parentage testing, is now commercially available from Thermo Fisher Scientific (Waltham, Massachusetts). Along with the published nomenclature system and the public database, the availability of the reagent kit has already promoted collaboration among different laboratories in a murder investigation. Similarly, to improve nonhuman forensic genetic testing and promote the use of this forensic evidence in civil and criminal investigations, genetic identification resources akin to the ones developed by Kanthaswamy et al. (2009) and Tom et al. (2010) have to be developed for other species as well. The loci in Canine Genotypes™ are presented in Table 12.1.

While having a comprehensive DNA database is often recommended following human criminal and population reference databases, most animal forensic laboratories do not have

Table 12.1 Locus Descriptions for Canine Genotypes™ Panel 2.1 Kit Microsatellites and Amelogenin Marker

Locus Name	Chromosome	Repeat Motif	Size Range (bp)[a]	Dye Color[b]
AHTk211	26	di	79–101	Blue
CXX279	22	di	109–133	Blue
REN169O18	29	di	150–170	Blue
INU055	10	di	190–216	Blue
REN54P11	18	di	222–244	Blue
INRA21	21	di	87–111	Green
AHT137	11	di	126–156	Green
REN169D01	14	di	199–221	Green
AHTh260	16	di	230–254	Green
AHTk253	23	di	277–297	Green
INU005	33	di	102–136	Black
INU030	12	di	139–157	Black
Amelogenin	X	—	174–218	Black
FH2848	2	di	222–244	Black
AHT121	13	di	68–118	Red
FH2054	12	tetra	135–179	Red
REN162C04	7	di	192–212	Red
AHTh171	6	di	215–239	Red
REN247M23	15	di	258–282	Red

Note: Canine Genotypes™ Panel 2.1 Manual; https://www.thermofisher.com/document-connect/document-connect.html?url=https%3A%2F%2Fassets.thermofisher.com%2FTFS-Assets%2FLSG%2Fmanuals%2FMAN0012410_Canine_Genotypes_Panel_2.1_UG.pdf&title=VXNlciBHdWlkZTogQ2FuaW5lIEdlbm90eXBlcyBQYW5lbCAyLjE=.

[a] Size ranges are based on information provided by ISAG and data generated by Thermo Fisher Scientific. The data represents a large selection of dog breeds. However, some breeds may have alleles outside the ranges provided.

[b] Dye colors are listed as they appear following electrophoresis with Filter Set G5.

the resources to establish large-scale population databases for forensic casework (Budowle et al. 2005; Kanthaswamy 2015). Population sampling efforts for building dog STR databases tend to be restricted to opportunistic sampling from local sources such as veterinary clinics and dog owners. Sampling of local dog samples has been shown to be representative of a much broader geographic sampling (Himmelberger et al. 2008), and also estimates of genetic diversity based on population samples proximate to the crime scene tend to exhibit a more similar genetic structure represented by the forensic sample than do those collected from different and diverse geographic regions (Himmelberger et al. 2008; Kanthaswamy et al. 2009). Therefore, opportunistic, local sampling of fewer than 50/50 mixed-breed dogs cannot produce relevant but also sufficient discriminatory and exclusionary and enhanced processing efficiency and reliability (Kanthaswamy et al. 2009) is most capable of distinctly identifying individual animals with high precision.

Recent court challenges to canine DNA analysis have demonstrated a need for animal forensic DNA testing that has been validated according to human forensic guidelines. The potential for widespread acceptance of animal DNA evidence was somewhat diminished by the September 2003 decision of the Washington Court of Appeals that excluded canine DNA evidence in the 1998 case *State of Washington v. Kenneth Leuluaialii*. The appellate court ruled that the trial court judge had erred by failing to establish "whether the scientific community generally accepted that the specific loci used by PE Zoogen in the present case were highly polymorphic and appropriate for forensic use." The court also stated that

> current canine DNA testing and mapping focuses on the goals of paternity testing, breed testing and cancer and research studies. There is little indication that polymorphic loci and alleles in canine DNA have been sufficiently studied such that probability estimates are appropriate for the forensic use applied in this case.

(Docket Number 43507-8-I)

Two additional examples of the use of animal forensic DNA typing in casework are described next. The DNA analysis for Case A was almost not performed because of the numerous media releases on the dismissal of canine DNA evidence in a 2003 appeals court hearing in the State of Washington. The author (Kanthaswamy) had to convince the prosecutor in Case A to allow the laboratory to continue with the DNA testing because the controversy with the DNA results in the Washington trial did not impact his case. The second case investigation was successfully performed because most of the issues raised by the court ruling had been put to rest by developments in the animal forensic DNA analysis community, including the study by Halverson and Basten (2005b).

12.3.1 Case A Details: Fatal Dog Attack

In 2003, Vivian Anthony, a mother of three and a grandmother of 17 from Columbus, Ohio, was severely mauled by two large rottweilers. She suffered massive trauma and blood loss and was on life support for two days before she died when her heart, lungs, and kidneys failed. A former physician was charged with manslaughter and/or murder because his dogs were implicated in the brutal attack. He denied it was his dogs, but DNA from dog saliva on the victim's clothing (see Figure 12.4) matched his dogs. Faced with this irrefutable evidence, the dogs' owner pled guilty and was incarcerated for six months for one count of reckless homicide and one count of assault involuntary manslaughter. Both his dogs were subsequently destroyed. This was the first case in Ohio where dog DNA was used in a trial.

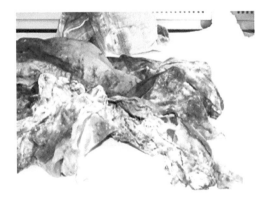

Figure 12.4 Victim's clothing from a fatal dog attack case. Dog saliva on the victim's clothing was tested for DNA. (File photo courtesy of VGL.)

12.3.2 Case B Details: Homicide

In 2009, two southwest London youths were attacked by rival gang members in a park. First, the boys were chased down and attacked with dogs, and then the dogs' owners kicked and punched the victims, and stabbed them with knives. One youth was stabbed 16 times and died, whereas another survived nine stab wounds. There were no eyewitnesses. Available CCTV footage, which recorded the attack, could not help identify the assailants with certainty because they had on hooded sweatshirts. In this case, the canine blood and saliva contributed to the physical evidence used by the British Crown Court to convict the defendants. The genetic identity testing kit Canine Genotypes™ 2.1 Multiplex STR Reagent Kit (see Kanthaswamy et al. 2009) was used to analyze the canine evidence, and the population genetic database used by the Crown's experts from LGC Forensics, UK, was composed of a sample set of UK dogs that represented 14 distinct dog breeds (R. Ogden, personal communication). An animal forensic expert from Questgen Forensics in Davis, California, and the author (Kanthaswamy) reviewed the results of LGC Forensics's DNA analysis and verified the veracity of the DNA analysis (J. Halverson, personal communication).

As a consequence of the State of Washington ruling, animal DNA evidence (canine in particular) was excluded from pending court cases in California and threatened the admissibility of this type of evidence in other states in the United States, including the previously stated case from Ohio (Case A). This ruling clearly underscored the need for validated nonhuman DNA analysis procedures that meet the rules of scientific acceptance and reliability. Recent developments of more informative genetic markers have demonstrated the polymorphic variability in nonhuman DNA analysis, leading to the same confidence of high discrimination as its human counterpart (Butler et al. 2010; Kanthaswamy et al. 2009). The second case example, (Case B) presented above used a canine forensic genetic testing kit, that is, the Canine Genotypes™ 2.1 Multiplex STR Reagent kit (Figure 12.5), which was developed and validated using a panel of canine-specific STRs using funding

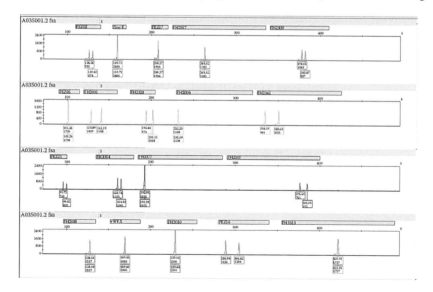

Figure 12.5 Canine forensic genetic testing kit, i.e., the Canine Genotypes™ 2.1 Multiplex STR Reagent Kit. GeneMapper electropherograms of amplified positive control dog (F-863) by the Canine 2.1 Multiplex STR Reagent Kit and the allelic distribution at each of the 19 loci. (From Kanthaswamy, S. et al., *J Forensic Sci*, 54, 829–840, 2009. With permission.)

from the US NIJ. Therefore, current nonhuman genetic testing techniques have now begun to address issues raised by the courts and are at the verge of adding another powerful tool of DNA typing methods to the forensic science arsenal.

12.4 Bovine Forensic DNA Testing

The recently sequenced bovine (*Bos taurus*) genome suggests a continued genetic exchange between wild cattle populations during their coexistence over a wide geographical range (Adelson 2008). At present, since the domestication event of cattle 8000 years ago, more than 50 distinct breeds are recognized. As a consequence of the close integration of cattle into the food chain of humans, forensically relevant cases involving cattle, such as identity forgery or cattle theft, are relatively common. Some of these examples have been published recently, demonstrating the relevance of this issue in forensic casework (van de Goor and van Haeringen 2007).

Microsatellites, also known as simple sequence repeats (SSRs) or short tandem repeats (STRs), represent specific sequences of DNA consisting of tandemly repeated units of one to six nucleotides, present as mono-, di-, tri-, tetra-, penta-, and hexanucleotide repeats, respectively. These sequences are abundant in prokaryotic and eukaryotic genomes, occurring in both coding and noncoding regions. Compared to other molecular markers, SSRs are uniquely characterized by their simplicity, abundance, ubiquity, variation, codominance, and multialleles among genomes (Powell et al. 1996). So, they have become a common tool broadly used in aspects of genetic mapping, molecular evolution, and systematic taxonomy in most genomes.

The conventional methods of generating microsatellite markers from genomic libraries (Weising et al. 2005) are being replaced rapidly by in silico mining of microsatellite sequences from DNA sequence databases (Jayashree et al. 2006; Korpelainen et al. 2007). Several search tools are available for mining microsatellite repeats in assembled genome sequences: BLASTN (Basic Local Alignment Search Tool; Temnykh et al. 2001), MISA (MIcroSAtellite; Thiel et al. 2003), Tandem-Repeats Finder (Benson 1999), and TROLL (tandem repeat occurrence locator; Castelo et al. 2002).

This approach has been used to identify microsatellite markers and investigate the type and distribution of repeat motifs in the expressed sequence tags of cattle. UniGene sequences of cattle were systematically searched for microsatellite markers using the "SSRFinder" Perl program. The dinucleotide repeat motifs were the most abundant SSRs in cattle, accounting for 54%, followed by 22%, 13%, 7%, and 4% for tri-, hexa-, penta-, and tetranucleotide repeats, respectively. Depending on the length of the repeat unit, the length of microsatellites varied from 14 to 86 bp. Among the di- and trinucleotide repeats, AC/TG (57%) and AGC (12%) were the most abundant types (Yan et al. 2008). These results indicated that the abundance of the different repeat motifs in the SSRs, as detected in UniGene sequences, was variable and thus unevenly distributed. These results are in agreement with previous reports on several animal species (Gupta et al. 2002; Ju et al. 2005; Perez et al. 2005; Rohrer et al. 2002; Serapion et al. 2004), which showed that the abundance of different repeats varied extensively depending on the species examined (Toth et al. 2000).

In 2006, the International Society for Animal Genetics (ISAG) recommended nine microsatellite loci (TGLA227, BM2113, ETH10, SPS115, TGLA126, TGLA122, INRA023,

ETH225, and BM1824) as the International Panel of Microsatellites for Cattle Parentage Testing (ISAG Panel), with a suggestion that other markers should be added to this panel to increase the efficacy in parentage testing (ISAG 2006). Recently, an additional three markers (BM1818, ETH3, and TGLA53) were suggested as candidate loci in cattle parentage analysis (ISAG 2008). At the moment, there are two firms—Finnzymes and Applied Biosystems—offering reagent kits for meat traceability and cattle parentage testing. Finnzymes, as part of Thermo Fisher Scientific, has three combinations of genotyping kits in its product portfolio: Bovine Genotypes™ Panels 1.2 (12 STR loci), 2.2 (6 STR loci), and 3.1 (18 STR loci). Applied Biosystems offers the StockMarks® for Cattle Bovine Parentage Typing Kit encompassing 11 STR loci. All herein mentioned reagent kits contain primers that amplify microsatellite markers recommended by ISAG for routine use in parentage testing and identification (Table 12.2). The alleles of all loci displayed the dinucleotide repeat structure.

In some developed countries, paternity identification using microsatellite markers is established for their cattle populations (Carolino et al. 2009; Cervini et al. 2006; Curi and Lopes 2002; Ozkan et al. 2009; Radko et al. 2005; Radko 2008; Řehout et al. 2006; Tian et al. 2008). In addition, microsatellites were used for the analyses of genetic diversity in cattle.

Table 12.2 30 Most Common Cattle STRs Recommended by the International Society for Animal Genetics (ISAG)

No	Markers	Chr	Primer Sequences (5′—3′) F&R (forward and reverse primers)		Reference
1	ETH225(D9S1)	9	GATCACCTTGCCACT ATTTCCT	ACATGACAGCC AGCTGCTACT	Steffen et al. (1993)
2	ETH152(D5S1)	5	TACTCGTAGGGCAGG CTGCCTG	GAGACCTCAGG GTTGGTGATCAG	Steffen et al. (1993)
3	HEL1(D15S10)	15	CAACAGCTATTTAACA AGGA	AGGCTACAGTC CATGGGATT	Kaukinen and Varvio (1993)
4	ILSTS005(D10S25)	10	GGAAGCAATGAAATC TATAGCC	TGTTCTGTGAG TTTGTAAGC	Brezinskyet al. (1993a)
5	HEL51(D21S15)	21	GCAGGATCACTTGTTA GGGA	AGACGTTAGTG TACATTAAC	Kaukinen and Varvio (1993)
6	INRA0052(D 12S4)	12	CAATCTGCATGAAGT ATAAATAT	CTT CAGGCATACC CTACACC	Vaiman et al. (1992)
7	INRA035(D16S11)	16	ATCCTTTGCAGCCTC CACATTG	TTGTGCTTTAT GACACTATCCG	Vaiman et al. (1994)
8	INRA063(D18S5)	18	ATTTGCACAAGCTAA ATCTAACC	AAACCACAGAA ATGCTTGGAAG	Vaiman et al. (1994)
9	MM8(D2S29)	2	CCCAAGGACAGAA AAGACT	CTCAAGATAAGAC CACACC	Mommens et al. (1994)
10	MM12(D9S20)	9	CAAGACAGGTGTTTCA ATCT	ATCGACTCTGG GGATGATGT	Mommens et al. (1994)
11	HEL9(D8S4)	8	CCCATTCAGTCTTCAG AGGT	CACATCCATGT TCTCACCAC	Kaukinen and Varvio (1993)
12	CSRM60(D10S5)	10	AAGATGTGATCCAAG AGAGAGGCA	AGGACCAGATC GTGAAAGG CATAG	Moore et al. (1994)
13	CSSM663(D14S31)	14	ACACAAATCCTTTCT GCCAGCTGA	AATTTAATGCA CTGAGGAG CTTGG	Barendse et al. (1994)
14	ETH185(D17S1)	17	TGCATGGACAGAGCA GCCTGGC	GCACCCCAACG AAAGCTCCCAG	Steffen et al. (1993)
15	HAUT24(D22S26)	22	CTCTCTGCCTTTGTCC CTGT	AATACACTTTA GGAGAAAAATA	Harlizius (comm.pers.)
16	HAUT27 (D26S21)	26	TTTTATGTTCATTTT TTGACTGG	AACTGCTGAAA TCTCCATCTTA	Harlizius (comm.pers.)
17	ETH3(D19S2)	19	GAACCTGCCTCTCCT GCATTGG	ACTCTGCCTGT GGCCAAGTAGG	Solinas Toldo et al. (1993)
18	ETH104(D5S3)	5	GTTCAGGACTGGCCCT GCTAACA	CCTCCAGCCCA CTTTCTCTTCTC	Solinas Toldo et al. (1993)
19	INRA0325(D11S9)	11	AAACTGTATTCTCTAAT AGCAC	GCAAGACATAT CTCCATTCCTTT	Vaiman et al. (1994)
20	INRA023(D3S10)	3	GAGTAGAGCTACAAG ATAAACTTC	TAACTACAGGG TGTTAGATG AACTCA	Vaiman et al. (1994)
21	BM2113(D2S26)	2	GCTGCCTTCTACCAAA TACCC	CTTAGACAACA GGGGTTTGG	Bishop et al. (1994)

(Continued)

Table 12.2 (Continued)   30 Most Common Cattle STRs Recommended by the International Society for Animal Genetics (ISAG)

No	Markers	Chr	Primer Sequences (5'—3') F&R (forward and reverse primers)		Reference
22	BM1818(D23S21)	23	AGCTGGGAATATAACC AAAGG	AGTGCTTTCAA GGTCCATGC	Bishop et al. (1994)
23	BM1824(D1S34)	1	GAGCAAGGTGTTTTTC CAATC	CATTCTCCAAC TGCTTCCTTG	Bishop et al. (1994)
24	HEL135(D11S15)	11	TAAGGACTTGAGATAA GGAG	CCATCTACCTC CATCTTAAC	Kaukinen and Varvio (1993)
25	ILSTS006(D7S8)	7	TGTCTGTATTTCTGCT GTGG	ACACGGAAGCG ATCTAAACG	Brezinskyet al. (1993b)
26	ILSTS030(D2S44)	2	CTGCAGTTCTGCATAT GTGG	CTTAGACAACA GGGGTTTGG	Kemp et al. (1995)
27	ILSTS0344(D5S54)	5	AAGGGTCTAAGTCCAC TGGC	GACCTGGTTTA GCAGAGAGC	Kemp et al. (1995)
28	ILSTS0332(D12S31)	12	TATTAGAGTGGCTCAG T GCC	ATGCAGACAGT TTTAGAGGG	Kemp et al. (1995)
29	ILSTS0113(D14S16)	14	GCTTGCTACATGGAAA GTGC	CTAAAATGCAG AGCCCTACC	Brezinskyet al. (1993c)
30	ILSTS0541(D21S44)	21	GAGGATCTTGATTTT GATGTCC	AGGGCCACTAT GGTACTTCC	Kemp et al. (1995)

12.5 Wildlife Forensic DNA Testing

Interpol recognized wildlife crime as the second most prevalent crime worldwide after drug trafficking (Linacre 2021). The value of the trade is often quoted as $20 billion per year (Alacs and Georges 2008; Linacre 2021), although the exact figure is unknown as very little illegal trade is ever detected. Furthermore, the low penalties and potential large financial gain led to extensive and lucrative trade in endangered species (Linacre and Ciavaglia 2017; Wilson-Wilde et al. 2010; Johnson et al. 2014). The scope and range of wildlife crime is wide and encompasses a variety of criminal activities such as poaching and illegal hunting of mammals (Zhang et al. 2020; Ewart et al. 2018), birds (Coghlan et al. 2012b; van Hoppe et al. 2016) and reptiles (Ciavaglia and Linacre 2018; Meganathan et al. 2010; Lee et al. 2009b), and the use of animal derivatives in traditional medicines (Byard 2016; Coghlan et al. 2012a).

It is the illegal trade in endangered species either as whole organisms, body parts, or as ornaments and medicines where international legislation can assist (Figures 12.6, 12.7, and 12.8). Forensic investigations can only be undertaken if legislation has been breached. The international trade in endangered species is monitored and regulated through recommendations made by CITES. These recommendations are enforced at a national level by legislation. Normally, this legislation stipulates the names of the species that are protected. It is for these reasons that much interest at a research level has focused on methods of species identification.

Genetic testing to enforce national legislation is rarely performed by the same operational laboratories that undertake human identification. This is primarily attributable to

Figure 12.6 The image shows the effect of poaching on a species where there may be fewer than 1000 Bengali tigers in the wild. (Image courtesy of the wildlife Institute of India.)

Figure 12.7 An ivory sculpture from which DNA can be isolated to determine if it is a CITES-listed species.

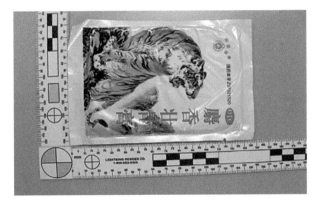

Figure 12.8 Illegal trade in endangered species as whole organisms, body parts, or as ornaments and medicines. A seized traditional East Asian medicine that was found to contain CITES-listed species.

time and cost constraints, and the fact that most operational laboratories are accredited to undertake standard operating procedures relevant to human identification only. Most wildlife forensic investigations are undertaken by universities on an ad hoc basis and performed by university academics who may not routinely undertake work for the criminal justice system. Good laboratory practice has been encouraged, and guidelines have been suggested for QC and QA within a wildlife forensic laboratory as much of the work is undertaken by academic institutions (Linacre et al. 2011). Notable exceptions where there are dedicated wildlife forensic science laboratories include the US Fish and Wildlife Laboratory in Ashland, Oregon; the Australian Museum in Sydney, Australia; and the Wildlife DNA Forensics laboratory, part of the Scottish Agricultural Science Agency (SASA), near Edinburgh in the United Kingdom. All three have attained ISO accreditation.

12.5.1 mtDNA in Species Testing

It may be that the seized material has been processed to create a sculpture, ornament, item of clothing, food, or supposed medicinal product. DNA can be obtained at trace levels from ivory (Ewart et al. 2020) where ivory is made into statues, or rhino horn (Ewart et al. 2018) where the rhino horn is processed to make knife handles. In so many instances DNA is at trace levels in wildlife trade such that mtDNA is the only genetic material available for processing to determine the species of origin (Linacre and Ciavaglia 2017).

Human identification has centered on a small section of mtDNA that does not encode a protein or an RNA molecule. This noncoding section is termed the hypervariable regions and comprises less than 1000 of the 16,569 bases that make up the human mitochondrial genome. The mitochondrial genome in eukaryotes encodes a total of 37 genes, 22 of which encode tRNA molecules, 2 encode rRNA molecules, and the other 13 encode proteins involved primarily with the process of oxidative respiration (Linacre and Tobe 2013). The number of genes on the mitochondrial genome is largely invariant for all vertebrate mitochondrial genomes, but the order of the genes may alter (Linacre and Tobe 2013); for instance, the order is different between avian and mammalian mitochondrial genomes. Vertebrate mtDNA has two strands of different densities: the heavy (or H-strand) and the light (or L-strand).

Although the use of mitochondrial loci in species testing has many benefits, there are a few disadvantages as well. Part of the benefit of mitochondrial loci is that they are inherited maternally and as a haplotype with no recombination. Maternal inheritance can be a problem if hybrid species occur, particularly if the mother is from a nonprotected species and the father is from a species that is protected (Amorim et al. 2020; González et al. 2020).

12.5.2 Species Identification Using Loci on Mitochondrial Genome

A classic characteristic for a locus used in species identification is that it should exhibit very little variation for any member of the same species such that all members of that species have the same, or very nearly the same, sequence; this is referred to as intraspecies variation. The second characteristic is that this same locus should exhibit sufficient differences from any member of the next closest species, being interspecies DNA sequence variation. The main locus used in taxonomic and phylogenetic studies until recently was cytochrome *b* (cyt *b*) (Kocher et al. 1989; Hsieh et al. 2001). This occurs between bases 14,747 and 15,887 in human mtDNA and encodes a protein 380 amino acids in length. The cyt *b* locus has been used extensively in taxonomic and forensic studies (Lee et al. 2009a; Lee et al. 2006; Lee et al. 2009b; Ewart et al. 2018; Meganathan et al. 2010).

The cytochrome *c* oxidase I (COI) mitochondrial gene locus was adopted by the Barcode for Life Consortium (BOLD) (www.boldsystems.org) (Hebert and Gregory 2005; Hebert et al. 2004; Ratnasingham and Hebert 2007). COI gene sequences, found between bases 5904 and 7445 in human mtDNA, were used initially in the identification of invertebrate species (Ball et al. 2005; Hajibabaei et al. 2006) and became the locus of choice in forensic entomology (Nelson and Melton 2007) before being adopted by BOLD.

The Commission for the International Society for Forensic Genetics (ISFG) recommended that there should be some comment as to which locus was used in any analysis (Linacre et al. 2011). Currently, there is no standardized locus in species testing. It is most likely that different loci will exhibit varying inter- and intraspecies similarity depending on whether examining insects, other invertebrates, fish, reptiles, birds, or mammals. Mammals are the only taxonomic group where a detailed studied has been performed to determine whether cyt *b* or COI have fewer false-positive and false-negative identifications (Tobe et al. 2010); in this study, cyt *b* was found to outperform COI.

Both COI and cyt *b* are large genes and currently the complete loci are not used in species identification—rather, sections of the loci are used; for the cyt *b* gene this is typically the first 400 bases (Hsieh et al. 2001) and for COI the first 645 bp (Hebert and Gregory 2005). The use of either section of the cyt *b* or COI loci is aided by the design of universal primers; these are primers that can be used in PCR to amplify a section of the gene from all the species for which the primers are designed.

12.5.3 Mitochondrial Sequence Analysis

Regardless of which primer pair has been used, amplification followed by sequencing is performed to obtain data for analysis. These sequences are then compared to a reference DNA sequence. Reference DNA sequences need to be obtained from known sources, and in this regard voucher specimens are ideal and were recommended by the ISFG Commission (Linacre et al. 2011). DNA sequence data can be otherwise found at repositories such as the National Center for Biotechnology Information (NCBI) database of genetic information called GenBank (www.ncbi.nlm.nih.gov/genbank/). Software including the Basic Local Alignment Search Tool (BLAST) compares the target sequence fragment to the data deposited on GenBank and produces a similarity score of the 100 most closely matched sequences. These are listed with the closest similarity at the top of the list. A similarity score of 100% to a known species, with the next closest being less than a similarity of 95%, would be ideal. This situation rarely occurs with a similarity score of 99% between the sequence tested and the nearest sequence being more typical. There is no consensus on how many differences and over what number of bases are due to either intra- or interspecies variation. It was reported that one base variation per 400 bp sequence is acceptable intraspecies variation (Linacre and Tobe 2013).

12.5.4 Conclusions on Animal Testing

Species testing using loci on the mitochondrial genome has become a standard method in conservation ecology and phylogenetic studies. These same processes were transferred to a forensic science arena after minimal validation. There remains no standardization as to which locus to use, nor which primer sets have been validated. Furthermore, many of the studies used in taxonomic research do not consider the opportunity for false inclusions or exclusions, although these factors are crucial and should be known before use of the test in the criminal justice system. It should be noted that many who undertake species testing in a forensic context come from a background where this type of testing is familiar, from taxonomic or evolutionary studies. Moreover, operational forensic science laboratories typically work to maximum capacity with examination of human DNA, and have little spare capacity to consider developing methods for species testing that may only be undertaken on occasion. With notable exceptions in Scotland, the US, and Australia, wildlife forensic science in many countries therefore remains the domain of university academics, where analysis can be performed to appropriate standards, provided that recommendations, such as those published by the ISFG and Society of Wildlife Forensic Science, are adhered to.

Acknowledgments

Felid section: The authors gratefully acknowledge the National Institute of Justice for funding of this research through interagency agreements to the National Cancer Institute's Laboratory of Genomic Diversity. We gratefully acknowledge the Cat Fanciers' Association and The International Cat association for their considerable help in facilitating sample collection from cat breeds. We also thank the hundreds of independent cat breeders who provided us with blood and buccal swab samples of their cats and pedigrees. Without the help of these individuals and organizations, this research would not have been possible. Amy Snyder, Leslie Wachter, and Nikia Coomber provided excellent technical assistance in this project. Grant Number 1999-IJ-R-A079 awarded by the National Institute of Justice, Office of Justice Programs, and US Department of Justice supported this project. Points of view in this document are those of the authors, and do not necessarily represent the official position or policies of the US Department of Justice. *Canine section:* The author (Kanthaswamy) thanks the VGL for allowing the review of the closed case forensic files for use in this research. Case A and the dog/cat case are from the period when the author headed the laboratory's forensic unit. The author was assisted by the staff of the forensic unit in the completion of work related to the cases, and had taken their contributions into account in the preparation of their case reports. This project was partially funded by the National Institute of Justice (NIJ Grant No. 2004-DN-BX-K007) awarded to the author. *Wildlife section:* The author wishes (Linacre) to acknowledge the Ministry of Justice, South Australia, for the ongoing support of forensic science at Flinders University.

References

Adelson, D.L. 2008. Insights and applications from sequencing the bovine genome. *Reprod Fertil Dev* 20(1):54–60.

Agronis, A. 2002. UC Davis Mag 19:3 (UC Davis).

Alacs, E.A., A. Georges, N.N. FitzSimmons, and J. Robertson. 2010. DNA Detective: A review of molecular approaches to wildlife forensics. *Forensic Sci Med Pathol.* 6(3):180–94.

Alacs, E., and A. Georges. 2008. Wildlife across our borders: A review of the illegal trade in Australia. *Aust J Forensic Sci* 40(2):147–60.

Amorim, A., F. Pereira, C. Alves, and O. Garcia. 2020. Species assignment in forensics and the challenge of hybrids. *Forensic Sci Int Genet* 48:102333.

Arcieri, M., G. Agostinelli, Z. Gray, A. Spadaro, L.A. Lyons, and K.M. Webb. 2016. Establishing a database of Canadian feline mitotypes for forensic use. *Forensic Sci Int Genet* 22:169–74.

Archibald, A.L., C.S. Haley, J.F. Brown, et al. 1995. The PIGMap consortium linkage map of the pig (Sus scrofa). *Mamm Genome* 6(3):157–75.

Ayres, K.L., and A.D.J. Overall. 1999. Allowing for within-subpopulation inbreeding in forensic match probabilities. *Forensic Sci Int* 103(3):207–16.

Ayres, K.L. 2000. Relatedness testing in subdivided populations. *Forensic Sci Int* 114(2):107–15.

Ball, S.L., P.D.N. Hebert, S.K. Burian, and J.M. Webb. 2005. Biological identifications of mayflies (Ephemeroptera) using DNA barcodes. *J North Am Benthol Soc* 24(3):508–24.

Barendse, W., S.M. Armitage, L.M. Kossarek, et al. 1994. A genetic linkage map of the bovine genome. *Nat Genet* 6(3):227–35.

Bensasson, D., D. Zhang, D.L. Hartl, and G.M. Hewitt. 2001. Mitochondrial pseudogenes: Evolution's misplaced witnesses. *Trends Ecol Evol* 16(6):314–21.

Benson, D.A., I. Karsch-Mizrachi, D.J. Lipman, J. Ostell, and E.W. Sayers. 2011. GenBank. *Nucleic Acids Res* 39(Database issue):D32–7.

Brooks, A., E.K. Creighton, B. Gandolfi, R. Khan, R.A. Grahn, and L.A. Lyons. 2016. SNP Miniplexes for individual identification of random-bred domestic cats. *J Foren Sci* 61(3):594–606.

Budowle, B., D.E. Adams, C.T. Comey, et al. 1990. Mitochondrial DNA: A possible genetic material suitable for forensic analysis. In: *Advances in Forensic Sciences*, ed. H.C. Lee and R.E. Gaensslen. Chicago: Medical Publishers.

Budowle, B., K.L. Monson, and R. Chakraborty. 1996. Estimating minimum allele frequencies for DNA profile frequency estimates for PCR-based loci. *Int J Legal Med* 108(4):173–6.

Budowle, B., P. Garofano, A. Hellmann et al. 2005. Recommendations for animal DNA forensic identity testing. *Int J Legal Med* 119(5):295–302.

Butler, J.M., P.M. Scheneider, and A. Carracedo. 2010. Journal update. *Forensic Sci Int* 4(3):143–4.

Byard, R. 2016. Traditional medicines and species extinction: Another side to forensic wildlife investigation. *Forensic Sci Med Pathol* 12(2):125–7.

Carolino, I., C.O. Sousa, S. Ferreira, N. Carolino, F.S. Silva, and L.T. Gama. 2009. Implementation of a parentage control system in Portuguese beef-cattle with a panel of microsatellite markers. *Genet Mol Biol* 32:306–11.

Carracedo, A., J.M. Prieto, L. Concheiro, and J. Estefanía. 1987. Isoelectric focusing patterns of some mamma-lian keratins. *J Forensic Sci* 32(1):93–9.

Cassidy, B.G., and R.A. Gonzales. 2005. DNA testing in animal forensics. *J Wildl Manag* 69(4):1454–62.

Castelo, A.T., W. Martins, and G.R. Gao. 2002. TROLL—Tandem repeat occurrence locator. *Bioinformatics* 18(4):634–6.

Ciavaglia, S., and A. Linacre. 2018. OzPythonPlex: An optimised forensic STR multiplex assay set for the Australasian carpet python (*Morelia spilota*). *Forensic Sci Int Genet* 34:231–48.

Cervini, M., F. Henrique-Silva, N. Mortari, and E. Matheucci Jr. 2006. Genetic variability of 10 microsatellite markers in the characterization of Brazilian Nellore cattle (Bos indicus). *Genet Mol Biol* 29:486–90.

Coghlan, M.L., J. Haile, J. Houston, et al. 2012a. Deep sequencing of plant and animal DNA contained within traditional Chinese medicines reveals legality issues and health safety concerns. *PLOS Genet* 8(4):e1002657.

Coghlan, M.L., N.E. White, L. Parkinson, J. Haile, P.B.S. Spencer, and M. Bunce. 2012b. Egg forensics: An appraisal of DNA sequencing to assist in species identification of illegally smuggled eggs. *Forensic Sci Int Genet* 6(2):268–73.

Coomber, N., V.A. David, S.J. O'Brien, and M. Menotti-Raymond. 2007. Validation of a short tandem repeat multiplex typing system for genetic individualization of domestic cat samples. *Croat Med J* 48(4):547–55.

Crawford, A.M., K.G. Dodds, A.J. Ede, et al. 1995. An autosomal genetic linkage map of the sheep genome. *Genetics* 140(2):703–24.

Culver, M., W.E. Johnson, J. Pecon-Slattery, and S.J. O'Brien. 2000. Genomic ancestry of the American puma (*Puma concolor*). *J Hered* 91(3):186–97.

Culver, M., M.A. Menotti-Raymond, and S.J. O'Brien. 2001. Patterns of size homoplasy at 10 microsatellite loci in pumas (*Puma concolor*). *Mol Biol Evol* 18(6):1151–6.

Curi, R.A. and C.R. Lopes. 2002. Evaluation of nine microsatellite loci and misidentification paternity frequency in a population of Gyr breed bovines. *Braz J Vet Res Anim Sci* 39:129–35.

D'Andrea, E., F. Fridez, and R. Conquoz. 1998. Preliminary experiments on the transfer of animal hair during simulated criminal behavior. *J Forensic Sci* 43(6):1257–8.

David, V.A., M. Menotti-Raymond, A.C. Wallace, et al. 2014. Endogenous retrovirus insertion in the KIT oncogene determines White and White spotting in domestic cats. *G3. Gen Geno Genet* 4(10):1881–91.

Dietrich, W., H. Katz, S.E. Lincoln, et al. 1992. A genetic map of the mouse suitable for typing intraspecific crosses. *Genetics* 131(2):423–47.

DNA Advisory Board (DAB). 1998. Quality assurance standards for forensic DNA testing laboratories. *Forensic Sci Commun* 2000:2.

Dog Bite Law, 16. http://www.dogbitelaw.com/plain-english-overview-of-dog-bite-law/when-dogs-bite-people.html.

Dreamin' Demon, 14. http://www.dreamindemon.com/forums/showthread.php?38050-Robert-Pettyjohn-animal-abusing-scum-faces-40-years-in-prison.

Drogemüller, C., S. Rufenacht, B. Wichert, and T. Leeb. 2007. Mutations within the FGF5 gene are associated with hair length in cats. *Anim Genet* 38(3):218–21.

Eizirik, E., J.H. Kim, M. Menotti-Raymond, P.G. Crawshaw, S.J. O'Brien, and W.E. Johnson. 2001. Phylogeography, population history and conservation genetics of jaguars (*Panthera onca*). *Mol Ecol* 10(1):65–79.

Eizirik, E., N. Yuhki, W.E. Johnson, M. Menotti-Raymond, S.S. Hannah, and S.J. O'Brien. 2003. Molecular genetics and evolution of melanism in the cat family. *Curr Biol* 13(5): 448–53.

Eizirik, E., V.A. David, Buckley-Beason, et al. 2010. Defining and mapping mammalian coat pattern genes: Multiple genomic regions implicated in domestic cat stripes and spots. *Genetics* 184(1):267–75.

Ellegren, H., B.P. Chowdhary, M. Johansson, et al. 1994. A primary linkage map of the porcine genome reveals a low rate of genetic recombination. *Genetics* 137(4):1089–100.

Ewart, K.M., G.J. Frankham, R. Mcewing, et al. 2018. An internationally standardized species identification test for use on suspected seized rhinoceros horn in the illegal wildlife trade. *Forensic Sci Int Genet* 32:33–9.

Ewart, K.M., A.L. Lightson, F.T. Sitam, J.J. Rovie-Ryan, N. Mather, and R. Mcewing. 2020. Expediting the sampling, decalcification, and forensic DNA analysis of large elephant ivory seizures to aid investigations and prosecutions. *Forensic Sci Int Genet* 44:102187.

Finnzymes Canine Genotypes™ Panel 1.1 Manual. 2012. http://diagnostics.finnzymes.fi/pdf/canine_genotypes_manual_1_1_low.pdf.

Fregeau, C.J., and R.M. Fourney. 1993. DNA typing with fluorescently tagged short tandem repeats: A sensitive and accurate approach to human identification. *BioTechniques* 15(1):100–19.

Fridez, F., S. Rochat, and R. Coquoz. 1999. Individual identification of cats and dogs using mitochondrial DNA tandem repeats? *Sci Justice* 39(3):167–71.

Gandolfi, B., C.A. Outerbridge, L.G. Beresford, et al. 2010. The naked truth: Sphynx and Devon Rex cat breed mutations in KRT71. *Mamm Genome* 21(9–10):509–15.

Gill, P., P.L. Ivanov, C. Kimpton, R. Piercy, N. Benson, G. Tully et al. 1994. Identification of the remains of the Romanov family by DNA analysis. *Nat Genet* 6(2):130–5.

Gilbert, D.A., C. Packer, A.E. Pusey, J.C. Stephens, and S.J. O'Brien. 1991. Analytical DNA fingerprinting in lions: Parentage, genetic diversity, and kinship. *J Hered* 82(5):378–86.

Gonzalez, B.A., A.M. Agapito, F. Novoa-Munoz, J. Vianna, W.E. Johnson, and J.C. Marin. 2020. Utility of genetic variation in coat color genes to distinguish wild, domestic and hybrid South American camelids for forensic and judicial applications. *Forensic Sci Int Genet* 45:102226.

Grahn, R.A., J.D. Kurushima, N.C. Billings, et al. 2011. Feline non-repetitive mitochondrial DNA control region database for forensic evidence. *Forensic Sci Int Genet* 5(1):33–42.

Gupta, P.K., S. Rustgi, S. Sharma, R. Singh, N. Kumar, and H.S. Balyan. 2002. Transferable EST-SSR markers for the study of polymorphism and genetic diversity in bread wheat. *Mol Genet Genomics* 270(4):315–23.

Hajibabaei, M., D.H. Janzen, J.M. Burns, W. Hallwachs, and P.D.N. Hebert. 2006. DNA barcodes distinguish species of tropical Lepidoptera. *Proc Natl Acad Sci U S A* 103(4):968–71.

Halverson, J.L., and C. Basten. 2005a. Forensic DNA identification of animal-derived trace evidence: Tools for linking victims and suspects. *Croat Med J* 46(4):598–605.

Halverson, J.L., and C. Basten. 2005b. A PCR multiplex and database for forensic DNA identification of dogs. *J Forensic Sci* 50(2):352–63.

Hauge, X.Y., and M. Litt. 1993. A study of the origin of 'shadow bands' seen when typing dinucleotide repeat polymorphisms by the PCR. *Hum Mol Genet* 2(4):411–5.

Hebert, P.D.N., and T.R. Gregory. 2005. The promise of DNA barcoding for taxonomy. *Syst Biol* 54(5):852–9.

Hebert, P.D.N., M.Y. Stoeckle, T.S. Zemlak, and C.M. Francis. 2004. Identification of birds through DNA barcodes. *PLOS Biol* 2(10):e312.

Higuchi, R., C.H. von Beroldingen, G.F. Sensabaugh, and H.A. Erlich. 1988. DNA typing from single hairs. *Nature* 332(6164):543–6.

Himmelberger, A.L., T.F. Spear, J.A. Satkoski, et al. 2008. Forensic utility of the mitochondrial hypervariable 1 region of domestic dogs, in conjunction with breed and geographic information. *J Forensic Sci* 53(1):81–9.

Holland, M.M., D.L. Fisher, L.G. Mitchell, et al. 1993. Mitochondrial DNA sequence analysis of human skeletal remains: Identification of remains from the Vietnam War. *J Forensic Sci* 38(3):542–53.

Holland, M.M., and T.J. Parsons. 1999. Mitochondrial DNA sequence analysis—Validation and use for forensic casework. *Forensic Sci Rev* 11(1):1–49.

Hsieh, H.M., H.L. Chiang, L.C. Tsai, et al. 2001. Cytochrome b gene for species identification of the conservation animals. *Forensic Sci Int* 122(1):7–18.

Humane Society of the United States. (2012). U.S. Pet ownership statistics 2011–2012. http://www .humanesociety.org/issues/pet_overpopulation/facts/pet_ownership_statistics.html.

Imes, D.L., L.A. Geary, R.A. Grahn, and L.A. Lyons. 2006. Albinism in the domestic cat (*Felis catus*) is associated with a tyrosinase (TYR) mutation. *Anim Genet* 37(2):175–8.

Inhuman.Org. http://www.inhumane.org/data/BEldred.html.

ISAG Conference. 2006. Porto Seguro, Brazil. Cattle Molecular Markers and Parentage Testing Workshop. http://www.isag.org.uk/ISAG/all ISAG2006_CMMPT.pdf.

ISAG Conference. 2008. Amsterdam, the Netherlands. Cattle Molecular Markers and Parentage Testing Workshop. http://www.isag.org.uk/ISAG/all/ISAG2008_CattleParentage.pdf.

Ishida, Y., V.A. David, E. Eizirik, et al. 2006. A homozygous single-base deletion in MLPH causes the dilute coat color phenotype in the domestic cat. *Genomics* 88(6):698–705.

Jacob, H.J., D.M. Brown, R.K. Bunker, et al. 1995. A genetic linkage map of laboratory rat, Rattus norvegicus. *Nat Genet* 9(1):63–9.

James, S.H., and J.J. Nordby (eds). 2003. *Forensic Science: An Introduction to Scientific and Investigative Techniques*. 2nd ed. Boca Raton, FL: CRC Press.

Jayashree, B., R. Punna, P. Prasad, et al. 2006. A database of simple sequence repeats from cereal and legume expressed sequence tags mined in silico: Survey and evaluation. *In Silico Biol* 6(6):607–20.

Jeffreys, A.J., V. Wilson, and S.L. Thein. 1985a. Individual-specific 'fingerprints' of human DNA. *Nature* 316(6023):76–9.

Jeffreys, A.J., J.F. Brookfield, and R. Semeonoff. 1985b. Positive identification of an immigration test- case using human DNA fingerprints. *Nature* 317(6040):818–9.

Jeffreys, A.J., and D.B. Morton. 1987. DNA fingerprints of dogs and cats. *Anim Genet* 18(1):1–15.

Johnson, R.N., L. Wilson-Wilde, and A. Linacre. 2014. Current and future directions of DNA in wildlife forensic science. *Forensic Sci Int Genet* 10:1–11.

Johnson, W.E., F. Shinyashiku, M. Menotti-Raymond, et al. 1999. Molecular genetic characterization of two insular Asian cat species, Bornean Bay Cat and Iriomote Cat. In: *Evolutionary Theory and Processes: Modern Perspectives*, ed. S.P. Wasser, 223–48. Netherlands: Kluwer Academic Publishers.

Jones, D.A. 1972. Blood samples: Probability of discrimination. *J Forensic Sci Soc* 12(2):355–9.

Ju, Z., M.C. Wells, A. Martinez, L. Hazlewood, and R.B. Walter. 2005. An in silico mining for simple sequence repeats from expressed sequence tags of zebrafish, Medak, Fundulus and Xiphophorus. *In Silico Biol* 5(5–6):439–63.

Kaelin, C., X. Xiao, L.Z. Hong, et al. 2012. Specifying and sustaining pigmentation patterns in domestic and wild cats. *Science* 337(6101):1536–41.

Kanthaswamy, S. 2009. The development and validation of a standardized canine STR panel for use in forensic casework (NIJ Grant No. 2004-DN-BX-K007 Final Report). National Institute of Justice (NIJ) National Criminal Justice Reference Service. http://www.ncjrs.gov/pdffiles1/nij/grants/226639.pdf.

Kanthaswamy, S. 2015. Review: domestic animal forensic genetics – biological evidence, genetic markers, analytical approaches and challenges. *Anim Genet* Oct; 46(5):473–84.

Kanthaswamy, S., B.K. Tom, A.M. Mattila, et al. 2009. Canine population data generated from a Multi-plex STR Kit for use in forensic casework. *J Forensic Sci* 54(4):829–40.

Kayser, M., and P. de Knijff. 2011. Improving human forensics through advances in genetics, genomics and molecular biology. *Nat Rev Genet* 12(3):179–92.

Kehler, J.S., V.A. David, A.A. Schäffer, et al. 2007. Four independent mutations in the feline fibroblast growth factor 5 gene determine the long-haired phenotype in domestic cats. *J Hered* 98(6):555–66.

Kocher, T.D., W.K. Thomas, A. Meyer, et al. 1989. Dynamics of mitochondrial DNA evolution in animals—Amplification and sequencing with conserved primers. *Proc Natl Acad Sci U S A* 86(16):6196–200.

Korpelainen, H., K. Kostamo, and V. Virtanen. 2007. Microsatellite marker identification using genome screening and restriction–ligation. *BioTechniques* 42(4):479–86.

Lee, J.C.I., L.C. Tsai, S.P. Liao, A. Linacre, and H.M. Hsieh. 2009a. Species identification using the cytochrome b gene of commercial turtle shells. *Forensic Sci Int Genet* 3(2):67–73.

Lee, J., H.M. Hsieh, and L.H. Huang. 2009b. Ivory identification by DNA profiling of cytochrome b gene. *Int J Legal Med* 123(2):117–21.

Lee, J.C.I., L.C. Tsai, C.Y. Yang, et al. 2006. DNA profiling of shahtoosh. *Electrophoresis* 27(17):3359–62.

Linacre, A. 2008. The use of DNA from non-human sources. *Forensic Sci Int Genet Suppl Ser* 1(1):605–6.

Linacre, A. (ed) 2009. *Forensic Science in Wildlife Investigations*. Boca Raton, FL: CRC Press, 178 pp

Linacre, A. 2021. Wildlife crime in Australia. *Emerg Top Life Sci* 5(3):487–494.

Linacre, A., and S.A. Ciavaglia. 2017. Wildlife forensic science. In: *Handbook of Forensic Genetics*, ed. A. Amorin and B. Budowle. London: World Scientific Publishing.

Linacre, A., L. Gusmao, W. Hecht, et al. 2011. ISFG: Recommendations regarding the use of nonhuman (animal) DNA in forensic genetic investigations. *Forensic Sci Int Genet* 5(5):501–5.

Linacre, A., and S. Tobo. 2013. *Wildlife DNA Analysis: Applications in Forensic Science*. Chichester, West Sussex, UK: John Wiley & Co.

Lopez, J.V., N. Yuhki, R. Masuda, W. Modi, and S.J. O'Brien. 1994. Numt, a recent transfer and tandem amplification of mitochondrial DNA to the nuclear genome of the domestic cat. *J Mol Evol* 39(5):174–90.

Lopez, J.V., S. Cevario, and S.J. O'Brien. 1996. Complete nucleotide sequence of the domestic cat (*Felis catus*) mitochondrial genome and a transposed mtDNA tandem repeat (Numt) in the nuclear genome. *Genomics* 33(2):229–46.

Lyons, L.A., D.L. Imes, H.C. Rah, and R.A. Grahn. 2005. Tyrosinase mutations associated with Siamese and Burmese patterns in the domestic cat (*Felis catus*). *Anim Genet* 36(2):119–26.

Lyons, L.A., R.A. Grahn, T.J. Kun, L.R. Netzel, E.E. Wictum, and J.L. Halverson. 2014. Acceptance of domestic cat mitochondrial DNA in a criminal proceeding. *Forensic Sci Int Genet* 13:61–7.

Meganathan, P.R., B. Dubey, K.N. Jogayya, N. Whitaker, and I. Haque. 2010. A novel multiplex PCR assay for the identification of Indian crocodiles. *Mol Ecol Resour* 10(4):744–7.

Mellersh, C.S., A.A. Langston, G.M. Acland, et al. 1997. A linkage map of the canine genome. *Genomics* 46(3):326–36.

Menotti-Raymond, M. et al. 2008b. *Nonhuman DNA Typing, Theory and Casework Applications*. ed. H.M. Coyle, Boca Raton, Fl: CRC Press. With permission from Taylor & Francis Group

Menotti-Raymond, M.A., V.A. David, and S.J. O'Brien. 1997a. Pet cat hair implicates murder suspect. *Nature* 386(6627):774.

Menotti-Raymond, M.A., V.A. David, J.C. Stephens, L.A. Lyons, and S.J. O'Brien. 1997b. Genetic individualization of domestic cats using feline STR loci for forensic applications. *J Forensic Sci* 42(6):1039–51.

Menotti-Raymond, M.A., V.A. David, L.A. Lyons, et al. 1999. A genetic linkage map of microsatellites in the domestic cat (*Felis catus*). *Genomics* 57(1):9–23.

Menotti-Raymond, M.A., V.A. David, and S.J. O'Brien. 2000. DNA yield from single hairs (wool and guard/shed and plucked), success rate in amplifying STR and mtDNA targets, estimating DNA yield using multicopy target. Eleventh International Symposium on Human Identification; Biloxi, MS. http://www.promega.com/geneticidproc/ussymp11proc/abstracts/menotti_raymond.pdf.

Menotti-Raymond, M.A., V.A. David, Z.Q. Chen, et al. 2003a. Second generation integrated genetic linkage/radiation hybrid maps of the domestic cat. *J Hered* 94:95–106.

Menotti-Raymond, M.A., V.A. David, L.L. Wachter, J.M. Butler, and S.J. O'Brien. 2005. An STR forensic typing system for the genetic individualization of domestic cat (*Felis catus*) samples. *J Forensic Sci* 50(5):1061–70.

Menotti-Raymond, M.A., V.A. David, A.A. Schäffer, et al. 2009. An autosomal genetic linkage map of the domestic cat, Felis silvestris catus. *Genomics* 93(4):305–13.

Menotti-Raymond, M.A., V.A. David, B.S. Weir, and S.J. O'Brien. 2012. A population genetic database of cat breeds developed in coordination with a domestic cat STR multiplex. *J Forensic Sci* 57(3):596–601.

Moore, J.E. 1988. A key for the identification of animal hairs. *J Forensic Sci Soc* 28(5–6):335–59.

Morin, P.A. 1992. Paternity exclusion using multiple hypervariable microsatellite loci amplified from nuclear DNA of hair cells. In: *Paternity in Primates: Genetic Tests and Theories*, ed. R.D. Martin, A.F. Dixson, and E.J. Wickings, 63–81. Basel: Karger.

Murphy, W.J., B. Davis, V.A. David, et al. 2007. A 1.5-Mb-resolution radiation hybrid map of the cat genome and comparative analysis with the canine and human genomes. *Genomics* 89(2):189–96.

Murphy, W.J., S. Sun, Z. Chen, et al. 2000. A radiation hybrid map of the cat genome: Implications for comparative mapping. *Genome Res* 10(5):691–702.

Nelson, L.A., J.F. Wallman, and M. Dowton. 2007. Using COI barcodes to identify forensically and medically important blowflies. *Med Vet Entomol* 21(1):44–52.

O'Brien, S.J., M. Menotti-Raymond, W.J. Murphy, and N. Yuhki. 2002. The feline genome project. *Annu Rev Genet* 36:657–86.

Ogden, R., N. Dawnay, and R. McEwing. 2009. Wildlife DNA forensics—Bridging the gap between conservation genetics and law enforcement. *Endang Species Res* 9(3):179–95.

Ozkan, E., M.I. Soysal, M. Ozder, E. Koban, O. Sahin, and I. Togan. 2009. Evaluation of parentage testing in the Turkish Holstein population based on 12 microsatellite loci. *Livest Sci* 124:101–6.

Pádár, Z., M. Angyal, B. Egyed, et al. 2001. Canine microsatellite polymorphisms as the resolution of an illegal animal death case in a Hungarian Zoological Gardens. *Int J Legal Med* 115(2):79–81.

Pádár, Z., B. Egyed, K. Kontadakis, et al. 2002. Canine STR analyses in forensic practice. Observation of a possible mutation in a dog hair. *Int J Legal Med* 116(5):286–8.

Peabody, A.J., R.J. Oxborough, P.E. Cage, and I.W. Evett. 1983. The discrimination of cat and dog hairs. *J Forensic Sci Soc* 23(2):121–9.

Perez, F., J. Ortiz, M. Zhinaula, C. Gonzabay, J. Calderon, and F.A. Volckaert. 2005. Development of EST-SSR markers by data mining in three species of shrimp, litopenaeus vannamei, litopenaeus stylirostris and Trachypenaeus birdy. *Mar Biotechnol (N Y)* 7(5):554–69.

Pontius, J.U., J.C. Mullikin, D.R. Smith, et al. 2007. Initial sequence and comparative analysis of the cat genome. *Genome Res* 17(11):1675–89.

Powell, W., G.C. Machray, and J. Provan. 1996. Polymorphism revealed by simple sequence repeats. *Trends Plant Sci* 1(7):215–22.

Radko, A., A. Zyga, T. Zabek, and E. Slota. 2005. Genetic variability among Polish Red, Hereford and Holstein-Friesian cattle raised in Poland based on analysis of microsatellite DNA sequences. *J Appl Genet* 46:89–91.

Radko, A. 2008. Microsatellite DNA polymorphism and its usefulness for pedigree verification of cattle raised in Poland. *Ann Anim Sci* 8:311–21.

Ratnasingham, S., and P.D.N. Hebert. 2007. BOLD: The Barcode of Life Data System (www.barcod inglife.org). *Mol Ecol Notes* 7(3):355–64.

Řehout, V., E. Hradecká, and J. Čítek. 2006. Evaluation of parentage testing in the Czech population of Holstein cattle. *Czech J Anim Sci* 51:503–9.

Rohrer, G.A., S.C. Fahrenkrug, D. Nonneman, N. Tao, and W.C. Warren. 2002. Mapping microsatellite markers identified in porcine EST sequences. *Anim Genet* 33(5):372–6.

Royal Canadian Mounted Police. 1996. *Interpretation of STR Profiles. Royal Canadian Mounted Police; Biology Section Methods Guide*. Ottawa, Ontario0

Savolainen, P., and J. Lunderberg. 1999. Forensic evidence based on mtDNA from dog and wolf hairs. *J Forensic Sci* 44(1):77–81.

Scharnhorst, G., and S. Kanthaswamy. 2011. An assessment of scientific and technical aspects of closed canine forensics DNA investigations—Case series from the University of California, Davis. *Croat Med J* 52(3):280–92.

Schmidt-Küntzel, A., E. Eizirik, S.J. O'Brien, and M. Menotti-Raymond. 2005. Tyrosinase and tyrosinase related protein 1 alleles specify domestic cat coat color phenotypes of the albino and brown loci. *J Hered* 96(4):289–301.

Schneider, P.M., Y. Seo, and C. Rittner. 1999. Forensic mtDNA hair analysis excludes a dog from having caused a traffic accident. *Int J Legal Med* 112(5):315–16.

Serapion, J., H. Kucuktas, J. Feng, and Z. Liu. 2004. Bioinformatics mining of type I microsatellites from expressed sequence tags of channel catfish (Ictalurus punctatus). *J Mar Biotechnol* 6(4):364–77.

Smalling, B.B., J.A. Satkoski, B.K. Tom, et al. 2010. The significance of regional and mixed breed canine mtDNA databases in forensic science. Bentham open-open. *Forensic Sci J* (Special Issue) 3:22–32.

Tampa Bay Online. 2010. http://www2.tbo.com/content/2010/jul/27/man-convicted-notorious -animal-cruelty-case-back-j/.

Tarditi, C.R., R.A. Grahn, J.J. Evans, J.D. Kurushima, and L.A. Lyons. 2011. Mitochondrial DNA sequencing of cat hair: An informative forensic tool. *J Forensic Sci* 56(Suppl 1):S36–46.

Temnykh, S., G. DeClerck, A. Lukashova, L. Lipovich, S. Cartinhour, and S. McCouch. 2001. Computational and experimental analysis of microsatellites in rice (*Oryza sativa* R.): Frequency, length variation, transposon associations, and genetic marker potential. *Genome Res* 11(8):1441–52.

Thiel, T., W. Michalek, R.K. Varshney, and A. Graner. 2003. Exploiting EST databases for the development and characterization of gene-derived SSR-markers in barley (*Hordeum vulgare* L.). *Theor Appl Genet* 106(3):411–22.

Tian, F., D. Sun, and Y. Zhang. 2008. Establishment of paternity testing system using microsatellite markers in Chinese Holstein. *J Genet Genom* 35:279–84.

Tobe, S.S., A. Kitchener, and A. Linacre. 2010. Reconstructing mammalian phylogenies: A detailed comparison of the cytochrome b and cytochrome oxidase subunit I mitochondrial genes. *PLOS ONE* 5(11):e14156.

Tom, B.K., M.T. Koskinen, M.R. Dayton, A.M. Mattila, E. Johnston, and D. Fantin. 2010. Development of a nomenclature system for a canine STR multiplex reagent kit. *J Forensic Sci* 55(3):597–604.

Toth, G., Z. Gaspari, and J. Jurka. 2000. Microsatellite in different eukaryotic genome, survey and analysis. *Genome Res* 10:1967–81.

Tresniowski, A., J.S. Podesta, and H. Breur. 2006. *People Mag*, January 23rd issue.

Van de Goor, L.H.P., and W.A. van Haeringen. 2007. *Identification of Stolen Cattle Using 22 Microsatellite Markers. Nonhuman DNA Typing: Theory and Casework Applications*, 122–3. Boca Raton, FL: CRC.

Van Hoppe, M.J.C., M.A.V. Dy, M. Van Der Einden, and A. Iyengar. 2016. A novel STR multiplex validated for forensic use in the hen harrier. *Forensic Sci Int Genet* 22:100–9.

Vella, C.M., L.M. Shelton, J.J. Mcgonagle, and T.W. Stanglein. 2003. *Robinson's Genetics for Cat Breeders and Veterinarians*. 4th ed. Edinburgh: Butterworth Heinemann.

Vigne, J.D., J. Guilaine, K. Debue, L. Haye, and P. Gérard. 2004. Early taming of the cat in Cyprus. *Science* 304(5668):259.

Weising, K., H. Nybom, K. Wolff, and G. Kahl. 2005. DNA Fingerprinting in Plants. Principles, Methods, and Applications. 2nd ed. Boca Raton, FL: CRC Press, Taylor & Francis Group.

Wilson-Wilde, L. 2010a. Combating wildlife crime. *Forensic Sci Med Pathol* 6(3):149–50.

Wilson-Wilde, L. 2010b. Wildlife crime: A global problem. *Forensic Sci Med Pathol* 6(3):221–2.

Wilson-Wilde, L., J. Norman, J. Robertson, S. Sarre, and A. Georges. 2010. Current issues in species identification for forensic science and the validity of using the cytochrome oxidase I (COI) gene. *Forensic Sci Med Pathol* 6(3):233–41.

Yan, Q., Y. Zhang, H. Li, et al. 2008. Identification of microsatellites in cattle unigenes. *J Genet Genomics* 35(5):261–6.

Zhang, H., G. Ades, M.P. Miller, F. Yang, K.W. Lai, and G.A. Fischer. 2020. Genetic identification of African pangolins and their origin in illegal trade. *Glob Ecol Conserv* 23:e01119.

Application of DNA-Based Methods in Forensic Entomology

13

JEFFREY D. WELLS AND VEDRANA ŠKARO

Contents

13.1 Introduction

Nonhuman animal DNA analysis plays an important role in forensic investigations. It includes invertebrates and vertebrates. The DNA analysis of vertebrates has equal application in both human and animal forensics, whereas DNA analysis of invertebrates has applications almost exclusively as an auxiliary tool in human forensics. Among the invertebrates, the most common sources of forensic evidence are arthropods. Hundreds of arthropod species, most importantly the insects, and primarily flies (Diptera) and beetles (Coleoptera), are attracted to corpses. Forensic entomology uses insects to help law enforcement determine the cause, location, and time of death of a human corpse as well as to detect poisons, drugs, and physical neglect and abuse (Pilli et al. 2016; Sharma et al. 2018; Wang et al. 2019; Al-Qahtni et al. 2020; Joseph et al. 2011; Verma and Paul 2013; Jales et al. 2020; de Carvalho et al. 2012). With information gathered at a crime scene and the study of insects found there (i.e., distribution, biology, and behavior), it may be possible to accurately determine how long a body has been at a specific location (equal to postmortem interval if death occurred at the site of discovery) or to connect the suspect to the crime scene (Anderson and Cervenka 2002; Catts and Goff 1992; Catts and Haskell 1990; Greenberg and Kunich 2002; Sharma et al. 2015a; Charabidze et al. 2017; Mona et al. 2019). However, the great majority of forensic entomological analyses are aimed at estimating the time of death. By conventional signs of postmortem decay (paleness or *pallor mortis*, hypostasis or *livor mortis*, coldness or *algor mortis*, stiffness or *rigor mortis*, etc.), time of death (postmortem interval [PMI]) can be accurately estimated for the first two to three days. Nevertheless, by forensic entomology (the study of the succession of necrophage

DOI: 10.4324/9780429019944-15

species and developmental stages of insects that feed on the corpse), the time elapsed from the moment of death to the preservation of insect specimens from the corpse or other entomological observations can be accurately calculated for the interval of one day to several weeks. Insect specimens should be considered physical evidence at the crime scene and processed as any other biological material, following the recommended procedures for quality assurance for collection, preservation, packaging, and transport in order to prevent contamination or destruction of evidence and to guarantee the chain of custody (Amendt et al. 2007; Amendt and Hall 2007; Budowle et al. 2005; Lord and Burger 1983; Randall et al. 1998; Ritz-Timme et al. 2000). The groups of invertebrates and their role in forensics are listed in Table 13.1, and the most important insects are listed in Table 13.2 (Gunn 2009).

Table 13.1   Groups of Invertebrates in Forensics

Invertebrates' Role in Forensics	Invertebrates Grouped According to Their Role in Forensics	Invertebrate Representatives
Forensic indicators in cases of murder or suspicious death	Those that are attracted to dead bodies	Detritivores (many different flies, dermestid beetles, pyralid and tineid moths, mites, earthworms, and miscellaneous invertebrates such as nematodes, slugs, snails, Collembola, Diplura, and millipedes)
		Carnivores and parasitoids (carabid and staphylinid beetles, burying beetles, histerid beetles, ants and wasps, centipedes and spiders)
	Those that leave dead bodies	Fleas and lice
	Those that become accidentally associated with the dead body and/or the crime scene	Miscellaneous insects
Cause of death	Those with direct response (anaphylactic reaction caused by stinging a person in the mouth or throat)	Honeybee, wasp, hornets
	Those with indirect response (causing distraction or panic attacks that result in a fatal accident—unexplained vehicle accidents)	Spiders, wasps and bees, bumblebees
Forensic indicators in cases of neglect and animal welfare	Those that infest wounds or which live on the body surface (possible consequences: pain and distress, fatal infection of the blood with bacteria)	Fleas, lice, and mites larvae of Diptera (the maggots of blowflies and flesh flies)
Food spoilage and hygiene litigation	Those that make a foodstuff unsuitable for human consumption (feces, cast skins, or bite marks) and transmit diseases by transferring pathogens such as bacteria, viruses, protozoa, and nematodes	Blowflies and other Diptera beetles, ants and wasps, butterflies and moths, cockroaches, mites, and many other insects
Subject of the illegal trade	Those that are captured above the allowed quota (results in population decline, habitat destruction, and possible extinction, and therefore some species become more valuable for illegal trade)	Mole cricket (*Gryllotalpa gryllotalpa*), large blue butterfly (*Maculinea arion*), swallowtail butterfly (*Papilio machaon*), medicinal leech (*Hirudo medicinalis*), *Triops cancriformis*, and food species (e.g., lobsters, oysters)

Table 13.2 **Most Important Insects in Forensics[a]**

Flies (order Diptera)	Blowflies (family Calliphoridae)
	Flesh flies (family Sarcophagidae)
	House flies (family Muscidae)
	Cheese flies (family Piophilidae)
	Scuttle flies (family Phoridae)
	Lesser corpse flies (family Sphaeroceridae)
	Lesser house flies (family Fanniidae)
	Black scavenger flies (family Sepsidae)
	Sun flies (family Heleomyzidae)
	Black soldier fly (family Stratiomyidae)
Beetles (order Coleoptera)	Rove beetles (family Staphylinidae)
	Hister beetles (family Histeridae)
	Carrion beetles (family Silphidae)
	Ham beetles (family Cleridae)
	Carcass beetles (family Trogidae)
	Leather/hide beetles (family Dermestidae)
	Scarab beetles (family Scarabaeidae)
	Sap beetles (family Nitidulidae)
Mites (Class Acari)	Family Tyroglyphidae
	Family Oribatidae
Moths (order Lepidoptera)	Clothing moths (family Tineidae)
	Grease moth (family Pyralinae)
Hymenoptera (order Hymenoptera)	Wasps (family Vespidae)
	Ants (family Formicidae)
	Bees (superfamily Apoidea)

[a] The list is incomplete because many other species can be found on a corpse and there may be significant differences with regard to climatic conditions.

Case study 1 (see later) describes a forensic case from 1948 in which the use of caddisfly casings was introduced as a tool for forensic investigation to estimate the time of death and link the forensic sample to the victim (Caspers 1952).

In forensic entomology, not only is the insect's life cycle important, but species identification is important as well. Recognition of morphological characteristics is often difficult and requires comprehensive knowledge of a taxonomic group. One reason for the complexity of this kind of recognition is that species of the class Insecta constitute more than 80% of the entire animal kingdom. Also, there are a limited number of experts able to identify forensically important insect larvae at the species level. Since some closely related species grow at very different rates, precise identification of species can be crucial to the analysis. Because of these difficulties, taxonomic misidentification of a specimen can lead to corresponding uncertainty in estimation of PMI which, depending on the species and temperature, could be miscalculated by more than a week. For some groups of insects, differentiating the morphological larvae stage is not possible. In these cases, an alternative would be rearing the larvae to adult. However, this method is slow, because of the duration of the process, and not reliable since rearing sometimes fails. An alternative approach is DNA analysis, not only to identify forensically important species in any life stage, but also to possibly identify human DNA that may be present in the insects' guts. In situations in which larvae are found in the absence of a corpse, it would be useful to know on what (and if it was a human, on whom) they were feeding. Molecular analysis allows identification through partial specimen rather than intact specimen as may be required in morphological identifications (Alacs et al. 2010). Forensic applications of entomological DNA typing are genetic fingerprints of insect specimen (Sharma and Singh 2017), insect species identification (Sharma et al. 2015a), and estimation of insects' age (Tarone et al. 2007; Zehner et al. 2009).

Molecular tests used in forensic entomology include mitochondrial DNA (mtDNA) and nuclear DNA (nDNA) analysis. Analysis of mtDNA can be effective for species identification. It is often preferred in cases of highly processed and degraded tissue because it is easier to type, since it is present in multiple copies per cell compared to one copy of nDNA from each parent. Also, in order to amplify an informative segment of mtDNA across a wide range of taxa, universal mtDNA primers are used. Therefore, the time needed for method development is typically less for mtDNA markers compared to nDNA. However, as with any method, the accuracy and reliability of an mtDNA-based species determination method can only be shown by empirical validation (Alacs et al. 2010).

Proposed forensic insect molecular genetics techniques include population genetic methods for inferring the postmortem relocation of a corpse (Picard and Wells 2012; Charabidze et al. 2017) and the use of gene expression levels to estimate specimen age (Tarone et al. 2007; Zehner et al. 2009; Boehme et al. 2013).

The different methods and technology have improved over the time and continue to evolve.

13.2 Methods of Insect DNA Analysis

The first step in the typical analysis of forensic insect material is DNA extraction followed by amplification by polymerase chain reaction (PCR), cycle sequencing, and comparison of obtained results with reference samples. An example of sequencing results is shown in Figure 13.1a. DNA extraction can be performed by one of the commonly used methods such as phenol/chloroform, cetyltrimethylammonium bromide, Chelex extraction, or by commercially available extraction kits (e.g., *Quick*-DNA Tissue/Insect Kits, DNeasy Blood & Tissue Kit, QIAamp® Fast DNA Tissue Kit, or DNAzol®). DNA analysis (mitochondrial or nuclear origin) enables insect species identification in each stage of its life cycle, and also in cases where only parts of the insect are available. To remain suitable for DNA extraction, entomological samples collected on and around the corpse are usually stored in 95%–100% alcohol and/or by freezing. Adult insects preserved by traditional pinning may be suitable for DNA extraction, but this is not optimal for genetic analysis. Contamination with foreign DNA is minimized by washing the surface of the specimen with ethanol or a weak bleach solution (Linville and Wells 2002), and by avoiding the gut when removing tissue for DNA extraction. Amplification and sequence analysis of DNA follows the usual protocols.

In order to avoid loss of DNA during the extraction procedure, direct PCR can be used as a cost-effective, fast, and highly efficient method. However, tissues used for direct PCR could not be reused for further genetic studies. To overcome this issue, direct PCR workflow has been optimized (Thongjued et al. 2019). Optimization included a dilution

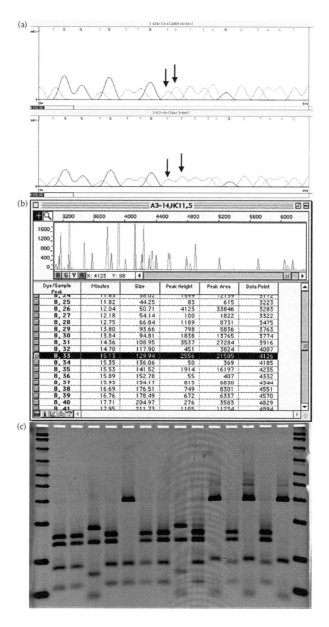

Figure 13.1 Two sequence electropherograms with arrows indicating a polymorphic site distinguishing the blowflies *Compsomyiops callipes* (a) and *Cochliomyia macellaria*. (b) An AFLP profile for the blowfly *Phormia regina*. (c) An acrylamide gel image showing heteroduplex profiles used to distinguish species of Sarcophagidae.

step using phosphate-buffered saline (PBS), a heating step, use of high-fidelity DNA polymerase, and additional primer pairs. The study showed that the workflow had high sensitivity and could be applied to dried samples, ethanol-preserved samples, and food samples with high success rates. Also, it can be applied to a wide range of insect taxa. The pre-PCR dilution part of the workflow may be combined with other amplification or detection techniques, such as next-generation sequencing, to resolve issues related to mixed DNA samples and decrease metabarcoding costs.

DNA sequencing is certainly the most common method used in forensic entomology, primarily because it is the most informative among the available methods of DNA analysis and because the availability of a large number of PCR primers makes it easy to analyze small overlapping DNA regions such as is necessary for degraded tissue. A variety of commercial kits, computer-operated genetic analyzers, and software-aided sequence analysis made sequencing widely available and relatively easy to perform. Furthermore, many companies now offer sequencing at a relatively affordable price, although using such a service is not appropriate for casework. The most common locus for specimen identification is some portion of the mtDNA genes for cytochrome *c* oxidase subunits one and two (COI and COII) (e.g., Wells and Williams 2007; Aly and Wen 2013; Sharma et al. 2015b). Because mtDNA is abundant in the cell and haploid, it is easier to sequence compared to nuclear DNA, and its use in forensic entomology preceded the term DNA barcoding (Hebert et al. 2003). Proposed alternatives to COI + II include different mtDNA genes (e.g., Zehner et al. 2004b, Gemmellaro et al. 2019; Tuccia et al. 2016) or nuclear DNA markers including internal transcribed spacer regions, noncoding ribosomal genes, period gene, other coding genes such as elongation factor 1α and specific targeting of introns (Nelson et al. 2008; Stevens and Wall 2001; Williams and Villet 2013; McDonagh and Stevens 2011; GilArriortua et al. 2015; Federico et al. 2018). A few publications have included more than one locus in a single analysis based on sequence data (Stevens 2003; Stevens et al. 2002; Wallman et al. 2005). Other authors have also suggested identification using the DNA fragment size genotypes produced by restriction digestion of PCR product (PCR-RFLP; e.g., Sperling et al. 1994; Schroeder et al. 2003), randomly amplified polymorphic DNA (RAPD; e.g., Benecke 1998; Sharma and Singh 2017), or amplified fragment length polymorphism (AFLP; e.g., Vos et al. 1995; Picard 2010; Picard and Wells 2012; Picard et al. 2017). An example of the AFLP profile is shown in Figure 13.1b.

The choice of DNA-based method that will be employed for the identification of the forensically important insects should rely on how well it distinguishes the most closely related species. Data interpretation should be considered in the context of the method employed since each has advantages and limits as well (Stevens et al. 2019).

One of the issues related to the methods used in forensic entomology is lack of official validation standards for DNA-based insect species determination (Stevens et al. 2019). Comparison of studies can be difficult when there are inconsistencies among gene choice and methods (Gemmellaro et al. 2019). In addition, if not successfully used on hundreds of specimens to produce correct determination, the method may not be reliable (Wells and Williams 2007). So, agreement is needed on standard best practices for this field.

Also, entomologists should have good knowledge of local carrion fauna well, as well as any trends in changes in distributions because it may help to narrow the choice to only the species found in the location at that time of year (Picard 2013; Owings and Picard 2018).

Whatever the genotype used, certain steps are required to design a reliable species-diagnostic test (see discussion in Wells and Stevens 2008). One is the assembly of a trusted and relevant database of reference genotypes. Many published genotypes are erroneous (Brunak et al. 1990), so an investigator must be confident that the reference specimens were correctly identified and that the record contains no artifact. Another common problem has been the use of an incomplete taxon sample for the reference database. A specimen could be misidentified by a given procedure if it is not a species represented in the database. Missing genotypes for all likely candidate species may lead to misidentification of the evidence specimen. Unique, trusted reference genotype data for candidate species could allow an unambiguous identification. The reference databases are improving. As mentioned earlier, with any forensic analytical procedure, a species-diagnostic test should be empirically validated.

The interpretation of a species-diagnostic test can vary according to the technology and requires specialized knowledge. One can expect there to be intraspecific genetic variation, especially in the loci useful for distinguishing close relatives. Phylogenetic analysis is a convenient way to handle large and complex genetic datasets with intraspecific variation, and identification is straightforward in taxa that show reciprocal monophyly for the locus (Wells and Williams 2007), as is not always the case (Will and Rubinoff 2004; Wells et al. 2007). Polymerase chain reaction-restriction fragment length polymorphism (PCR-RFLP) techniques depend on an exact match between members of a species. Therefore, PCR-RFLP would be unreliable to the extent that there was intraspecific variation in the relevant restriction enzyme sites. However, as with any method, the measure of reliability is empirical validation rather than theoretical concerns. Random amplified polymorphic DNA (RAPD) is problematic in that the results have been shown to be extremely sensitive to experimental conditions (MacPherson et al. 1993; Perez et al. 1998).

The difference between intraspecific sequence variation and the genetic distance between close relatives has been shown for some comparisons. For *Chrysomya rufifacies* and *Chrysomya albiceps*, it was detected that there was a less than 1% mismatch within a species and about 3% mismatch between species. Similar levels of variation were found for flesh flies (family Sarcophagidae) *Sarcophaga argyrostoma* and *Sarcophaga crassipalpis* (within-species variation, 1%; between-species variation, about 3%) and blowflies *C. vicina* and *C. vomitoria* (within-species variation, <1%; between-species variation, about 5%). An example of heteroduplex profiles used to distinguish species of Sacrophagidae is shown in Figure 13.1c. Although variation within species is generally less than the variation between species, a careful interpretation is necessary because data on sequences and variations are still not complete. An additional problem is the possible breeding between different species and major differences in sequences of the same species from geographically distant locations (Wells and Sperling 2001; Wells et al. 2001; Zehner et al. 2004b). More research needs to be published in order to make clear whether all populations of a given species can be identified by a particular gene sequence, and whether there are problems caused by subspecies or cryptic species (Gunn 2009; Wells and Stevens 2008).

Geographic genetic variation might be useful to reconstruct the postmortem movement of a corpse (Picard and Wells 2012). However, a false conclusion is possible in cases when geographic population genetic structure is not well understood (Picard et al. 2017). Using SNPs for population genetic analysis allows relatively simple identification of both neutral markers and loci under selection, even for organisms without a well-annotated genome. It should also be considered if populations differ in any other forensically important aspect of their biology, for example, development rate (Gallagher et al. 2010; Tarone et al. 2011; Owings et al. 2014). Though the life cycles are predictable, certain environmental factors can shorten or prolong the time of maturation of the insects. The most significant variables are temperature and species (Greenberg and Kunich 2002. Also, carrion-feeding insects can consume and accumulate the non-metabolized drugs ingested by the deceased person, since some drugs stay in the body systems for longer period of time. These drugs alter the metabolism of the insects and consequently affect their rate of growth and size (Goff and Lord 2001). Cocaine and methamphetamine in the cadaver can accelerate the larval development (Joseph et al. 2011; Verma and Paul 2013; de Carvalho et al. 2012).

Organophosphate poisons in the carrion can delay insect colonization (Joseph et al. 2011; Jales et al. 2020). The drugs stored in insect flesh can be detected and even if the drug cannot be quantified, just the fact that it is detected is enough to make the diagnosis (Verma and Paul 2013).

Although RAPD has been used for species determination (see case study 2), a result may be difficult to reproduce. However, RAPD profiles can be generated from any well-preserved tissue sample, and the method does produce complex genotypes that are more suitable for population genetic analysis than the data from a typical sequencing experiment (Stevens and Wall 2001).

For animal studies, the role of RAPD profiles has been replaced by the more repeatable genotypes produced by AFLP (Vos et al. 1995). This included investigations of calliphorids (Baudry et al. 2003; Alamalakala et al. 2009; Picard and Wells 2009). Although AFLP has been proposed for forensic insect specimen identification (Alamalakala et al. 2009; Picard and Wells 2012; Picard et al. 2017), the method requires a large amount of nondegraded specimen DNA, and the protocol is more complicated than sequencing. Therefore, it is not likely to be used for this purpose except for taxa that cannot be distinguished by standard sequence-based methods. The extreme polymorphism of AFLP genotypes makes them useful for applications such as population assignment, which in theory could be used to reconstruct the postmortem transportation of a corpse (Picard and Wells 2010).

In many scientific disciplines, the use of microsatellite loci has now become established as the gold standard for revealing population-level genetic variation. One of the key advantages of using microsatellite loci is the ability to reliably recognize alleles and to interpret the resulting genetic data using powerful conventional Mendelian/Hardy–Weinberg-based analyses. Microsatellite typing methods have been developed for some carrion-feeding Calliphoridae (Florin and Gyllenstrand 2002; Torres and Azeredo-Espin 2008; Rodrigues et al. 2009; Farncombe et al. 2014). However, we are not aware of them having been applied in a forensic context.

Species-diagnostic methods, such as pyrosequencing (Balitzki-Korte et al. 2005; Hsieh et al. 2020), heteroduplex analysis (Boakye et al. 1999), real-time PCR (Wetton et al. 2004), or microarrays (Belosludtsev et al. 2004), are used for non-insect taxa, and it can be concluded that such methods can play a greater role in forensic entomology. These technologies allow simultaneous analysis of >1 locus in order to overcome problems of DNA identification of closely related species. Real-time PCR is a reliable method for multilocus analysis (Jang et al. 2019), but the number of different genotypes that can be simultaneously screened is limited. Real-time PCR is used not only for multilocus analysis but for analysis of differential gene expression as well in order to predict insects' developmental stages (Tarone and Foran 2011; Tarone et al. 2007; Zehner et al. 2009).

Multilocus analysis using a variety of oligonucleotide probes can be placed on a single microarray chip; a single microarray could be produced that is capable of performing virtually every genotyping task needed for a forensic insect specimen (Wells and Stevens 2008). However, it is necessary to think through about making such a chip that would work for all combinations of insects in all geographical regions and, at the same time, to make it affordable to end users.

When DNA is degraded, which is usually the case in forensic case samples, full-length barcode sequences are not easily amplified. Mini-barcodes of short COI and 16S rDNA fragments may overcome this problem (Hsieh et al. 2020) and have higher success rates than full-length sequences with little compromise in accuracy. The region of an insect's mitochondrial 16S rDNA gene exhibits high interspecies divergence and low intraspecies diversity. Sanger sequencing is the most common method used to obtain DNA sequences, but it involves multiple steps and is time-consuming. By contrast, as an alternative method for direct sequencing of DNA templates, pyrosequencing is a time-efficient method and effective in sequencing short fragments with high degree of polymorphism (Chen et al. 2018). Such fragments contain sufficient information for species identification (Balitzki-Korte et al. 2005). Pyrosequencing uses a series of enzymatic reactions to detect visible light emitted during the synthesis of DNA. Combined with the GenBank database, pyrosequencing makes an extendable species identification system (Hsieh et al. 2020). Since only short fragments of <500 bp of DNA can be sequenced with pyrosequencing methods, highly variable and informative regions need to be targeted in order to overcome the limitation of application of this method in forensics (Alacs et al. 2010).

Another approach that may be suitable for forensic entomology is massively parallel sequencing, often referred to as next-generation sequencing (NGS). An NGS method that generates longer reads is typically more useful for de novo genome assembly in non-model organisms (without a reference genome to use as a scaffold during assembly). A method that generates short reads typically provides greater depth of coverage, allowing robust exploration of heterozygosity and population-level variation. A disadvantage of short versus long reads is an increase in error rates, as well as cost per base (costs are expected to decrease over time). Although currently there may not be many forensic insect genomes that are publicly available (Andere et al. 2016), it is realistic to expect in the near future that more genomes become available. This will likely lead to new, more discriminatory molecular markers for identification and also to a better understanding of their biochemistry and physiology (Stevens et al. 2019), or the genetic basis of variation in development rate (Tarone et al. 2016).

Relatively inexpensive low coverage NGS analysis can be used to reliably sequence high copy number loci such as mtDNA, a process known as "genome skimming" (Coissac et al. 2016). Combining high-throughput sequencing techniques and a mini-barcode primer set such as one for 16S rDNA (Hsieh et al. 2020) will be helpful for species discovery and forensic identification. Rapid genotyping of a large number of immature specimens may be facilitated after a panel or multiplex of mitochondrial and/or nuclear loci is developed. This could be done relatively cheaply, as NGS cost has decreased over the years, although it is not yet as cost-effective as in the field of forensic human identification that benefits from greater economies of scale (Stevens et al. 2019). An NGS approach can be applied to low template and/or degraded DNA samples, and can complement classical entomological analyses. One example is a case of human identification from DNA extracted from the gastrointestinal tract of lice found at the crime scene. The results were used for identification and matching purposes to confirm the victim's state of neglect prior to death (Pilli et al. 2016).

Prior to the wide use of new genotyping techniques, they must be understood and accepted by the courts. Therefore, existing guidelines for entomological data collection and processing needed for use in forensic cases must be constantly revised.

Regular updates on advances in the laboratory techniques provided by medicolegal centers and laboratories and their collaboration with forensic entomologists are very important for proper DNA analysis and identification based on entomological data (Sharma et al. 2015a; Singh and Bala 2009).

13.3 Analysis of Human DNA Extracted from Insects

An insect can also be a useful source of vertebrate DNA. It has been shown that it is possible to obtain mtDNA, AFLP, and STR data from adult insect blood meals that can lead to individualization of the ingested host blood (Lord et al. 1998; Vos et al. 1995; Mukabana et al. 2002; Spitaleri et al. 2006; Pilli at al. 2016; Mukherjee et al. 2019). Forensic analysis of DNA recovered from a larvae's digestive tract can be used to identify what the larvae was feeding on (Wells et al. 2001; Zehner et al. 2004a). This approach has been used in casework to identify a corpse too badly burned to directly genotype (Chávez-Briones et al. 2013) and to associate human remains from separate locations (Li et al. 2011). Case study 3 describes a laboratory experiment demonstrating that larval crops may be a suitable source of DNA for identification of both the insect and its gut contents, and therefore mtDNA analysis may potentially be used to associate a larva with a human corpse, even if physical contact between the two is not observed (Wells et al. 2001). Potential practical applications of such gut content analysis include (1) resolving a chain-of-evidence dispute because the source of the larvae's food is disputed, (2) determining if the human corpse was present at a site at which only larvae are found, (3) detecting that a larva on a corpse had moved to the victim from another food source, and (4) identifying sperm DNA that was otherwise unretrievable because of larval feeding (Wells et al. 2001; Clery 2001).

Primer specificity is very important for determining the host DNA in the insect gut. Identification of vertebrates is typically based on the mtDNA cytochrome b (cyt b) gene, and less often on the noncoding D-loop. An investigator can select PCR primers that will amplify only vertebrate or only insect mtDNA, thus avoiding sequencing problems that result from a mixture. Because of the importance of genetic analysis of the contents of an insect's gut in crime cases, it is important to follow the recommendations for the collection, storage, and further handling of the specimen. If the larvae are too small for dissection or the larva's crop is no longer visible, DNA extraction may be less successful. Although some limitations of the technique still have to be explored, the results of studies so far suggest that the larvae's foregut, the crop, is the best source of larval gut contents, and that the mature larvae (third-instar larvae: 15–20 mm long, 4–6 days old) actively feeding on the corpse can be considered the best source of human DNA (Campobasso et al. 2005; Zehner et al. 2004a). Live larvae no longer on the corpse must be quickly located and preserved before the gut contents are digested (Linville and Wells 2002; Campobasso et al. 2005). Case study 4 describes a forensic case that showed that it is possible to successfully obtain a complete human genetic profile even if the DNA is recovered from small and biologically contaminated traces such as mosquito blood trace on the wall (Spitaleri et al. 2006). In order to use mosquitoes as a source of the perpetrator's DNA, the selection of appropriate mosquito individuals should be optimized (Trájer 2018).

Other studies have also shown that the contents of a larvae crop can be used to obtain a human or nonhuman mtDNA haplotype (Wells et al. 2001; Linville and Wells 2002) or a human Y-STR profile (Clery 2001; Chamoun et al. 2020.). It is even possible to successfully analyze human hypervariable regions using beetle larvae that fed on human bone (DiZinno et al. 2002). DiZinno and colleagues (2002) presented arguments for the potential forensic utility of carrion insect gut analysis. Nevertheless, for a successful analysis, carefully controlled conditions are required as well as using relatively sensitive mtDNA techniques and/or undegraded tissue. It is still not certain if a less sensitive but more informative STR analysis is likely to work given the typically degraded state of human corpses during actual death investigations (Njau et al. 2016).

13.3.1 Case Study 1: A Caddisfly Casing in Service of Criminalistics

Caddisflies are moth-like insects that have aquatic larvae that make protective cases of silk decorated with gravel, sand, twigs, or other debris (Caspers 1952; in English described by Benecke 2001). In a case from 1948, a caddisfly (most likely *Limnephilus flavicornis* L.) casing was used to suggest a possible PMI. Also, because of the fibers it was built of, it could also be connected to the socks of a dead woman. A few years later, Hubert Caspers described the case and since then the use of caddisfly casings was introduced as a tool for forensic investigation.

The naked body of a dead woman had been found in a moat of a windmill. She wore only a pair of red socks and she was wrapped in a sack. On one sock a caddisfly casing was found and it contained red fibers on its top and bottom. Sock fibers had obviously been used to build the casing. The conclusion was that the larva had mostly built its casing before entering the sack in which the deceased was wrapped in. The larva then finished the casing (fibers on top) and attached it to the red sock (fibers on bottom). According to the duration of the attachment procedure, which takes at least a few days, it was estimated that the body had been in the water for at least one week.

13.3.2 Case Study 2: Identity of Maggots Found on Outside and Inside of Body Bag

In October 1997, a severely decomposed body was analyzed for insect evidence in order to determine the PMI (Benecke 1998). The average size of maggots found on the corpse and on the outside of the closed body bag was 9 mm. The question was whether the maggots were of the same or different species. In one case, they might have squeezed themselves through tiny holes and came from outside the bag (e.g., to find a place for pupation). The other option suggested that a second oviposition of another species might have taken place after the body was put into the bag. Moreover, pupae were found on the floor beneath the corpse. They could not be directly related to any specific body in the morgue, that is, they may have fallen down from several unknown corpses. Because of the different developing time of fly species, PMI estimation is only possible if the insect species is known and can thus be used as a time since death indicator. It is difficult to identify the species of maggots, particularly in the early stage of their life cycle. Therefore, RAPD, a quick and inexpensive DNA test known to help in cases dealing with a variety of species, was used. The method was used to determine if the maggots that were found on the inside of a body bag were identical (a) with maggots found on the outside of the bag and (b) with pupae found on the floor under the corpse. Four single maggots found at the body bag were compared to a pupa found near the corpse, a blowfly from another case, and an adult silphid carrion beetle (negative control). The RAPD DNA profile showed that the profiles of maggots collected at the body bag were of the same genus. It was also found that the pupa was not a different (later) stage of one of the maggots from the actual corpse. There was no similarity between the DNA profiles of the two fly species and between the profiles produced by the different primers. The DNA profile of the beetle (negative control) did not show any similarities to the blowfly profiles.

Although this was not a serious evaluation of the utility of a method for specimen identification, it has been shown that RAPD results could be transferred from research to practical applications such as some urgent forensic investigations. In this example, RAPD profiles were used to distinguish *Lucilia sericata*, *Calliphora vicina* (=*erythrocephala*), and a beetle.

13.3.3 Case Study 3: Human and Insect mtDNA Analysis from Maggots

Usually, the estimate of the PMI is based on the age of a fly larva (maggot), if all of their development and feeding occurred on the victim (Wells et al. 2001). This is the only case when the age of the larva is relevant to PMI. However, there is a variety of situations where an alternative method for associating a maggot with a corpse would have been useful for forensic investigation: (1) the investigators discover maggots but no corpse; (2) the maggots are not found directly on a corpse, and an alternative food source is nearby; and (3) the maggots are found on a corpse, but may have come from somewhere else. In this study, it was demonstrated that mtDNA sequence data can be obtained from both the gut contents and the gut itself from maggots that had fed on human tissue. These data can be used to identify both the human corpse upon which the maggot was feeding and the species of the maggot itself.

Eggs were obtained from wild *Cynomya cadaverina* collected near San Antonio, Texas, and the resulting larvae were raised on human liver tissue that had been removed during an organ transplant procedure. Third larval instars were preserved in 70% ethanol at about 12 mm in length and stored at −20°C until used for DNA extraction. Other larvae were allowed to complete development to the adult stage. In order to produce reference haplotypes to be compared to human and fly DNA from the maggot crops, liver donor donated blood as well and the thoracic muscle from one of the emerged adult flies was used. The DNA of each sample type was extracted following the protocols from Qiagen DNA extraction kits. PCR analysis was done for a region of the fly cytochrome oxidase subunit one (COI) and a region of the human hypervariable region II (HVII).

13.3.3.1 Insect mtDNA

Successful amplification of a fragment of COI was obtained in all tested larval crops and the adult fly. Three of the larvae differed from the adult due to the silent T to A substitution, whereas two produced COI haplotypes identical to the adult fly. The COI primers failed to amplify DNA from the reference blood sample, as expected.

13.3.3.2 Human mtDNA

The HV2 primers failed to amplify DNA from the reference adult fly, as expected. A fragment of HV2 from the blood reference sample and from three of five crop extracts was successfully amplified. All HV2 sequences were identical and differed from the standard human sequence by the presence of a guanine at position 263 and by a cytosine at position 315.1.

Evidently, maggot crops can be a suitable source of DNA for identification of both the insect and its gut contents showing that mtDNA analysis may potentially be used to associate a maggot with a human corpse, even if physical contact between the two is not observed.

13.3.4 Case Study 4: Genotyping of Human DNA Recovered from Mosquitoes Found at a Crime Scene

Often, a mosquito will rest very close to the location of its recent blood meal, thereby potentially being a source of blood from a person who is no longer present (Spilateri et al. 2006). The possibility of amplifying human DNA from blood meals of Diptera species was demonstrated in a casework that occurred in Sicily. The corpse of a transsexual prostitute was discovered in a vegetated area near a beach. There were marks on the body suggesting strangulation as the cause of death. A suspect, a distinguished businessman, was identified because his car had been seen in the area around the time of the murder. A search of the suspect's home, which was located inland and far from the beach, did not yield any conventional biological evidence. There were, however, the remains of a blood-fed mosquito that had apparently been swatted against a wall, leaving both a bloodstain and the damaged insect specimen. Both the stain and insect specimen were collected, along with samples of the suspect's clothing that contained traces of sand grains and vegetation fragments. An optimized DNA extraction was carefully carried out in order to yield DNA. PCR amplification and STR profiling at 15 human genetic loci was then performed on the extracted DNA, using the AmpFLSTR Identifiler (Applied Biosystems). A nearly complete 15-loci profile was obtained either from the blood on filter paper or from the blood debris obtained by scraping the stain on the wall. So, it was confirmed that each signal corresponded to a given human STR amplicon and not to unspecific amplification of unknown DNA. The obtained profile fully matched the victim's profile, confirming her presence at least in the neighborhood of the suspect's home. Entomological examination attributed the mosquito to the species whose members, under normal conditions, do not travel the distance between the beach and the suspect's home. Electron microscopy and chemical x-ray microanalysis of the sand collected from the soles of the sneakers revealed that the grains found in the suspect's sneakers were consistent with the ones sampled at the beach. Leaf fragments found on the suspect's clothing were from the same plant species whose bushes were hiding the victim's body. In addition, this plant was not reported to flourish in any medium other than sandy soil. If taken alone, all of the aforementioned occurrences could not have served as an unquestionable link between the suspect and the murder. However, when taken into consideration together with police observation and significant scientific data, the extreme rarity of such joined events can lead the jury toward a conviction of second-degree murder.

Results of this case report showed that it is possible to successfully obtain a complete genetic profile even if DNA is recovered from poor and biologically contaminated traces. It is crucial to treat the evidence properly and to evaluate it with respect to other factors revealed by a thorough approach to the crime scene. The applied analytical strategy has proved to be a powerful tool for the investigation, which enabled the authorities to obtain the DNA profile of the major suspect of the murder.

References

Alacs, E.A., A. Georges, N.N. FitzSimmons, and J. Robertson. 2010. DNA detective: A review of molecular approaches to wildlife forensics. *Forensic Sci Med Pathol* 6(3):180–194.

Alamalakala, L., S.R. Skoda, and J.E. Foster. 2009. Amplified fragment length polymorphism used for inter- and intraspecific differentiation of screwworms (Diptera: Calliphoridae). *Bull Entomol Res* 99(2):139–149.

Al-Qahtni, A.H., M.S. Al-Khalifa, and A.M. Mashaly. 2020. Two human cases associated with forensic insects in Riyadh, Saudi Arabia. *Saudi J Biol Sci.* 27(3):881–886.

Aly, S.M., and J. Wen. 2013. Applicability of partial characterization of cytochrome oxidase I in identification of forensically important flies (Diptera) from China and Egypt. *Parasitol Res* 112(7):2667–2674. doi: 10.1007/s00436-013-3449-5.

Amendt, J.R., C.P. Campobasso, E. Gaundry, C. Reiter, H. LeBlance, and M. Hall. 2007. Best practice in forensic entomology: Standards and guidelines. *Int J Legal Med* 121(2):90–104.

Amendt, J.R., and M. Hall. 2007. Forensic entomology—Standards and guidelines. In *Proceedings of the 6th International Congress of the Baltic Medico-Legal Association—New Technologies in Forensic Medicine. Forensic Sci Int* 169(1):S27.

Andere, A.A., R.N. Platt, D.A. Ray, and C.J. Picard. 2016. Genome sequence of *Phormia regina* Meigen (Diptera: Calliphoridae): Implications for medical, veterinary and forensic research. *BMC Genomics* 17(1):842. doi: 10.1186/s12864-016-3187-z.

Anderson, G.S., and V.J. Cervenka. 2002. Insects associated with the body: Their use and analyses. In: *Advances in Forensic Taphonomy—Method, Theory and Archaeological Perspectives*, eds. W.D. Haglund and M.H. Sorg, 173–200. Boca Raton, FL: CRC Press.

Balitzki-Korte, B., K. Anslinger, C. Bartsch, and B. Rolf. 2005. Species identification by means of pyrosequencing the mitochondrial 12S rRNA gene. *Int J Leg Med* 119(5):291–294.

Baudry, E., J. Bartos, K. Emerson, T. Whitworth, and J.H. Werren. 2003. Wolbachia and genetic variability in the birdnest blowfly *Protocalliphora sialia. Mol Ecol* 12(7):1843–1854.

Belosludtsev, Y.Y., D. Bowerman, R. Weil, N. Marthandan, R. Balog, K. Luebke, et al. 2004. Organism identification using a genome sequence-independent universal microarray probe set. *BioTechniques* 37(4):654–658, 660.

Benecke, M. 1998. Random amplified polymorphic DNA (RAPD) typing of necrophageous insects (Diptera, Coleoptera) in criminal forensic studies: Validation and use in praxi. *Forensic Sci Int* 98(3):157–168.

Benecke, M. 2001. A brief history of forensic entomology. *Forensic Sci Int* 120(1–2):2–14.

Boakye, D.A., J. Tang, P. Truc, A. Merriweather, and T.R. Unnasch. 1999. Identification of blood-meals in haematophagous Diptera by cytochrome B heteroduplex analysis. *Med Vet Entomol* 13(3):282–287.

Boehme, P., P. Spahn, J.R. Amendt, and R. Zehner. 2013. Differential gene expression during metamorphosis: A promising approach for age estimation of forensically important *Calliphora vicina* pupae (Diptera: Calliphoridae). *Int J Legal Med* 127(1):243–249.

Brunak, S., J. Englebrecht, and S. Knudsen. 1990. Cleaning up gene databases. *Nature* 343(6254):123.

Budowle, B., P. Garofano, A. Hellman, M. Ketchum, S. Kanthaswamy, W. Parson, W. van Haeringen, S. Fain, and T. Broad. 2005. Recommendations for animal DNA forensic and identity testing. *Int J Legal Med* 119(5):295–302.

Campobasso, C.P., J.G. Linville, J.D. Wells, and F. Introna. 2005. Forensic genetic analysis of insect gut contents. *Am J Forensic Med Pathol* 26(2):161–165.

Caspers, H. 1952. Ein Köcherfliegen-Gehäuse im Dienste der Kriminalistik [A caddis-fly casing in the service of criminalistics]. *Arch Hydrobiol* 46:125–127, 155.

Catts, E.P., and M.L. Goff. 1992. Forensic entomology in criminal investigations. *Annu Rev Entomol* 37:253–272.

Catts, E.P., and N.H. Haskell. 1990. *Entomology and Death—A Procedural Guide.* Clemson: Joyce's Print Shop.

Chamoun, C.A., M.S. Couri, R.G. Garrido, R.S. Moura-Neto, and J. Oliveira-Costa. 2020. Recovery & identification of human Y-STR DNA from immatures of *Chrysomya albiceps* (Diptera: Calliphoridae). Simulation of sexual crime investigation involving victim corpse in state of decay. *Forensic Sci Int* 310:110239. doi: 10.1016/j.forsciint.2020.110239.

Charabidze, D., M. Gosselin, and V. Hedouin. 2017. Use of necrophagous insects as evidence of cadaver relocation: Myth or reality? *PeerJ* 5:e3506. doi: 10.7717/peerj.3506.

Chávez-Briones, M.L., R. Hernández-Cortéz, P. Díaz-Torres, A. Niderhauser-García, J. Ancer-Rodríguez, G. Jaramillo-Rangel, and M. Ortega-Martínez. 2013. Identification of human remains by DNA analysis or gastrointestinal contents of fly larvae. *J Forensic Sci* 58(1): 248–250.

Chen, W., Y. Shang, L. Ren, K. Xie, X. Zhang, C. Zhang, S. Sun, Y. Wang, L. Zha, and Y. Guo. 2018. Developing a MtSNP-based genotyping system for genetic identification of forensically important flesh flies (Diptera: Sarcophagidae). *Forensic Sci Int* 290:178–188. doi: 10.1016/j.forsciint.2018.07.012.

Clery, J.M. 2001. Stability of prostate specific antigen (PSA) and subsequent Y-STR typing of *Lucilia* (*Phaenicia*) *sericata* (Meigen) (Diptera: Calliphoridae) maggots reared from a simulated postmortem sexual assault. *Forensic Sci Int* 120(1–2):72–76.

Coissac, E., P.M. Hollingsworth, S. Lavergne, and P. Taberlet. 2016. From barcodes to genomes: Extending the concept of DNA barcoding. *Mol Ecol* 25(7):1423–1428.

de Carvalho, L.M., A.X. Linhares, and F.A. Badan Palhares. 2012. The effect of cocaine on the development rate of immatures and adults of *Chrysomya albiceps* and *Chrysomya putoria* (Diptera: Calliphoridae) and its importance to postmortem interval estimate. *Forensic Sci Int* 220(1–3):27–32. doi: 10.1016/j.forsciint.2012.01.023.

DiZinno, J.A., W.D. Lord, M.B. Collins-Morton, M.R. Wilson, and M.L. Goff. 2002. Mitochondrial DNA sequencing of beetle larvae (Nitidulidae: *Omosita*) recovered from human bone. *J Forensic Sci* 47(6):1337–1339.

Farncombe, K.M., D. Beresford, and C.J. Kyle. 2014. Characterization of microsatellite loci in *Phormia regina* towards expanding molecular applications in forensic entomology. *Forensic Sci Int* 240:122–125. doi: 10.1016/j.forsciint.2014.04.024.

Federico, C., D. Lombardo, N. La Porta, A.M. Pappalardo, V. Ferrito, F. Lombardo, and S. Saccone. 2018. Rapid molecular identification of necrophagous Diptera by means of variable-length intron sequences in the wingless gene. *J Forensic Leg Med* 56:66–72. doi: 10.1016/j.jflm.2018.03.003.

Florin, A.B., and N. Gyllenstrand. 2002. Isolation and characterization of polymorphic microsatellite markers in the blowflies *Lucilia illustris* and *Lucilia sericata*. *Mol Ecol Notes* 2(2): 113–116.

Gallagher, M.B., S. Sandhu, and R. Kimsey. 2010. Variation in developmental time for geographically distinct populations of the common green bottle fly, *Lucilia sericata* (Meigen). *J Forensic Sci* 55(2):438–442. doi: 10.1111/j.1556-4029.2009.01285.x.

Gemmellaro, M.D., G.C. Hamilton, and J.L. Ware. 2019. Review of molecular identification techniques for forensically important Diptera. *J Med Entomol* 56(4):887–902. doi: 10.1093/jme/tjz040.

GilArriortua, M., M.I. Saloña-Bordas, L.M. Cainé, F. Pinheiro, and M.M. de Pancorbo. 2015. Technical note. Mitochondrial and nuclear DNA approaches for reliable identification of Lucilia (Diptera, Calliphoridae) species of forensic interest from Southern Europe. *Forensic Sci Int* 257:393–397. doi: 10.1016/j.forsciint.2015.10.010.

Goff, M.L., and W.D. Lord. 2001. Entomotoxicology: Insects as toxicological indicators and the impact of drugs and toxins on insect development. In: *Forensic Entomology. The Utility of Arthropods in Forensic Investigations*, 1st ed., eds. J.H. Byrd and J.L. Castner, 331–340. Boca Raton, FL: CRC Press.

Greenberg, B., and J.C. Kunich. 2002. *Entomology and the Law—Flies as Forensic Indicators*. Cambridge: Cambridge University Press.

Gunn, A. 2009. *Essential Forensic Biology*. Hong Kong: Wiley-Blackwell.

Hebert, P.D., A. Cywinska, S.L. Ball, and J.R. deWaard. 2003. Biological identifications through DNA barcodes. *Proc Biol Sci* 270(1512):313–321.

Hsieh, C.H., C.G. Huang, W.J. Wu, and H.Y. Wang. 2020. A rapid insect species identification system using mini-barcode pyrosequencing. *Pest Manag Sci* 76(4):1222–1227. doi: 10.1002/ps.5674.

Jales, J.T., T.M. Barbosa, L.C. Dos Santos, V.P.S. Rachetti, and R.A. Gama. 2020. Carrion decomposition and assemblage of necrophagous dipterans associated with terbufos (organophosphate) intoxicated rat carcasses. *Acta Trop* 212:105652. doi: 10.1016/j.actatropica.2020.105652.

Jang, H., S.E. Shin, K.S. Ko, and S.H. Park. 2019. SNP typing using multiplex real-time PCR assay for species identification of forensically important blowflies and fleshflies collected in South Korea (Diptera: Calliphoridae and Sarcophagidae). *BioMed Res Int* 2019:6762517. doi: 10.1155/2019/6762517.

Joseph, I., D.G. Mathew, P. Sathyan, and G. Vargheese. 2011. The use of insects in forensic investigations: An overview on the scope of forensic entomology. *J Forensic Dent Sci* 3(2):89–91. doi: 10.4103/0975-1475.92154.

Li, X., J.F. Cai, Y.D. Guo, F. Xiong, L. Zhang, H. Feng, F.M. Meng, Y. Fu, J.B. Li, and Y.Q. Feng. 2011. Mitochondrial DNA and STR analyses for human DNA from maggot crop contents: A forensic entomology case from central-southern China. *Trop Biomed* 28(2): 333–338.

Linville, J.G., and J.D. Wells. 2002. Surface sterilization of a maggot using bleach does not interfere with mitochondrial DNA analysis of crop contents. *J Forensic Sci* 47(5):1055–1059.

Lord, W.D., and J.F. Burger. 1983. Collection and preservation of forensically important entomological materials. *J Forensic Sci* 28(4):936–944.

Lord, W.D., J.A. DiZinno, M.R. Wilson, B. Budowle, D. Taplin, and T.L. Meinking. 1998. Isolation, amplification, and sequencing of human mitochondrial DNA obtained from human crab louse, *Pthirus pubis* (L.), blood meals. *J Forensic Sci* 43(5):1097–1100.

MacPherson, J.M., P.E. Eckstein, G.J. Scoles, and A.A. Gajadhar. 1993. Variability of the random amplified polymorphic DNA assay among thermal cyclers, and effects of primer and DNA concentration. *Mol Cell Probes* 7(4):293–299.

McDonagh, L.M., and J.R. Stevens. 2011. The molecular systematics of blowflies and screwworm flies (Diptera: Calliphoridae) using 28S rRNA, COX1 and EF-1α: Insights into the evolution of dipteran parasitism. *Parasitology* 138(13)(v):1760–1777. doi: 10.1017/S0031182011001089.

Mona, S., M. Jawad, S. Noreen, S. Ali, and A. Rakha. 2019. Forensic entomology: A comprehensive review. *Adv Life Sci* 6(2):48–59.

Mukabana, W.R., W. Takken, P. Seda, G.F. Killeen, W.A. Hawley, and B.G. Knols. 2002. Extent of digestion affects the success of amplifying human DNA from blood meals of *Anopheles gambiae* (Diptera: Culicidae). *Bull Entomol Res* 92(3):233–239.

Mukherjee, S., P. Singh, F. Tuccia, J. Pradelli, G. Giordani, and S. Vanin. 2019. DNA characterization from gut content of larvae of *Megaselia scalaris* (Diptera, Phoridae). *Sci Justice* 59(6) (v):654–659. doi: 10.1016/j.scijus.2019.06.006.

Nelson, L.A., J.F. Wallman, and M. Dowton. 2008. Identification of forensically important *Chrysomya* (Diptera: Calliphoridae) species using the second ribosomal internal transcribed spacer (ITS2). *Forensic Sci Int* 177(2–3):238–247.

Njau, D.G., E.K. Muge, P.W. Kinyanjui, C.O.A. Omwandho, and S.Mukwana. 2016. STR analysis of human DNA from maggots fed on decomposing bodies: Assessment of the time period for successful analysis. *Egypt J Forensic Sci* 6(3):261–269. doi: 10.1016/j.ejfs.2015.04.002.

Owings, C.G., and C.J. Picard. 2018. New distribution record for *Lucilia cuprina* (Diptera: Calliphoridae) in Indiana, United States. *J Insect Sci* 18(4):8. doi: 10.1093/jisesa/iey071.

Owings, C.G., C. Spiegelman, A.M. Tarone, and J.K. Tomberlin. 2014. Developmental variation among *Cochliomyia macellaria* Fabricius (Diptera: Calliphoridae) populations from three ecoregions of Texas, USA. *Int J Legal Med* 128(4):709–717. doi: 10.1007/s00414-014-1014-0.

Perez, T., J. Albornozm, and A. Dominguez. 1998. An evaluation of RAPD fragment reproducibility and nature. *Mol Ecol* 7(10):1347–1357.

Picard, C.P., M.H. Villet, and J.D. Wells. 2012. Amplified fragment length polymorphism confirms reciprocal monophyly in *Chrysomya putoria* and *Chrysomya chloropyga*: A correction of reported shared mtDNA haplotypes. *Med Vet Entomol* 26(1):116–119.

Picard, C.J. 2010. Analysis of Carrion Fly (Diptera: Calliphoridae) Population Genetics Using Amplified Fragment Length Polymorphism (AFLP) Profiles [PhD dissertation]. Morgantown, WV: Department of Biology, West Virginia University.

Picard, C.J., J.D. Wells, A. Ullyot, and K. Rognes. 2017. Amplified fragment length polymorphism analysis supports the valid separate species status of Lucilia caesar and L. illustris (Diptera: Calliphoridae). *Forensic Sci Res* 3(1):60–64. doi: 10.1080/20961790.2017.1398286.

Picard, C.J. 2013. First record of *Chrysomya megacephala* Fabricius (Diptera: Calliphoridae) in Indiana, U.S.A. *Proc Entomol Soc Wash* 115(3):265–267.

Picard, C.J., and J.D. Wells. 2009. Survey of the genetic diversity of *Phormia regina* (Diptera: Calliphoridae) using amplified fragment length polymorphisms. *J Med Entomol* 46(3):664–670.

Picard, C.J., and J.D. Wells. 2010. The population genetic structure of North American *Lucilia sericata* (Diptera: Calliphoridae), and the utility of genetic assignment methods for reconstruction of postmortem corpse relocation. *Forensic Sci Int* 195(1–3):63–67.

Picard, C.J., and J.D. Wells. 2012. A test for carrion fly full siblings: A tool for detecting postmortem relocation of a corpse. *J Forensic Sci* 57(2):535–538.

Pilli, E., A. Agostino, D. Vergani, E. Salata, I. Ciuna, A. Berti, D. Caramelli, and S. Lambiase. 2016. Human identification by lice: A next generation sequencing challenge. *Forensic Sci Int* 266:e71–e78. doi: 10.1016/j.forsciint.2016.05.006.

Randall, B.B., M.F. Fierro, and R.C. Froede. 1998. Practice guideline for forensic pathology. *Arch Pathol Lab Med* 122(12):1056–1064.

Ritz-Timme, S., G. Rochholz, H.W. Schutz, M.J. Collins, E.R. Waite, C. Cattaneo, and H.J. Kaatsch. 2000. Quality assurance in age estimation based on aspartic acid racemisation. *Int J Legal Med* 114(1–2):83–86.

Rodrigues, R.A., A.M.L. Azeredo-Espin, and T.T. Torres. 2009. Microsatellite markers for population genetic studies of the blowfly *Chrysomya putoria* (Diptera: Calliphoridae). *Mem Inst Oswaldo Cruz* 104(7):1047–1050.

Schroeder, H., H. Klotzbach, S. Elias, C. Augustin, and K. Pueschel. 2003. Use of PCR-RFLP for differentiation of calliphorid larvae (Diptera, Calliphoridae) on human corpses. *Forensic Sci Int* 132(1):76–81.

Sharma, A., M. Bala, and N. Singh. 2018. Five case studies associated with forensically important entomofauna recovered from human corpses from Punjab, India. *J Forensic Sci Crim Inves* 7(5):55572. doi: 10.19080/JFSCI.2018.07.555721.

Sharma, R., R.K. Garg, and J. Gaur. 2015a. Various methods for the estimation of the post mortem interval from Calliphoridae: A review. *Egypt J Forensic Sci* 5(1):1–12.

Sharma, M., D. Singh, and A.K. Sharma. 2015b. Mitochondrial DNA based identification of forensically important Indian flesh flies (Diptera: Sarcophagidae). *Forensic Sci Int* 247:1–6. doi: 10.1016/j.forsciint.2014.11.017.

Sharma, M., and D. Singh. 2017. Utility of random amplified polymorphic DNA (RAPD) in forensic entomology. *Int J Sci Res Sci Technol.* 3(1):56–58.

Singh, D., and M. Bala. 2009. The effect of starvation on the larval behavior of two forensically important species of blow flies (Diptera: Calliphoridae). *Forensic Sci Int* 193(1–3):118–121.

Sperling, F.A.H., G.S. Anderson, and D.A. Hickey. 1994. A DNA-based approach to the identification of insect species used for postmortem interval estimation. *J Forensic Sci* 39(2):418–427.

Spitaleri, S., C. Romano, E. Di Luise, E. Ginestra, and L. Saravo. 2006. Genotyping of human DNA recovered from mosquitoes found on a crime scene. In *International Society for Forensic Genetics; Progress in Forensic Genetics 11: Proceedings of the 21st International ISFG Congress, International Congress Series* 1288:574–557.

Stevens, J.R. 2003. The evolution of myiasis in blowflies (Calliphoridae). *Int J Parasitol* 33(10):1105–1113.

Stevens, J., and R. Wall. 2001. Genetic relationships between blowflies (Calliphoridae) of forensic importance. *Forensic Sci Int* 120(1–2):116–123.

Stevens, J.R., C.J. Picard, and J.D. Wells. 2019. Molecular genetic methods for forensic entomology. In: *Forensic Entomology. The Utility of Arthropods in Legal Investigations*, 3rd ed., eds. J.H. Byrd and J.K. Tomberlin, 253–268. Boca Raton, FL: CRC Press.

Stevens, J.R., R. Wall, and J.D. Wells. 2002. Paraphyly in Hawaiian hybrid blowfly populations and the evolutionary history of anthropophilic species. *Insect Mol Biol* 11(2):141–148.

Tarone, A.M., K.C. Jennings, and D.R. Foran. 2007. Aging blow fly eggs using gene expression: A feasibility study. *J Forensic Sci* 52(6):1350–1354.

Tarone, A.M., and D.R. Foran. 2011. Gene expression during blow fly development: Improving the precision of age estimates in forensic entomology. *J Forensic Sci* 56(Suppl 1):S112–122. doi: 10.1111/j.1556-4029.2010.01632.x.

Tarone, A.M., C.J. Picard, and S.-H. Sze. 2016. *Genomic Tools to Reduce Error in PMI Estimates Derived from Entomological Evidence*. National Institute of Justice Grant Report 2012-DN-BX-K024.

Thongjued, K., W. Chotigeat, S. Bumrungsri, P. Thanakiatkrai, and T. Kitpipit. 2019. A new cost-effective and fast direct PCR protocol for insects based on PBS buffer. *Mol Ecol Resour* 19(3):691–701. doi: 10.1111/1755-0998.13005.

Torres, T.T., and A.M.L. Azeredo-Espin. 2008. Characterization of polymorphic microsatellite markers for the blowfly *Chrysomya albiceps* (Diptera: Calliphoridae). *Mol Ecol Resour* 8(1):208–210.

Trájer, A.J. 2018. Which mosquitoes (Diptera: Culicidae) are candidates for DNA extraction in forensic practice? *J Forensic Leg Med* 58:183–191. doi: 10.1016/j.jflm.2018.07.002.

Tuccia, F., G. Giordani, and S. Vanin. 2016. A general review of the most common COI primers for Calliphoridae identification in forensic entomology. *Forensic Sci Int Genet* 24:e9–e11. doi: 10.1016/j.fsigen.2016.07.003.

Verma, K., and R. Paul. 2013. Assessment of post mortem interval, (PMI) from forensic entomotoxicological studies of larvae and flies. *Entomol Ornithol Herpetol* 2(104). doi: 10.4172/2161-0983.1000104.

Vos, P., R. Hogers, M. Bleeker, M. Reijans, T. van de Lee, M. Hornes, A. Frijters, J. Pot, J. Peleman, and M. Kuiper. 1995. AFLP: A new technique for DNA fingerprinting. *Nucleic Acids Res* 23(21):4407–4414.

Wallman, J.F., R. Leys, and K. Hogendoom. 2005. Molecular systematics of Australian carrion breeding blowflies (Diptera: Calliphoridae) based on mitochondrial DNA. *Invertebr Syst* 19(1):1–15.

Wang, M., J. Chu, Y. Wang, F. Li, M. Liao, H. Shi, Y. Zhang, G. Hu, and J. Wang. 2019. Forensic entomology application in China: Four case reports. *J Forensic Leg Med* 63:40–47. doi: 10.1016/j.jflm.2019.03.001.

Williams, K., and M.H. Villet. 2013. Ancient and modern hybridization between *Lucilia sericata* and *L. cuprina* (Diptera: Calliphoridae). *Eur J Entomol* 110(2):187–196.

Wells, J.D., F. Introna Jr., G. DiVella, C.P. Campobasso, J. Hayes, and F.A. Sperling. 2001. Human and insect mitochondrial DNA analysis from maggots. *J Forensic Sci* 46(3):685–687.

Wells, J.D., T. Pape, and F.A.H. Sperling. 2001. DNA-based identification and molecular systematics of forensically important Sarcophagidae (Diptera). *J Forensic Sci* 46(5):1098–1102.

Wells, J.D., and F.A.H. Sperling. 2001. DNA-based identification of forensically important Chrysomyinae (Diptera: Calliphoridae). *Forensic Sci Int* 120(1):110–115.

Wells, J.D., and J.R. Stevens. 2008. Application of DNA-based methods in forensic entomology. *Annu Rev Entomol* 53:103–120.

Wells, J.D., R. Wall, and J.R. Stevens. 2007. Phylogenetic analysis of forensically important Lucilia flies based on cytochrome oxidase I: A cautionary tale for forensic species determination. *Int J Legal Med* 121(3):229–233.

Wells, J.D., and D.W. Williams. 2007. Validation of a DNA-based method for identifying Chrysomyinae (Diptera: Calliphoridae) used in a death investigation. *Int J Legal Med* 121(1):1–8.

Wetton, J.H., C.S. Tsang, C.A. Roney, and A.C. Spriggs. 2004. An extremely sensitive species-specific ARMs PCR test for the presence of tiger bone DNA. *Forensic Sci Int* 140(1):139–135.

Will, K.W., and D. Rubinoff. 2004. Myth of the molecule: DNA barcodes for species cannot replace morphology for identification and classification. *Cladistics* 20(1):47–55.

Zehner, R., J. Amendt, and P. Boehme. 2009. Gene expression analysis as a tool for age estimation of blowfly pupae. *Forensic Sci Int Genet Suppl S* 2(1):292–293.

Zehner, R., J. Amendt, and R. Krettek. 2004a. STR typing of human DNA from fly larvae fed on decomposing bodies. *J Forensic Sci* 49(2):337–340.

Zehner, R., J. Amendt, S. Schütt, J. Sauer, R. Krettek, and D. Povolný. 2004b. Genetic identification of forensically important flesh flies (Diptera: Sarcophagidae). *Int J Legal Med* 118(4):245–247.

Forensic Botany Plants as Evidence in Criminal Cases and as Agents of Bioterrorism

14

HEATHER MILLER COYLE, HENRY C. LEE, AND TIMOTHY M. PALMBACH

Contents

14.1 Introduction

This chapter provides a general overview of forensic botany and highlights some of the specialty areas, such as palynology (pollen identification), drug enforcement, and bioterrorism, in criminal casework. Plants as criminal evidence are highly varied in species but are prevalent in homicide as well as physical and sexual assaults as trace evidence. Almost every forensic science laboratory has examples of plant evidence, either intact or as fragments in their trace analysis section (e.g., grass species, tree species, algae, diatoms). Typically, a species identification is made first, macroscopically or microscopically, followed by further individualizing tests by chemistry or DNA. In addition, drug and toxicology laboratories house many different examples of plant-derived evidence. Although this chapter focuses on criminal case applications, there is almost an equal area of civil case application that includes plants and plant products. Those areas in the United States are under the jurisdiction of the Food and Drug Administration and commercial and private plant breeders who seek to protect the quality of their plant-based products or new genetic varieties under the Plant Variety Protection Act and the Plant Patent Act (Miller Coyle 2005).

The number of ways that plants can present as biological evidence is innumerable but a few examples include:

- Grass stains on a dress after sexual assault or botanical samples such as leaves at a homicide scene.

- Plant leaves and stems caught in the undercarriage of a vehicle at a secondary crime scene where a victim's body is dumped.
- Seeds caught in the pants cuffs of a burglar who is fleeing the scene of a home invasion.
- Stomach contents with vegetable matter from a victim's last meal to support time of death or an alibi to a location.
- Pollen to date the season of burial of skeletal remains in a mass grave.
- Geographic profiling to determine possible regions where a plant sample could have originated either via the plant itself or its packaging; also, for linking suspect/ victim to vehicle or location.

14.2 Evidence Collection

Crime scene investigators need to consider the potential for locating botanical evidence and modify their search methods and search locations to maximize the likelihood of locating this type of evidence. Before releasing a crime scene report, a review of the entire crime scene should be made to determine if any botanical evidence might have been overlooked for collection (Miller Coyle 2005; Lee et al. 2001). In particular, one should search for the presence of small plant leaves, seeds, pollen grains, algae, grass, plant-based drugs, wood chips, and stems in or on a body, clothing, and at the scene. Known reference samples and control samples also need to be collected for later comparison, and with plant evidence, one sample is not typically sufficient. For example, if a holly leaf is found clutched in a victim's hand at a secondary scene, and if three holly bushes are present at the suspect's residence, then at least three samples (one from each bush) must be collected for later comparison. There are many different genetic varieties of the same plant species; therefore, for later individualization (either by distinct morphological features or by genetic analysis), a known reference sample from each must be collected for comparison to one at the scene. If plant samples are small, investigators may have to obtain reference samples from a wide variety of potential plant donors located in or adjacent to the scene (Miller Coyle 2005). If no reference samples are found at the scene, then herbarium or collection plant reference samples can be used as comparative references for species identification (Miller Coyle 2005).

One of the most critical and time-consuming factors in forensic botany is performing the appropriate comparisons. From a single piece of evidence, it is often possible to make an identification of the plant species; however, to perform individualization back to a single source, often a comparative reference database must be created. For human identification (HID) by DNA methods, forensic scientists use a preexisting population reference database to assess the individuality of the sample (Lee et al. 1998). With most plant cases, a reference population database will need to be constructed from (a) the same geographic region as for the crime scene or (b) a representative sampling of the plant species if it is very mobile and widely transportable (e.g., crops or drugs). Unfortunately, with very few exceptions, comparative databases do not exist, and part of the overall examination process may require the creation of a specific database. Some places to look for reference databases are academic research institutions already performing population genetic studies on the plant of interest or private seed and plant technology companies that are involved in breeding genetic varieties of plant species for profit (Miller Coyle

2005). Herbariums and botanical gardens may also allow for their private collections to be used as references.

The actual process of collecting plant evidence follows traditional forensic collection methods including crime scene documentation, photography, individual packaging of samples in paper envelopes, flat pressing of herbarium specimens between cardboard, or pollen collection on swabs and microscope slides for sterile containers (Miller Coyle 2005; De Forest et al. 1983; Miller Coyle et al. 2001; Lee et al. 2001). Each item of evidence must be wrapped separately. And a chain of custody and evidence seals must be maintained. For any unusual forms of plant evidence (e.g., stomach contents, vomit, fecal matter), preservation may be a bit different, such as collection in a plastic cup and storage at 4°C since drying may destroy the features of the plant cells such that species identification will be difficult. For plant DNA testing and appropriate archival of the sample for long-term storage, fast technology for analysis of nucleic acids (FTA) cards can be used in a similar fashion as for HID testing (Allgeier et al. 2011). When in doubt, a consultation with the forensic botanist or forensic science laboratory would be recommended.

Crime scene documentation should include intermediate and close-up photos of the different plant sources in the crime scene (Lee et al. 2001). Investigators should carefully evaluate the entire scene such as to collect reference samples from any plants that appear to be of the same species as the questioned botanical material (Miller Coyle 2005). Before proceeding to court with the plant evidence after it has been compared and tested, one should consider the following issues that may affect admissibility in court (Miller Coyle 2005):

- An appropriate review of the evidence and testing procedures has been made from the crime scene through forensic laboratory testing.
- A chain of custody has been documented and maintained.
- Has plant evidence ever been admitted in this particular court system?
- What are the qualifications of the examiners performing the tests and writing the reports?
- Was it possible to perform a blind comparison for identification purposes?
- Is a representative population database of samples or DNA profiles available for comparative purposes? Was it created specifically for this case?
- Is the genetic reproductive strategy of the plant species known? For example, is it clonally propagated and likely to have more than one match genetically? Or is it seed propagated and, therefore, individual in its genetic profile?
- Are all of the participants in the case apprised of the benefits and limitations of the botanical evidence to avoid discrepancies in testimony?
- What is the known and potential error rate related to this type of analysis?
- Are the procedures based on "generally accepted" scientific principles?

Botanical evidence is inherently variable because of the great number of plant species found on land and in water. A comprehensive understanding among crime scene personnel, forensic scientists, academic practitioners, and attorneys can lend clarity to a judge and jury when plant evidence is presented in a court of law. Currently, botanical evidence is only rarely used in criminal trials, but it is used investigatively somewhat frequently to verify or refute alibis and to provide information about geographic locations for crime scene reconstruction.

14.3 Overview of Techniques

14.3.1 Microscopy

Simple, cost-effective procedures such as light microscopy and use of a comparison microscope are time-honored methods in forensic science. Most forensic science laboratories have a trace evidence section that collects fibers (many of which are of plant origin: cotton, jute), hairs, paint chips, soil, and pollen grains off of evidence using a vacuuming or scrape-down procedure. The vacuuming process uses a modified handheld vacuum with a specialized capture filter to collect microscopic pollen (Lee et al. 2001). Commercially available micro vacuum systems such as the wet vacuum system M-Vac have been highly successful in solving cold cases by enhancing the collection of trace biologicals for DNA testing. Surfaces such as rough rocks used as weapons have been vacuumed to recover trace DNA from fingerprints by concentrating the DNA (https://www.m-vac.com). It would be useful for pollen grain recovery as well. Larger fragments of plant evidence are collected with forceps and placed in individual paper envelopes to preserve the color, shape, and surface features such as trichomes, glands, stomata, and other important features useful for species identification. The scrape-down procedure uses a spatula to physically scrape items such as clothing and carpets to loosen small fragments of plant material for collection and storage in paper envelopes (Lee et al. 2001). Resins such as hashish can be scraped into a druggist fold and placed in a secondary envelope for transport. FTA collection cards are usually used at the scene and are sealed in a secondary envelope before being sent to the laboratory for analysis (Allgeier et al. 2011).

Microscopy magnifies the physical features of a plant fragment so that vein patterns, stomata patterns, trichome shapes, and other features that lend uniqueness to a plant can be more readily identified. Typically, a botanical key is used to search a reference database for the plant species that possess those characteristics and enables species identification (Miller Coyle 2005). A comparison microscope is useful if one is trying to determine a physical match between two plant fragments. Often, plant evidence is small and to the unaided eye, not very unique. Under magnification, however, if both fragments are placed under the same magnification on a comparative microscope, the images can be overlaid and matched by morphology and identified to known references. The leaf surface patterns of stomata and trichomes can provide additional information to the comparison as well as having a direct physical match between two edges of torn leaf fragments. Be wary of using microscopy as a formal means of identification on some plant species because many appear morphologically similar and are not easy to distinguish with this method. For those morphologically similar species or for fragmented plant material, DNA identification is the optimal technique (Kress et al. 2009; Wiltshire 2006; Kress et al. 2005).

14.3.2 Species Identification

The proper identification of a plant to the species level is useful if one is trying to establish a connection to a primary or secondary crime scene (Miller Coyle 2005; Lee et al. 1998). Frequently, victims are killed at one location and relocated to a secondary location in an attempt to hide the crime. Many cases include plant evidence found clutched in the victim's hand, attached to the victim's clothing, on a blanket, or in a vehicle that was used to transport the victim. In other cases, ask whether wood fragments collected from a vehicle's undercarriage can be matched back to a tree at the scene of the crime; if pine needles found inside the barrel of a gun can be linked to the tree where the attacker was crouched and waiting for the victim; if leaf samples from a serial killer's body-dumping site match to each other; or if drug samples from a suitcase match those recovered from a suspected thief. These may sound like odd forms of associative evidence, but these are based on real cases where plant individualization by DNA was requested. Some additional cases where pollen is used investigatively include the profiling of species geographically from vehicles to obtain an investigative lead for missing person cases to find a link to a body dump site. Pollen can be collected from suspect shoes to match to a gas or brake pedal, air filter, upholstery, or clothing. At the US Customs and Border Protection (CBP) Laboratories and Scientific Services, pollen is even used as a tracer system to track drugs such as fentanyl to the country of origin in shipment seizures at US borders (https://qz.com/1635897/the-us -is-using-pollen-to-track-illegal-drugs-like-fentanyl/). Plant species identification answers the question, What type of plant is this and what type of region does it grow in?, but cannot provide individualization to answer the question, Did this fragment come from this tree and only this tree? Plant species identification is a necessary first classification step in trying to establish a forensic linkage and then individualization is performed if sufficient sample quantity and quality is available for subsequent DNA testing (amplified fragment length polymorphisms [AFLP]; Figure 14.1). Other types of DNA test methods include random amplified polymorphic DNA (RAPD) used in the Arizona Maricopa County homicide (*State of Arizona vs. Mark Bogan*) where *Palo verde* seed pods recovered from a truck bed were used to link a vehicle to the homicide scene. In this case, a plant reference DNA database was constructed from other area trees to show that the only one that matched was at the crime scene (https://apnews.com/article/326da16edd9e677ce6033700e470e9ae).

Figure 14.1 Pollen from drug samples can be used to try to determine a geographic origin if mixed with pollen of indigenous species grown in the same location. The presence of cystolith hairs on *Cannabis* leaves is considered one form of botanical identification by microscopy. All botanical evidence should be properly photographed, documented, and packaged in clean paper envelopes with a chain of custody maintained.

The simplest form of species identification is to observe the morphology and compare the distinctive features to an identification key. This can be done visually or the use of image recognition software can be helpful to render an identification. Some examples of image recognition software include LeafSnap (https://www.appurse.com/leafsnap), PlantSnap (https://www.plantsnap.com), and Virginia Tech Tree ID (Vtree App, https://dendro.cnre.vt.edu/dendrology/vtree.htm). Species identification can be made by assessing the characteristic physical features of the plant evidence; however, in many instances, the fragment is too small or lacks sufficient detail to make a conclusive identification microscopically. When this situation occurs, DNA can also be used to make a species identification (Kress et al. 2009; Wiltshire 2006; Kress et al. 2005; Bruni et al. 2010). Using commercially available plant DNA extraction kits, plant DNA can be obtained in less than two hours. A portion of the recovered plant DNA is then amplified using specific polymerase chain reaction (PCR) primers to selectively target a region that is specific to that plant species. There are hundreds of plant sequences that can be used for identification (nuclear genes: *18S, ITS1, ITS2*; chloroplast genes: *rbcL, atpB, ndhF*), and many searchable databases (e.g., http://www.ncbi.gov) are available for confirmation of sequence information (Miller Coyle 2005; Kress et al. 2009; Wiltshire 2006; Kress et al. 2005; Bruni et al. 2010). A recent publication by Bruni et al. (2010) established that the matK plastid DNA marker was the most useful for rapid identification of poisonous plant species by DNA barcoding. This is especially important for use at poison control centers where rapid identification is a must to help ascertain best treatment for ingested poison victims. This particular approach can also be used for the species identification of trace plant particulates found on outer packaging for drugs as one example or on outer layers of clothing collected from a suspect as another example. The ability to effectively perform this analysis using DNA from trace particulates may be invaluable for geosourcing samples that are being trafficked across borders and can then be regulated by federal and international restrictions. In addition, many grass species are challenging to identify microscopically because of the extreme similarity of their morphological features. In Australia, a DNA identification system was developed as a tool for progressive identification of grass samples to different taxonomic levels (Ward et al. 2005, 2009).

14.3.3 DNA Individualization

14.3.3.1 Amplified Fragment Length Polymorphism

The AFLP methodology was first described by Vos et al. (1995) and represents a method for genotyping single-source plant samples for a large number of markers to generate a highly discriminating DNA band pattern. The AFLP technique is based on the detection of genomic restriction fragments after PCR amplification. Band patterns are produced without prior sequence knowledge, using a limited set of amplification primers. The AFLP method permits the inspection of polymorphisms at a large number of loci within a short period of time and requires only a small amount of DNA (about 10 ng). Although other DNA typing techniques exist and have been developed for specific applications, AFLP analysis is the method of choice since it has general applicability to any single-source plant sample regardless of the species (Miller Coyle 2005; Miller Coyle et al. 2001; Vos et al. 1995). It is similar to mitochondrial DNA sequencing, however, in that if two or more samples of the same species or if two different species are present, then the AFLP profile will appear as a mixture and will be difficult to interpret because of its complexity even with known reference samples for comparisons (Miller Coyle 2005).

Several commercial kits for AFLP analysis are available and all follow the general procedure of

- One-step digestion–ligation reaction where short DNA adaptors are added to each end of a DNA fragment.
- PCR preselective amplification of a subset of DNA fragments with one primer set to generate a pool of fragments.
- PCR selective amplification and fluorescent labeling of the subset of DNA fragments to generate an individualizing "barcode."
- Detection of the DNA fragments with a DNA sequencer and genetic analysis with computerized software to convert the "barcode" to a numeric code for comparisons.

Figure 14.2 illustrates a typical AFLP profile that differs from short tandem repeat (STR) analysis in that discrete alleles are not typed. AFLP is most analogous to the original HID method developed by Alec Jeffreys known as multilocus typing by RFLP (restriction fragment length polymorphisms) and used bins to score the DNA fragments even if the fragments showed slight differences in mobility during gel electrophoresis (De Forest et al. 1983; Miller Coyle et al. 2001; Lee and Ladd 2001; Lee et al. 2001, 1998; Vos et al. 1995). The binning process involves creating "software bins" based on the size of DNA fragments and are arbitrary in the sense that the size accommodation of the bins can be anywhere from 30 to 100 bases, and as long as the scoring process is consistent, then the presence or absence of a peak is either 1 or 0, respectively. The greater the number of DNA fragments scored, the higher the degree of certainty of a match between the evidentiary sample and the known reference comparative sample (Miller Coyle 2005). AFLP marker systems are dominant in inheritance patterns.

14.3.3.2 Short Tandem Repeat

STRs, also known as simple sequence repeats (SSRs) or minisatellites, are highly individualizing and excellent for the detection of mixed samples once they have been developed and validated for the plant species of interest. If a panel of STR markers is already in existence or the time and resources required are not a limitation, then they are a good method to use (Newton and Vo 2003; Craft et al. 2007; Howard et al. 2009; Štambuk et al. 2007; Howard et al. 2008). Often, however, a case may occur where an unusual plant species is recovered but there is no interest in developing a panel of STR markers for a single case since that may take approximately two years to complete and validate. In those cases, AFLP is a good choice since the technology is applicable to any plant species as long as it is a single-source sample. As technology advances, automation and chip-based technology for STR markers

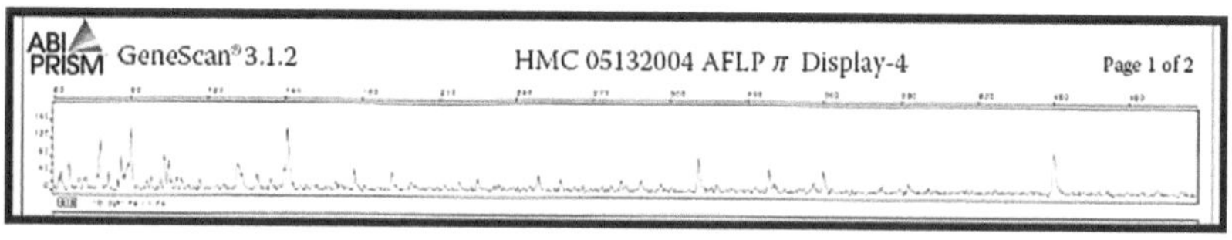

Figure 14.2 An example of an AFLP profile from Sutera leaves. Relative fluorescence units (RFU) are represented on the y-axis (0–1000); size (0–460 nucleotide bases) is represented by the x-axis.

or single-nucleotide polymorphisms may become a rapid and viable alternative for genetic individualization of plant samples and rapid construction of plant databases.

STR markers are usually multiplexed into a single reaction tube for PCR amplification and then detected by a DNA sequencer as discrete DNA fragments of a specific size when compared to an allelic ladder. Unlike human identity testing in forensics, each plant species may not have an allelic ladder available, and, therefore, sizing of the DNA fragment would be based on the software analysis and fragment mobility without a commercial reference. Computer software can generate fixed bins for fragment scoring and allele assignment. Nevertheless, many different animal and plant species have had STR marker panels developed, and some kits are even commercially available (Miller Coyle 2005; Ward et al. 2005; Newton and Vo 2003; Craft et al. 2007). For plant species that are highly inbred, AFLPs have been suggested when STR markers may fail to distinguish between individual plants because of high levels of genetic similarity (Miller Coyle 2005; Miller Coyle et al. 2001; Vos et al. 1995). This marker system is a co-dominant system.

14.3.3.3 Random Amplified Fragment Polymorphism (RAPD)

Random amplified fragment polymorphisms are generated by the random recognition of DNA sequences in the template by short arbitrary primers of 8–12 nucleotides that if sufficiently close will amplify as a detectable PCR product. These sequences will vary from species to species and within individuals in the same species. Although an older technique, it is not species dependent and will work on any botanical evidence to generate randomized patterns. It does not work well with highly degraded samples, as random breakage in the template will cause nonreproducibility. RAPD analysis was used in the first criminal case in Arizona as associative plant evidence and has the added benefit of being cost-effective and simple to perform. This marker system detects dominant fragments typically.

14.3.3.4 Single-Nucleotide Polymorphism (SNP)

Single-nucleotide polymorphisms are single base changes also known as mutation events that are heritable from generation to generation. If the mutation occurs in a noncoding region, then it is harmless. If it occurs in a coding region for gene function, then it may be detected by differences in product. A good example of this is the protein Tetrahydrocannabinolic acid synthase (THCAS) "null" form that affects the production of cannabigerolic acid (CBGA) production (https://pubmed.ncbi.nlm.nih.gov/33557333/). Studies on biosynthetic pathways in *Cannabis* are being elucidated by examining the effect of point mutations (SNPs) in the tetrahydrocannabinol (THC) and CBD pathways. Any region of DNA can be sequenced for the order of nucleotide bases to examine SNP markers and their usefulness to discriminate between plant samples of any species. SNP testing is used in plant genetics and breeding to generate assays to quickly screen breeding stocks for desirable traits. It takes thousands of SNP markers to fully classify a plant genome compared to STR/SSR technology, but with high-throughput sequencing technologies and robotics it is simple to perform.

14.4 Examples of Plants with Bioterrorist Potential

Two of our best-known medicines are commonly called aspirin and Taxol. Both are classified as miracle medicines, however, at the wrong dosage these could also be used as bioterrorist or poison agents. They are simple to acquire and simple to administer. For these reasons, one should place a careful watch on quantities being purchased. In 1763, the Royal Society reported on willow bark in powder form being useful to cure fevers. This report is attributed to a vicar from Oxfordshire named Edward Stone. Its original discovery may date as far back as 1500 BC in which Egyptian medical texts refer to it as an anti-inflammatory pain reliever from papyrus. In 400 BC, historical literature references the use of willow tea by Hippocrates in Greece for pain in childbirth. The recent history of aspirin dates to the 1890s when Felix Hoffman, a scientist at Bayer Corporation in Germany, discovered the substance as a chemist and gave some to his father for his rheumatism. Shortly thereafter, it was sold in powder form to physicians so they could treat their patients. The chemical name for the active ingredient in aspirin is acetylsalicylic acid, the derivation of the word aspirin is "A" for aspirin, "spir" for *Spiracea ulmaria* (meadowsweet), which like willow also contains salicin. In 1950, aspirin was entered into the *Guinness Book of World Records* for the most frequently sold painkiller. Aspirin overdoes can result in seizures, and renal and respiratory failure as well. Any natural product substance can be a medicine or a poison depending on the quantity and manner in which it is administered.

Paclitaxel is derived from the Pacific yew tree (*Taxus brevifolia*) and sold under the brand name Taxol as the most well-known natural source anticancer therapeutic. It is commonly used as a medicine for ovarian, breast, and lung cancers, and Kaposi sarcoma. The Taxol substance blocks cell division by disruption of microtubules and its development into a formal medicine was funded by the National Cancer Institute (NCI). Taxol was discovered by Dr Mansukh Wani and Monroe Wall. The research was part of a collaborative agreement with the US Department of Agriculture, with the NCI to screen for natural plant products and health benefits. Between the years of 1960 and 1981, over 300,000 samples were processed under this amazing collaboration. Today, the World Health Organization lists Taxol as an essential medicine to support the worldwide healthcare system. If used inappropriately, neutropenia or lack of neutrophils in the immune system develops leading to increased susceptibility of infections.

However, unlike medicinal plants or therapeutic cancer agents that inhibit cell division, the natural plant products that are considered the most likely potential bioterrorist agents fall generally into two categories: (a) those that rapidly inhibit protein synthesis and cell division with no known cure, and (b) hemorrhagic agents that respond poorly to therapy. Three examples are described here out of the hundreds of thousands of plant species on our planet that may contain toxic substances at certain dosages. Digitalis is derived from *Digitalis purpurea* and is known to aid in muscle contraction of a weak or diseased heart. Its common name is foxglove and it is grown as a cottage landscaping plant for its multitude of colors and bell-shaped flowers. It is a native plant to Europe, West Asia, and northwest Africa. Although it is considered medicinal, it can also be highly toxic to humans and animals due to the active component of cardiac and steroidal glycosides (https://www .aaas.org/digitalis-flower-drug-poison). Symptoms include headache, nausea, vomiting, diarrhea, cardia arrhythmia, delirium, hallucination, convulsion, and xanthopsia. Death from toxicity is possible. The whole plant is toxic and accidental overdose has been reported in children drinking water from foxglove in a vase so not much is required for toxic effect. Drying the plant does not reduce the toxicity and, on occasion, it has been mistaken for harmless comfrey tea (*Symphytum* spp.).

Abrin is a toxic albumin found in the seeds of the jequirity pea, *Abrus precatorius*. The toxic dose for humans ranges between 10 and 1000 micrograms per kilogram when ingested; 3.3 micrograms per kilogram when inhaled. The seeds are not available through wide-scale production, meaning that limitation prevents them from being used on a large population (https://pubmed.ncbi.nlm.nih.gov/15181663/). Ricin is an even more toxic substance from castor bean, *Ricinus communis*, a common landscaping plant. It is a perennial flowering plant in the Euphorbiaceae family. Ricin has a history of use as a bioterrorist agent. In the 1940s, the US military experimented with this substance as a possible warfare agent. It is only transmitted by contact, or by inhalation or ingestion. Ricin is simple to make and can come in powder, mist, or pellet form, or as a weak acid dissolved in water, and can be exposed to a large group of people through air, food, or water contamination. The toxic dose is four to eight seeds or the equivalent. Symptoms include pain, nausea, internal bleeding, and death within three to five days. In 1978, Georgi Markov, a writer and dissident, was attacked and killed by a poisonous metal pellet that contained microscopic amounts of ricin. Ricin rapidly prevents protein production, and causes organ failure and death (https://emergency.cdc.gov/agent/ricin/facts.asp). There is no antidote to ricin poisoning.

14.5 Summary

This chapter provides a brief overview of plant associative evidence and explains how it is useful in criminal casework and as a tracer substance to provide investigative leads. Of concern, however, are the poisonous plants that are in our environment that could be used as bioterrorist agents due to their toxicity. These plant species are currently unregulated and sold for other purposes such as landscaping ornamentals and herbal medicines and remedies. Even food substances can be toxic if not prepared correctly and include such edible plants as nutmeg spice powder, which contains myristicin (toxic with just two teaspoons); apple seeds, cherry pits, elderberries, and almonds, which contain trace cyanide, but if ingested in large quantities can be poisonous; raw kidney beans with toxic lectins; mango and cashew with the irritant urushiol; and *Ephedra sinica* and Chinese ma huang, which in large doses can yield serious side effects and even death. Proper labeling and careful administration have left these plants as toxic or necessarily medically administered for thousands of years based on historical records. In the wrong hands, these plants could be used to poison or terrorize populations, and many have no known antidotes. This cautionary chapter is written for information and educational purposes only, and to raise awareness of the potential benefits and misuse of plant substances.

References

Allgeier, L., J. Hemenway, N. Shirley, T. LaNier, and H. Miller Coyle. 2011. Field testing of DNA collection cards for *Cannabis sativa* with a single hexanucleotide marker. *J. Forensic Sci.* 56(5): 1245–1249.

Bruni, I., F. De Mattia, and A. Galimberti. 2010. Identification of poisonous plants by DNA barcoding approach. *Int. J. Legal Med.* 124(6): 595–603.

Craft, K.J., J.D. Owens, and M.V. Ashley. 2007. Application of plant DNA markers in forensic botany: Genetic comparison of *Quercus* evidence leaves to crime scene trees using microsatellites. *Forensic Sci. Int.* 165(1): 64–70.

De Forest, P.R., R.E. Gaensslen, and H.C. Lee. 1983. *Forensic Science: An Introduction to Criminalistics.* New York: McGraw-Hill.

Howard, C., S. Gilmore, J. Robertson, and R. Peakall. 2008. Developmental validation of a *Cannabis sativa* STR multiplex system for forensic analysis. *J. Forensic Sci.* 53(5): 1061–1067.

Howard, C., S. Gilmore, J. Robertson, and R. Peakall. 2009. A *Cannabis sativa* STR genotype database for Australian seizures: Forensic applications and limitations. *J. Forensic Sci.* 54(3): 556–563.

Kress, W.J., D.L. Erickson, F.A. Jones, et al. 2009. Plant DNA barcodes and a community phylogeny of a tropical forest dynamics plot in Panama. *Proc. Natl. Acad. Sci. U. S. A.* 106(44): 18621–18626.

Kress, W.J., K.J. Wurdack, E.A. Zimmer, L.A. Weigt, and D.H. Janzen. 2005. Use of DNA barcodes to identify flowering plants. *Proc. Natl. Acad. Sci. U. S. A.* 102(23): 8369–8374.

Lee, H.C., and C. Ladd. 2001. Preservation and collection of DNA evidence. *Croat. Med. J.* 42(3): 225.

Lee, H.C., C. Ladd, C.A. Scherczinger, and M.T. Bourke. 1998. Forensic applications of DNA typing, collection and preservation of DNA evidence. *Am. J. Forensic Med. Pathol.* 19(1): 10.

Lee, H.C., T.M. Palmbach, and M.T. Miller. 2001. *Henry Lee's Crime Scene Handbook*. Boston, MA: Academic Press.

Miller Coyle, H. 2005. *Forensic Botany: Principles and Applications to Criminal Casework*. Boca Raton, FL: CRC Press.

Miller Coyle, H., C. Ladd, T. Palmbach, and H.C. Lee. 2001. The Green Revolution: Botanical contributions to forensics and drug enforcement. *Croat. Med. J.* 42(3): 340–345.

Miller Coyle, H., C.L. Lee, W.Y. Lin, H.C. Lee, and T.M. Palmbach. 2005. Forensic botany: Using plant evidence to aid in forensic death investigation. *Croat. Med. J.* 46(4): 606–612.

Newton, C., and T. Vo. 2003. Application of microsatellite markers in plant forensics. *Can. Soc. Forensic Sci. J.* 36: 57.

Štambuk, S., D. Sutlović, P. Bakarić, Š. Petričević, and Š. Andjelinović. 2007. Forensic botany: Potential usefulness of microsatellite-based genotyping of Croatian olive (*Olea europaea* L.) in forensic casework. *Croat. Med. J.* 48(4): 556–562.

Vos, P., R. Hogers, M. Bleeker, et al. 1995. AFLP: A new technique for DNA fingerprinting. *Nucleic Acids Res.* 23(21): 4407.

Ward, J., R. Peakall, S.R. Gilmore, and J. Robertson. 2005. A molecular identification system for grasses: A novel technology for forensic botany. *Forensic Sci. Int.* 152(2–3): 121–131.

Ward, J., S.R. Gilmore, J. Robertson, and R. Peakall. 2009. A grass molecular identification system for forensic botany: A critical evaluation of the strengths and limitations. *J. Forensic Sci.* 54(6): 1254–1260.

Wiltshire, P.E. 2006. Hair as a source of forensic evidence in murder investigations. *Forensic Sci. Int.* 163(3): 241–248.

Recent Developments and Future Directions in Human Forensic Molecular Biology

III

Forensic Body Fluid and Tissue Identification

15

CORNELIUS COURTS

Contents

DOI: 10.4324/9780429019944-18

15.1 A Shift of Focus: From Individualization to Contextualization

In 1930, the French criminologist Edmond Locard formulated what today is known as *Locard's exchange principle*, which may be paraphrased as "every contact leaves a trace" (Locard 1930). Applied to traces in forensic biology, this basic principle points to the fact that biological agents will leave small amounts of DNA containing material down to single cells by any contact with objects or other biological agents. As biological agents may be *individualized* by simultaneous analysis of a sufficient number of DNA polymorphisms, the presence of DNA-containing material from a specific individual may be proven beyond rational doubt given the power of contemporary forensic DNA analysis.

However, this DNA-based *individualization* does not inform on the mode of contact or interaction causing the biological material to be transferred to/from where a sample was collected, i.e., it does not contain information as to the activity level (Gill et al. 2020) and hence does not allow for *contextualization* of the trace material and to reconstruct how it had been deposited. This context, though, may be critical to assess the actual modus of involvement of an individual in the crime in question, especially when in court a contesting defense hypothesis has been presented that would explain how DNA that undoubtedly (not even by the defense) originates from a suspect/defendant had been conveyed to a crime site or victim in an innocent manner, e.g., via DNA transfer (Gosch and Courts 2019).

Consequently, the question as to *how* a particular biological sample came to be from where it was recovered or what different components it comprises or whether it contains a certain component, e.g., a specific body fluid, is of increasing forensic interest and will often be essential to assess the evidentiary value of a given stain or trace and hence weigh the probabilities of alternative hypotheses in court.

For example, in a potential case of alleged vaginal penetration of an unconsenting victim with a bottle, the defense hypothesis could hold that the victim merely drank from the bottle, so her DNA that had been detected on the bottle's mouth would be explained. If, however, traces of vaginal secretion could be confirmed, this would strongly support the penetration hypothesis. In another example, it could be required to determine the true source of a trace of blood spatter, e.g., to discern whether it dropped from a cut wound or had been expectorated. In both cases, DNA results would be identical and not helpful, but in the latter case the detection of an admixture of saliva would lend context to the trace (Park et al. 2015).

15.2 Forensic Identification of Body Fluids and Organ Tissues

In the aforementioned examples and many other analogous cases *proving the identity of a body fluid/tissue* or that a specific body fluid/tissue is part of a mixed stain would be of decisive importance. Also, in addition to the aspect of contextualization mentioned earlier, knowing the composition of a stain or trace can also inform downstream procedures, e.g., which DNA extraction method will be best suited, if a differential lysis step should be applied, and so forth. Consequently, available methods and their intricacies, areas of applications, limitations, and new developments have been and are being discussed and reviewed in the forensic science community on a regular basis (Virkler and Lednev 2009a; Virkler and Lednev 2009b; An et al. 2012; Sijen 2015; Harbison and Fleming 2016; Sakurada et al. 2020; Williams 2020).

In summary, forensic body fluid identification (BFI) and organ tissue identification (OTI) are very important elements of the process of the investigation of forensically relevant biological material, and sensitive, reliable, and confirmatory methods need to be accessible and constantly be improved and extended. Therefore, the "classical" tests for BFI that will be discussed in Section 15.3 are increasingly being complemented by more modern, accurate, versatile, and efficient methods.

15.2.1 The "Big Five" and Then Some

Theoretically, any body fluid and tissue can be of forensic interest in contextualizing a stain or trace, and hence all methods to reliably identify a body fluid or tissue are of forensic value. Five body fluids, i.e., peripheral and menstrual blood, semen, saliva, and vaginal secretion, are, however, of particular importance as they are very frequently spilled, spattered, smeared, or left in different kinds of serious crimes and hence encountered at many crime scenes and on persons and objects related to the crime.

15.2.1.1 Peripheral Blood

Peripheral blood consists of plasma (fluid portion), red and white blood cells, and platelets (cellular portion). While red blood cells (erythrocytes) have no nucleus and hence contain no DNA, they do comprise heme, a blood-specific component. On the other hand, white blood cells (leukocytes) lack heme but do contain nuclei filled with DNA. Obviously, peripheral blood is of considerable forensic interest, as in many instances its presence at a scene points to violence and inflicted injury. As blood is well visible, even small amounts can easily be spotted by the unaided eye at crime scenes. Therefore, to conceal that a crime has happened, perpetrators will often try to clean up traces, droplets, or spatters of blood. In many cases, however, even such concealed traces can be detected when very sensitive presumptive tests like Luminol are applied (Table 15.1).

15.2.1.2 Vaginal Secretion

Vaginal secretion is produced by the cells of the vaginal mucosa and the cervix and the Bartholin's glands. It is composed of cells and bacteria, suspended in a liquid to lubricate and protect the vagina from where it exits through the vaginal opening. However, unlike other body fluids, vaginal secretion contains no characteristic protein component, resulting in a lack of classical tests to detect it (Table 15.1). In forensic investigations, particularly in cases of sexual assault, the detection of vaginal secretion, e.g., on a swab taken from a suspect's penis or an object that the victim claims has been introduced vaginally without her consent, can be essential for an evidence-based reconstruction of the course of events.

15.2.1.3 Menstrual Blood

Menstrual discharge or "menstrual blood" consists of blood, cervical mucus, and fragments of endometrial tissue that sloughs monthly during menstruation and ranges from 10 to 80 ml in volume. As the discharge passes the vaginal tract and exits the vaginal opening, it will also have admixed vaginal secretion rendering it a complex multicomponent body fluid. While menstrual blood does contain peripheral blood and will be detected by all tests devised to detect the latter, it also contains specific components originating from the cervix and uterus that are unique for menstrual blood. Differentiating menstrual from peripheral blood in, e.g., a forensic blood stain from an alleged sexual assault thus can yield

Table 15.1 Forensic Body Fluid and Tissue Identification Tests

Body Fluid	Detected Component	Name of Test	Test Principle	SoE
Peripheral blood	Heme	Kastle-Meyer	Colorless phenolphthalein reacts with H_2O_2 catalyzed by heme, reaction product is pink	P
		Leucomalachite Green (LMG)	Colorless LMG reacts with H_2O_2 catalyzed by heme, reaction product is blue-green	P
		Tetramethylbenzidine (TMB); Hemastix®	Colorless reduced ortho-Tolidine reacts with H_2O_2 catalyzed by heme, reaction product is blue	P
		Luminol	Luminol reacts with H_2O_2 catalyzed by heme, reaction creates a visible glow	P
		Fluorescein	Fluorescein reacts with H_2O_2 catalyzed by heme, reaction product creates a visible glow when exposed to light (425–485 nm)	P
	Heme derivatives	Hemochromogen crystal assay (Takayama test)	Pyridine and glucose react with heme to form crystals of pyridine ferriprotoporphyrin crystals detectable microscopically	C
		Hematin crystal assay (Teichmann test)	Glacial acetic acid and salts react with heme when heated to form crystals of hematin chloride crystals detectable microscopically	C
	Human hemoglobin	Bluestar OBTI (quick test cassette)	Immunochromatographic detection of hemoglobin	C
	Human glycophorin A antigen	RSID-Blood (quick test cassette)	Immunochromatographic detection of glycophorin A	C
Menstrual blood	D-dimer	Seratec® PMB Test, menstrual blood test (quick test cassette)	Immunochromatographic detection of D-dimers (and human hemoglobin)	C
Vaginal secretion	Vaginal acid phosphatase (VAP)	Ablett test	Isoelectric focusing under optimal conditions can separate VAP from SAP	P
	Nucleated squamous epithelial cells (NSEC)	Lugol's iodine assay	Iodine-based dyeing of glycogen containing NSEC; identification by microscopy	C
Semen	Semen fluid, dried	Lighting	Lighting a stain with an ALS, laser, or UV light will cause semen stains to emit fluorescence	P
	Seminal acid phosphatase (SAP)	Colorimetric assays	A colorless reagent (e.g., α-naphtyl phosphate) has its phosphate group removed by the SAP and then together with a diazonium salt (e.g. Brentamine Fast Blue B) will form a colored product	P

(Continued)

Table 15.1 (Continued) Forensic Body Fluid and Tissue Identification Tests

Body Fluid	Detected Component	Name of Test	Test Principle	SoE
		Fluorometric assays	A reagent (e.g., 4-methylumbelliferone phosphate) has its phosphate group removed by the SAP, which under UV light generates fluorescence	P
	Sperm cells*	Christmas Tree Staining	Specific dyeing of sperm cell components (nuclei, neck and tail, acrosome); identification by microscopy	C
		Sperm Hy-Liter™	Specific immunofluorescent dyeing of sperm cells; identification by fluorescence microscopy	C
	Semenogelin	RSID-Semen (quick test cassette)	Immunochromatographic detection of human semenogelin antigen	C
	Prostate-specific antigen (PSA)	Seratec® PSA Semiquant	Immunochromatographic detection of human PSA	C
Saliva	Saliva, dried	Lighting	Lighting a stain with light of 470 nm will cause saliva stains to emit fluorescence, visible with goggles	P
	α-Amylase	Starch-iodine assay (radial diffusion assay)	Amylose in starch reacts with iodine and forms a blue complex; if amylase is present, starch is broken down so that on adding iodine no colored product is made	P
		Phadebas amylase test (colorimetric assay)	A water-insoluble, colorless, cross-linked starch polymer carrying a blue dye is hydrolyzed by α-amylase to form water-soluble blue fragments	P
		RSID-Saliva (quick test cassette)	Immunochromatographic detection of human semenogelin antigen	C
	Buccal epithelial cells	HE staining	Dyeing of buccal epithelial cells; identification by microscopy	P
Urine	Urine, dried	Lighting	Lighting a stain with an ALS will cause urine stains to emit fluorescence	P
	Creatinine	Jaffe test	Picric acid reacts with creatinine and a weak base to form a deep orange-red color	P
	Urea	DMAC test	Para-dimethylaminocinnamaldehyde reacts with urea	P
	Tamm–Horsfall glycoprotein	RSID-Urine (quick test cassette)	Immunochromatographic detection of Tamm–Horsfall protein	C
Sweat	Sweat-specific antigen	Sagawa test	Immunological assay using monoclonal antibody	C

Notes: SoE, strength of evidence P, presumptive; C, confirmatory; ALS: alternate light source.

* Ineffective in semen from sterilized or azoospermic men.

important contextualizing information, as the presence of menstrual blood does not imply violence or injury. However, it has to be kept in mind, that a positive result for menstrual blood does not exclude the admixture of peripheral blood originating, e.g., from a wound. Currently, there are very few classical tests to detect menstrual blood (Table 15.1).

15.2.1.4 Semen

Semen is the male reproductive fluid. It is produced by the seminal vesicles, the bulbourethral gland, and the prostrate and consists of a viscous liquid that in addition to sperm cells contains fructose, carnitine, glycerylphosphorylcholine, acid phosphatase, and several other components. The ejaculate fluid of a healthy fertile male has a volume of 1.6–5.5 ml and will contain 17 million–192 million sperm cells per milliliter each harboring a haploid male genome (Gamblin and Morgan-Smith 2019). Ejaculated semen will in most cases be indicative of some form of male sexual activity, hence its detection is of central importance in the investigation of sexual assault cases, and there is a plethora of classical tests to detect semen (Table 15.1). When applying these tests, it has to be taken heed that semen fluid does not necessarily contain sperm cells (e.g., in vasectomized or azoospermic males).

15.2.1.5 Saliva

Saliva is a watery, slightly viscous liquid produced by three pairs of major glands and numerous minor salivary glands located in the oral cavity to keep the mouth wet. It contains nitrogenous products, mucins, immunoglobulins, and enzymes such as α-amylase that help to prepare food for digestion. Also, it is heavily fraught with oral microbiota. Saliva stains are frequently encountered at crime scenes due to the common habit of spitting but will also typically be present in cases of violent and sexual assault, and are of particular importance for trace contextualization when oral rape but also kissing or licking of body parts without consent has been alleged. Most classical tests to detect saliva depend on its content of α-amylase (Table 15.1).

15.2.1.6 Other Body Fluids

Apart from these "big five" there are other body fluids the identification or differentiation of which may in particular case constellations be of forensic interest. These include sweat, urine, pre-ejaculate, nasal mucosa, and rectal mucosa, for some of which classical tests are available (Table 15.1).

The detection of *sweat*, especially after having dried on a garment, may help to contextualize how or by whom it has been worn. Also, it may be of forensic relevance to differentiate dried stains of saliva from sweat. Sweat is a clear, watery, slightly acidic liquid (pH 6.3) that contains minerals and metals, lactate, and urea and that may be identified by the Sagawa test (Table 15.1).

Urine may be involved, for instance, in cases of sexual assault that include debasement of the victim. Detecting stains of urine is difficult, however, as they are typically pale, diffuse, and may be spread over large areas. Classical tests to detect urine focus on components like urea, creatinine, and Tamm–Horsfall protein (Table 15.1).

Pre-ejaculate is a clear liquid originating from the bulbourethral gland that functions as a natural lubricant during intercourse and exits the penis on sexual arousal but before proper ejaculation. It may or may not contain viable sperm cells and may or may not produce a positive result for semen (Kelly 2020; Zukerman 2003; Killick 2011). Hence, there is no test to reliably differentiate pre-ejaculate from ejaculated semen.

The detection of *nasal mucosa* will only very seldom be of forensic interest. It can, however, serve as a proxy to detect nasal blood in which the former will be contained, e.g., to help in differentiating the exact source of a blood stain. No classical tests to detect nasal mucosa are available.

To identify *rectal mucosa* can be relevant if, e.g., in a case of sexual assault or rape penile swabs of a suspect have been taken and anal penetration needs to be proven or differentiated from vaginal penetration. Currently, there is no classical test to identify rectal mucosa.

15.2.1.7 Organ Tissues

In the reconstruction of violent crimes, the identification of *skin and organ tissues* can provide crucial information complementing the DNA-based source attribution. Proving that a tissue sample recovered, for instance, from the blade of a knife or a bullet secured at a crime scene originates from an internal organ can contextualize the sample and produces evidence that a serious injury has been inflicted with the weapon, while the mere presence of the victim's DNA established by standard short tandem repeat (STR) profiling may be argued to result from a superficial cut or innocuous graze. Also, in cases involving postmortal dismemberment to enable covert disposal of a corpse, the identification of even mangled and degraded tissue recovered, for instance, from sewers or waste pipes may be required.

There are some conventional immunological, histological, and/or enzymatic techniques for the assignment of tissue origin (Kimura et al. 1995; Takahama 1996; Seo et al. 1997; Takata et al. 2004). However, these may pose problems in terms of specificity and sensitivity, especially when only trace amounts of material are present, and may interfere with DNA profiling.

15.3 Classical Tests for Forensic Body Fluid Identification

"Classical" tests to detect forensically relevant body fluids have been in use for decades, and utilize chemical, colorimetric, and immunological principles and processes to demonstrate the presence of (a specific component of) a specific body fluid. Though non-comprehensive, Table 15.1 lists widely used classical tests for the detection of forensically relevant body fluids.

Many of these classical tests are still widely used and do retain current relevance as they are cost-efficient and most of them are quick and easy to perform even by non-experts, e.g., at the crime site, rendering their use appropriate, e.g., in cases where it is necessary to search for "removed" or cleaned-up traces and when sufficient trace material is available. They do have a number of disadvantages and caveats, though, that need to be taken into account when deciding if and how the bodily origin of a stain or trace has to be determined. Many presumptive tests lack specificity and are prone to false-positive (due to cross reactivity) and false-negative results. Also, some test chemicals when applied to, e.g., blood stains may be detrimental to DNA (de Almeida 2011). Many confirmative tests exhibit limited sensitivity and, in most cases, consume scarce sample material that will not be available for downstream DNA analysis. Moreover, classical tests can only detect one kind of body fluid at once. So, if complex mixed stains, as frequently encountered in sexual assault cases, have to be analyzed, the amount of trace material required to perform all tests deemed appropriate may become a limiting factor. Also, there is no reliable classic test for vaginal secretion, which is of high contextual interest in forensic casework analysis, and there are no classical tests at all for organ tissues.

Fortunately, progress in forensic science facilitated the emergence of novel physical, biochemical, and molecular methods for sensitive, specific, non-consumptive identification of the cellular or bodily provenience of biological stains and traces found at crime scenes that can be performed in parallel and complementary to standard forensic DNA analysis.

15.4 RNA-Based Approaches to Forensic Body Fluid and Tissue Identification

The most promising and most thoroughly tested and validated "non-classic" method for forensic BFI is centered on RNA expression analysis (Lynch and Fleming 2020). The origin of forensic RNA analysis may be traced back to 1994 when Phang et al. were the first to use RT-PCR to analyze RNA from postmortem tissues in a forensic setting. Since then and with the availability of commercial qPCR devices (see Section 15.4.4.1), the interest in highly versatile RNA analysis surged within the forensic and medicolegal community (Courts and Madea 2012). In parallel to this, long-held concerns that RNA may be too unstable and prone to degradation to be of any forensic use have been relativized by numerous studies reporting successful analysis of RNA from an array of tissues and body fluids with extended postmortem intervals or storage times (Bahar et al. 2007; Zubakov et al. 2009; King et al. 2001; Koppelkamm et al. 2011; Kohlmeier and Schneider 2012). Thus, while almost all >200 distinct cell types in an adult human body (except gametes, red blood cells, platelets, and squamous cells) in all tissues possess *the same diploid genome* and therefore cannot be discriminated DNA-wise, the common principle of all RNA-based approaches to the identification of forensically relevant cell-containing biological material is that functional and/or organizational differences between different kinds of cells and hence tissues and body fluids will be *represented by differently constituted "RNAomes,"* i.e., the entirety of molecules of a specific RNA species (e.g., mRNA) in a temporospatial context. This is possible, because although only about 2% of the about 21,000 human genes are protein coding, probably more than 80% of the genome is transcribed to produce a plethora of RNAomes comprising many different species of RNA (Uhlén et al. 2010; The ENCODE Project Consortium 2012; The Genotype-Tissue Expression [GTEx] project 2013; Uhlén et al. 2015). Spatial and temporal differences in gene expression result from complex regulatory interactions of transcription factors, effects of chromatin state, enhancer/silencer regions in the DNA, and epigenetic regulation. As a result, each specialized cell type expresses a highly specific subset of possible RNAs tailored to its needs, and it is mainly for technical or methodological reasons that currently mostly only one species of RNA is analyzed simultaneously.

15.4.1 Messenger RNA

Primary transcription generates messenger RNA (mRNA), and a fully processed mRNA after splicing has an average half-life of 10 h in human cells. This allows a cell to rapidly adapt protein synthesis to its changing needs. Hence, mRNA transcription has been shown to be the "master sculptor" of cellular identity and specificity (Melé et al. 2015) rendering the mRNA transcriptome a highly plastic molecular depiction or "snapshot" of a cell's phenotype and also its condition and tissue provenience. This well of information has been exploited in forensic mRNA-based BFI since in 2002 when Bauer and Patzelt were first to report the detection of menstrual blood due to its specific mRNA expression. To date, an ample body of literature is available on the application of mRNA analysis in the detection of a range of body fluids and organ tissues (Sijen 2015; Lynch and Fleming 2020), for some of which, like nasal mucosa and rectal mucosa, RNA analysis is the only existing test today (Sakurada et al. 2012; Chirnside et al. 2020; Bamberg et al. 2021). Multiplex assays (e.g., Bamberg et al. 2021; Lindenbergh et al. 2012; Xu et al. 2014; Albani and Fleming 2019; Lindenbergh et al. 2013) are available and numerous international trials have already and successfully been executed, employing capillary electrophoresis (Section 15.4.4.1) (Haas et al. 2010; Haas et al. 2012; Haas et al. 2013; Haas et al. 2014; van den Berge 2014; Haas et al. 2015; Salzmann et al. 2021; Carnevali et al. 2017) or massively parallel sequencing (MPS; see Section 15.4.4.3) (Ingold et al. 2018; Ingold et al. 2020a; Ingold et al. 2020b) as the method of detection, respectively. Consequently, today, mRNA-based methods are routinely used in forensic BFI and OTI by accredited forensic laboratories and the results are accepted by courts in several countries.

15.4.2 Micro-RNA (miRNA)

Discovered only in 1993, micro-RNA (miRNA) is a class of small noncoding RNA (ncRNA) molecules with a length of 18–24 nucleotides that play an essential regulative role for many cellular processes. Like mRNA, miRNA also exhibit a specific expression spectrum in different body fluids (Weber et al. 2010), and their high tissue specificity reported in biomedical studies led to the suggestion that miRNAs could be utilized as markers for body fluids in forensic investigations with specific body fluids expected to exhibit specific miRNA expression profiles or "signatures." Consequently, their potential as forensic markers has been validated and have been discussed and explored since Hanson et al. in 2009 published the first study on forensic BFI employing qRT-PCR to assess miRNA expression (Courts and Madea 2010; Silva et al. 2015; Glynn 2020; Hanson and Ballantyne 2013).

MiRNA analysis is of particular interest for forensic investigations because miRNA profiling has some advantages as compared to mRNA profiling. Most important, due to their tiny size, mature miRNAs are more stable than mRNAs rendering them less susceptible against fractionation by chemical or physical strain (Jung et al. 2010; Zubakov et al. 2010; Lux et al. 2014). Sauer et al. even successfully analyzed miRNA from and identified blood that had been aged for more than 36 years (Sauer et al. 2016), as well as putrefied organ tissues after 28 days of environmental exposition (Sauer et al. 2017).

Since 2009, several qRT-PCR assays for the identification of body fluids (Sauer et al. 2016; Wang et al. 2013; Park et al. 2014; Sirker et al. 2017; Mayes et al. 2018) and organ tissues (Sauer et al. 2017) have been presented. Since 2014, MPS has also been employed in forensic miRNA-based BFI (Petersen et al. 2014; Seashols-Williams et al. 2016; Wang et al. 2016; Dørum et al. 2019). The data available so far indicates without doubt the potential of miRNA markers for the identification of forensically relevant body fluids and organ tissues. In contrast to mRNA, however, there is little agreement between studies as to which (combinations of) miRNA markers should be used to specifically detect a particular body fluid or organ tissue. Therefore, more validation experiments, standardization. as well as interlaboratory trials employing sets of ubiquitously used markers will be required to establish reliable assays and hence reach broad acceptance and allow for introduction to routine use of forensic miRNA analysis.

15.4.3 Other RNA species

With the ongoing expansion of the "RNA zoo" (Morris and Mattick 2014), hence with more and more different RNA species being discovered and their cellular functions being described, several groups have now focused on the forensic potential of some of these "new" RNA species.

15.4.3.1 Piwi-Interacting RNA (piRNA)

Piwi-interacting RNA (piRNA) is a large class of small (21–35 nt) noncoding RNA molecules expressed in animal cells. PiRNAs are named from their interaction in RNA protein with piwi-subfamily Argonaute proteins. These piRNA complexes have mostly protective functions being involved in the silencing of transposable elements but can also be involved in the regulation of other genetic elements in germ line cells (Ozata et al. 2019). PiRNA expression is highly regulated resulting in a cell and tissue specific expression pattern that may be exploited for forensic BFI and OTI, and like miRNA (Courts et al. 2010), due to its short length, piRNA is expected to be resistant against degradation.

In a proof-of-concept study, Wang et al. in 2019 were first to examine the expression levels of four preselected piRNAs in venous blood, saliva, semen, menstrual blood and vaginal secretion employing qRT-PCR and could show that piR-55521 can differentiate semen from other body fluids and is detectable in samples with as little as 200 pg of total RNA (Wang et al. 2019). They then expanded their work and utilized MPS and partial least squares-discriminant analysis (PLS-DA) to study global piRNA expression in venous blood, menstrual blood, saliva, semen, vaginal secretions, and skin, and selected for differentially regulated candidates. From this, by qRT-PCR verification they distilled panels of three piRNAs to discriminate venous blood and menstrual blood, and two piRNAs to discriminate saliva and vaginal secretions (Wang et al. 2019). This recent initial work indicates the potential of piRNA as another class of small RNA for forensic BFI.

15.4.3.2 Circular RNA (circRNA)

Circular RNA (circRNA) that are produced by the infrequent backsplicing of exons in pre-mRNA have only in 2012 been discovered to be the predominant transcript isoform from hundreds of human genes in diverse cell types (Salzman et al. 2012) and shown to exhibit cell-type specific features (Salzman et al. 2013). CircRNA have post-transcriptional regulatory and "meta-regulatory" (regulating regulators) functions, an example for the latter being circRNA that act as "miRNA sponges" in that they strongly suppress the activity of target miRNAs, resulting in turn in increased levels of the miRNAs targets (Hansen et al. 2013).

Specific distributions of circRNA in saliva and blood had already been described in the non-forensic literature in 2015 (Memczak et al. 2015; Bahn et al. 2015). Moreover, circRNA have high abundance in many cell and tissue types, and due to their lack of free ends and potential super coiling exhibit remarkable stability, in summary recommending them to forensic purposes.

In 2017, Zhang et al. were first to include circRNAs (the major circRNAs of ALAS2 and MMP7) in an mRNA profiling procedure for BFI and could show that this significantly improved detection sensitivity and stability. Liu et al. in 2019 also described the inclusion of circRNAs in the mRNA-based detection of biomarkers for blood, menstrual blood, vaginal secretion, semen, and urine, and recently presented a multiplex assay for forensic BFI including 14 mRNA markers with circRNAs expression that could discriminate these body fluids, even when all linear transcripts had been completely erased (i.e., mimicking intense degradation) (Liu et al. 2020). These initial studies indicate the value of integrating circRNA detection into mRNA-based BFI, warranting further research.

15.4.4 Methods of Forensic RNA Analysis

15.4.4.1 Multiplex PCR and CE
The method most widely used, as of today, for standard forensic DNA analysis, *capillary electrophoresis (CE)*, that will be accessible for most forensic genetics laboratories, is also most commonly applied in and most rigorously validated for mRNA-based BFI and OTI. Its advantages are its relative time- and cost-efficiency and its high multiplexing capacity, allowing for the simultaneous analysis of 20-plus RNA markers. However, quantification from peak heights of relative input amounts of the enclosed markers and dynamic range are limited, which complicates interpretation and may require multiple replicate runs.

15.4.4.2 Quantitative Reverse Transcription PCR (qRT-PCR)
Since in 1996, when real-time PCR instruments became commercially available, qRT-PCR quickly rose to being considered the gold standard for quantitative gene expression analysis because its accuracy and reproducibility when applied correctly, i.e., *adhering to the MIQE guidelines* (Courts et al. 2019; Bustin et al. 2009), are still unsurpassed. Moreover, qRT-PCR devices are affordable and will be available in most forensic genetic routine labs.

However, due to the quite limited capacity of qRT-PCR for multiplexing, each body-fluid-specific marker has to be amplified in a separate well. Given the methodological requirement of technical replicates this renders the simultaneous detection of two or more body fluids in a sample a labor- and cost-intensive task. Thus, qRT-PCR is only seldomly applied in mRNA-based forensic BFI (Visser et al. 2011; Park et al. 2013).

For forensic miRNA analysis, in contrast, there is only very preliminary research on applying CE in the detection of specific markers (van der Meer et al. 2013), hence most published efforts reporting on methods for miRNA-based BFI relied on qRT-PCR.

15.4.4.3 Massively Parallel Sequencing (MPS)
The most modern and promising method for RNA-based forensic BFI and OTI is MPS (Haas et al. 2021) applied in forensic RNA-seq, and although degradation and low amounts of sample material typical for the forensic setting had been assumed to considerably constrict forensic RNA-Seq, several studies produced proof to the contrary (Bruijns et al. 2018; Ballard et al. 2020), even managing to sequence whole transcriptomes (Lin et al. 2015; Weinbrecht et al. 2017; Salzmann et al. 2019).

MPS has a huge multiplexing capacity and can easily accommodate hundreds of markers in a single targeted sequencing assay. Also, mRNA sequencing can be combined with standard STR analysis (Zubakov et al. 2015). Consequently, the amount of available data on the forensic application of MPS to sequencing mRNA and miRNA in body fluids (Dørum et al. 2019; Salzmann et al. 2019; Hanson et al. 2018; Dørum et al. 2018; Wang et al. 2017) and organ tissues (Hanson et al. 2017; Javan et al. 2015) is constantly increasing, and first collaborative exercises on sequencing mRNA are being executed (Ingold et al. 2018; Ingold et al. 2020a; Ingold et al. 2020a). The most distinguishing unique feature of MPS in forensic RNA analysis, however, is its potential to not only quantify transcripts but also capture their base sequences, making it possible to detect and analyze sequence variants, and hence, to assign transcripts from mixed stains, e.g., blood from two persons, to their relative donors (Ingold et al. 2020a; Ingold et al. 2020b; Hanson et al. 2019; Liu et al. 2021), which bears enormous potential for mixed stain analysis in forensic casework.

As a downside, MPS and RNA sequencing in particular still is comparably expensive, labor intensive, and requires additional bioinformatic expertise and infrastructure to deliver reliable results. These limitations notwithstanding and given further research and validation, it is plausible that MPS will become the dominant and most comprehensive method for RNA-based forensic BFI and OTI in the future.

15.4.4.4 Other Methods

Apart from CE, RT-qPCR, and MPS, some miscellaneous other methods for RNA-based BFI have been described that represent creative and versatile new approaches to forensic RNA analysis.

Park et al. (2013) described how they adopted the *hybridization-based nCounter system by "Nanostring"* for forensic BFI to, using a logistic regression model, identify blood, semen, saliva, and vaginal secretion.

Hanson et al. (2013) showed that RNA-based BFI can be performed without fluorescence detection by applying duplex and triplex *high-resolution melt assays (HRM)* with unlabeled primers to detect markers for vaginal secretion, skin, blood, menstrual blood, saliva, and semen. Each body-fluid-specific marker could be identified by the presence of a distinct melt peak. Usually, HRM assays are used to detect variants or isoforms for a single gene target.

Su et al. (2015) applied *real-time reverse transcription loop-mediated isothermal amplification (real-time RT-LAMP)* to show that a blood-specific marker could only be detected in blood but not in other body fluids. Compared to standard multiplex RT-PCR, RT-LAMP is more rapid and requires only a simple heating block, which could make RNA-based BFI more flexible.

Donfack et al. (2015) used matrix-assisted laser desorption/ionization–time of flight mass spectrometry (MALDI-TOF MS) to successfully identify specific mRNA-derived cDNA from venous blood, saliva, and semen. They could show that MALDI-TOF MS with some caveats may be a potential fluorescent dye-free alternative method for RNA-based forensic BFI.

And recently, Jackson et al. (2020) applied *LAMP* in the simultaneous detection of mRNA markers from whole blood, semen, saliva, and vaginal fluid using a *smartphone* for optical detection and analysis.

15.5 DNA-Based Approaches to Forensic Body Fluid and Tissue Identification

15.5.1 Forensic Epigenetics: Methylation Analysis

Epigenetics is a field of study that analyzes changes in gene expression that occur *without alteration to the DNA sequence*. The most widely studied epigenetic modification in homo sapiens is DNA methylation. which usually occurs on cytosine nucleotides embedded in so-called CpG islands. CpG methylation, which is associated with condensed chromatin structure, is an important epigenetic regulatory mechanism normally resulting in repression of transcription (Nustad et al. 2018).

Consequently, numerous regions exhibiting differential methylation have been discovered by genome-wide methylation analyses from which cell-type and tissue-specific methylation patterns could be inferred (Ghosh et al. 2010). Ten years ago, Frumkin et al. (2011) were first to apply this finding to a forensic setting when they used methylation-sensitive restriction endonuclease (MSRE) followed by multiplex amplification of differentially methylated regions (DMRs) on 50 DNA samples from blood, saliva, semen, and skin epidermis, and successfully identified the body fluid or tissue in all cases. Since then, research on the application of methylation analysis in forensic BFI has surged, and today an extensive body of literature is available (Kader et al. 2020), introducing assays for single body fluids and tissues like blood, saliva, semen (Silva et al. 2016; Watanabe et al. 2016; Wasserstrom et al. 2013; Watanabe et al. 2018; Liu et al. 2020; Holtkötter et al. 2018), and brain tissue (Samsuwan et al. 2018) as well as multiplex assays (Forat et al. 2016; Holtkötter et al. 2017; Gauthier et al. 2019; Xie et al. 2020). Moreover, a first collaborative exercise on DNA methylation-based BFI has been performed (Jung et al. 2016).

Taken together, there are promising approaches of forensic methylation analysis. Notably, DNA methylation is a comparably stable epigenetic modification, and unlike temporary expression states of RNA and RNAomes, which are highly dynamic, DNA methylation is static, thus allowing for quantitative analysis of older samples. Also, individualization (STR analysis) and contextualization (BFI/OTI) of a trace can be performed from the same sample preparation, the same molecule even (DNA). On the other hand, DNA methylation has been shown to be influenced by various external and environmental factors such as nutrition, smoking, aging, exposure to pollutants, and so forth, which has to be taken into account in forensic validation of DNA methylation-based methods for routine casework. Another limiting factor in the progress of forensic epigenetics is the amount and quality of DNA typically recovered from crime scenes.

15.5.1.1 *Methods of Forensic Methylation Analysis*

There are several approaches to analyze differential DNA methylation on a global, gene-specific as well as epigenome-wide scale (Kurdyukov and Bullock 2016). In forensic genetics, there are reports on methylation analysis using mostly sequencing (minisequencing, i.e., single-base extension sequencing [SBE]), Sanger sequencing, pyrosequencing and MPS), microarrays (such as "Infinium" by Illumina), high-resolution melting curve analysis (HRMA), and methylation-specific PCR (MS PCR) *after bisulfite treatment* as well as methylation-sensitive restriction enzyme PCR (MSRE PCR), *which does not require prior bisulfite conversion.*

Briefly, under bisulfite treatment genomic DNA is denatured and treated with *sodium bisulfite*, leading to deamination of unmethylated cytosines, i.e., a "conversion" into uracils, while methylated cytosines (both 5-methylcytosine and 5-hydroxymethylcytosine) remain unchanged. Upon PCR amplification, uracil is then converted to thymine. This process requires a relatively large amount of DNA, however, and also is not fully effective, i.e., leaving a small percentage of cytosines unconverted. Also, DNA loss and especially DNA fragmentation may be troublesome. Therefore, choosing the appropriate kit or method (Kint et al. 2018) for the material at hand and controlling for DNA damage and conversion efficiency is essential.

Some of the methods for forensic epigenetics are technically challenging and analysis of DNA methylation is not a straightforward task. No single technique is "the best" and can generate unambiguous data on DNA methylation and which method to choose when applying epigenetics to forensic BFI and OTI is nontrivial and highly dependent on available equipment and expertise, but also the quantity and quality of DNA extracted from forensic material as technologies for the analysis of methylation status of multiple CpG in a quantitative single-nucleotide resolution and high-throughput manner require a large number of epigenetic markers and high-grade quality DNA that will not normally be encountered in forensic trace material. Also, it has to be considered that sample material that has been used for methylation analysis, depending on the chosen method, may not be usable for standard forensic STR analysis anymore.

15.5.2 Copy-Number Variations

Copy-number variations (CNV) are repeated sections of the genome with the number of repeats in the genome varying between individuals (McCarroll and Altshuler 2007). Hence, CNV are structural variations of the DNA derived from duplication or deletion events that affect a considerable number of bases, and 4.8%–9.5% of the human genome can be classified as CNV (Zarrei et al. 2015). In 2018, Zubakov et al. were first to investigate the potential of CNV analysis in forensic BFI. In a genome-wide CNV screening followed by targeted qPCR confirmation they identified DNA markers specific for blood and semen, respectively. Advantages of CNV analysis are that it does not require prior DNA treatment, is highly sensitive, and that CNV–PCR product can be used as input material for standard forensic DNA analysis, hence allowing the use of the same DNA aliquot for both tissue and individual identification.

15.6 Other Approaches for Forensic Body Fluid and Tissue Identification

This chapter briefly introduces other non-classical but neither RNA- nor DNA-based approaches for forensic body fluid identification.

15.6.1 Raman Spectrometry

Named after the Indian physician C.V. Raman, Raman spectrometry is a nondestructive test that relies upon the scattering of low-intensity infrared laser light (785 nm) by compounds including biological materials (Butler et al. 2016). This produces spectra that are complex and from which unique spectroscopic signatures of the molecular structures of each fluid can be derived by advanced statistical analysis. Unique Raman spectroscopic signatures have been determined for blood (Boyd et al. 2011), saliva (Virkler and Lednev 2010), semen (Virkler and Lednev 2009a; Virkler and Lednev 2009b), vaginal secretion (Sikirzhytskaya et al. 2012), and sweat (Sikirzhytskaya et al. 2012) that are caused by and correspond to specific components in the respective body fluids. Thus, forensically relevant body fluids can be differentiated by visual comparison of their Raman spectra (Virkler and Lednev 2008).

15.6.2 Fourier Transform Infrared Spectrometry

Fourier transform infrared (FT-IR) spectroscopy is widely used to identify characteristic spectral signatures determined for various biological materials in a nondestructive manner (Baker et al. 2014). Eventually, this sparked interest in the forensic community and some groups successfully applied FT-IR in the detection and differentiation of forensically relevant body fluids (Orphanou et al. 2015). However, FT-IR is somewhat limited in forensic settings as it is typically used to analyze fresh body fluids in comparably large quantities on even, clean, nonabsorbent surfaces without environmental influence or contamination. Very recently, therefore, Sharma et al. (2021) used attenuated total reflectance FT-IR (ATR-FTIR) to detect stains of blood, and they specifically analyzed the effect of substrates, washing/chemical treatment, aging, and dilution limits, and could show that ATR-FTIR is promising for the detection of blood traces even under forensic casework conditions.

15.6.3 Microbiome Characterization

Some relevant body fluids, including saliva, vaginal secretion, and rectal mucosa, contain large numbers of microbiota, and the quantitative classification of the components of an unknown body fluid's microbiome has been shown to be suitable for its identification (Clarke et al. 2017) based on findings that the human microbiome consists of distinct bacterial populations at different locations in and on the body (The Human Microbiome Project Consortium, Structure, function and diversity of the healthy human microbiome 2012). In 2010, Power et al. and Donaldson et al. applied this principle to show that blood had been expirated and hence contained traces of saliva by detecting salivary bacteria. In 2012, before MPS was widely available, Giampaoli et al., Doi et al. and others used RT-PCR to identify samples of vaginal secretion, saliva, and fecal matter (Giampaoli et al. 2012; Doi et al. 2014; Nakanishi et al. 2009; Nakanishi et al. 2013), with Giampoli's assay later being validated in an interlaboratory trial (Giampaoli et al. 2014). With MPS becoming more widely accessible, forensic microbiomics (Hampton-Marcell et al. 2017) has surged and various groups applied MPS of the microbiome in forensic BFI, e.g., to detect saliva on skin (Hanssen et al. 2017) and vaginal secretion (Giampaoli et al. 2017). Recently, to investigate whether exposition to temporal and environmental effects would affect the bacterial composition of body fluid stains, hence impacting forensic microbiomic assessment, Dobay et al. (2019) exposed samples of saliva, skin, vaginal fluid, and menstrual blood to indoor conditions for 30 days. Applying MPS to a total of 46 control and exposed samples, they could show that both groups of samples could be assigned to the correct body site, and also that vaginal and menstrual samples cannot be distinguished using bacterial markers due to shared microbial signatures. In 2020, Díez Lopez et al. successfully subclassified blood samples by applying forensic microbiomics. They used a taxonomy-independent deep neural network approach based on microbiome MPS data to assign forensic blood samples according to their bodily origin, such as menstrual, nasal, fingerprick, and venous blood.

In conclusion, while some challenges of exploring and using microbiome data including correct use of databases and bioinformatics tools need to be addressed and adapted to forensic conditions and standards, it is safe to say that microbiomics based on MPS bears considerable potential for future applications of forensic BFI.

15.6.4 Protein Analysis and Proteomics

Proteins contain information about an individual's genotype (in the form of single amino acid polymorphisms, the result of nonsynonymous SNPs) and by their expression level also reflect aspects of its phenotype. Proteomics employing mass spectrometry (MS), therefore, can both identify biological samples and characterize the conditions that produced them. Moreover, MS proteomics can be used in cases where nucleic acids are absent, degraded, or uninformative. This, combined with the technical advancement of MS instrumentation, the improvement of protein sequence databases and the maturing of data analysis methods motivate interest in proteomics as a potential tool for forensic BFI and OTI (Merkley et al. 2019; Parker et al. 2021).

In fact, protein analysis to identify brain for forensic OTI was used as early as 2004 (Takata et al. 2004). In the following years, various reports on the application of MS to detect protein candidates to identify different body fluids and materials from forensic stains had been published (Yang et al. 2013; van Steendam et al. 2013; Igoh et al. 2015; de Beijer et al. 2018).

After valid candidates had been identified, targeted assays were designed to detect and quantitate proteins with low abundance against a background of other non-target molecules and have been evaluated in casework-type samples but also authentic casework samples (Pieri et al. 2019; Legg et al. 2014; Dammeier et al. 2016). While most of these studies used QQQ mass spectrometers operating in MRM mode, Riman et al. (2018), in contrast, used *proximity ligation real-time PCR (PLiRT-PCR)* with two polyclonal antibodies for the identification of semen and sperm proteins. The method relies on protein recognition by pairs of proximity probes (antibody–DNA conjugates) that give rise to a ligated DNA strand.

In summary, forensic proteomics has come a long way from its beginnings and today bears considerable potential for application in forensic BFI and OTI, especially in challenged material where nucleic acids are degraded beyond analysis.

15.7 Outlook

Acknowledging and internalizing the forensic shift of focus toward trace contextualization and assessing activity level propositions, forensic body fluid and organ tissue identification has grown into a prolific, highly active and creative field of application-oriented research ("from bench to crime site").

While classic tests for BFI will remain in use for probably still quite some time, future research may focus on broadening the scope of body fluids, tissues, and mixtures thereof that can reliably be identified by available current methods, and on integrating BFI and OTI in standard analysis pipelines in forensic and criminal investigation but also on further validation and improvement of existing methods and techniques.

BFI and OTI approaches based on nucleic acid sequencing, be they analysis of DNA methylation or RNA expression, are promising and will without doubt benefit from employing and further validating current MPS technologies to perform more collaborative exercises and to work toward their routine applicability seems highly advisable.

In addition, to render such approaches more mobile and flexible, assessing the use of portable devices like the "MinION," which utilizes nanopore technology in "single-molecule sequencing" (sometimes dubbed "third-generation sequencing") (Plesivkova et al. 2019) and has already been shown to be suitable for forensic purposes (; Vasiljevic et al. 2021) and also RNA sequencing (Seki et al. 2019), should be considered.

Reference

Albani, P.P., R. Fleming. 2019. Developmental validation of an enhanced mRNA-based multiplex system for body fluid and cell type identification. *Sci Justice* 59(3) 217–227.

An, J.H., K.J. Shin, W.I. Yang, H.Y. Lee. 2012. Body fluid identification in forensics. *BMB Rep* 45 545–553.

Bahar, B., F.J. Monahan, A.P. Moloney, O. Schmidt, D.E. MacHugh, T. Sweeney. 2007. Long-term stability of RNA in post-mortem bovine skeletal muscle, liver and subcutaneous adipose tissues. *BMC Mol Biol* 8:108.

Bahn, J.H.Q. et al. 2015. The landscape of microRNA, piwi-interacting RNA, and circular RNA in human saliva. *Clin Chem* 61(1) 221–230.

Baker, M.J., et al. 2014. Using Fourier transform IR spectroscopy to analyze biological materials. *Nat Protoc* 9(8) 1771–1791.

Ballard, D., J. Winkler-Galicki, J. Wesoły. 2020. Massive parallel sequencing in forensics: Advantages, issues, technicalities, and prospects. *Int J Legal Med* 134(4) 1291–1303.

Bamberg, M., L. Dierig, G. Kulstein, S.N. Kunz, M. Schmidt, T. Hadrys, P. Wiegand. 2021. Development and validation of an mRNA-based multiplex body fluid identification workflow and a rectal mucosa marker pilot study. *Forensic Sci Int Genet* 54:102542.

Bauer, M., D. Patzelt. 2002. Evaluation of mRNA markers for the identification of menstrual blood. *J Forensic Sci* 47(6) 1278–1282.

Boyd, S., M.F. Bertino, S.J. Seashols. 2011. Raman spectroscopy of blood samples for forensic applications. *Forensic Sci Int* 208(1–3) 124–128.

Bruijns, B., R. Tiggelaar, H. ardeniers. 2018. Massively parallel sequencing techniques for forensics: A review. *Electrophoresis* 39(21):2642–2654.

Bustin, S.A., et al. 2009. The MIQE guidelines: Minimum information for publication of quantitative real-time PCR experiments. *Clin Chem* 55(4) 611–622.

Butler, H.J., et al. 2016. Using Raman spectroscopy to characterize biological materials. *Nat Protoc* 11(4) 664–687.

Carnevali, E., et al. 2017. A GEFI collaborative exercise on DNA/RNA co-analysis and mRNA profiling interpretation. *Forensic Sci Int Genet Suppl S* 6 e18–e20.

Chirnside, O., A. Lemalu, R. Fleming. 2020. Identification of nasal mucosa markers for forensic mRNA body fluid determination. *Forensic Sci Int Genet* 48 102317.

Clarke, T.H., A. Gomez, H. Singh, K.E. Nelson, L.M. Brinkac. 2017. Integrating the microbiome as a resource in the forensics toolkit. *Forensic Sci Int Genet* 30 141–147.

Courts, C., B. Madea. 2010. Micro-RNA—A potential for forensic science? *Forensic Sci Int* 203(1–3) 106–111.

Courts, C., B. Madea. 2012. Ribonucleic acid. Importance in forensic molecular biology. *Rechtsmedizin* 22(2) 135–144.

Courts, C., M.W. Pfaffl, E. Sauer, W. Parson. 2019. Pleading for adherence to the MIQE-guidelines when reporting quantitative PCR data in forensic genetic research. *Forensic Sci Int Genet* 42:e21-e24.

Dammeier, S., S. Nahnsen, J. Veit, F. Wehner, M. Ueffing, O. Kohlbacher. 2016. Mass-spectrometry-based proteomics reveals organ-specific expression patterns to be used as forensic evidence. *J Proteome Res* 15(1) 182–192.

de Almeida, J.P., N. Glesse, C. Bonorino. 2011. Effect of presumptive tests reagents on human blood confirmatory tests and DNA analysis using real time polymerase chain reaction. *Forensic Sci Int* 206(1–3) 58–61.

de Beijer, R.P., et al. 2018. Identification and detection of protein markers to differentiate between forensically relevant body fluids. *Forensic Sci Int* 290 196–206.

Díez López, C., D. Montiel González, C. Haas, A. Vidaki, M. Kayser. 2020. Microbiome-based body site of origin classification of forensically relevant blood traces. *Forensic Sci Int Genet* 47 102280.

Dobay, A., et al. 2019. Microbiome-based body fluid identification of samples exposed to indoor conditions. *Forensic Sci Int Genet* 40 105–113.

Doi, M., S. Gamo, T. Okiura, H. Nishimukai, M. Asano. 2014. A simple identification method for vaginal secretions using relative quantification of Lactobacillus DNA. *Forensic Sci Int Genet* 12C 93–99.

Donaldson, A.E., M.C. Taylor, S.J. Cordiner, I.L. Lamon. 2010. Using oral microbial DNA analysis to identify expirated bloodspatter. *Int J Legal Med* 124(6) 569–576.

Donfack, J., A. Wiley. 2015. Mass spectrometry-based cDNA profiling as a potential tool for human body fluid identification. *Forensic Sci Int Genet* 16 112–120.

Dørum, G., et al. 2019. Predicting the origin of stains from whole miRNome massively parallel sequencing data. *Forensic Sci Int Genet* 40 131–139.

Dørum, G., S. Ingold, E. Hanson, J. Ballantyne, L. Snipen, C. Haas. 2018. Predicting the origin of stains from next generation sequencing mRNA data. *Forensic Sci Int Genet* 34 37–48.

Forat, S., et al. 2016. Methylation markers for the identification of body fluids and tissues from forensic trace evidence. *PLOS ONE* 11(2) e0147973.

Frumkin, D., A. Wasserstrom, B. Budowle, A. Davidson. 2011. DNA methylation-based forensic tissue identification. *Forensic Sci Int Genet* 5(5) 517–524.

Gamblin, A.P., R.K. Morgan-Smith. 2019. The characteristics of seminal fluid and the forensic tests available to identify it, WIREs. *Forensic Sci* 8 7.

Gauthier, Q.T., S. Cho, J.H. Carmel, B.R. McCord. 2019. Development of a body fluid identification multiplex via DNA methylation analysis. *Electrophoresis* 40(18–19) 2565–2574.

Ghosh, S., A.J. Yates, M.C. Frühwald, J.C. Miecznikowski, C. Plass, D. Smiraglia. 2010. Tissue specific DNA methylation of CpG islands in normal human adult somatic tissues distinguishes neural from non-neural tissues. *Epigenetics* 5(6) 527–538.

Giampaoli, S., et al. 2012. Molecular identification of vaginal fluid by microbial signature. *Forensic Sci Int Genet* 6(5) 559–564.

Giampaoli, S., et al. 2014. Forensic interlaboratory evaluation of the ForFLUID kit for vaginal fluids identification. *J Forensic Leg Med* 21 60–63.

Giampaoli, S.,. E. DeVittori, F. Valeriani, A. Berti, V. Romano Spica. 2017. Informativeness of NGS analysis for vaginal fluid identification. *J Forensic Sci* 62(1) 192–196.

Gill, P., et al. 2020. DNA commission of the International society for forensic genetics: Assessing the value of forensic biological evidence—Guidelines highlighting the importance of propositions. Part II: Evaluation of biological traces considering activity level propositions. *Forensic Sci Int Genet* 44 102186.

Glynn, C.L. 2020. Potential applications of microRNA profiling to forensic investigations. *RNA* 26(1) 1–9.

Gosch, A., C. Courts. 2019. On DNA transfer: The lack and difficulty of systematic research and how to do it better. *Forensic Sci Int Genet* 40 24–36.

Haas, C., et al. 2010. mRNA profiling for the identification of blood—Results of a collaborative EDNAP exercise. *Forensic Sci Int Genet* 5(1) 21–26.

Haas, C., et al. 2012. RNA/DNA co-analysis from blood stains—Results of a second collaborative EDNAP exercise. *Forensic Sci Int Genet* 6(1) 70–80.

Haas, C., et al. 2013. RNA/DNA co-analysis from human saliva and semen stains—Results of a third collaborative EDNAP exercise. *Forensic Sci Int Genet* 7(2) 230–239.

Haas, C., et al. 2014. RNA/DNA co-analysis from human menstrual blood and vaginal secretion stains: Results of a fourth and fifth collaborative EDNAP exercise. *Forensic Sci Int Genet* 8(1) 203–212.

Haas, C., et al. 2015. RNA/DNA co-analysis from human skin and contact traces—Results of a sixth collaborative EDNAP exercise. *Forensic Sci Int Genet* 16 139–147.

Haas, C., J. Neubauer, A.P. Salzmann, E. Hanson, J. Ballantyne. 2021. Forensic transcriptome analysis using massively parallel sequencing. *Forensic Sci Int Genet* 52 102486.

Hampton-Marcell, J.T., J.V. Lopez, J.A. Gilbert. 2017. The human microbiome: An emerging tool in forensics. *Microb Biotechnol* 10(2) 228–230.

Hansen, T.B., et al. 2013. Natural RNA circles function as efficient microRNA sponges. *Nature* 495(7441) 384–388.

Hanson, E., J. Ballantyne. 2017. Human organ tissue identification by targeted RNA deep sequencing to aid the investigation of traumatic injury. *Genes (Basel)* 8(11):319.

Hanson, E., S. Ingold, G. Dorum, C. Haas, R. Lagace, J. Ballantyne. 2019. Assigning forensic body fluids to DNA donors in mixed samples by targeted RNA/DNA deep sequencing of coding region SNPs using ion torrent technology. *Forensic Sci Int Genet Suppl S* 7(1) 23–24.

Hanson, E., S. Ingold, C. Haas, J. Ballantyne. 2018. Messenger RNA biomarker signatures for forensic body fluid identification revealed by targeted RNA sequencing. *Forensic Sci Int Genet* 34 206–221.

Hanson, E.K., H. Lubenow, J. Ballantyne. 2009. Identification of forensically relevant body fluids using a panel of differentially expressed microRNAs. *Anal Biochem* 387(2) 303–314.

Hanson, E.K., J. Ballantyne. 2013a. Rapid and inexpensive body fluid identification by RNA profiling-based multiplex High Resolution Melt (HRM) analysis. *F1000 Res* 2:281.

Hanson, E.K., J. Ballantyne. 2013b. Circulating microRNA for the identification of forensically relevant body fluids. *Methods Mol Biol* 1024 221–234.

Hanssen, E.N., E. Avershina, K. Rudi, P. Gill, L. Snipen. 2017. Body fluid prediction from microbial patterns for forensic application. *Forensic Sci Int Genet* 30 10–17.

Harbison, S., R. Fleming. 2016. Forensic body fluid identification: State of the art. *RRFMS* 6 11–23.

Holtkötter, H., et al. 2017. Independent validation of body fluid-specific CpG markers and construction of a robust multiplex assay. *Forensic Sci Int Genet* 29 261–268.

Holtkötter, H., K. Schwender, P. Wiegand, H. Pfeiffer, M. Vennemann. 2018. Marker evaluation for differentiation of blood and menstrual fluid by methylation-sensitive SNaPshot analysis. *Int J Legal Med* 132(2) 387–395.

Igoh, A., Y. Doi, K. Sakurada. 2015. Identification and evaluation of potential forensic marker proteins in vaginal fluid by liquid chromatography/mass spectrometry. *Anal Bioanal Chem* 407(23) 7135–7144.

Ingold, S., et al. 2020a. Body fluid identification using a targeted mRNA massively parallel sequencing approach - Results of a EUROFORGEN/EDNAP collaborative exercise. *Forensic Sci Int Genet* 34 105–115.

Ingold, S., et al. 2020b. Body fluid identification and assignment to donors using a targeted mRNA massively parallel sequencing approach—Results of a second EUROFORGEN/EDNAP collaborative exercise. *Forensic Sci Int Genet* 45 102208.

Ingold, S., G. Dørum, E. Hanson, J. Ballantyne, C. Haas. 2020c. Assigning forensic body fluids to donors in mixed body fluids by targeted RNA/DNA deep sequencing of coding region SNPs. *Int J Legal Med* 134(2):473–485.

Jackson, K.R., et al. 2020. A novel loop-mediated isothermal amplification method for identification of four body fluids with smartphone detection. *Forensic Sci Int Genet* 45 102195.

Javan, G.T., I. Can, S.J. Finley, S. Soni. 2015. The apoptotic thanatotranscriptome associated with the liver of cadavers. *Forensic Sci Med Pathol* 11(4) 509–516.

Jung, M., et al. 2010. Robust microRNA stability in degraded RNA preparations from human tissue and cell samples. *Clin Chem* 56(6) 998–1006.

Jung, S.E., et al. 2016. A collaborative exercise on DNA methylation-based body fluid typing. *Electrophoresis* 37(21):2759-2766.

Kader, F., M. Ghai, A.O. Olaniran. 2020. Characterization of DNA methylation-based markers for human body fluid identification in forensics: A critical review. *Int J Legal Med* 134(1) 1–20.

Kelly, M.C. 2020. Pre-ejaculate fluid in the context of sexual assault: A review of the literature from a clinical forensic medicine perspective. *Forensic Sci Int* 318:110596.

Killick, S.R., C. Leary, J. Trussell, K.A. Guthrie. 2011. Sperm content of pre-ejaculatory fluid. *Hum Fertil (Camb)* 14(1) 48–52.

Kimura, A., H. Ikeda, S. Yasuda, K. Yamaguchi, T. Tsuji. 1995. Brain tissue identification based on myosin heavy chain isoforms. *Int J Legal Med* 107(4) 193–196.

King, K., F.A. Flinter, P.M. Green. 2001. Hair roots as the ideal source of mRNA for genetic testing. *J Med Genet* 38(6) E20.

Kint, S., W. de Spiegelaere, J. de Kesel, L. Vandekerckhove, W. van Criekinge. 2018. Evaluation of bisulfite kits for DNA methylation profiling in terms of DNA fragmentation and DNA recovery using digital PCR. *PLOS ONE* 13(6) e0199091.

Kohlmeier, F., P.M. Schneider. 2012. Successful mRNA profiling of 23 years old blood stains. *Forensic Sci Int Genet* 6(2) 274–276.

Koppelkamm, A., B. Vennemann, S. Lutz-Bonengel, T. Fracasso, M. Vennemann. 2011. RNA integrity in post-mortem samples: Influencing parameters and implications on RT-qPCR assays. *Int J Legal Med* 125(4) 573–580.

Kurdyukov, S., M. Bullock. 2016. DNA methylation analysis: Choosing the right method. *Biology (Basel)* 5(1):3.

Legg, K.M., R. Powell, N. Reisdorph, R. Reisdorph, P.B. Danielson. 2014. Discovery of highly specific protein markers for the identification of biological stains. *Electrophoresis* 35(21–22) 3069–3078.

Lin, M.H., D.F. Jones, R. Fleming. 2015. Transcriptomic analysis of degraded forensic body fluids. *Forensic Sci Int Genet* 17 35–42.

Lindenbergh, A., et al. 2012. A multiplex (m)RNA-profiling system for the forensic identification of body fluids and contact traces. *Forensic Sci Int Genet* 6(5) 565–577.

Lindenbergh, A., et al. 2013. Development of a mRNA profiling multiplex for the inference of organ tissues. *Int J Legal Med* 127 891–900.

Liu, B., et al. 2019. Characterization of tissue-specific biomarkers with the expression of circRNAs in forensically relevant body fluids. *Int J Legal Med* 133(5) 1321–1331.

Liu, B., et al. 2020. Development of a multiplex system for the identification of forensically relevant body fluids. *Forensic Sci Int Genet* 47 102312.

Liu, J., et al. 2021. Identification of coding region SNPs from specific and sensitive mRNA biomarkers for the deconvolution of the semen donor in a body fluid mixture. *Forensic Sci Int Genet* 52 102483.

Liu, K.-L., et al. 2020. Identification of spermatozoa using a novel 3-plex MSRE-PCR assay for forensic examination of sexual assaults. *Int J Legal Med* 134(6) 1991–2004.

Locard, E. 1930. The analysis of dust traces. Part I. *Am J Police Sci* 1(3) 276.

Lux, C., C. Schyma, B. Madea, C. Courts. 2014. Identification of gunshots to the head by detection of RNA in backspatter primarily expressed in brain tissue. *Forensic Sci Int* 237 62–69.

Lynch, C., R. Fleming. 2020. RNA-based approaches for body fluid identification in forensic science. *WIREs Forensic Sci* 3:e1407.

Mayes, C., S. Seashols-Williams, S. Hughes-Stamm. 2018. A capillary electrophoresis method for identifying forensically relevant body fluids using miRNAs. *Leg Med (Tokyo)* 30 1–4.

McCarroll, S.A., D.M. Altshule. 2007. Copy-number variation and association studies of human disease. *Nat Genet* 39(7) Supplement S37–S42.

Melé, M., et al. 2015. Human genomics. The human transcriptome across tissues and individuals. *Science* 348(6235) 660–665.

Memczak, S., P. Papavasileiou, O. Peters, N. Rajewsky, S. Pfeffer. 2015. Identification and characterization of circular RNAs as a new class of putative biomarkers in human blood. *PLOS ONE* 10(10) e0141214.

Merkley, E.D., D.S. Wunschel, K.L. Wahl, K.H. Jarman. 2019. Applications and challenges of forensic proteomics. *Forensic Sci Int* 297 350–363.

Morris, K.V., J.S. Mattick. 2014. The rise of regulatory RNA. *Nat Rev Genet* 15(6) 423–437.

Nakanishi, H., et al. 2009. A novel method for the identification of saliva by detecting oral streptococci using PCR. *Forensic Sci Int* 183(1–3) 20–23.

Nakanishi, H., et al. 2013. Identification of feces by detection of Bacteroides genes. *Forensic Sci Int Genet* 7(1) 176–179.

Nustad, H.E., M. Almeida, A.J. Canty, M. LeBlanc, C.M. Page, P.E. Melton. 2018. Epigenetics, heritability, and longitudinal analysis. *BMC Genet* 19(Suppl 1) 77.

Orphanou, C.-M., L. Walton-Williams, H. Mountain, J. Cassella. 2015. The detection and discrimination of human body fluids using ATR FT-IR spectroscopy. *Forensic Sci Int* 252 e10-6.

Ozata, D.M., I. Gainetdinov, A. Zoch, D. O'Carroll, P.D. Zamore. 2019. PIWI-interacting RNAs: Small RNAs with big functions. *Nat Rev Genet* 20(2) 89–108.

Park, H.Y., B.N. Son, Y.I. Seo, S.K. Lim. 2015. Comparison of four saliva detection methods to identify expectorated blood spatter. *J Forensic Sci* 60(6) 1571–1576.

Park, J.-L., et al. 2013. Forensic body fluid identification by analysis of multiple RNA markers using NanoString technology. *Genomics Inform* 11(4) 277–281.

Park, J.L., et al. 2014. Microarray screening and qRT-PCR evaluation of microRNA markers for forensic body fluid identification. *Electrophoresis* 5(21–22):3062–8.

Park, S.M., et al. 2013. Genome-wide mRNA profiling and multiplex quantitative RT-PCR for forensic body fluid identification. *Forensic Sci Int Genet* 7(1) 143–150.

Parker, G.J., H.E. McKiernan, K.M. Legg, Z.C. Goecker. 2021. Forensic proteomics. *Forensic Sci Int Genet* 54:102529.

Petersen, C., B. Hjort, T. Tvedebrink, L. Kielpinski, J. Vinther, N. Morling. 2014. Body fluid identification of blood, saliva and semen using second generation sequencing of micro-RNA. *Forensic Sci Int Genet Supplser* 4 e204–e205.

Phang, T.W., C.Y. Shi, J.N. Chia, C.N. Ong. 1994. Amplification of cDNA via RT-PCR using RNA extracted from postmortem tissues. *J Forensic Sci* 39(5): 391275–391279.

Pieri, M., et al. 2019. Mass spectrometry-based proteomics for the forensic identification of vomit traces. *J Proteomics* 209 103524.

Plesivkova, D., R. Richards, S. Harbison. 2019. A review of the potential of the MinION™ single-molecule sequencing system for forensic applications. *Wires Forensic Sci* 1(1) e1323.

Power, D.A., S.J. Cordiner, J.A. Kieser, G.R. Tompkins, J. Horswell. 2010. PCR-based detection of salivary bacteria as a marker of expirated blood. *Sci Justice* 50(2) 59–63.

Riman, S., C.H. Shek, M.A. Peck, J. Benjamin, D. Podini. 2018. Proximity ligation real-time PCR: A protein-based confirmatory method for the identification of semen and sperm cells from sexual assault evidence. *Forensic Sci Int Genet* 37 64–72.

Sakurada, K., K. Watanabe, T. Akutsu. 2020. Current methods for body fluid identification related to sexual crime: Focusing on saliva, semen, and vaginal fluid. *Diagnostics (Basel)* 10(9):693.

Sakurada, K., T. Akutsu, K. Watanabe, M. Yoshino. 2012. Identification of nasal blood by real-time RT-PCR. *Leg Med (Tokyo)* 14(4) 201–204.

Salzman, J., C. Gawad, P.L. Wang, N. Lacayo, P.O. Brown. 2012. Circular RNAs are the predominant transcript isoform from hundreds of human genes in diverse cell types. *PLOS ONE* 7(2) e30733.

Salzman, J., R.E. Chen, M.N. Olsen, P.L. Wang, P.O. Brown, J.V. Moran. 2013. Cell-type specific features of circular RNA expression. *PLOS Genet* 9(9) e1003777.

Salzmann, A.P., et al. 2021. mRNA profiling of mock casework samples: Results of a FoRNAP collaborative exercise. *Forensic Sci Int Genet* 50 102409.

Salzmann, A.P., G. Russo, S. Aluri, C. Haas. 2019. Transcription and microbial profiling of body fluids using a massively parallel sequencing approach. *Forensic Sci Int Genet* 43 102149.

Samsuwan, J., T. Muangsub, P. Yanatatsaneejit, A. Mutirangura, N. Kitkumthorn. 2018. Combined bisulfite Restriction Analysis for brain tissue identification. *Forensic Sci Int* 286 42–45.

Sauer, E., A. Extra, P. Cachee, C. Courts. 2017. Identification of organ tissue types and skin from forensic samples by microRNA expression analysis. *Forensic Sci Int Genet* 28 99–110.

Sauer, E., A.K. Reinke, C. Courts. 2016. Differentiation of five body fluids from forensic samples by expression analysis of four microRNAs using quantitative PCR. *Forensic Sci Int Genet* 22 89–99.

Seashols-Williams, S., et al. 2016. High-throughput miRNA sequencing and identification of biomarkers for forensically relevant biological fluids. *Electrophoresis* 37(21):2780–2788.

Seki, M., et al. 2019. Evaluation and application of RNA-Seq by MinION. *DNA Res* 26(1) 55–65.

Seo, Y., E. Kakizaki, K. Takahama. 1997. A sandwich enzyme immunoassay for brain S-100 protein and its forensic application. *Forensic Sci Int* 87(2) 145–154.

Sharma, S., R. Chophi, J.K. Jossan, R. Singh. 2021. Detection of bloodstains using attenuated total reflectance-Fourier transform infrared spectroscopy supported with PCA and PCA-LDA. *Med Sci Law* 61(4):292-301. 258024211010926.

Sijen, T. 2015. Molecular approaches for forensic cell type identification: On mRNA, miRNA, DNA methylation and microbial markers. *Forensic Sci Int Genet* 18 21–32.

Sikirzhytskaya, A., V. Sikirzhytski, I.K. Lednev. 2012. Raman spectroscopic signature of vaginal fluid and its potential application in forensic body fluid identification. *Forensic Sci Int* 216(1–3) 44–48.

Sikirzhytski, V., A. Sikirzhytskaya, I.K. Lednev. 2012. Multidimensional Raman spectroscopic signature of sweat and its potential application to forensic body fluid identification. *Anal Chim Acta* 718 78–83.

Silva, D.S.B.S., J. Antunes, K. Balamurugan, G. Duncan, C.S. Alho, B. McCord. 2016. Developmental validation studies of epigenetic DNA methylation markers for the detection of blood, semen and saliva samples. *Forensic Sci Int Genet* 23 55–63.

Silva, S.S., C. Lopes, A.L. Teixeira, M.J. Sousa, R. Medeiros. 2015. Forensic miRNA: Potential biomarker for body fluids? *Forensic Sci Int Genet* 14C 1–10.

Sirker, M., R. Fimmers, P.M. Schneider, I. Gomes. 2017. Evaluating the forensic application of 19 target microRNAs as biomarkers in body fluid and tissue identification. *Forensic Sci Int Genet* 27 41–49.

Su, C.W., et al. 2015. A novel application of real-time RT-LAMP for body fluid identification: Using HBB detection as the model. *Forensic Sci Med Pathol* 11(2):208-15.

Takahama, K. 1996. Forensic application of organ-specific antigens. *Forensic Sci Int* 80(1–2) 63–69.

Takata, T., S. Miyaishi, T. Kitao, H. Ishizu. 2004. Identification of human brain from a tissue fragment by detection of neurofilament proteins. *Forensic Sci Int* 144(1) 1–6.

The ENCODE Project Consortium. 2012. An integrated encyclopedia of DNA elements in the human genome. *Nature* 489(7414) 57–74.

The Genotype-Tissue Expression (GTEx) Project. 2013. The Genotype-Tissue Expression (GTEx) project. *Nat Genet* 45(6) 580–585.

The Human Microbiome Project Consortium. 2012. Structure, function and diversity of the healthy human microbiome. *Nature* 486(7402) 207–214.

Uhlen, M., et al. 2010. Towards a knowledge-based Human Protein Atlas. *Nat Biotechnol* 28(12) 1248–1250.

Uhlén, M., et al. 2015. Proteomics. Tissue-based map of the human proteome. *Science* 347(6220) 1260419.

van den Berge, M., et al. 2014. A collaborative European exercise on mRNA-based body fluid/skin typing and interpretation of DNA and RNA results. *Forensic Sci Int Genet* 10 40–48.

van der Meer, D., M.L. Uchimoto, G. Williams. 2013. Simultaneous analysis of micro-RNA and DNA for determining the body fluid origin of DNA profiles. *J Forensic Sci* 58(4) 967–971.

van Steendam, K., M. de Ceuleneer, M. Dhaenens, D. van Hoofstat, D. Deforce. 2013. Mass spectrometry-based proteomics as a tool to identify biological matrices in forensic science. *Int J Legal Med* 127(2) 287–298.

Vasiljevic, N., et al. 2021. Developmental validation of Oxford nanopore Technology MinION sequence data and the NGSpeciesID bioinformatic pipeline for forensic genetic species identification. *Forensic Sci Int Genet* 53 102493.

Virkler, K., I.K. Lednev. 2008. Raman spectroscopy offers great potential for the nondestructive confirmatory identification of body fluids. *Forensic Sci Int* 181(1–3) e1–e5.

Virkler, K., I.K. Lednev. 2009a. Raman spectroscopic signature of semen and its potential application to forensic body fluid identification. *Forensic Sci Int* 193(1–3) 56–62.

Virkler, K., I.K. Lednev. 2009b. Analysis of body fluids for forensic purposes: From laboratory testing to non-destructive-destructive rapid confirmatory identification at a crime scene. *Forensic Sci Int* 188(1–3) 1–17.

Virkler, K., I.K. Lednev. 2010. Forensic body fluid identification: The Raman spectroscopic signature of saliva. *Analyst* 135(3) 512–517.

Visser, M., D. Zubakov, K.N. Ballantyne, M. Kayser. 2011. mRNA-based skin identification for forensic applications. *Int J Legal Med* 125(2) 253–263.

Wang, S., et al. 2019a. Expression profile analysis of piwi-interacting RNA in forensically relevant biological fluids. *Forensic Sci Int Genet* 42 171–180.

Wang, S., Z. Wang, R. Tao, G. He, J. Liu, C. Li, Y. Hou. 2019b. The potential use of piwi-interacting RNA biomarkers in forensic body fluid identification: A proof-of-principle study. *Forensic Sci Int Genet* 39 129–135.

Wang, Z., et al. 2016. Characterization of microRNA expression profiles in blood and saliva using the Ion Personal Genome Machine((R)) System (Ion PGM System). *Forensic Sci Int Genet* 20 140–146.

Wang, Z., J. Zhang, H. Luo, Y. Ye, J. Yan, Y. Hou. 2013. Screening and confirmation of microRNA markers for forensic body fluid identification. *Forensic Sci Int Genet* 7(1) 116–123.

Wang, Z., X. Zhao, Y. Hou. 2017. Exploring of microRNA markers for semen stains using massively parallel sequencing. *Forensic Sci Int Genet Suppl S* 6 e107–e109.

Wasserstrom, A., D. Frumkin, A. Davidson, M. Shpitzen, Y. Herman, R. Gafny. 2013. Demonstration of DSI-semen-A novel DNA methylation-based forensic semen identification assay. *Forensic Sci Int Genet* 7(1) 136–142.

Watanabe, K., K. Taniguchi, T. Akutsu. 2018. Development of a DNA methylation-based semen-specific SNP typing method: A new approach for genotyping from a mixture of body fluids. *Forensic Sci Int Genet* 37 227–234.

Watanabe, K., T. Akutsu, A. Takamura, K. Sakurada. Evaluation of a blood-specific DNA methylated region and trial for allele-specific blood identification from mixed body fluid. *DNA Leg Med (Tokyo)* 22 49–53.

Weber, J.A., et al. 2010. The microRNA spectrum in 12 body fluids. *Clin Chem* 56(11) 1733–1741.

Weinbrecht, K.D., J. Fu, M. Payton, R. Allen. 2017. Time-dependent loss of mRNA transcripts from forensic stains. *RRFMS* 7(7) 1–12.

Williams, G.A. 2020. Body fluid identification: A case for more research and innovation. *Forensic Sci Int Reports* 2 100096.

Xie, B., F. Song, S. Wang, K. Zhang, Y. Li, H. Luo. 2020. Exploring a multiplex DNA methylation-based SNP typing method for body fluids identification: As a preliminary report. *Forensic Sci Int* 313 110329.

Xu, Y., et al. 2014. Development of highly sensitive and specific mRNA multiplex system (XCYR1) for forensic human body fluids and tissues identification. *PLOS ONE* 9(7) e100123.

Yang, H., B. Zhou, H. Deng, M. Prinz, D. Siegel. 2013. Body fluid identification by mass spectrometry. *Int J Legal Med* 127(6) 1065–1077.

Zarrei, M., J.R. MacDonald, D. Merico, S.W. Scherer. 2015. A copy number variation map of the human genome. *Nat Rev Genet* 16(3) 172–183.

Zhang, Y., et al. 2017. Evaluation of the inclusion of circular RNAs in mRNA profiling in forensic body fluid identification. *Int J Legal Med* 188 1.

Zubakov, D., et al. 2015. Towards simultaneous individual and tissue identification: A proof-of-principle study on parallel sequencing of STRs, amelogenin, and mRNAs with the Ion Torrent PGM. *Forensic Sci Int Genet* 17 122–128.

Zubakov, D., et al. 2018. Introducing novel type of human DNA markers for forensic tissue identi-fication: DNA copy number variation allows the detection of blood and semen. *Forensic Sci Int Genet* 36 112–118.

Zubakov, D., A.W. Boersma, Y. Choi, P.F. van Kuijk, E.A. Wiemer, M. Kayser. 2010. MicroRNA markers for forensic body fluid identification obtained from microarray screening and quan-titative RT-PCR confirmation. *Int J Legal Med* 124(3) 217–226.

Zubakov, D., M. Kokshoorn, A. Kloosterman, M. Kayser. 2009. New markers for old stains: Stable mRNA markers for blood and saliva identification from up to 16-year-old stains. *Int J Legal Med* 123(1) 71–74.

Zukerman, Z., D.B. Weiss, R. Orvieto. 2003. Does preejaculatory penile secretion originating from Cowper's gland contain sperm? *J Assist Reprod Genet* 20(4) 157–159.

Evolving Technologies in Forensic DNA Analysis

16

HENRY ERLICH, GUNMEET K. BALI,
AND CASSANDRA D. CALLOWAY

Contents

16.1 Introduction

Polymerase chain reaction (PCR)-based genetic analysis has revolutionized the practice of forensics since the first application of the HLA-DQA1 test in the *Pennsylvania vs. Pestinikis* case in 1986 (Mihalovich et al. 1992). In recent years, the focus of forensics genetic analysis has been primarily on length polymorphisms, panels of short tandem repeat (STR) markers, and, to some extent, on sequence polymorphism of the mitochondrial DNA (mtDNA) hypervariable regions (HVI and HVII) (Butler 2004; Holland and Parsons 1999; Sullivan, Hopgood, and Gill 1992; Just, Irwin, and O'Callaghan 2004). The STR markers have been analyzed by capillary electrophoresis and the mtDNA HVI and HVII sequence markers by dideoxy chain termination Sanger sequencing or by various probe-based methods, such as the linear array (Melton et al. 2005; Gabriel et al. 2003; Divne et al. 2005; Roberts and Calloway 2007; Date Chong et al. 2005). Recently, novel technologies and new genetic markers have been developed that promise to have a significant impact on the future of forensic genetic analysis. In this chapter, we discuss some of the most significant developments and their applications. Next-generation sequencing (NGS) technologies (also known as massively parallel sequencing or MPS) provide a single, very high-throughput platform for analyzing sequence and tandem repeat polymorphisms, and provide a digital readout for analyzing forensics mixtures based on the clonal sequencing feature. The other significant recent development is the implementation of Rapid DNA instrumentation, small instruments for automating DNA extraction, PCR amplification, capillary electrophoresis, and analysis of a panel of STR markers. Both of these technologies promise to have a major impact on how forensic DNA analysis is carried out in the future.

DOI: 10.4324/9780429019944-19

16.2 NGS/MPS

Many different commercial NGS or MPS systems are available today, all with two properties in common: (1) they are massively parallel, with millions of simultaneous sequencing reactions and (2) each reaction is initiated with a single DNA molecule or from a clonal population generated from a single DNA molecule. The two most commonly used platforms (Illumina® and Ion Torrent®, Thermo-Fisher) carry out clonal sequencing by limiting dilution so that the amplification and sequencing starts from a single DNA molecule. This clonal sequencing property of NGS facilitates the analysis of mixtures because the individual components of a mixture can be sequenced separately. Another benefit of NGS technology is that it allows analysis of different categories of genetic markers (e.g., STRs, SNPs, mitochondrial DNA polymorphisms) on the same instrument.

In addition, the ability to detect sequence variants of STR alleles of the same size and, thus, indistinguishable by capillary electrophoresis, provides higher discrimination power, increased mixture resolution and more accurate results.

16.2.1 Introduction to NGS Technology

There are several different commercial NGS platforms that use different sequencing chemistries with unique characteristics and limitations (Erlich, Calloway, and Lee 2020), but they are all capable of producing vast amounts of sequencing data in a cost effective and timely manner (McCombie, McPherson, and Mardis 2018). The common initial step for all MPS platforms is the creation of a "library," that is, the conversion of DNA from the evidence or reference sample to a form that can be sequenced in each of the parallel sequencing reactions for a given NGS platform. For many platforms, this step involves the addition of short DNA sequences, "adapters," to the ends of all the DNA molecules in the sample in a ligation reaction. These adapters contain sequences that allow each of the individual DNA fragments to be PCR amplified and to be bound to a location where the sequencing reaction will occur. These added sequences also contain short DNA sequences that function as "bar codes" (or "indices"), and the specific pair of bar codes indicates which DNA sample was the source of the library. Identifying each library with a unique pair of bar codes allows the sequencing of multiple libraries in the same NGS run. The mixture of many millions of sequence reads generated in the run can be deconvoluted and assigned by bioinformatic software to an individual sample (Lee, Varlaro, and Holt 2016; McCombie, McPherson, and Mardis 2018).

MPS libraries can either be "targeted" or "shotgun." Shotgun libraries are constructed from all the DNA present in the sample; this is the strategy for generating whole genome sequences (WGS). Targeted libraries are focused on particular regions of interest; for forensic applications involving individual identification, the targeted regions are typically the highly polymorphic regions of the genome, such as the current standard STR panels, panels of SNPs, or the mitochondrial DNA genome. Since less than 1% of the human genome is polymorphic, the vast majority of forensic NGS applications involve targeted libraries.

Targeted libraries can be prepared in one of two ways. PCR amplification can be used to target specific regions of DNA and specific genetic markers. The PCR primers are designed to flank the region(s) of interest; amplification results in a targeted PCR product that contains the primers. In some methods, the adapter sequences are directly attached to the ends of PCR primers along with their bar codes. These adapter sequences then bind to complementary sequences tethered to the sequencing platform, typically beads or a slide (or flow cell) on which sequencing reactions occur. Alternatively, the adapters are connected in a ligation reaction to the amplified DNA.

An alternative strategy for creating targeted libraries does not rely on selective PCR amplification but is based on the selective "capture" of the regions of interest from a shotgun library (Bose et al. 2018; Shih et al. 2018). The specific capture is accomplished by a set of probes, short synthetic DNA fragments that hybridize to specific sequences in the shotgun library. These captured targeted regions can then be recovered and subsequently sequenced. This probe capture strategy, while more cumbersome than the PCR-targeted library system, offers some advantages for analyzing forensic samples. Many forensic samples contain highly degraded DNA; many of these short DNA fragments may not contain both of the intact PCR primer binding sites and therefore cannot be amplified by PCR. The probe-capture-based library system, however, is often effective in recovering short DNA sequences from such samples (Bose et al. 2018; Shih et al. 2018). Another advantage of the probe capture strategy is that a given set of genetic markers (e.g., STRs and SNPs) can be selectively recovered from a shotgun library, sequenced, and then, with a different set of probes, the mitochondrial genome can then be captured and sequenced from the same library. These target enrichment strategies using probe capture have been recently applied to hair samples (Shih et al. 2018; Wisner et al. 2021). These capabilities are particularly valuable for forensic samples with limited DNA.

Most current commercial NGS or MPS platforms rely on sequencing a clonal population generated by some form of DNA amplification from a single DNA molecule (Mardis 2008; de Knijff 2019). The target DNA to be sequenced serves as a template for a DNA synthesis reaction. The sequence of a clonal population of DNA molecules is determined by detecting which of the four nucleotides is incorporated into a growing new DNA strand. The Ion Torrent systems flow one of the four nucleotides in succession over the parallel sequencing reactions and determines which nucleotide is incorporated into a specific site in the new DNA strand by detecting a hydrogen ion (a proton) that is released during the incorporation event, with a highly sensitive pH meter.[*] The Illumina system uses four different fluorescent labels for each of the four nucleotides to determine which nucleotide is incorporated at each position of the growing DNA strand using a camera to detect the fluorescence. Both the Illumina and Ion Torrent systems provide relatively short sequence reads, typically less than 300 (Illumina) or 400 (Ion Torrent) bases. Both strands of the double-stranded DNA template are sequenced; these sequence reads are termed "forward" or "reverse." In the Illumina system, these "paired-end" reads from a single DNA molecule can be "merged" so that a DNA molecule of around 500 bp can be sequenced completely with sequence reads in both directions. Sequence read length limitations may create problems for STR genotyping by NGS; determining the sequence of the entire tandem repeat region and the unique flanking DNA sequences in very large STR alleles may not be possible due to the sequence read length constraint. For NGS analysis of STRs, some STR loci, which have long alleles, have therefore been excluded from some STR panels designed for NGS.

---

[*] The first NGS system, the 454 platform, introduced in 2005, used a similar strategy by detecting a light flash that was generated during the nucleotide incorporation into the new DNA strand. This system is no longer available.

Newer and less widely used NGS platforms, such as Pacific BioSciences technology and Oxford Nanopore technology, do not require the clonal amplification step and sequence the individual DNA molecule. Unlike the Illumina platform, these two sequencing technologies are capable of producing very long sequence reads. The Oxford Nanopore technology uses a small portable instrument and promises potential use in the field. Another more recent DNA sequencing technology, the MinION (Oxford Nanopore Technologies), also does not require amplification or fluorescent labels. Instead, it detects ion currents generated when a charged DNA molecule passes through a nanoscale pore in a membrane separating two chambers. It is capable of producing very long sequence reads and has the unique advantage of being light and portable, offering the potential for use in the field. For potential applications in the field, the small, compact MinION will have to be used in conjunction with a system for extracting DNA from the crime scene evidence sample, as has been done in the Rapid DNA technology. However, the current MinIon system has the disadvantage of high error rates, especially for long homopolymer stretches (Rang 2018). A recent study demonstrated that STRs could be sequenced with the MinIon instrument (Cornelis et al. 2018). For all these platforms, NGS has the potential to analyze both length and sequence polymorphisms and the ability to deconvolute mixtures.

16.3 NGS, Mixtures, and Mitochondrial DNA

Until the introduction of NGS/MPS technology, forensic DNA sequencing was carried out by the Sanger method, performed on automated capillary electrophoresis instruments using fluorescently labeled DNA fragments generated during a DNA synthesis reaction. Forensic sequencing was carried out primarily on the mitochondrial DNA since the length polymorphism of the STR loci could be analyzed faster and more cheaply by electrophoresis. The development of massively parallel and clonal sequencing systems (NGS) represented a true paradigm shift for all DNA sequencing, and for forensic sequencing in particular, it offered the potential for much more effective analysis of mixed samples.

16.3.1 Mitochondrial DNA and Mixtures

The analysis of forensics specimens that contain multiple contributors remains one of the most problematic issues in forensics from both a technical and statistical perspective. Analysis of the haploid mtDNA genome has the potential for effective deconvolution of mixtures. mtDNA has served as a valuable forensics genetic target because of the high copy number per cell, allowing the analysis of minute amounts of specimen DNA that could not be analyzed with chromosomal genetic markers (Higuchi et al. 1988; Holland and Parsons 1999; Sullivan, Hopgood, and Gill 1992). It is the haploid nature of the mtDNA genome (excluding the issue of heteroplasmy) that is particularly valuable in the analysis of forensic mixtures, a specimen category notably difficult to analyze and interpret with chromosomal markers. If each individual contributor to a mixture is represented by one mtDNA sequence, a forensics mixture with n different sequences would be expected to contain at least n different contributors. Although less informative than a panel of chromosomal markers, mtDNA analysis can thus provide a useful estimate of the number of different contributors and, thus, facilitate and simplify the analysis of chromosomal marker data, if available, from the same mixture. Resolving the number of different mtDNA sequences in a mixture can be challenging with Sanger sequencing. However, the clonal sequencing aspect of NGS platforms addresses this issue and allows a digital readout based on the number of sequence reads for each individual mtDNA sequence.

The HV (hypervariable) regions of the human mitochondrial genome have high sequence diversity and, accordingly, have been the primary area of study for identification purposes. Some HVI/HVII types are relatively common among all populations, even though the overall distribution of mtDNA HVI/HVII sequences is highly skewed toward rare types. Approximately 7% of Caucasians share the most common HVI/HVII sequence differing from the rCRS at 263G, and 13 additional HV sequences are shared among >0.5% of the population (Parsons and Coble 2001). The power of discrimination is limited for all population groups because of the presence of a few common HVI/II sequences in all populations tested (Melton et al. 2005; Date Chong et al. 2005; Gabriel et al. 2003). Therefore, additional sequence polymorphisms outside the HVI/HVII regions need to be targeted to increase the power of discrimination of mtDNA analysis. NGS technology has made analysis of the entire mtDNA genome and deconvolution of mixtures more feasible.

mtDNA mixture analysis by Sanger sequencing was not typically attempted because the peak heights in electropherograms do not always correlate with mixture proportions and, while the signal defines what bases are present at each position, the electropherogram does not provide linkage information (Calloway, Stuart, and Erlich 2009). Results were, therefore, most often deemed inconclusive. The development of next-generation sequencing (NGS), a massively parallel clonal technique which sequences the individual components of the mixture, addresses these shortcomings of Sanger sequencing (Moorthie, Mattocks, and Wright 2011). Churchill et al. (2017) attempted mtDNA mixture deconvolution of three-person mixtures utilizing whole mtDNA sequencing and the Torrent Suite software v5.2.1, which generates haplotype calls as well as assesses the relative quantity of each contributor. However, with more complex three-person mixtures where two minor contributors have relatively similar contribution amounts, identification of a second minor haplotype failed. Moreover, NGS libraries generated by targeted PCR amplification, while highly specific, are not ideal for the sequence analysis of short DNA fragments since PCR requires two intact priming sites. The probe capture strategy of targeted enrichment from shotgun DNA libraries allows the analysis of the short DNA fragments found in many forensic samples. The massively parallel aspect of NGS technology allows the entire mitochondrial genome to be sequenced, increasing the discrimination power, and given the large number of reads for each genomic region, provides sensitivity in detecting a minor contributor and statistical confidence in estimating the proportions of the various contributors (Shih et al. 2018). The clonal aspect of NGS allows for greater accuracy and precision in frequency interpretation by counting individual sequence reads. Each contributor can then be associated with a haplotype assembled from the sequence reads, which include variants outside the hypervariable regions. As noted earlier, sequencing the entire mtDNA genome is critical for individual identification because analyzing only the HVI/HVII regions provides a limited power of discrimination (POD). mtDNA analysis has an intrinsically lower POD than nuclear markers due to the absence of recombination between polymorphic sites, and so the POD is determined by the frequency of the mtDNA haplotype in a population database. Consequently, analyzing the entire mtDNA genome rather than just the HVI/HVII regions is valuable for the interpretation of mixtures, provided the short reads generated by NGS can be assembled to form discrete haplotypes.

One of the potential confounding factors in the interpretation of mtDNA sequences by NGS is the presence of homologous sequences in the nuclear genome, known as NUMTs (nuclear sequences of mitochondrial origin) (Maude et al. 2019). Analysis of the short sequence reads can be complicated by the incorrect alignment of reads generated from NUMT sequences to the mtDNA genome. In mixtures, these low-level NUMT reads could be interpreted incorrectly as an additional trace mtDNA contributor.

As noted earlier, the probe capture strategy of generating NGS libraries is more likely to yield sequence reads from the short DNA fragments found in many forensic specimens than is PCR-based enrichment. However, this choice involves trade-offs. PCR enrichment is more specific, faster, and cheaper. Another advantage of the probe capture technology is that a single shotgun library can be prepared from a forensic specimen and then targeted with different probe panels for nuclear genes (SNPs and STRs) as well as mtDNA. Analysis of mtDNA libraries targeted with our probe capture system provides even coverage over the entire genome even at DNA inputs of 1 pg (Figure 16.1).

16.3.2 Software Deconvolution of mtDNA Mixtures

For deconvolution of mtDNA mixtures by NGS analysis, the primary challenge is to reconstruct entire mtDNA sequences by assembling the short reads generated by the NGS technology. Whether mtDNA NGS libraries are targeted by probe capture or PCR, interpreting these short reads as specific haplogroups or haplotypes can be challenging. Haplotype identification and estimation of contributor proportions can be achieved with the software GeneMarker®HTS, which uses the Revised Cambridge Reference Sequence (rCRS) as the reference sequence and allele frequency tables to rank the most likely genotype combinations and resolve contributors into a major and minor donor (Holland, Pack, and McElhoe 2017). Mixture interpretation is only unambiguous when resolving simple mtDNA mixtures of two contributors with unequal proportions above trace amounts. Frequency-based approaches to mtDNA mixture deconvolution, therefore, are not practical for more than two contributors or mixtures where the contributions are nearly equal or at trace amounts where low-level sequencing noise confounds interpretation. The software program Mixemt

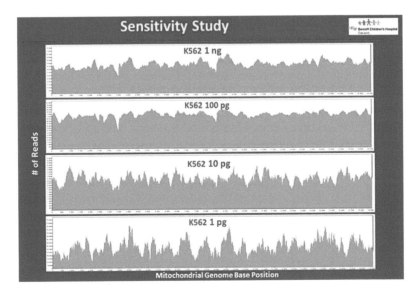

Figure 16.1 Sequence read coverage of the mtDNA genome at low DNA inputs. K562 is a control cell line.

utilizes a phylogenetic approach for mixture deconvolution to address this problem. It employs information on variants and haplotypes from mitochondrial diversity surveys using the Reconstructed Sapiens Reference Sequence (RSRS) as the reference to co-estimate the mixture proportions and the source haplogroup for each read (Vohr et al. 2017). Mixemt was initially tested using 336 in silico mixtures of two persons ranging from 50:50 to 95:5 ratios; 132 in silico mixtures of three persons of a 75:20:5 ratio; and 3 in vitro mixtures of two persons, one of equal proportions and two of distinct proportions. For mixtures containing more than two contributors of various proportions, phylogenetic information can successfully be used for interpretation when compared to frequency-based approaches because they are able to link variants and more readily resolve the mixtures (Vohr et al. 2017).

A recent study sought to determine whether utilizing both Mixemt and GeneMarker®HTS software programs can create a more comprehensive analysis technique to interpret mtDNA mixtures. GeneMarker®HTS and Mixemt software were utilized to analyze in vitro mixtures made from two and three persons, as well as with simulated case type samples of hairs mixed with various bodily fluids (Wisner et al. 2021). The results of analyzing two-person mixtures with approximately equal contributions (50:50), 95:5, and three-person mixtures (33:33:33) are shown in Figure 16.2 (Wisner et al. 2021).

Both software programs used here in the analysis of mtDNA mixtures have benefits and limitations. GeneMarker®HTS can resolve two-person mixtures when the minor contributor ranges from approximately 15% to 40% contribution. Mixemt can overcome the limitations of GeneMarker®HTS in interpreting mixtures with more than one contributor and mixtures with low-level contributor proportions. For the resolution of mixtures with more than two contributors, GeneMarker®HTS analysis alone proved difficult if the contributing haplotypes are not known. Mixemt can also identify trace contributors and contamination since the software considers all phylogenetically informative sites. There is, as expected, a trade-off between the sensitivity to detect a minor variant (contributor) and the background noise, requiring careful setting of thresholds.

For analysis of mtDNA sequences, one of the confounding factors is the presence of NUMTs (nuclear mitochondrial DNA), homologous sequences in the nuclear genome. By focusing only on phylogenetically informative sites, Mixemt can disregard some sequence noise because it does not use non-phylogenetic variants for phasing. The current version of the Mixemt program, however, has some practical limitations. The command line interface can be difficult, and analysis of one sample takes one day, on average. Additionally, it does not assign unlinked private mutations, indels, or some fast sites. Finally, the Reconstructed Sapiens Reference Sequence is used as the reference sequence rather than the traditional Revised Cambridge Reference Sequence (rCRS). Therefore, cross-checking to the rCRS is required. In general, analysis of mtDNA mixtures is most successful when *both* frequency and phylogenetic information is used. In addition, modification of Mixemt to include frequency analysis for the assignment of the unlinked private mutations and indels, and to incorporate probabilistic genotyping will be very valuable in the interpretation of forensic mixtures.

16.4 NGS and STRs

As noted earlier, one of the advantages of creating shotgun libraries from forensic specimens is that, in principle, sequence files for SNPs, STRs, and mtDNA could all be generated using different capture probe panels for analysis of the same sample. Targeted sequencing

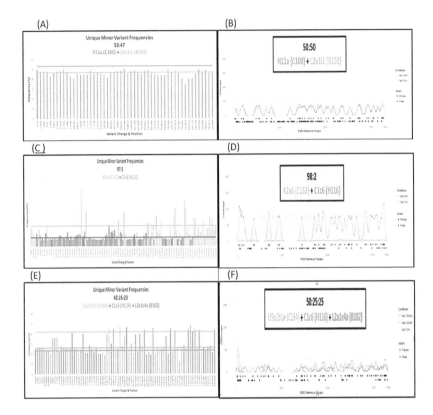

Figure 16.2 Unique minor variant frequencies manually obtained from GeneMarker®HTS in A, C, and E. Horizontal lines depict the average frequencies for each contributor. Colors reflect the colors generated by Mixemt to designate each contributor. B, D, and F are coverage curves with called variants and haplogroups obtained from Mixemt; the gray-shaded area denotes the number of reads at each base position. Circles denote phylogenetically informative sites defined by Phylotree build 17 that were detected for a haplogroup assignment; triangles denote private mutations detected, but not used in haplogroup assignment. Only phylogenetically informative sites are considered by Mixemt and therefore are reflected in the colored haplogroup lines. Areas with no phylogenetically informative sites will result in no colored lines in this area, despite there being ample coverage. (A) Unique minor variant frequencies manually obtained from GeneMarker®HTS in expected 50:50 mixtures. Six variants of major contributor C100 (avg 53.09%) appear in the minor report. (B) Coverage curve with called variants and haplogroups obtained from Mixemt of expected 50:50 mixtures. Mixture is estimated as 50.01% C100 (turquoise) and 49.99% B102 (salmon). (C) Unique minor variant frequencies in a 95:5 mixture (GeneMarker®HTS). At a variant filter setting of 1%, 106 unexpected variants (avg 1.04%) appeared in the minor report and at variant filter setting of 2%, six unexpected variants (avg 2.8%) appeared in the minor report. (D) Coverage curve with variants and haplogroups of 95:5 mixtures (Mixemt). Mixture estimated at 98.14% C163 (salmon) and 1.86% H116 (turquoise). (E) Unique minor variant frequencies estimated by GeneMarker®HTS in a 33:33:33 mixture. All contributors appeared in the minor report and varied in frequencies, making deconvolution by frequency alone impossible. (F) Coverage curves with variants and haplogroups obtained by Mixemt analysis of NGS reads from a contrived 3 person mixture with an estimated proportions of 33%, 33%, and 33%. Based on the the observed data, the contributor proportions are estimated as 49.98% C164 (salmon), 24.77% B182 (green), and 25.25% H116 (turquoise).

by the probe capture method also overcomes many limitations of capillary electrophoresis (CE) where genotyping is based on detection of fragment length. NGS enables the identification of STR alleles that are identical by size but different by sequence. Additionally, NGS analysis allows identification of variations that occur in regulatory, intronic, and coding regions of DNA, not possible in CE analysis. A variety of software for interpreting STRs by NGS have been reported like the ForenSeq Universal Analysis software, STRscan, NextGene, STRetch, LobSTR, HipSTR, RepeatSeq, and MaSTR (Wang et al. 2020).

We have previously reported analyses of libraries captured with SNP and mtDNA panels (Bose et al. 2018; Shih et al. 2018). Analyzing STRs by these strategies requires the design of probes that flank the repeat regions and sequence reads that span the repeat region, extending into the unique flanking regions. Consequently, we have not included probes in our STR probe panel for STRs with long alleles. Recently, we have analyzed sequence files from targeted libraries captured with probe panels for a set of STR markers, using a web-browser-based software called toaSTR, developed by LABCON-OWL, for genotyping NGS STRs (Ganschow et al. 2018). toaSTR is an alignment-free software that utilizes upstream and downstream recognition elements (REs) to identify the STR markers. The toaSTR software currently has 44 STRs available for analysis, including 30 autosomal STRS, 8 Y STR, 5 X STRs, and the Amelogenin marker. Penta E and Penta D markers were included in the toaSTR software but not in our STRv1.0 panel due to the larger amplicon sizes of the loci (>400 bps), which would exceed the sequencing length (300 bps) on the MiSeq with the MiSeq v3 reagents. toaSTR is a versatile software that utilizes direct graphical user interface to mentor through the STR genotyping workflow and supporting data from various NGS platforms. The software allows users to define locus-specific stutter thresholds and create custom sets of STR markers for analysis. toaSTR displays data visualization with interactive diagrams and a productive tabular analysis of detected sequences. The software sequence-based stutter model predicts and detects common stutter variants. The algorithm differentiates biological (iso-)alleles from stutter and other artifacts to aid in mixture interpretation.

As illustrated in Figure 16.3, toaSTR provides valuable information including electropherogram visualization of the allele calls, sequences of the STR alleles and stutters, and coverage for each allele (and the stutter). As an example, allele calls 10 and 10b were made for the D2S441 locus and the data was visualized in the form of an electropherogram, histogram, and sequences. Information on the coverage per allele was also available. The stutter threshold was set at 20% because stutter fragments are usually less than 15% of the true tetranucleotide allele repeats. The expected stutter was calculated by multiplying the number of total reads per allele by the stutter threshold percentage. If the coverage of the suspected allele falls below the expected stutter reads, then the allele is called a stutter (yellow), as seen in this case (Figure 16.3). One expectation for a stutter fragment is that the length is equal to $n - 1$ or $n + 1$, that the length should be within a base of the true allele (with n repeats). If the sequence read coverage of the suspected allele exceeds the coverage of the expected stutter, then the allele is categorized as an artifact or contaminant (Figure 16.3). In rare cases, a true allele was characterized by the software as stutter and vice versa.

This example shows the advantage of NGS over CE to detect isoalleles. The CE for this sample at D2S441 locus detects a homozygous 10 but NGS shows sequence variation in allele 10 and detects it as 10 ([TCTA]8 TCTG TCTA) and 10b ([TCTA]10).

An example of mixture analysis using toaSTR is shown in Figure 16.4. The mixture control sample NIST SRM 2391c-D control DNA is a 3:1 mixture of control samples NIST

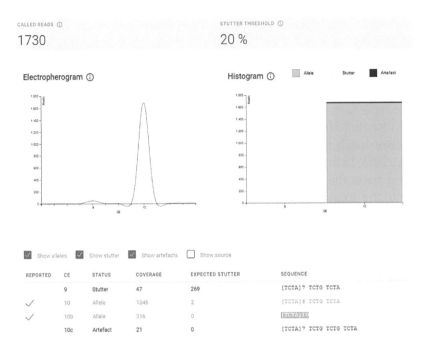

Figure 16.3 A screenshot of toaSTR analysis of D2S441 locus from NIST SRM 2391c-A control DNA. toaSTR analysis yielded allele calls 10 and 10b. A 9 stutter allele (n–1) of 1 true allele 10 was detected at 47 reads per locus coverage. A 10c artifact derived from allele 10 was also detected at 21 read coverage.

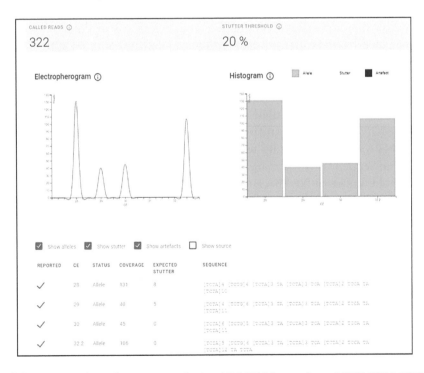

Figure 16.4 A screenshot of toaSTR analysis of D21S11 locus from NIST SRM 2391c-D control DNA. NIST SRM 2391c-D control DNA is a 3:1 mixture of NIST SRM 2391c-A control DNA and NIST SRM 2391c-C control DNA toaSTR analysis yielded allele calls 28 and 32.2 (NIST SRM 2391c-A) and alleles 29 and 30 (NIST SRM 2391c-C).

SRM 2391c-A and NIST SRM 2391c-C. This library was prepared at 10 ng DNA input; all four alleles from the two contributors were detected. Alleles 28 and 32.2 come from NIST SRM 2391c-A, and alleles 29 and 30 are from the contributor NIST SRM 2391c-C. The sequence read coverage provides an approximation of the mixture proportions. The alleles from the major contributor (NIST SRM 2391c-A) are threefold higher in coverage than the minor contributor (NIST SRM 2391c-C). It is also worth noting that longer alleles are successfully detected in mixture samples. In additional testing of more mixture samples at lower DNA input, the analysis of sequence read coverage was not always quantitative. In addition, alleles from the major contributor were detected in almost all cases, but the alleles from the minor contributor were detected in some but not all cases. Analysis of mixtures will require consideration of missing data, i.e., allele dropout, and caution in interpreting proportions from read coverage.

16.5 Rapid DNA

Unlike NGS, the commercial Rapid DNA technology does not represent a technical innovation allowing the analysis of novel genetic markers or the more effective genotyping of complex mixtures; instead, it automates an existing system, the characterization of STRs by electrophoresis. Rapid DNA instruments can obtain a CODIS core STR profile from a reference cheek swab in less than 90 minutes (Grover 2017; Shackleton et al. 2019; Buscaino et al. 2018). The practical consequences of this automation and speed are very significant for law enforcement practices. These compact instruments can be deployed in police stations and border patrol stations, and have the potential to be used in the field (Morrison et al. 2018).

Starting in 2012, several different companies developed integrated microfluidic devices to automate DNA extraction from a cheek swab, PCR amplification of a panel of STR loci, electrophoresis of the PCR products, and data analysis (Erlich, Calloway, and Lee 2020). The companies offering these fully integrated devices (swab in, STR profile out) include ParaDNA, which amplifies 5 STRs and amelogenin directly from human and casework samples; ANDE, which uses PowerPlex 16 chemistry or a new 27 locus STR system, FlexPlex, an automated allele calling expert system and RFID sample tracking; and RapidHit200 with PowerPlex 16 HS chemistry. An updated version, RapidHIT200, with the GlobalFiler Express runs 8 swabs simultaneously.

All of these commercial compact instruments are capable of providing an STR profile in less than 90 minutes. This short turnaround time is achieved, in part, by rapid PCR thermal cycling. Gibson-Daw, Crenshaw, and McCord (2017) were able to reduce the total time to amplify a seven loci multiplex consisting of six STRs and Amelogenin, to 6.5 minutes. The speed of these instruments and newly developed chemistries means that, in principle, these instruments could be deployed in police stations so that STR profiles of arrestees could be obtained and screened against criminal and evidence databases while the suspect is still at the booking station. Rapid DNA results that matched the profile of an evidence sample or a database profile would, presumably, lead to the arrestee being detained; rapidly generated profiles could also potentially prevent the detention of the wrongfully accused. These instruments could also be used at the border to investigate claimed familial relationships in immigration cases; these applications, however, raises issues of both consent and privacy.

These two applications have in common the analysis of high-quality DNA from blood or a cheek swab unlike the degraded DNA typically found in forensic crime scene samples or in the human remains used in missing person cases. In validation studies (Morrison et al. 2018; Shackleton et al. 2019), the performance of most of these Rapid DNA instruments, which have been optimized for reference samples, is quite good for reference samples but is somewhat less reliable for forensic evidence samples (Mapes et al. 2019; Alshehhi and Roy 2015). In some of these validation studies, some degree of manual interpretation of the profiles generated from the instrument was necessary to achieve the correct result (i.e., concordant with a known sample profile). A recent validation study (Salceda et al. 2017) evaluated the RapidHIT ID system using the Global Filer Express chemistry on reference samples and reported a 98% success rate for a concordance data set. Another evaluation of the same instrument system (Date Chong et al. 2005) found that some variability in the STR results was observed and that some samples required manual editing of the profiles; they reported that 50% of the cheek swab samples would have been successfully genotyped without any need for manual review or editing.

One group has developed a system that concentrates the DNA from low template samples (Turingan et al. 2016) and has reported improved rapid DNA results from blood, buccal cells, chewing gum, bone, cell phones, cigarette butts, and other low template samples.

US Federal Bureau of Investigation (FBI) guidelines for application of STR genotyping by Rapid DNA specifies that it has been validated for reference samples, not casework evidence; accordingly, reference sample profiles can be uploaded to CODIS, but currently not casework profiles. There is a growing body of research to improve results on crime scene samples and to provide decision support guidance for rapid DNA end users (Mapes et al. 2019).

16.6 Conclusions

The development and application of NGS (or MPS) technology to forensic individual identification promises to have a significant impact on the criminal justice system. Here, we have reviewed two specific applications, namely, the deconvolution of forensic mixtures using mtDNA and a phylogenetics-based software, Mixemt; and the analysis of STRs using a web-based software, toaSTR. The other major recent technical innovation in forensic DNA technology, Rapid DNA instruments that have automated DNA extraction, PCR amplification, and capillary electrophoresis for analysis of STR markers, promises to transform when and where genetic analyses can be carried out. We can anticipate that the evolution of DNA technology that has characterized forensic genetics since the mid-1980s will continue, introducing faster, more sensitive, and more powerful strategies for individual identification.

Acknowledgments

We are grateful to Mary Wisner for the analysis of mitochondrial DNA mixtures.

Sections of this chapter were originally published in *Silent Witness*, edited by Erlich, Stover, and White, and have been reproduced by permission of Oxford University Press (http://global.oup.com/academic). For permission to reuse this material, please visit http://global.oup.com/academic/rights.

References

Alshehhi, A., and R. Roy. 2015. Generating rapid DNA profiles from crime scene samples commonly encountered in the United Arab Emirates. *J Forensics Res* 06(04): 296.

Bose, N., K. Carlberg, G. Sensabaugh, H. Erlich, and S. Calloway. 2018. Target capture enrichment of nuclear SNP markers for massively parallel sequencing of degraded and mixed samples. *Forensic Sci Int Genet* 34: 186–196.

Buscaino, J., A. Barican, L. Farrales, B. Goldman, J. Klevenberg, M. Kuhn, and D. King. 2018. Evaluation of a rapid DNA process with the RapidHIT® ID system using a specialized cartridge for extracted and quantified human DNA. *Forensic Sci Int Genet* 34: 116–127.

Butler, J. M. 2004. Short tandem repeat analysis for human identity testing. *Curr Protoc Hum Genet.* Chapter 14:Unit 14.8.

Calloway, C. D., S. M. Stuart, and H. A. Erlich. 2009. Development of a multiplex PCR and linear array probe assay targeting informative polymorphisms within the entire mitochondrial genome. https://www.ncjrs.gov/App/Publications/abstract.aspx?ID=250297.

Churchill, J. D., M. Stoljarova, J. L. King, and B. Budowle. 2017. Parsing apart the contributors of mitochondrial DNA mixtures with massively parallel sequencing data. *Forensic Sci Int Genet Suppl S* 6: e439–e441.

Cornelis, S., S. Willems, C. V. Neste, O. Tytgat, J. Weymaere, A. V. Plaetsen, and F. V. Nieuwerburgh. 2018. Forensic STR profiling using Oxford Nanopore Technologies MinION sequencer. https://doi.org/10.1101/433151.

Date Chong, M., C. D. Calloway, S. B. Klein, C. Orrego, and M. R. Buoncristiani. 2005. Optimization of a duplex amplification and sequencing strategy for the HVI/HVII regions of human mitochondrial DNA for forensic casework. *Forensic Sci Int* 154(2): 137–148.

de Knijff, Peter. 2019. From next generation sequencing to now generation sequencing in forensics. *Forensic Sci Int Genet* 38: 175–180.

Divne, A. M., M. Nilsson, C. Calloway, R. Reynolds, H. Erlich, and M. Allen. 2005. Forensic casework analysis using the HVI/HVII mtDNA linear array assay. *J Forensic Sci* 50(3): 548–554.

Erlich, H., C. Calloway, and S. B. Lee. 2020. Recent developments in forensic DNA technology. In: *Silent Witness: Forensic DNA Analysis in Criminal Investigations and Humanitarian Disasters*, edited by Henry Erlich, Eric Stover and Thomas J. White, 105–127. New York: Oxford University Press.

Gabriel, M. N., C. D. Calloway, R. L. Reynolds, and D. Primorac. 2003. Identification of human remains by immobilized sequence-specific oligonucleotide probe analysis of mtDNA hypervariable regions I and II. *Croat Med J* 44(3): 293–298.

Ganschow, S., J. Silvery, J. Kalinowski, and C. Tiemann. 2018. toaSTR: A web application for forensic STR genotyping by massively parallel sequencing. *Forensic Sci Int Genet* 37: 21–28.

Gibson-Daw, G., K. Crenshaw, and B. McCord. 2017. Optimization of ultrahigh-speed multiplex PCR for forensic analysis. *Anal Bioanal Chem* 410(1): 235–245.

Grover, R., H. Jiang, R. S. Turingan, J. L. French, E. Tan, and R. F. Selden. 2017. FlexPlex27—Highly multiplexed rapid DNA identification for law enforcement, kinship, and military applications. *Int J Legal Med* 131(6): 1489–1415.

Higuchi, R., C. H. von Beroldingen, G. F. Sensabaugh, and H. A. Erlich. 1988. DNA typing from single hairs. *Nature* 332(6164): 543–546.

Holland, M., and T. Parsons. 1999. Mitochondrial DNA sequence analysis—Validation and use for forensic casework. *Forensic Sci Rev* 11(1): 21–50.

Holland, M. M., E. D. Pack, and J. A. McElhoe. 2017. Evaluation of GeneMarker® HTS for improved alignment of mtDNA MPS data, haplotype determination, and heteroplasmy assessment. *Forensic Sci Int Genet* 28: 90–98.

Just, R. S., J. A. Irwin, and J. E. O'Callaghan. 2004. Toward increased utility of mtDNA in forensic identifications. *Forensic Sci Int* 146(Suppl): S147–S149.

Lee, S. B., J. Varlaro, and C. Holt. 2016. Stepping into the future of forensic genomics: Developmental validation of a next-generation sequencing forensic DNA sample-to-answer system. *Forensic Magazine.* https://www.forensicmag.com/article/2016/07/future-forensic-genomics-developmental-validation-ngs.

Mapes, A. A., R. D. Stoel, C. J. De Poot, P. Vergeer, and M. Huyck. 2019. Decision support for using mobile Rapid DNA analysis at the crime scene. *Sci Justice* 59(1): 29–45.

Mardis, E. R. 2008. Next-generation DNA sequencing methods. *Annu Rev Genomics Hum Genet* 9: 387–402.

Maude, H., M. Davidson, N. Charitakis, L. Diaz, W. Bowers, E. Gradovich, T. Andrew, and D. Huntley. 2019. NUMT confounding biases mitochondrial heteroplasmy calls in favor of the reference allele. *Front Cell Dev Biol* 7: 201.

McCombie, W. R., J. D. McPherson, and R. Mardis. 2018. Next-generation sequencing technologies. *Cold Spring Harb Perspect Med* a036798. https://doi.org/10.1101/cshperspect.a036798.

Melton, T., G. Dimick, B. Higgins, L. Lindstrom, and K. Nelson. 2005. Forensic mitochondrial DNA analysis of 691 casework hairs. *J Forensic Sci* 50(1): 73–80.

Mihalovich, J., E. Blake, R. Higuchi, P. S. Walsh, and H. Erlich. 1992. Polymerase chain reaction (PCR) amplification and human leukocyte antigen (HLA)-DQ alpha oligonucleotide typing on biological evidence samples: Casework experience. *J Forensic Sci* 37(3): 700–726.

Moorthie, S., C. J. Mattocks, and C. F. Wright. 2011. Review of massively parallel DNA sequencing technologies. *HUGO J* 5(1–4): 1–12.

Morrison, J., G. Watts, G. Hobbs, and N. Dawnay. 2018. Field-based detection of biological samples for forensic analysis: Established techniques, novel tools, and future innovations. *Forensic Sci Int* 285: 147–160.

Parsons, T. J., and M. D. Coble. 2001. Increasing the forensic discrimination of mitochondrial DNA testing through analysis of the entire mitochondrial DNA genome. *Croat Med J* 42(3): 304–309.

Roberts, K. A., and C. Calloway. 2007. Mitochondrial DNA amplification success rate as a function of hair morphology. *J Forensic Sci* 52(1): 40–47.

Salceda, S., A. Barican, J. Buscaino, B. Goldman, J. Klevenberg, M. Kuhn, and D. King. 2017. Validation of a rapid DNA process with the RapidHIT® ID system using GlobalFiler® Express chemistry, a platform optimized for decentralized testing environments. *Forensic Sci Int Genet* 28: 21–34.

Shackleton, D., N. Gray, L. Ives, S. Malsom, and D. Vanhinsbergh. 2019. Development of RapidHIT® ID using NGMSElect™ Express chemistry for the processing of reference samples within the UK criminal justice system. *Forensic Sci Int* 295: 179–188.

Shih, S. Y., N. Bose, A. Gonçalves, H. A. Erlich, and C. D. Calloway. 2018. Applications of probe capture enrichment next generation sequencing for whole mitochondrial genome and 426 nuclear SNPs for forensically challenging samples. *Genes* 9(1): 49.

Sullivan, K. M., R. Hopgood, and P. Gill. 1992. Identification of human remains by amplification and automated sequencing of mitochondrial DNA. *Int J Legal Med* 105(2): 83–86.

Turingan, R. S., S. Vasantgadkar, L. Palombo, C. Hogan, H. Jiang, E. Tan, and R. F. Selden. 2016. Rapid DNA analysis for automated processing and interpretation of low DNA content samples. *Investig Genet* 7(1): 2.

Vohr, S. H., R. Gordon, J. M. Eizenga, H. A. Erlich, C. D. Calloway, and R. E. Green. 2017. A phylogenetic approach for haplotype analysis of sequence data from complex mitochondrial mixtures. *Forensic Sci Int Genet* 30: 93–105.

Wang, D., R. Tao, Z. Li, D. Pan, Z. Wang, L. Chengtao, and Y. Shi. 2020. STRsearch: A new pipeline for targeted profiling of short tandem repeats in massively parallel sequencing data. *Hereditas* 157(1): 8.

Wisner, M., S. Shelly, H. Erlich, and C. Calloway. 2021. Resolution of mitochondrial DNA mixtures using a probe capture next generation sequencing system and phylogenetic-based software. *Forensic Sci Int Genet* 53: 102531.

Prediction of Physical Characteristics, such as Eye, Hair, and Skin Color, Based Solely on DNA

17

SUSAN WALSH AND MANFRED KAYSER

Contents

17.1 Introduction

The gold standard in forensic DNA analysis is forensic DNA profiling, which is the generation of short tandem repeat (STR) data at multiple STR loci in an effort to obtain a comparable match for identification purposes. Either the STR data generated from the crime scene sample matches a suspect involved in the case whose sample had been collected during the investigation, or the DNA profile matches that of an individual within an operational and regulated national forensic DNA database (e.g., CODIS in the United States). This match is determined with a statistically calculated degree of certainty that allows the identification of the individual that had deposited the crime scene sample. Whether the DNA-identified sample donor is the perpetrator is up to the court to decide, often with the help of non-DNA evidence. This comparable identification approach is the most widely accepted form of DNA identification used in legislation throughout the world. Research in improving the effectiveness of this method has seen increases in the number of STR loci included in commercial kits, alongside a significant technological improvement that has resulted in higher sensitivity of forensic STR testing so that only picograms of DNA are sufficient and the move from fragment length analysis with inferred allele repeat number information to massively parallel sequence (MPS) analysis for determining the repeat number directly. In addition, developments in more accurately determining match statistics, i.e., additional population frequency data being made available, and global agency database cooperation have further helped in this comparable profiling methodology.

However, the limitation of this method will always be its dependency on a comparison with the STR profile of an already known person. Without a match with a case suspect

DOI: 10.4324/9780429019944-20

or someone in the forensic DNA database to calculate a statistical match probability that allows individual identification, the DNA data generated is noninformative. Taking this into consideration, researchers have begun to explore other aspects of information that can be gleaned from DNA material, and over the last 15 years, appearance prediction using DNA, also known as the field of forensic DNA phenotyping (FDP), has emerged and grown with great potential. FDP is a relatively new field within the realm of forensic DNA analyses and although it will never replace the gold standard of DNA identification, which is and will remain forensic DNA profiling, it does have the capacity to complement intelligence-driven applications toward finding previously unknown individuals and thus making them available for STR profile comparison. Predicting one's physical appearance, i.e., externally visible characteristics solely from DNA, not only assists in providing intelligence information about individuals that have left a crime scene sample to locate a potential suspect, but moreover it also allows the generation of characteristics helpful in missing persons and mass disaster victim identification scenarios without knowledge on the putative identity of the person and the relatives, so that these individuals have a better chance at being identified using DNA. Its application also expands into the realm of anthropology and human history, including the field of ancient DNA (aDNA) analysis, as generating appearance information from DNA provides more information on how the deceased person had looked rather than what can be deduced from skeletal remains. It also complements fellow intelligence-driven DNA applications such as biogeographical ancestry inference (BGA) and forensic genetic genealogy (FGG). Performing appearance prediction from DNA alongside DNA-based BGA and DNA methylation-based estimation of a person's age increases the investigative value of FDP, which in a wider view encompasses all three aspects. This is due to increased reliability, given that some appearance traits depend on BGA and age, in addition to BGA providing further specifications. For example, brown-eyed, black-haired, light-skinned individuals are found in a large geographical area covering native populations from three continents—Europe, Asia, and the Americas—while DNA-based BGA allows the inference of these three continental regions separately, thus providing additional investigative information on top of DNA-predicted eye, hair, and skin color. FDP utilizes sets of carefully ascertained single-nucleotide polymorphisms (SNPs), combined for each appearance trait being predicted (with other sets used for BGA). Each SNP has a given weight toward the predicted outcome using training and test data from thousands of individuals that have been used to build and validate the trait-specific prediction models. Here we summarize the current state of appearance prediction from DNA with regards to pigmentation traits, its history, and development, and allude to further developments within the field of FDP to expand visible trait prediction from DNA beyond pigmentation. The other two elements of wider-sense FDP—DNA-based BGA and DNA methylation-based age estimation—will not be covered in this chapter.

17.2 Complex Traits: Pigmentation

The ability to predict a phenotype using genetic data falls within the framework of complex genetics because all normal human appearance traits are polygenic. Examples such as pigmentation, height, and facial structure are highly heritably but influenced by hundreds of genes, in addition to environmental factors such as age and health which vary in influence on the final trait appearance. Significant progress has been made exploring the fundamental aspects of complex genetics, not only in medical genetics, since all common heritable diseases are complex traits, but also in nonmedical human genetics such as unearthing DNA markers associated with normal phenotypic traits and exploring their function, interactions, and pathways. Unveiling the genetic basis of normal appearance without direct medical relevance is rather a niche area relative to the large field of human disease genetics and has therefore also etched its way into some applied disciplines such as forensic genetics, anthropological genetics, and consumer genetics. Pigmentation genetics as a whole is seen as a complex trait with the most success achieved over the last two decades through focused explorations of eye, hair, and skin color phenotypes (and more recently other pigmentation phenotypes such as freckles and eyebrow color) that have unearthed genes with mostly overlapping impact due to interrelating function and interacting pathways that are typically shared between global human populations. Melanin pigments are packaged into melanosomes, which are specialized subcellular compartments that are instructed to plant themselves within adjacent keratinocytes (Barsh 2003). Pigmentation variation arises from differing numbers of melanosomes within the cell, as well as their size, composition, and distribution (Lin and Fisher 2007). Two forms of melanin can be produced—eumelanin (brown-black) and pheomelanin (red-yellow)—and it is their differing ratios and placement that are responsible for the degree of variation that is observed across eye, hair, and skin color phenotypes (Sturm and Frudakis 2004) (see Figure 17.1 for a cellular overview). In the early days of exploring human pigmentation genetics, pigmentation disorders in combination with a comparative genomics approach (Jackson 1997; Brilliant 2001) led to developments concerning the mechanisms and genes involved in the color spectrum of its variation. Animal pigmentation models revealed the first few genes, with the most dramatic examples being gene inactivation. Disease-classified albinism phenotypes of oculocutaneous albinism *OCA1* and *OCA2*, which result in a complete loss of all pigment within the iris, hair follicles, and skin cells of an individual, were thought to represent the most informative genes/pathways for human pigments. However, later on, many additional loci were identified through the characterization of naturally occurring or artificially induced mutations affecting melanin amount and distribution, such as genes *TYR*, *OCA2*, *TYRP1*, and *SLC45A2 (MATP)* (Sturm and Larsson 2009).

Improved understanding of the melanin pathway also led to the melanocortin coupled receptor gene 1 (*MC1R*), highly influential in not only pigmentation genetics but also cancer genetics, due to its critical role in the switch of melanin type from eumelanin to pheomelanin (Valverde et al. 1995) in melanocytes. The role of this gene in normal human pigmentation was substantiated by studies showing an association of variants in *MC1R* with red hair and fair skin (Valverde et al. 1995; Bastiaens et al. 2001; Naysmith et al. 2004). The agouti protein (*ASIP*) was also shown to act as an inverse agonist of *MC1R* in mouse melanocyte cell culture studies (Le Pape et al. 2009), providing support for its allelic associations within lighter or darker pigmentation phenotypes in humans (Kanetsky et al. 2002; Bonilla et al. 2005; Sulem et al. 2008). A similar approach found *SLC24A5*, the homologue of the zebrafish golden gene mutation (Lamason et al. 2005), implicated in human skin color through the analysis of its allelic distribution worldwide. The *KITLG* gene was implicated in human pigmentation variation by the association of its ancestral allele with skin color, and strong linkage for the derived allele within Europeans and East Asians (Biswas and Akey 2006; Voight et al. 2006; Sabeti et al. 2007). An intronic SNP (rs12203592) in the *IRF4* gene is another pigmentation variant that displays unusual variation in allele frequencies in Europe, as the T allele tends to lighten skin and eye phenotypes but darkens hair (Sulem et al. 2007; Duffy et al. 2010). Transgenic models have confirmed

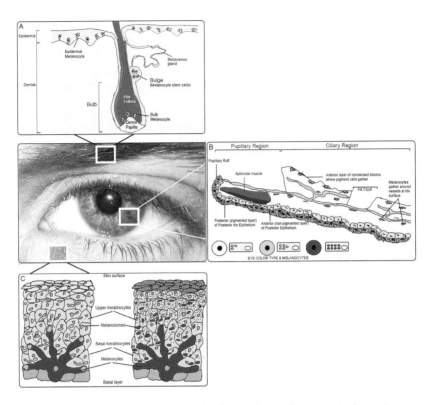

Figure 17.1 A schematic overview of the biology of eye, hair, and skin pigmentation. (A) Visual representation of a hair follicle adapted from (Picardo et al. 2015). The origin of pigment production for the hair shaft occurs at the hair bulb, due to the presence of melanocytes that contain mature melanosomes that secret melanin granules into the hair shaft. The hair bulge contains hair follicle stem cells necessary for growth and pigment deposition, which migrate to the hair bulb when instructed. The dermal papilla is instrumental in the formation of new hair follicles and plays a pivotal role in regulating hair follicle pigmentation. (B) Visual representation of an iris cellular structure (adapted from Hughes and Luce 2020). The iris can be divided into two regions, a pupillary region and ciliary region. Iris stromal melanocytes in the anterior layer are derived from the neural crest, whereas the posterior pigment epithelium forms the posterior layer, which is neuroectodermal in origin (Rennie 2012). Light reflection within the walls of iris vessels against the dark brown pigment of the posterior iris pigment layer creates apparent variation. A thinly pigmented iris may appear blue, while a thin stroma allows color to be seen from the posterior portion of the iris mimicking green or hazel. A highly pigmented anterior layer absorbs light and is visibly brown. Distinct folds like an iris crypt create structure that expose the color of the posterior layer. (C) Visual representation of the epidermis of the light and dark skin (adapted from OpenStax 2013). Melanocytes embedded in the epidermis produce melanosomes, which are transferred into surrounding keratinocytes. These organelles are responsible for synthesizing and depositing melanin pigment (represented by the dots in this image).

that a 400 bp fragment of the *IRF4* intron that includes rs12203592 is an enhancer and that the T allele reduces fragment activity (Praetorius et al. 2013).

As the determinants of human skin, eye, and hair color fall into the quasi-Mendelian inheritance pattern of a polygenic trait, only a few major genes show high effects, such as those mentioned earlier, but each trait also has many additional modifier genes with rather small effects (Sturm et al. 2001) and are therefore usually difficult to identify. However, the application of genome-wide association studies (GWAS) to pigmentation traits has opened the door to discovering hundreds of additional pigmentation genes. The GWAS approach searches through wide-ranging SNP databases using large numbers of individuals with phenotype information for SNPs with genome-wide significant trait associations, which must be replicated in independent datasets. The success of GWAS in unveiling the polygenic nature of heritable complex traits largely depends on the sample size. Over the last 15 years, GWAS for all human pigmentation traits separately were done with ever-increasing sample size, thus increasing statistical power to newly identify underlying genetic loci with small effects, while confirming those with larger effects discovered in previously smaller-sized GWAS. Extremely large GWAS involving hundreds of thousands of individuals were performed for hair (Hysi et al. 2018), eye (Simcoe et al. 2021), and skin color (Visconti et al. 2018), which demonstrated the large genetic complexity of human pigmentation traits, while in the past they were assumed, and by mistake as we now know, to be genetically simple traits with one or two genes only. Although several genes do overlap within the pigmentation pathway system, there are typically noted differences in gene penetrance for each pigment trait, therefore it is best to explore eye, hair, and skin color independently when searching for association with common and low-frequency variants.

17.2.1 Eye Color Genetics

Eye color in its simplest definition refers to a perceived color within the ocular aperture that is observed at a distance. If one were up close, it is easier to pick up on the assortment of structures and pigment spread that relay this overall color. The color arises from melanocytes in the anterior pigment epithelium (anterior border layer) that predominantly carry eumelanin and the stromal melanocytes that carry both eumelanin and pheomelanin in the iris. Melanocyte density is constant across irises of a different color; therefore, variation arises from the density and content of the melanin within the melanosome (Figure 17.1). For example, blue or green eyes contain large amounts of pheomelanin, a 14:1 pheomelanin:eumelanin ratio; while brown eyes contain higher amounts of melanin with a 1:1 pheomelanin:eumelanin ratio (Duffy 2015).

Concentrated studies have led to a vast increase in the genetic understanding of human eye color within the last decade. Essentially, genome-wide association and linkage analysis or candidate gene studies (Kanetsky et al. 2002; Sulem et al. 2008; Sulem et al. 2007; Kayser et al. 2008; Eiberg et al. 2008; Han et al. 2008; Sturm et al. 2008; Duffy et al. 2007; Zhu et al. 2004; Posthuma et al. 2006; Frudakis et al. 2003) encapsulated the majority of successful gene mapping techniques for eye color. The OCA2 gene on chromosome 15 that was originally thought to be the most informative human eye color gene, due to its association with the human P protein in its role for the processing of melanosomal proteins (Rebbeck et al. 2002) and resulting mutations leading to pigmentation disorders such as albinism (Brilliant 2001), was surpassed in importance. Further exploration led to its neighboring gene HERC2, as it was more significantly associated with eye color variation than variants in OCA2 (Sulem et al. 2008; Kayser et al. 2008; Eiberg et al. 2008; Han et al. 2008; Sturm et al. 2008). A subtle case of gene penetrance was found, where one of the most significant nonsynonymous SNPs associated with eye color, rs1800407 located in exon 12 of the OCA2 gene, simply acts as a modifier of rs12913832 in HERC2 and to a much lesser extent displays independent association with its variation (Sturm et al. 2008). It was proposed that the genetic variation occurring at HERC2 acted as a functional regulator of the adjacent OCA2's gene activity (Kayser et al. 2008; Eiberg et al. 2008; Sturm et al. 2008), which was then further substantiated in a functional genetics study demonstrating that the sequence including rs12913832 in *HERC2* functions as a long-distance enhancer for the *OCA2* promotor, working as a kind of on/off switch for *OCA2* expression (Visser et al. 2012). This study showed that the ancestral *HERC2* rs12913832-A (or T) allele promotes melanin production, ultimately leading to dark pigmentation by recruiting transcription factors, and in combination with increased chromatin looping to the nearby promoter of the pigmentation gene *OCA2*, facilitates OCA2 expression. In contrast, these are all reduced in lightly pigmented melanocytes carrying the derived rs12913832 G (or C) allele (Visser et al. 2012). While the *HERC2/OCA2* region harbors a large genetic proportion of blue and brown eye color information, there are also other genes noted for their contribution to eye color variation through their involvement in melanin biosynthesis and transport, such as *SLC24A4*, *SLC45A2 (MATP)*, *TYRP1*, *TYR*, *ASIP*, and *IRF4*, although to a much lesser extent (Kanetsky et al. 2002; Sulem et al. 2008; Sulem et al. 2007; Han et al. 2008; Frudakis et al. 2003).

There has been a shift toward exploring the quantitative nature of iris color as opposed to the simple categorical classification. Although using categorical trait information, most often a three-category scale of blue, green or intermediate, and brown eye color (Sulem et al. 2008; Sulem et al. 2007; Sturm et al. 2008; Duffy et al. 2007) have been successful in finding genetic variants associated with eye color, it is known that in reality iris color exists on a more continuous spectrum from the lightest shades of blue to the darkest of brown or black. The use of categorized information from continuous traits can oversimplify the true quantitative nature of the trait. Additional genes such as *LYST* and *DSCR9*, previously unidentified using the simpler categorical approach, were found using quantitative approaches based on a color spectrum of hue (H) and saturation (S), thus showing the success of alternative phenotyping (Liu et al. 2010).

More recently, expanding the numbers used in GWAS to increase power has substantially improved genetic discovery related to eye color, albeit on a categorical scale. Using an $N = 194,622$ for discovery and replication, researchers revealed strong associations in 61 discrete regions of the genome related to categorical eye color phenotypes in the largest eye color GWAS to date, collectively explaining 53.2% of eye color variation (Simcoe et al. 2021). Novel genes associated with eye color that were discovered in this study include *GPR157*, *HIVEP3*, *DTL*, *PRKCE*, *INO80D*, *WNT10A*, *FARSB*, *TRAF3IP1*, *CCDC13*, *MITF*, *PPARGC1A*, *TIGD2*, *PDCD6-AHRR*, *DAB2*, *ZNF608*, *ADRB2*, *GCNT2*, *TULP4*, *IGFBP3*, *SEMA3A*, *LONRF1*, *AP3M2*, *ZBTB10* , *MOB3B*, *IER5L*, *TPCN2*, *SOX5*, *BTG1*, *KLF12*, *DCT*, *ZFP36L1*, *SLC24A5*, *UBE2I*, *SMG6*, *MAP2K6*, *ZNF358*, *SLC23A2*, *SIK1*, *WNT7B*, *SSX1*, and *TMEM255A*. Ideally, future explorations of the genetic association of eye color toward increased numbers of participants and continuous color spectrum classifications will substantiate many of these genes and enable future exploration of their functional significance within iris color development. That there are more than these recently discovered 61 genetic loci involved in human eye color is indicated by the high but still limited eye color variance that can be explained by these genetic loci, together with the high heritability of eye color. More direct evidence comes from the outcome of an even larger GWAS on hair color, which doubled the number of genes of which most, if not all, are expected to also contribute to eye color (Hysi et al. 2018).

Most genetic eye color work was done in individuals of complete or at least partial European descent, which makes complete sense as strong eye color variation between the lightest blue and the darkest brown only exists in people of at least partial European descent. However, there also is a subtle variation of brown eye color in Asians. Simcoe et al. (2021) used 1636 Asians (Han Chinese and Indians) to test the genetic loci they had identified in almost 200,000 Europeans and found that 70% had the same direction of effect in Asians as in Europeans,, while 11% showed significant eye color associations also found in Asians. This suggests that the genetic makeup for eye color in Europeans and Asians is rather similar, although the degree of phenotypic eye color variation is much smaller in Asians than it is in Europeans.

17.2.2 Hair Color Genetics

Hair follicles are complex and independent, with multiple cell cycles from growth to regression. Keratinocytes make up the majority of a hair follicle cellular structure, with dermal papilla cells, which are specialized mesenchymal cells, residing at the bottom controlling phases of its development (Driskell et al. 2011). These cells interact with melanocytes embedded in the follicular epithelium and therefore allow pigment to be deposited on the forming shaft (Figure 17.1).

The genetic basis of human hair color variation has also been explored quite considerably over the last few years. Estimates from twin studies suggest upward of 97% heritability for hair color (Pavan and Sturm 2019). Valverde et al. (1995) found that red hair color is mainly associated with polymorphisms in *MC1R*, a single small gene. This information has also been confirmed in concurrent studies that have been performed using various population samples (Sulem et al. 2007; Han et al. 2008; Kanetsky et al. 2004; Box et al. 1997; Harding et al. 2000; Flanagan et al. 2000; Pastorino et al. 2004; Rana et al. 1999). Using individuals from an Icelandic population produced additional information on red hair color inheritance, revealing two SNPs within the *ASIP* gene that represent an *MC1R* antagonist that are significantly associated with red hair color (Sulem et al. 2008). Further candidate gene approaches found that a position in the 3¢-UTR of the *ASIP* gene was also associated with dark hair color in European populations (Kanetsky et al. 2002; Voisey et al. 2006). Two non-synonymous SNPs in *SLC45A2* (MATP) were also found to be associated with dark hair color with several confirmation studies (Branicki et al. 2008; Fernandez et al. 2008; Graf et al. 2005).

GWAS have contributed a lot of knowledge toward understanding genetic variants associated with hair color variation. One of the earliest hair color GWAS (Sulem et al. 2008; Sulem et al. 2007) found variant associations previously implicated in human pigmentation such as *MC1R*, *HERC2-OCA2*, and *KITLG* including the *SLC24A4* gene, a member of the *SLC45A5* family (Lamason et al. 2005). Functional tests exploring *KITLG*, which was significantly associated more specifically with common blond hair color in Northern Europeans (Sulem et al. 2007), found that the region contains a regulatory enhancer that drives expression in hair follicle development, a particular SNP rs12821256, and alters binding the lymphoid enhancer-binding factor 1 (*LEF1*) transcription factor, reducing *LEF1* responsiveness and enhancer activity (Guenther et al. 2014). An intergenic region between *IRF4* and *EXOC2* was also identified (Sulem et al. 2007). Another independent GWAS substantiated that *IRF4* and *SLC24A4* were also linked with hair color (Han et al. 2008), as well as confirming the roles of SNPs within the *HERC2-OCA2*, *MC1R*, and *SLC45A2* loci. Sulem et al. (2008) found via GWAS two coding variants within the *TPCN2* gene, which impacts blond versus brown hair color. Other associated genes found through multiple GWAS include *TYR, TYRP1, KIF26A*, and *OBSCN* (Sulem et al. 2008; Sulem et al. 2007; Han et al. 2008). Branicki et al. (2009) reexamined the previously discovered

HERC2-OCA2 region, not only confirming the association of *HERC2* rs12913832 with eye color, but they also showed that this SNP was significantly associated with skin and hair coloration. In 2018, Hysi et al. (2018) performed the largest GWAS meta-analyses on hair color, using over 300,000 individuals from the 23andMe database and the UK Biobank cohort study, with replication provided by additional data from the Visible Trait Genetics (VisiGen) Consortium. They identified 124 discrete genetic loci significantly associated with hair color. Of those, several genes were noted to be involved in melanocyte development: *EDNRB* (rs1279403), *MITF* (rs9823839), *HPS5* (rs201668750), *FGF5* (rs7681907), *SOX5* (rs9971729), and *TWIST2* (rs11684254). Also in 2018, Morgan et al. further explored the UKBB identifying a total of 163 distinct genes (93 overlapping with Hysi et al.) but focused specifically on exploring *MC1R* and its contribution toward hair color, in addition to identifying measured epistatic interactions in the *PKHRD1* gene. In 2021, Lona-Durazo et al. explored hair color using fine-mapping analyses on a Canadian cohort of over 12,741 individuals with European ancestry, in addition to data mining of expression and methylation quantitative trait loci (QTLs) of cultured melanocytes to further substantiate *EDNRB* and *CDK10* roles in hair color development. More recently, researchers have begun exploring eyebrow hair color (Peng et al. 2019), again finding known genes related to head hair color variation as well as new candidates. Further functional analyses may find out why it is possible to have differences in human head hair variation versus facial hair (eyebrow/beard) in the same individual, perhaps related to pigment transport within the hair follicle at differing facial sites.

Overall, the majority of work in recent years focused on categorical hair color definitions, including, but very limited, hair graying (Adhikari et al. 2016), to unearth associated genes but with increased power. The mechanism of action (Commo et al. 2012) in hair graying or whitening is quite different from other hair color changes that occur from a child to adult in certain individuals. Research over the last few years into the genes associated with hair graying have shown that it is linked with the self-maintenance of melanocyte stem cells, and graying is accelerated by bcl-2 deficiency performed on mouse models (Nishimura et al. 2005), but many more GWAS on hair color loss should be performed. Only a handful of studies have explored quantitative hair color genetic association (Candille et al. 2012) with little novelty, as they merely substantiated categorical hits. This is certainly an area that could be further explored with adequate cohort sizes and reflectance or color space (CIELAB, HSV, RGB) phenotypes.

Although most GWAS on hair color were done in people of European or partial European descent, which makes sense as hair color variation is largely a European phenomenon, in contrast to eye color this is not completely so. Some Pacific Islanders from Papua New Guinea, Vanuatu, and the Solomon Islands have blond hair together with the same dark eyes and dark skin as their black-haired population group. A small-scale hair color GWAS in Solomon Islanders that included blond-haired individuals highlighted a missense mutation in *TYRP1*, a gene also involved in blond hair in Europeans, as a single source for blond hair in the Solomon Islands (and expectedly in the neighboring Pacific region), which does not exist outside Oceania (Kenny et al. 2012).

17.2.3 Skin Color Genetics

Similar to the deposit of melanin on hair shafts, pigment-containing melanosome granules affect skin coloration by their transfer to surrounding keratinocytes in the epidermis.

The melanocytes located in the basal layer of the epidermis transfer these melanosomes to adjacent keratinocytes through dendritic structures, and eventually the keratinocytes migrate to reach the upper layers of the epidermis (Rees 2003) where we perceive skin color tones (see Figure 17.1).

In contrast to eye and hair color, variation in skin color exists globally, which provides difficulties in finding skin color genes via GWAS, which can only be done in homogeneous populations, thus within continental groups, where skin color variation is smaller than between groups. Therefore, some genes involved in skin color have been found through studies on positive selection (Jablonski and Chaplin 2010; Wilde et al. 2014; de Gruijter et al. 2011) between continental populations concerning UV protection (Rees 2003). Many of the genetic studies on human skin pigmentation have assessed skin color via a categorical approach using a scale such as the Fitzpatrick scale (Fitzpatrick 1988), with some success in GWAS, however, less than the level of eye and hair color associations due to the lower degree of variation in skin color within continental populations. One of the first successful GWAS (Han et al. 2008) revealed associated SNPs within a European population in relation to skin color from the *IRF4* gene. Using a South Asian population, Stokowski et al. (2007) also revealed skin color associations with variants from three genes, *SLC45A2*, *TYR*, and *SLC24A5*. Support from a previous study (Lamason et al. 2005) on African Americans confirmed that skin color-associated variants were located in the *SLC24A5* gene. Studying skin color pigmentation effects between populations, Myles et al. (2007) measured allele frequencies in Europeans, Chinese, and Africans for 24 pigmentation genes from two large-scale SNP datasets. This study revealed that the *DCT* gene was strongly associated with pigmentation control in Asians. The *ASIP* and *OCA2* genes appeared to play important roles in European and Asian patterns, while the *SLC45A2 (MATP)* and *TYR* genes affect were seen only in Europeans (Myles et al. 2007).

Ascertaining skin color genes associated with its quantitative (continuous) color scale using a comprehensive candidate gene study (Jacobs et al. 2013) highlighted new skin color genes *UGT1A* and *BNC2* in determining continuous skin color variation within Europeans. In 2017, GWAS searches of more genetically diverse populations began to reveal additional factors that contributed towards the genetic basis of skin pigmentation. Using Africans from multiple countries, Crawford et al. (2017) recorded light reflectance measures and identified new loci: *MFSD12*, and a region surrounding *DDB1* and *THEM138*, as well as confirming the already established skin color gene *SLC24A5*. Unlike *SLC24A5*, which was more than likely introduced into East Africa by gene flow via non-African populations (Liu et al. 2018), the identification of *MFSD12* by using a more diverse non-European population allowed exploration of the more locally adaptive genes within the region that were associated with skin color as opposed to the broad continental evolutionary trajectories that can stifle lower signals. Derived alleles of the two SNPs with the highest probability within the region, rs56203814˙T and rs10424065˙T, are associated with darker pigmentation; while the ancestral alleles at two other loci, rs6510760˙G and rs112332856˙T, are associated with lighter pigmentation (Crawford et al. 2017), with backing from cell line and mouse model knockdowns at these loci. In 2019, Adhikari et al. focused on exploring a large cohort of Latin American individuals with high native American ancestry from multiple populations. Typical European skin color lightening genes were highlighted in this study—*TYR*, *TYRP1*, *OCA2/HERC2*, *SLC45A2*, *SLC24A5*, and *IRF4*—in addition to a novel region close by *EXM2*, which was also linked with skin color tanning response from other studies (Visconti et al. 2018). Within this population cohort, the known skin color gene *MFSD12,* previously found in an African-associated search, rose as a top contributor toward skin color variation and has also been determined to have undergone significant evolutionary selection in East Asians (Adhikari et al. 2019). Visconti et al. (2018) also explored tanning response in a GWAS based on almost 177,000 Europeans, unearthing 20 genes, most of them were novel. They include *SLC45A2* (rs16891982), *IRF4* (rs12203592), *TYR* (rs1126809), *OCA2/HERC2* (rs12913832), *MC1R* (rs369230), *RALY/ASIP* (rs6059655), *PDE4B* (rs1308048), *RIPK5* (rs12078075), *PA2G4P4* (rs9818780), *PPARGC1B* (rs251464), *AHR/AGR3* (rs117132860), *TRPS1* (rs2737212), *TYRP1* (rs1326797), *BNC2* (rs10810650), *EMX2* (rs35563099), *TPCN2* (rs72917317), *DCT* (rs9561570), *ATP11A* (rs1046793), *SLC24A4* (rs746586), and *KIAA0930* (rs11703668).

17.3 Developing Genetic Prediction Systems for Eye, Hair, and Skin Color

For the practical aspect of predicting phenotypes from genotypes, two elements are required: (1) a laboratory assay or DNA test that generates the genetic data for the selected DNA predictors from a DNA sample and (2) a statistical tool that generates trait probabilities based on training data (both genetic and phenotypic) that form the basis of the underlying statistical prediction models.

For consideration of the laboratory tool design, DNA tests for SNP analysis from low-quality and low-quantity DNA samples typically confronted within a forensic setting, high-throughput SNP genotyping tools such as SNP microarrays used in GWAS typically do not work. What matters most are technologies that allow high sensitivity while using low amounts of DNA in a multiplex fashion to cater for as many SNPs as possible in parallel. Most previously developed noncommercial DNA tests for appearance prediction were based on the principle of PCR followed by single base extension (SBE) using commercial SNaPshot chemistry (Thermo Fisher), which is highly sensitive, allowing successful DNA analysis on a picogram level, is comparable with commercial STR kits, and facilitates multiplex designs of up to two dozen SNPs in a single reaction. With the growing number of predictive DNA markers in recent years, there is a mounting need for having sensitive SNP genotyping technologies with larger multiplexing capacity. Currently, targeted MPS is the method of choice for multiplexing hundreds of SNPs with high sensitivity and is starting to be used for FDP purposes (Breslin et al. 2019; Xavier et al. 2020; Palencia-Madrid et al. 2020).

To develop an optimal statistical model for predicting a phenotype from genotype data, it is first advised to gather all known DNA variants associated with the trait in question, this allows an exhaustive exploration of the contribution of these variants toward prediction during the model development stage. Notably, the dataset used for marker discovery needs to be independent from the datasets used for prediction modeling. Typically, a systematic study then explores these markers in a rank-based fashion by investigating their contribution to the trait. This can be done simplistically via a linear categorical approach, i.e., 1, 2, 3, or by quantitively measuring the phenotype and assessing the correlation of the variant with the trait. Additional measures that allow the exploration of independence, while also adjusting for the biggest contributors, as well as exploring epistatic interactions and effects that may be present within the candidate list can also be generated. The most important element here is filtering of valid contributors toward prediction, as noise from the use of ineffective contributors can decrease the accuracy of the model's overall performance.

Once an appropriate set of predictive DNA markers has been selected, the choice of statistical model building approach must be determined. This choice is highly reflective of the definition of the phenotype, whether it is defined as categorical or continuous, and through the years, various model building approaches have been tried and tested to optimally predict appearance traits. Using categorical eye color as an example, methods explored include ordinal regression, multinomial logistic regression, and classification trees (Liu et al. 2009) to Bayesian naïve classifier (Ruiz et al. 2013), support vector machines, random forest, and artificial neural networks (Katsara et al. 2021; Palmal et al. 2021). Importantly, the dataset used for prediction model building needs to be independent from those previously used for discovery of the DNA predictors used in the model, and at best also independent from the dataset used for validating the prediction model.

Next, the development of any statistical predictive model that is based on an internal reference dataset of genotypes and phenotypes must be assessed by statistical validation of the model, yielding parameters of test accuracy that are comparable between model systems using an independent model testing or model validation dataset. Overall test accuracy reflects the reliability of the DNA test and can be expressed using several different parameters: AUC (area under the receiver operating curve; where 1 is perfect prediction and 0.5 is random chance), sensitivity, specificity, positive predictive value (PPV), and negative predictive value (NPV). Caliebe et al. (2017, 2018) proposed the use of PPV and NPV more specifically in DNA phenotyping accuracy estimates, as they would be the same in every population, irrespective of the prevalence of the phenotype in question, that is if all causal factors were accounted for by the prediction model, which is not always the case. PPV can be seen as the percentage of correctly predicted color type among the predicted positives and NPV as the percentage of correctly predicted non-color type among the predicted negatives (Liu et al. 2019). In case data restrictions do not allow independent datasets for model building and model validation, the same dataset may be used with a leave-one-out cross-validation (LOOCV) approach. In addition, it is advised to perform additional tests on independent datasets not involved in model training and testing to truly assess the precision of any prediction test. With the vast multitude of new variants being found through association and candidate studies, and improved phenotype measures and classification, these systematic quality control assessments and model building methodologies will only increase in performance accuracy, thus providing even more efficient ways of predicting phenotypes, especially in the age of artificial intelligence and ease of using deep learning approaches (van Hilten et al. 2021). At present, there are several categorical prediction systems available and used worldwide for the prediction of eye, hair, and skin color that were built using various approaches and are discussed next. More detailed information on maintained prediction systems that have web tool availability, such as the variants used and overlap, can be found as an overview in Table 17.1.

17.3.1 Genetic Prediction Systems for Eye Color

In 2008, the development of an eye color prediction model showed potential due to the discovery of a particular variant: rs12913832 (*HERC2*) with a large phenotypic effect (Kayser et al. 2008; Eiberg et al. 2008). Therefore, in 2009, a systematic study was performed investigating the predictive effect of 37 eye-color-associated SNPs from eight genes in 6168 Dutch

Table 17.1 Overview of Current and Maintained Prediction Systems (Webtools) for Eye, Hair and Skin Color from DNA

System	Number	Variants	Genes	Model Type	Training Dataset Size (*N*)	Training Data Major Populations
IrisPlex [SWAGDAM 2017][Xavie et al. 2020]	6 SNPs	rs12913832 rs1800407	HERC2 OCA2	Categorical MLR prediction 3-category eye color system:	9466	Europe, North and South America, Africa, Asia
		rs12896399 rs16891982 rs1393350 rs12203592	LOC105370627 SLC45A2 TYR IRF4	blue, intermediate, brown Prediction probabilities https://hirisplex.erasmusmc.nl/		
All overlap with snipper						
Snipper [Praetorius et al. 2013]	7 SNPs	All of the above plus rs1129038	HERC2	Categorical Bayesian Naïve classifier 3-category eye color system:	256	Europe
	13 SNPs	All of the above plus rs11636232 rs1667394 rs7183877	HERC2 HERC2 HERC2	blue, green/hazel, brown Likelihood ratio (LR) or Prediction probabilities http://mathgene.usc.es/snipper/eyeclassifier.html		

(Continued)

Table 17.1 (Continued) Overview of Current and Maintained Prediction Systems (Webtools) for Eye, Hair and Skin Color from DNA

System	Number	Variants	Genes	Model Type	Training Dataset Size (N)	Training Data Major Populations
		rs4778232	OCA2			
		rs4778241	OCA2			
		rs8024968	OCA2			
	23 SNPs	All of the above plus				
		rs1015362	ASIP			
		rs6058017	ASIP			
		rs12592730	HERC2			
		rs916977	HERC2			
		rs1375164	OCA2			
		rs7495174	OCA2			
		rs4778138	OCA2			
		rs1408799	TYRP1			
		rs683	TYRP1			
		rs26722	SLC45A2			
HIrisPlex [Voisey et al. 2006][Xavie et al. 2020]	21 SNPs 1 INDEL	rs312262906 rs11547464 rs885479 rs1805008 rs1805005	MC1R MC1R MC1R MC1R MC1R	Categorical MLR prediction 4-category hair color system: brown, blond, red, black Prediction probabilities https://hirisplex.erasmusmc.nl/	1878	Europe, North and South America, Africa, Asia

(Continued)

Table 17.1 (Continued) Overview of Current and Maintained Prediction Systems (Webtools) for Eye, Hair and Skin Color from DNA

System	Number	Variants	Genes	Model Type	Training Dataset Size (N)	Training Data Major Populations
		rs1805006	MC1R			
		rs1805007	MC1R			
		rs201326893	MC1R			
		rs2228479	MC1R			
		rs1110400	MC1R			
		rs1805009	TUBB3			
		rs28777	SLC45A2			
		rs16891982	SLC45A2			
		rs12821256	KITLG			
		rs4959270	LOC105374875			
		rs12203592	IRF4			
		rs1042602	TYR			
		rs683	TYRP1			
		rs1800407	OCA2			
		rs2402130	SLC24A4			
		rs12913832	HERC2			
		rs2378249	PIGU			
7 overlap with snipper						
Snipper [Walsh et al. 2011]	12 SNPs	rs1129038	HERC2	Categorical Bayesian Naïve classifier	230	Europe
		rs12913832	HERC2	4-category hair color system:		
		rs11547464	MC1R	brown, blond, red, black		
		rs1805006	MC1R	Likelihood ratio (LR) or		
		rs1805007	MC1R	Prediction probabilities		
		rs1805008	MC1R	http://mathgene.usc.es/snipper/hairclassifier.html		

(Continued)

Table 17.1 (Continued) Overview of Current and Maintained Prediction Systems (Webtools) for Eye, Hair and Skin Color from DNA

System	Number	Variants	Genes	Model Type	Training Dataset Size (N)	Training Data Major Populations
		rs1805009	MC1R			
		rs12931267	FANCA			
		rs28777	SLC45A2			
		rs35264875	TPCN2			
		rs4778138	OCA2			
		rs7495174	OCA2			
HIrisPlex-S [Wollstein et al. 2017][Xavie et al. 2020]	36 SNPs	rs1426654 rs12203592	SLC24A5 IRF4	Categorical Bayesian Naïve classifier 5-category skin color system:	1423	Europe, North and South America, Africa, Asia
4 overlap with snipper		rs1805007	MC1R	very pale, pale, intermediate,		
		rs1805008	MC1R	dark, dark to black		
		rs11547464	MC1R	Prediction probabilities		
		rs885479	MC1R	https://hirisplex.erasmusmc.nl/		
		rs228479	MC1R			

(Continued)

Table 17.1 (Continued) Overview of Current and Maintained Prediction Systems (Webtools) for Eye, Hair and Skin Color from DNA

System	Number	Variants	Genes	Model Type	Training Dataset Size (N)	Training Data Major Populations
		rs1805006	MC1R			
		rs1110400	MC1R			
		rs1126809	MC1R			
		rs3212355	MC1R			
		rs1800414	OCA2			
		rs1800407	OCA2			
		rs12441727	OCA2			
		rs1470608	OCA2			
		rs1545397	OCA2			
		rs16891982	SLC45A2			
		rs28777	SLC45A2			
		rs1667394	HERC2			
		rs2238289	HERC2			
		rs1129038	HERC2			
		rs12913832	HERC2			
		rs6497292	HERC2			
		rs1042602	TYR			
		rs1393350	TYR			
		rs6059655	RALY			
		rs8051733	DEF8			
		rs2378249	PIGU			
		rs6119471	ASIP			
		rs2402130	SLC24A4			

(Continued)

Table 17.1 (Continued)   Overview of Current and Maintained Prediction Systems (Webtools) for Eye, Hair and Skin Color from DNA

System	Number	Variants	Genes	Model Type	Training Dataset Size (N)	Training Data Major Populations
		rs17128291	SLC24A4			
		rs12896399	SLC24A4			
		rs683	TYRP1			
		rs12821256	KITLG			
		rs3114908	ANKRD11			
		rs10756819	BNC2			
Snipper	10 SNPs	rs10777129	KITLG	Categorical Bayesian Naïve classifier	118	Europe, America, Africa, Asia
		rs13289	SLC45A2	3-category system:		
[Wilde et al. 2014]		rs1408799	TYRP1	white, intermediate, black		
		rs1426654	SLC24A5	Likelihood ratio (LR) or		
		rs1448484	OCA2	prediction probabilities		
		rs16891982	SLC45A2	http://mathgene.usc.es/snipper/skinclassifier.html		
		rs2402130	SLC24A4			
		rs3829241	TPCN2			
		rs6058017	ASIP			
		rs6119471	ASIP			

Europeans (Liu et al. 2009). This study demonstrated that using 15 SNPs from eight genes, blue and brown eye colors can be predicted with >90% prevalence-adjusted population-based accuracy. This accuracy, as expressed by the AUC, found that most eye color information is provided by a subset of just six SNPs from six genes (Liu et al. 2009): rs12913832 (*HERC2*), rs1800407 (*OCA2*), rs12896399 (*SLC24A4*), rs16891982 (*SLC45A2* [*MATP*]), rs1393350 (*TYR*), and rs12203592 (*IRF4*). Termed IrisPlex, a predictive system, including both a DNA test and a statistical prediction model, was developed that included both a six-variant genetic test for data generation and a mathematical algorithm for predicting the three categories of eye color: blue, brown, and intermediate (Walsh, Lindenbergh, Zuniga et al. 2011). The prediction model built using a multinomial logistic regression (MLR) approach now consists of >9000 individuals from multiple countries to train the model coefficients.

Performance metrics from cross-validation at present can be found on the IrisPlex site web tool located at https://hirisplex.erasmusmc.nl/. In short, the AUC, PPV, NPV of the three categories are, for blue, 0.94, 0.90, 0.90; brown 0.95, 0.77, 0.96; and intermediate 0.74, 0.09, 0.96, respectively. While highly accurate for the prediction of blue and brown eye color, this system is not so accurate for the prediction of non-blue and non-brown eye color, lumped into the intermediate category. Notably, this intermediate category is not only green, which is very rare, but includes eyes that have both blue and brown components, and are therefore difficult to predict from DNA with current knowledge, not only using the IrisPlex system. The IrisPlex assay that was designed to work in conjunction with this prediction model and completes the IrisPlex system consists of a six-SNP multiplex design (Walsh, Lindenbergh, Zuniga et al. 2011). It is a developmentally validated robust assay that can produce complete SNP profiles with only 31 pg of DNA, approximately six human diploid cell equivalents (Walsh, Lindenbergh, Zuniga et al. 2011), and has been rigorously tested according to the guidelines suggested by the Scientific Working Group on DNA Analysis Methods (SWGDAM 2017).

Since publication of the IrisPlex system in 2011, there have been discussions on ways to improve the predictability of eye color from DNA using alternative DNA markers and alternative prediction methods, especially concerning the intermediate/green eye color category, in addition to population bias during model development. Several groups have challenged its effectiveness in this type of modeling approach used and its prediction accuracy, especially with regard to its accuracy in admixed populations (Dembinski and Picard 2014; Yun et al. 2014). There have been several different statistical approaches proposed and published over the last few years for the prediction of phenotypic traits from genetic data. Mengel-From et al. (2009) took a likelihood ratio (LR) approach to prediction. Using four SNPs—rs12913832 (*HERC2*), rs1800407 (*OCA2*), rs1129038 (*HERC2*), and rs11636232 (*HERC2*)—they reach an LR of 29.3 in discriminating between light and dark eye color in Europeans. Spichenok et al. (2011) used a scenario-based exclusion approach with six SNPs for eye color prediction: rs12913832 (*HERC2*), rs16891982 (*SLC45A2*), rs12203592 (*IRF4*), rs1545397 (*OCA2*), rs6119471 (*ASIP*), and rs885479 (*MC1R*). After genotyping the most highly associated eye color SNP—rs12913832—to indicate either a proposed blue or non-blue eye color, there is a stepwise inclusion of SNP data from the rest of the set to come to a final conclusion. However, there is no measurement of its accuracy or explanation of variance other than the list of samples the scenario-based approach was performed on (Spichenok et al. 2011; Pneuman et al. 2012; Hart et al. 2013). Valenzuela et al. (2010) found that a combination of three SNPs—rs12913832 (*HERC2*), rs16891982 (*SLC45A2*), and rs1426654 (*SLC24A5*)—were able to explain 76% of eye color variation.

Pospiech et al. (2014) proposed a Bayesian network model approach using the six highest-ranked SNPs from six genes in the Liu et al. (2009) study. This LR-type approach uses two models: the first model assumes eye color categories of blue, brown, green, and hazel; and the second, light and dark, although they could not improve Liu et al.'s (2009) results using this method. Following on with Bayesian approaches, Ruiz et al. (2013) proposed a similar Bayesian likelihood-based classification approach using their snipper tool. Overall, the aim of the study was to show that this LR-based approach including seven additional SNPs improved eye color prediction, particularly in the intermediate category. Their analysis on the six core SNPs from Liu et al. (2009) did not show an improvement in prediction, and only a slight increase is seen with the addition of the other SNPs from the Mengel-From et al. study (2010). However, the largest increase in prediction was observed by the addition of several other SNPs, with rs1129038, rs1667394, and rs7183877 from *HERC2* displaying the largest improvement in prediction. In fact, Eriksson et al. (2010) previously identified both rs1667394 and rs7183877 as being independently significant for eye color prediction of intermediates. At present, the snipper suites eye color classifier currently uses a 7-set, 13-set, and 23-set marker model for eye color prediction that is built using 256 individuals from European populations. Performance metrics using their 7-set classifier were recorded from Ruiz et al. (2013) on this small set and are reported as PPV, NPV for blue 0.80, 0.99; green/hazel 0.55, 0.93; and brown 0.51, 0.96. However, definitions of intermediate versus green/hazel mean there could be phenotype classification issues. A real check on the performance must be in an assessment of the same individuals using all prediction systems available to truly get an accurate comparison. This, in addition to the fact that these systems contain both known causal as well as non-causal DNA markers, may be affected by population frequencies in model building; therefore, it is advised to train models using a more global population to circumvent this issue.

The limitations of the current DNA prediction systems for eye color lie in the inadequate prediction of non-blue and non-brown eyes that are classified into the intermediate category, and these problems exist for all populations. Obviously, this limitation is larger in populations with increased frequencies of these intermediate eye colors, which most notably are populations of European–non-European mixed ancestry. As this inadequacy in prediction occurs using all currently developed iris color prediction models, future research needs to show if this problem can be overcome by including many more DNA markers and/ or moving from a categorical to a quantitative phenotype definition to correctly classify these intermediate individuals; technologically, this would require a move toward targeted MPS for genotyping effectiveness using small sample input. In practice and until then, however, current tools used to predict these non-blue and non-brown eyes are recognized by the tool in some fashion, albeit typically not with a high probability of the intermediate category, but instead with similar frequencies for both the blue and the brown category, which can be used to suggest the presence of intermediate eye color.

At present when predicting eye color on a categorical level, given that there are no valid continuous models available that significantly surpass the accuracies currently seen in categorical eye color prediction, the IrisPlex system using probability estimates and the snipper suite eye color classifier using likelihood ratio estimates are best used in tandem, with more specific reporting strategies following their published suggestions. However, researchers are seeing the validity of moving from categorical eye color prediction toward quantitative eye color prediction with the production of new phenotyping methods (Andersen et al. 2013; Wollstein et al. 2017) that can be utilized in developing quantitative prediction models. This could avoid minor problems involved with user interpretation of the prediction output in addition to more accurately predicting colors of admixed populations where there are combinations of European and non-European alleles that drive the final color.

17.3.2 Genetic Prediction Systems for Hair Color

With regard to head hair color prediction from DNA, preliminary attempts made this trait one of the first externally visible traits to be predicted from DNA, albeit at least one of the categories of this trait. An early red hair prediction protocol based on a combination of nonsynonymous variants in the *MC1R* gene that incur the red hair phenotype was developed for forensic use over 20 years ago (Grimes et al. 2001) with an accuracy of 84% in the prediction of red hair individuals. Also while assessing potential candidates for pigmentation traits, Sulem et al. (2007) developed a hair color prediction tool that was capable of excluding red and either blond or brown hair color in its prediction for many of their individuals. In 2010, Valenzuela et al. performed an exhaustive assessment of 75 SNPs from 24 genes previously implicated in hair, skin, and eye color in samples of various biogeographic origins (Europe and elsewhere) and found that using three of them, i.e., rs12913832 (*HERC2*), rs16891982 (*SLC45A2*), and rs1426654 (*SLC24A5*), combined gave the best prediction for light and dark hair color. Around this time, Branicki et al. (2009) also examined the *HERC2-OCA2* region, not only confirming the association of *HERC2* rs12913832 with eye color but also found that variants within this gene provided the best prediction for light and dark hair color as it explained 76% of total hair melanin using a multilinear regression modeling approach. The evaluation by Branicki et al. in 2011 of 46 known hair color-associated SNPs from 13 genes for model-based population-wise (Polish) hair color prediction was influential in understanding the levels at which hair color prediction could be achieved. The aim of the study was to rank the SNPs according to the most useful for hair color prediction, and, in turn, they found that 13 DNA markers, 2 *MC1R* compound markers, and 11 single DNA markers from 11 genes contained the most predictive information for hair color. These genes include *MC1R, HERC2, OCA2, SLC45A2 (MATP), KITLG, EXOC2, TYR, SLC24A4, IRF4, PIGU/ASIP,* and *TYRP1*.

From this information, and in combination with eye color prediction and the IrisPlex system, the HIrisPlex system was developed (Walsh et al. 2013), which includes a single multiplex genotyping tool for 24 variants, including the six IrisPlex SNPs, and two prediction models (i.e., the IrisPlex model for eye color prediction and the HIrisPlex model for hair color prediction). The HIrisPlex assay is robust and sensitive, as it was designed to cope with low amounts of DNA including highly degraded DNA, with sensitivity testing assessments providing full DNA profiles down to 63 pg DNA input (Walsh et al. 2013). The assay has also been assessed according to SWGDAM guidelines (SWGDAM 2017). These guidelines include species testing, simulated casework that satisfies 88% of samples tested, in addition to concordance testing performed by five independent forensic laboratories, which display consistent reproducible results on varying sample types (Walsh et al. 2014).

The HIrisPlex system as a whole is currently capable of predicting both eye and hair color on a categorical level with 3 out of 4 individuals correctly predicted for both traits on a test set of 119 individuals (Walsh et al. 2014). To cater for large numbers and even incomplete profiles (23 DNA variants and less) often found in casework and aDNA samples, the HIrisPlex prediction tool is also available online at https://hirisplex.erasmusmc.nl/

and is one of the most reliable eye and hair color prediction system currently available. Performance metrics at present are based on a set of 1878 individuals from global populations. In short, the AUC, PPV, NPV for the four categories are blond 0.8, 0.63, 0.79; brown 0.72, 0.57, 0.72; red 0.92, 0.73, 0.97; and black 0.83, 0.7, 0.91. However, it is important to note is that these measures are based upon the highest probability representing the final prediction, which is not the case in practice for hair color prediction, as it is recommended to use prediction guides according to the HIrisPlex publications.

One aspect to mention in terms of the performance of the HIrisPlex system is the inability to predict blond hair to a high degree of accuracy in comparison to the other category levels, brown, black, and red (Walsh et al. 2013). This is due to age-related changes that occur in a low percentage of individuals. This phenomenon leads to blond prediction results as the individuals are born with blond hair, however, the color changes to brown, even black during adolescence (ages 12–13), and sometimes earlier in age (ages 2–3). For these individuals, an incorrect blond prediction is given using the HIrisPlex system that is based on all current knowledge of hair-color-associated and predictive SNPs/genes. Therefore this shows that this area of prediction needs improvement and where more research is needed to understand the mechanisms/processes involved in this phenomenon beyond what has already been performed (Kukla-Bartoszek et al. 2018). Notably, the previously mentioned blond hair mutation in *TYRP1* found in Pacific Islanders is not part of the HIrisPlex system, due to its restricted geographic distribution in the Western Pacific.

Again, utilizing a Bayesian approach, in 2015, the team at the University of Santiago de Compostela proposed a hair color likelihood-based classification approach using their modified snipper tool. Söchtig et al. (2015) developed the LR-based approach that included 12 pigmentation variants from literature (8 that overlap with HIrisPlex) using 230 individuals from European populations. In short, Snipper uses a naïve Bayes classification system for single or multiple SNP profiles by estimating membership likelihood to one of several populations (phenotypes) defined by their allele frequencies, which are estimated from a predefined training set. After the likelihoods are ranked, Snipper assigns the profile to a population from the ratio of the two largest likelihoods (for more details, see http://mathgene.usc.es/snipper/hairclassifier.html). Performance metrics such as PPV and NPV are not available for this set, however, AUC performance has been ascertained as 0.86, 0.64, 0.94, and 0.84 for blond, brown, red, and black, respectively, using this small subset.

In 2020, Pospiech et al. (2020) explored the prediction of hair graying using 10 variants from 10 genes found using whole-exome and targeted NGS data to develop a binary model of graying versus no graying that achieved a prediction accuracy estimate of AUC=0.873. Further exploration of its incorporation into hair color prediction tools would be beneficial. As iterated for eye color prediction performance, a similar dataset assessment should be performed and models evaluated, however, given the different approaches used (i.e., generation of prediction probability and Likelihood ratio), it may be valuable to generate predictions using both tools for a thorough assessment of the unknown. Until we move closer to a quantitative prediction tool with more variants included that can catch the more subtle shade differences in hair, this would currently be the best course of action for hair color prediction.

17.3.3 Genetic Prediction Systems for Skin Color

Of the three traits that make up our pigmentation profile, skin color prediction was a more difficult trait to develop prediction models due to the variation that exists between different populations and not within a population to a high degree, in addition to the difficulty of using current GWAS approaches to find these variants. Due to the convergent evolution of this trait, there are multiple genetic pathways involved in the evolution of this trait, therefore there needs to be an overlap in all of the different genes that have evolved within multiple populations to cater for skin color prediction on a global scale. Early prediction models focused on European skin color variants that evolved from an African ancestral state only. Valenzuela et al. (2010) obtained skin reflectance measures with an R2 value of 45.7% simply using three SNPs, while Spichenok et al. (2011) utilized seven variants for predicting non-white, not dark, and not-white/not-dark skin color, although with such limitations it meant they were unable to produce prediction outcomes in 28% of their samples (Spichenok et al. 2011; Hart et al. 2013). Liu et al. (2015) also focused on European-derived skin color variants and their use in prediction using nine variants; however, these variants did not prove as useful in Asians, Native Americans, and sub-Saharan Africans.

Combining skin color variants from literature, Maroñas et al. (2014) published the first extensive study on skin color prediction using 59 candidate pigmentation variants and three skin color categories: white, intermediate, and black. This study identified a subset of ten markers from eight genes that were most predictive for skin color variation in their small 118 individual subset, and achieved AUC values of .999 for white, 0.996 for black, and 0.803 for intermediate skin color (Maroñas et al. 2014). Aiming to improve on this knowledge using a more global set of individuals (N = 2025) for model training and testing, Walsh et al. (2017) assessed 77 pigmentation variants from 37 genetic loci and systematically ranked their prediction contribution towards a dermatological skin color scale called the Fitzpatrick scale (Fitzpatrick 1988). From this information, and in combination with eye and hair color prediction, the HIrisPlex-S system was developed (Walsh et al. 2013). The HIrisPlex-S system consists of two multiplex genotyping assays to generate genotype information on 41 DNA variants, the previous HIrisPlex assay and a second assay of 17 SNPs, together with three prediction models—the IrisPlex model for eye color, the HIrisPlex model for hair color, and the HIrisPlex-S model for skin color. In forensic developmental validation testing, the novel 17-plex assay part of HIrisPlex-S performed in full agreement with SWGDAM guidelines (Chaitanya et al. 2018), as previously shown for the 24-plex assay. Sensitivity testing also revealed complete SNP profiles from as little as 63 pg of input DNA, in addition to successful typing of simulated forensic casework samples such as blood, semen, saliva stains, inhibited DNA samples, low-quantity touch (trace) DNA samples, and artificially degraded DNA samples, equaling the previously demonstrated performance of the 24-plex HIrisPlex assay.

The HIrisPlex-S model skin color prediction was designed for five skin color categories using 36 SNPs from 16 genes. In a direct comparison of the Maroñas et al. (2014) model and the HIrisPlex-S model, 194 individuals from 17 different populations not involved in prediction model training or assessment were predicted revealing improvement to overall skin color prediction using this new HIrisPlex-S model (Walsh et al. 2017). HIrisPlex-S performance metrics at present are based on a set of 1878 individuals from global populations with the five categories producing values for AUC, PPV, NPV at very pale 0.74, 0.4, 0.94; pale 0.72, 0.6, 0.94; intermediate 0.73, 0.6, 0.73; dark 0.88, 0.34, 0.98; and dark to black 0.96, 0.81, 0.99. Some additional prediction attempts have been made since for singular populations. A Polish study examined categorical skin color and skin tanning ability prediction using random forest and other machine learning approaches (Zaorska et al. 2019), albeit not achieving significant improvements to skin color prediction in the cohort.

More recently a dataset consisting of admixed ancestry from Latin America (Palmal et al. 2021) was used to measure and compare the performance of well-known eye, hair, and skin color prediction tools in this admixed cohort. More specifically it compared the CAN dataset (including 56, 101, and 120 SNPs for eye, hair, and skin color prediction, respectively) to the HIrisPlex-S SNP set (including 6, 22, and 36 SNPs for eye, hair, and skin color prediction, respectively) on both a categorical and a quantitative scale. Although the performance was comparable between both sets of variants in the prediction of this admixed cohort, particularly for eye and hair color, it was noted that skin color prediction by HIrisPlex-S could be improved by expanding the training dataset to capture haplotypes specific to Latin American populations, otherwise, the marker list seemed appropriate to cover categorical predictions of pigmentation traits (Palmal et al. 2021). The success observed by using an admixed population in model design for this CANDELA study (Palmal et al. 2021) is a good example of how moving from categorical to quantitative pigmentation prediction could improve those definitions of intermediate pigment classes that prove the most difficult to predict using categorical systems. Not to mention exploring and including samples of multiethnic populations that can complement models by including skin color genes that cater to their genetic architecture.

Notably, there is a significant overlap in the fields of molecular anthropology and forensic genetics, as the predictive methods that have been developed by the forensic community and highlighted in this chapter are often used in ancient DNA (aDNA) research, from the identification of historical figures to insights on the appearance of our ancestors (Olalde et al. 2014; Keller et al. 2012; Bogdanowicz et al. 2009; King et al. 2014). The power of these tools is also the fact that the models exist independently of the assays, therefore alternative DNA technologies can be used, such as capture sequencing, to generate the genotype information necessary to predict appearance from these aDNA remains.

17.4 Future for Pigmentation Prediction and Forensic DNA Phenotyping

We have already witnessed a change in genotyping technology with regard to forensic DNA phenotyping. In the past, biological assays that generated markers needed for prediction systems such as HIrisPlex-S and snipper suite were designed and several validated with a specific task in mind, which was the generation of genotype profiles of the required markers needed for input into the prediction models. This idea of an all-in-one system that was catered for low-input, low-quality samples passing quality control metrics that would allow their use on forensic casework samples exemplifies translating research into practical applied tools. This has not been eliminated per se; it is still quite valuable when there is limited sample available, but certainly there has been a drive toward massive parallel sequencing over capillary electrophoresis, with the idea that not only biallelic markers would be genotyped but also STR fragments could be sequenced in tandem that would allow much higher resolution for intelligence and identity testing. The ability to generate genotype information on variants beyond the capacity of 25 markers (which was the limitation of standard capillary electrophoresis chemistries such as SNaPshot) have propelled the potential to multiplex traits for their prediction from DNA. Now, for example, markers required for eye, hair, and skin color prediction using the HIrisplex-S system are contained in one system by use of targeted MPS for use on both benchtop sequencers from Illumina

(MiSeq) and Thermo Fisher (Ion Torrent) and have been validated for this purpose (Breslin et al. 2019). Not to mention the strew of commercial assays that have since been produced, which is of great benefit to the community: Illumina's ForenSeq™ DNA Signature Prep Kit (HIrisPlex 24 markers only); Thermo Fisher's Ion Ampliseq DNA Phenotyping Panel (HIrisPlex 24 markers only); and the Ion Ampliseq VISAGE-Basic Tool Research Panel (HIrisPlex-S 41 marker set in addition to BGA markers) (Xavier et al. 2020). The latter two are not full-scale commercial tools, but simply represent the AmpliSeq primer pool, respectively, while all other reagents needed are commercially available. What does this technological development toward the use of targeted MPS mean for prediction model designs; it allows even more DNA predictors to be used in the assays and models, which will provide increased prediction accuracy and greater degrees of differentiation for pigmentation variation. That the increase of DNA predictors leads to an increase in prediction accuracy is seen in the work by Hysi et al. (2018) where they increased the number of hair color predictors used to 258 SNPs and directly compared performance with the original HIrisPlex method of 24. This led to an increase in AUC for all hair color categories except for red, albeit in a minor way, relative to the increase in SNP number. In particular, the tenfold increase of SNP predictors only led to an increase in AUC between 0.04 and 0.12 depending on the category and dataset. This exemplifies the fact that the inclusion of more predictive markers will lead to improved prediction, but the multiplex capacity of the assay technology must allow this. For a robust assay based on current MPS technology, multiplexing of hundreds of SNPs is possible, but not thousands. With >250 SNPs simply for hair color alone, we must also be aware that current MPS technology may still not have enough room to include many other appearance traits for which predictive markers are already available in the development of an all-in-one genotyping assay.

Moreover, it is finally time to move toward quantitative prediction of pigmentation traits using color space classifications such as RBG (red, blue, green), HSV (hue, saturation, value), or melanin index (MI) with deep learning or other artificial intelligence approaches. This avenue of development would also require many more numbers of SNPs in comparison to categorical pigmentation prediction needs. Also, epistatic interactions, in particular, have been proposed in both categorical (Ruiz et al. 2013; Pośpiech et al. 2014) and quantitative (Liu et al. 2009) approaches to improve prediction. A further look at these contributions and their inclusion into pigmentation prediction models for green/intermediate eye color and age-related hair color changes and global skin color variation may lead to fruitful results when combined with new independent SNP effects.

In terms of forensic DNA phenotyping as a whole, the ability to combine multiple appearance traits into one MPS-based assay with multiple separate prediction models is the future. As we know so much about pigmentation, it is a clear leader in the amount of fundamental genetics knowledge required to build a model and successfully predict a trait. Other traits are improving in our genetic understanding of their development, with several more recently generating their first prediction models. Additional pigmentation traits that would allow at least a partial marker overlap with eye, hair, and skin color are eyebrow color (Peng et al. 2019) and freckling (Kukla-Bartoszek et al. 2019; Hernando et al. 2018) prediction. Male-pattern baldness research over the last ten years has led to the development of models to predict early-onset cases with a 14-SNP model that shows accuracy levels in the AUC 0.74 range (Liu et al. 2016), while another predictor developed by Hagenaars et al. (2017) can discriminate individuals with no signs of hair loss from those with severe baldness with similar degrees of accuracy. Along the lines of hair prediction, but regarding shape, significant progress has been made over the last few years in understanding the genetic contributors of curly versus straight hair (Medland et al. 2009). In 2018, a large study was conducted using approximately 12,000 samples of varying ethnicities to explore and design an effective binomial logistic prediction model capable of predicting straight versus non-straight hair using 32 informative variants from 26 loci (Liu et al. 2018). A range in AUC performance accuracy was obtained: from 0.66 in Europeans to 0.79 in non-Europeans. Height, a trait often recorded by medical professionals, thus leading to the availability of thousands of phenotypic collections and large GWAS, alas is a much more difficult trait to predict than others due to its high degree of environmental influences and suspected heterogeneity effects. Over the years, studies have explored the use of 54 DNA markers to predict extreme height (Aulchenko et al. 2009) with AUCs reaching 0.65, to minimal improvement of up to 0.75 with 180 markers (Liu et al. 2014) and 0.79 AUC (Liu et al. 2019) but at the cost of generating data on at least 697 variants (Wood et al. 2014) to capture the tall extreme phenotype. For other identifying traits such as facial structure, we are still in the very early days of exploration of the genes that develop this two-tier (bone and tissue) structure. The very few prediction studies that have been performed (Claes et al. 2014; Lippert et al. 2017; Qiao et al. 2018; Sero et al. 2019) only emphasize the fact that we have still a lot to learn before we can actually take into account all of the environmental, not to mention biological effects that combine to give us our visual identity.

Hence, for the near and likely distant future, further research in appearance genetics combined with expected advances in parallel SNP genotyping technologies will likely allow FDP to describe individuals on a finer and more detailed scale. This will allow the reduction of large numbers of unknown suspects, which are typical in cases without leads, and where FDP is applied to do its job to help find unknown perpetrators through visual descriptions. However, concluding a complete and individual-specific depiction using FDP, thus making subsequent gold-standard forensic DNA profiling and forensic DNA databases obsolete, would require individualized facial prediction (among these other traits) from crime scene DNA, and that is still too far-fetched for the foreseeable future. Yes, we know from the very close resemblance of monozygotic twins that appearance in its totality is (besides identical twins) individual-specific, unveiling the underlying genetic architecture of the face in combination with individual-specific prediction for all the other traits mentioned in this chapter is still not complete. Therefore, the very idea of individualized appearance prediction from DNA still has many years of fundamental research and application development to work through, especially given the rather limited research funding available at present. With increased support, however, perhaps that day could be sooner than we think given how it could revolutionize the very meaning of DNA profiling.

Acknowledgments

We would like to thank all our colleagues in the human genetics and forensic genetics fields who contributed their hard work, which has led us to where we are today in our genetic understanding of human appearance and its predictability from DNA, including forensic DNA. In addition, a huge thank you to all participants; without your contribution to research, these developments would not have been possible. Last, a thank you to all our funding agencies past and present; your continued support allows us to perform this research.

References

Adhikari, K., et al. 2016. A genome-wide association scan implicates DCHS2, RUNX2, GLI3, PAX1, and EDAR in human facial variation. *Nat Commun* 7(1):11616.

Adhikari, K., et al. 2019. A GWAS in Latin Americans highlights the convergent evolution of lighter skin pigmentation in Eurasia. *Nat Commun* 10(1):358.

Andersen, J. D., et al. 2013. Genetic analyses of the human eye colours using a novel objective method for eye colour classification. *Forensic Sci Int Genet* 7(5):508–15.

Barsh, G. S. 2003. What controls variation in human skin color? *PLOS Biol* 1(1):E27.

Bastiaens, M. T., et al. 2001. Melanocortin-1 receptor gene variants determine the risk of nonmelanoma skin cancer independently of fair skin and red hair. *Am J Hum Genet* 68(4):884–94.

Biswas, S. and J. M. Akey. 2006. Genomic insights into positive selection. *Trends Genet* 22(8): 437–46.

Bonilla, C., et al. 2005. The 8818G allele of the agouti signaling protein (ASIP) gene is ancestral and is associated with darker skin color in African Americans. *Hum Genet* 116(5):402–6.

Box, N. F., J. R. Wyeth, L. E. O'Gorman, N. G. Martin and R. A. Sturm. 1997. Characterization of melanocyte stimulating hormone receptor variant alleles in twins with red hair. *Hum Mol Genet* 6(11):1891–7.

Branicki, W., U. Brudnik, J. Draus-Barini, T. Kupiec and A. Wojas-Pelc. 2008. Association of the SLC45A2 gene with physiological human hair colour variation. *J Hum Genet* 53(11–12):966–71.

Branicki, W., U. Brudnik and A. Wojas-Pelc. 2009. Interactions between HERC2, OCA2 and MC1R may influence human pigmentation phenotype. *Ann Hum Genet* 73(2):160–70.

Branicki, W., et al. 2011. Model-based prediction of human hair color using DNA variants. *Hum Genet* 129(4):443–54.

Breslin, K., et al. 2019. HIrisPlex-S system for eye, hair, and skin color prediction from DNA: Massively parallel sequencing solutions for two common forensically used platforms. *Forensic Sci Int Genet* 43:102152.

Brilliant, M. H. 2001. The mouse p (pink-eyed dilution) and human P genes, oculocutaneous albinism type 2 (OCA2), and melanosomal pH. *Pigment Cell Res* 14(2):86–93.

Caliebe, A., M. Krawczak and M. Kayser. 2018. Predictive values in Forensic DNA phenotyping are not necessarily prevalence-dependent. *Forensic Sci Int Genet* 33:e7–e8.

Caliebe, A., S. Walsh, F. Liu, M. Kayser and M. Krawczak. 2017. Likelihood ratio and posterior odds in forensic genetics: Two sides of the same coin. *Forensic Sci Int Genet* 28:203–10.

Candille, S. I., et al. 2012. Genome-wide association studies of quantitatively measured skin, hair, and eye pigmentation in four European populations. *PLOS ONE* 7(10):e48294.

Chaitanya, L., et al. 2018. The HIrisPlex-S system for eye, hair and skin colour prediction from DNA: Introduction and forensic developmental validation. *Forensic Sci Int Genet* 35:123–135.

Commo, S., K. Wakamatsu, I. Lozano, S. Panhard, G. Loussouarn, B. A. Bernard and S. Ito. 2012. Age-dependent changes in eumelanin composition in hairs of various ethnic origins. *Int J Cosmet Sci* 34(1):102–7.

Crawford, N. G., et al. 2017. Loci associated with skin pigmentation identified in African populations. *Science* 358(6365):eaan8433.

de Gruijter, J. M., et al. 2011. Contrasting signals of positive selection in genes involved in human skin-color variation from tests based on SNP scans and resequencing. *Investig Genet* 2(1):24.

Dembinski, G. M. and C. J. Picard. 2014. Evaluation of the IrisPlex DNA-based eye color prediction assay in a United States population. *Forensic Sci Int Genet* 9:111–7.

Driskell, R. R., C. Clavel, M. Rendl and F. M. Watt. 2011. Hair follicle dermal papilla cells at a glance. *J Cell Sci* 124(8):1179–82.

Duffy, D. L. 2015. *Genetics of Eye Colour*. Chichester: John Wiley & Sons Ltd.

Duffy, D. L., et al. 2007. A three-single-nucleotide polymorphism haplotype in intron 1 of OCA2 explains most human eye-color variation. *Am J Hum Genet* 80(2):241–52.

Eiberg, H., J. Troelsen, M. Nielsen, A. Mikkelsen, J. Mengel-From, K. W. Kjaer and L. Hansen. 2008. Blue eye color in humans may be caused by a perfectly associated founder mutation in a regulatory element located within the HERC2 gene inhibiting OCA2 expression. *Hum Genet* 123(2):177–87.

Eriksson, N., et al. 2010. Web-based, participant-driven studies yield novel genetic associations for common traits. *PLOS Genet* 6(6):e1000993.

Fernandez, L. P., R. L. Milne, G. Pita, J. A. Avilés, P. Lázaro, J. Benítez and G. Ribas. 2008. SLC45A2: A novel malignant melanoma-associated gene. *Hum Mutat* 29(9):1161–7.

Fitzpatrick, T. B. 1988. The validity and practicality of sun-reactive skin types I through VI. *Arch Dermatol* 124(6):869–71.

Flanagan, N., et al. 2000. Pleiotropic effects of the melanocortin 1 receptor (MC1R) gene on human pigmentation. *Hum Mol Genet* 9(17):2531–7.

Frudakis, T., et al. 2003. Sequences associated with human iris pigmentation. *Genetics* 165(4):2071–83.

Graf, J., R. Hodgson and A. van Daal. 2005. Single nucleotide polymorphisms in the MATP gene are associated with normal human pigmentation variation. *Hum Mutat* 25(3):278–84.

Grimes, E. A., P. J. Noake, L. Dixon and A. Urquhart. 2001. Sequence polymorphism in the human melanocortin 1 receptor gene as an indicator of the red hair phenotype. *Forensic Sci Int* 122(2–3):124–9.

Guenther, C. A., B. Tasic, L. Luo, M. A. Bedell and D. M. Kingsley. 2014. A molecular basis for classic blond hair color in Europeans. *Nat Genet* 46(7):748–52.

Han, J., et al. 2008. A genome-wide association study identifies novel alleles associated with hair color and skin pigmentation. *PLOS Genet* 4(5):e1000074.

Harding, R. M., et al. 2000. Evidence for variable selective pressures at MC1R. *Am J Hum Genet* 66(4):1351–61.

Hart, K. L., S. L. Kimura, V. Mushailov, Z. M. Budimlija, M. Prinz and E. Wurmbach. 2013. Improved eye- and skin-color prediction based on 8 SNPs. *Croat Med J* 54(3):248–56.

Hughes, M. O. and C. A. Luce. 2020. Iris, Limbus and Sclera. Artificial eye clinic. Available from: http://wordpress.artificialeyeclinic.com/eye-anatomy/iris-limbus-and-sclera/.

Hysi, P. G., et al. 2018. Genome-wide association meta-analysis of individuals of European ancestry identifies new loci explaining a substantial fraction of hair color variation and heritability. *Nat Genet* 50(5):652–6.

Jablonski, N. G. and G. Chaplin. 2010. Colloquium paper: Human skin pigmentation as an adaptation to UV radiation. *Proc Natl Acad Sci U S A* 107(Supplement 2):8962.

Jackson, I. J. 1997. Homologous pigmentation mutations in human, mouse and other model organisms. *Hum Mol Genet* 6(10):1613–24.

Jacobs, L., et al. 2013. Comprehensive candidate gene study highlights UGT1A and BNC2 as new genes determining continuous skin color variation in Europeans. *Hum Genet* 132(2): 147–58.

Kanetsky, P. A., et al. 2004. Assessment of polymorphic variants in the melanocortin-1 receptor gene with cutaneous pigmentation using an evolutionary approach. *Cancer Epidemiol Biomarkers Prev* 13(5):808–19.

Kanetsky, P. A., J. Swoyer, S. Panossian, R. Holmes, D. Guerry and T. R. Rebbeck. 2002. A polymorphism in the agouti signaling protein gene is associated with human pigmentation. *Am J Hum Genet* 70(3):770–5.

Katsara, M.-A., W. Branicki, S. Walsh, M. Kayser and M. Nothnagel. 2021. Evaluation of supervised machine-learning methods for predicting appearance traits from DNA. *Forensic Sci Int Genet* 53:102507.

Kayser, M., et al. 2008. Three genome-wide association studies and a linkage analysis identify HERC2 as a human iris color gene. *Am J Hum Genet* 82(2):411–23.

Keller, A., et al. 2012. New insights into the Tyrolean Iceman's origin and phenotype as inferred by whole-genome sequencing. *Nat Commun* 3(1):698.

Kenny, E. E., et al. 2012. Melanesian blond hair is caused by an amino acid change in TYRP1. *Science* 336(6081):554.

Kukla-Bartoszek, M., et al. 2018. Investigating the impact of age-depended hair colour darkening during childhood on DNA-based hair colour prediction with the HIrisPlex system. *Forensic Sci Int Genet* 36:26–33.

Lamason, R. L., et al. 2005. SLC24A5, a putative cation exchanger, affects pigmentation in zebrafish and humans. *Science* 310(5755):1782–6.

Le Pape, E., T. Passeron, A. Giubellino, J. C. Valencia, R. Wolber and V. J. Hearing. 2009. Microarray analysis sheds light on the dedifferentiating role of agouti signal protein in murine melanocytes via the Mc1r. *Proc Natl Acad Sci U S A* 106(6):1802.

Lin, J. Y. and D. E. Fisher. 2007. Melanocyte biology and skin pigmentation. *Nature* 445(7130):843–50.

Liu, F., K. van Duijn, J. Vingerling, A. Hofman, A. Uitterlinden, A. Janssens and M. Kayser. 2009. Eye color and the prediction of complex phenotypes from genotypes. *Curr Biol* 19(5):192–3.

Liu, F., et al. 2010. Digital quantification of human eye color highlights genetic association of three new loci. *PLOS Genet* 6(5):e1000934.

Liu, F., et al. 2015. Genetics of skin color variation in Europeans: Genome-wide association studies with functional follow-up. *Hum Genet* 134(8):823–35.

Liu, F., et al. 2018. Meta-analysis of genome-wide association studies identifies 8 novel loci involved in shape variation of human head hair. *Hum Mol Genet* 27(3):559–75.

Liu, F., K. Zhong, X. Jing, A. G. Uitterlinden, A. E. J. Hendriks, S. L. S. Drop and M. Kayser. 2019. Update on the predictability of tall stature from DNA markers in Europeans. *Forensic Sci Int Genet* 42:8–13.

Lona-Durazo, F., et al. 2021. A large Canadian cohort provides insights into the genetic architecture of human hair colour. *Commun Biol* 4(1):1253.

Maroñas, O., et al. 2014. Development of a forensic skin colour predictive test. *Forensic Sci Int Genet* 13:34–44.

Mengel-From, J., C. Borsting, J. J. Sanchez, H. Eiberg and N. Morling. 2010. Human eye colour and HERC2, OCA2 and MATP. *Forensic Sci Int Genet* 4(5):323–8.

Mengel-From, J., T. Wong, N. Morling, J. Rees and I. Jackson. 2009. Genetic determinants of hair and eye colours in the Scottish and Danish populations. *BMC Genet* 10(1):88.

Morgan, M. D., et al. 2018. Genome-wide study of hair colour in UK Biobank explains most of the SNP heritability. *Nat Commun* 9(1):5271.

Myles, S., M. Somel, K. Tang, J. Kelso and M. Stoneking. 2007. Identifying genes underlying skin pigmentation differences among human populations. *Hum Genet* 120(5):613–21.

Naysmith, L., et al. 2004. Quantitative measures of the effect of the melanocortin 1 receptor on human pigmentary status. *J Investig Dermatol* 122(2):423–8.

Nishimura, E. K., S. R. Granter and D. E. Fisher. 2005. Mechanisms of hair graying: Incomplete melanocyte stem cell maintenance in the niche. *Science* 307(5710):720–4.

Olalde, I., et al. 2014. Derived immune and ancestral pigmentation alleles in a 7,000-year-old Mesolithic European. *Nature* 507(7491):225–8.

Openstax College. 2013. Anatomy & physiology. OpenStax. Available from: https://openstax.org/books/anatomy-and-physiology/pages/5-1-layers-of-the-skin.

Palencia-Madrid, L., et al. 2020. Evaluation of the VISAGE basic tool for appearance and ancestry prediction using PowerSeq chemistry on the MiSeq FGx system. *Genes* 11(6):708.

Palmal, S., et al. 2021. Prediction of eye, hair and skin colour in Latin Americans. *Forensic Sci Int Genet* 53:102517.

Pastorino, L., et al. 2004. Novel MC1R variants in *Ligurian melanoma* patients and controls. *Hum Mutat* 24(1):103.

Pavan, W. J. and R. A. Sturm. 2019. The genetics of human skin and hair pigmentation. *Annu Rev Genomics Hum Genet* 20(1):41–72.

Peng, F., et al. 2019. Genome-wide association studies identify multiple genetic loci influencing eyebrow color variation in Europeans. *J Investig Dermatol* 139(7):1601–5.

Picardo, M., M. L. Dell'Anna, K. Ezzedine, I. Hamzavi, J. E. Harris, D. Parsad and A. Taieb. 2015. Vitiligo. *Nat Rev Dis Primers* 1(1):15011.

Pneuman, A., Z. M. Budimlija, T. Caragine, M. Prinz and E. Wurmbach. 2012. Verification of eye and skin color predictors in various populations. *Leg Med (Tokyo)* 14(2):78–83.

Pośpiech, E., et al. 2014. The common occurrence of epistasis in the determination of human pigmentation and its impact on DNA-based pigmentation phenotype prediction. *Forensic Sci Int Genet* 11:64–72.

Pośpiech, E., et al. 2020. Exploring the possibility of predicting human head hair greying from DNA using whole-exome and targeted NGS data. *BMC Genomics* 21(1):538.

Posthuma, D., et al. 2006. Replicated linkage for eye color on 15q using comparative ratings of sibling pairs. *Behav Genet* 36(1):12–7.

Praetorius, C., et al. 2013. A polymorphism in IRF4 affects human pigmentation through a tyrosinase-dependent MITF/TFAP2A pathway. *Cell* 155(5):1022–33.

Rana, B. K., et al. 1999. High polymorphism at the human melanocortin 1 receptor locus. *Genetics* 151(4):1547–57.

Rebbeck, T. R., et al. 2002. P gene as an inherited biomarker of human eye color. *Cancer Epidemiol Biomarkers* 11(8):782–4.

Rees, J. L. 2003. Genetics of hair and skin color. *Annu Rev Genet* 37:67–90.

Rennie, I. G. 2012. Don't it make my blue eyes brown: Heterochromia and other abnormalities of the iris. *Eye (Lond)* 26(1):29–50.

Ruiz, Y., et al. 2013. Further development of forensic eye color predictive tests. *Forensic Sci Int Genet* 7(1):28–40.

Sabeti, P. C., et al. 2007. Genome-wide detection and characterization of positive selection in human populations. *Nature* 449(7164):913–8.

Scientific Working Group on DNA Analysis Methods (SWGDAM). 2017. SWGDAM Interpretation Guidelines for Autosomal STR Typing by Forensic DNA Testing Laboratories. Available from: www.swgdam.org.

Simcoe, M., et al. 2021. Genome-wide association study in almost 195,000 individuals identifies 50 previously unidentified genetic loci for eye color. *Sci Adv* 7(11):eabd1239.

Sochtig, J., et al. 2015. Exploration of SNP variants affecting hair colour prediction in Europeans. *Int J Legal Med* 129(5):963–75.

Spichenok, O., et al. 2011. Prediction of eye and skin color in diverse populations using seven SNPs. *Forensic Sci Int Genet* 5(5):472–8.

Stokowski, R. P., et al. 2007. A genomewide association study of skin pigmentation in a South Asian population. *Am J Hum Genet* 81(6):1119–32.

Sturm, R. A., et al. 2008. A single SNP in an evolutionary conserved region within intron 86 of the HERC2 gene determines human blue-brown eye color. *Am J Hum Genet* 82(2):424–31.

Sturm, R. A. and T. N. Frudakis. 2004. Eye colour: Portals into pigmentation genes and ancestry. *Trends Genet* 20(8):327–32.

Sturm, R. A. and M. Larsson. 2009. Genetics of human iris colour and patterns. *Pigment Cell Melanoma Res* 22(5):544–62.

Sturm, R. A., R. D. Teasdale and N. F. Box. 2001. Human pigmentation genes: Identification, structure, and consequences of polymorphic variation. *Gene* 277(1):49–62.

Sulem, P., et al. 2007. Genetic determinants of hair, eye and skin pigmentation in Europeans. *Nat Genet* 39(12):1443–52.

Sulem, P., et al. 2008. Two newly identified genetic determinants of pigmentation in Europeans. *Nat Genet* 40(7):835–7.

Valenzuela, R. K., et al. 2010. Predicting phenotype from genotype: Normal pigmentation. *J Forensic Sci* 55(2):315–22.

Valverde, P., E. Healy, I. Jackson, J. Rees and A. Thody. 1995. Variants of the melanocyte-stimulating hormone receptor gene are associated with red hair and fair skin in humans. *Nat Genet* 11(3):328–30.

Van Hilten, A., et al. 2021. GenNet framework: Interpretable deep learning for predicting phenotypes from genetic data. *Commun Biol* 4(1):1094.

Visconti, A., et al. 2018. Genome-wide association study in 176,678 Europeans reveals genetic loci for tanning response to sun exposure. *Nat Commun* 9(1):1684.

Visser, M., M. Kayser and R. J. Palstra. 2012. HERC2 rs12913832 modulates human pigmentation by attenuating chromatin-loop formation between a long-range enhancer and the OCA2 promoter. *Genome Res* 22(3):446–55.

Voight, B. F., S. Kudaravalli, X. Wen and J. K. Pritchard. 2006. A map of recent positive selection in the human genome. *PLOS Biol* 4(3):e72.

Voisey, J., M. del C. Gomez-Cabrera, D. J. Smit, J. H. Leonard, R. A. Sturm and A. van Daal. 2006. A polymorphism in the agouti signalling protein (ASIP) is associated with decreased levels of mRNA. *Pigment Cell Res* 19(3):226–31.

Walsh, S., et al. 2014. Developmental validation of the HIrisPlex system: DNA-based eye and hair colour prediction for forensic and anthropological usage. *Forensic Sci Int Genet* 9:150–61.

Walsh, S., et al. 2017. Global skin colour prediction from DNA. *Hum Genet* 136(7):847–63.

Walsh, S., A. Lindenbergh, S. Zuniga, T. Sijen, P. de Knijff, M. Kayser and K. Ballantyne. 2011. Developmental validation of the IrisPlex system: Determination of blue and brown iris colour for forensic intelligence. *Forensic Sci Int Genet* 5(5):464–71.

Walsh, S., et al. 2013. The HIrisPlex system for simultaneous prediction of hair and eye colour from DNA. *Forensic Sci Int Genet* 7(1):98–115.

Wilde, S., et al. 2014. Direct evidence for positive selection of skin, hair, and eye pigmentation in Europeans during the last 5,000 y. *PNAS* 111(13):4832.

Wollstein, A., et al. 2017. Novel quantitative pigmentation phenotyping enhances genetic association, epistasis, and prediction of human eye colour. *Sci Rep* 7:43359.

Xavier, C., et al. 2020. Development and validation of the VISAGE AmpliSeq basic tool to predict appearance and ancestry from DNA. *Forensic Sci Int Genet* 48:102336.

Yun, L., Y. Gu, H. Rajeevan and K. K. Kidd. 2014. Application of six IrisPlex SNPs and comparison of two eye color prediction systems in diverse Eurasia populations. *Int J Legal Med* 128(3):447–53.

Zaorska, K., P. Zawierucha and M. Nowicki. 2019. Prediction of skin color, tanning and freckling from DNA in Polish population: Linear regression, random forest and neural network approaches. *Hum Genet* 138(6):635–47.

Zhu, G., et al. 2004. A genome scan for eye color in 502 twin families: Most variation is due to a QTL on chromosome 15q. *Twin Res Off J Int Soc Twin Stud* 7(2):197–210.

Molecular Autopsy

18

FREDERICK R. BIEBER, ZORAN M.
BUDIMLIJA, AND ANTTI SAJANTILA

Contents

18.1 Molecular Autopsy

Multidisciplinary investigation of death involves forensic pathology, forensic toxicology, and forensic genetics, which are crucial for public safety and legal medicine. The goals of the forensic autopsy are to determine both the cause of death (CoD; e.g., blunt trauma, infection, asphyxia, cardiac arrest) and the legal manner of death (MoD; homicide, suicide, natural, accidental, undetermined). The complete forensic autopsy may require evaluation of the location and position of the body, evidence attached to the body or found at the scene of death, and reconstruction of the scene itself. Death investigations often include a psychological autopsy component based on medical records, historical data, or notes left by the decedent. The forensic autopsy is best performed by designated professionals—forensic pathologists or qualified coroners, forensic pathologists in both cases—to determine if the death was a consequence of natural or unnatural causes (e.g., crime, accident).

It is necessary to determine whether the death was a result of criminal act per se, or if foul play was involved in cases of alleged suicides, accidents, or "natural deaths." In the case of accidental deaths, it is important to determine the mechanism of occurrence to ensure that the accident itself caused the death and to help to prevent future accidents of the same nature. In addition, forensic autopsies must be performed to ensure that the cause of death in a single case (e.g., rare infectious agent, toxin) is not a concern for public health, as well as to aid in the identification of unidentified human remains.

Forensic autopsies are typically unique and reliable sources of valuable data to investigate unexpected deaths. They also serve as data sources for characterization of health and

DOI: 10.4324/9780429019944-21

way of life at a population level. Information obtained from postmortem investigations, when properly archived and analyzed, can reveal patterns and trends that, ultimately, can be used by decision-makers to implement preventive actions and to develop key public health policies. More sophisticated information can be obtained from molecular autopsies, including postmortem pharmacogenetics.

18.1.1 Molecular Autopsy: Definition(s)

Molecular autopsy emerges from molecular pathology, a study of biochemical and bio-physical cellular mechanisms as the basic factors in diseases. So far, the molecular autopsy has been described by many as a "number of genetic tests performed on the tissues taken during the autopsy to confirm the presence of suspect gene mutations linked to the sud-den, unexpected death without visible morphological substrate." This chapter deals with principles and standards of molecular genetics and their application in the cases of sudden, unexpected deaths.

18.2 Molecular Genetics

18.2.1 Genetics and Genomics in Sudden Natural Death

18.2.1.1 Introduction

In addition to investigating death as a result of violence, accident, or suicide, foren-sic pathologists also determine the CoD as a consequence of natural disease processes. Among natural deaths that forensic pathologists investigate, the majority are sudden and unexpected (Gill and Scordi-Bello 2010). Such deaths are often associated with diseases affecting the cardiovascular system (Barsheshet et al. 2011; Ilina et al. 2011), metabolic disorders, and sudden death in epilepsy.

The ability to identify underlying causes of death, especially genetic disorders, not only allows for the determination of CoD in a decedent but may also help avoid a recurring tragedy within a family. Recently, molecular studies of common disorders have shifted from examining a single or a few candidate genes in a few families (genetics) to the whole human genome level in a population (genomics), where the involvement and interaction of many genes and pathways are looked at simultaneously (Podgoreanu and Schwinn 2005). Understanding the genetic components of the common diseases, coupled with the use of molecular genetic testing, has become an increasingly important tool in the forensic pathologist's armamentarium in determining CoD or contributing factors to death.

A macroscopic or microscopic autopsy finding can give forensic pathologists clues to the choice of genetic test. A negative autopsy finding, on the other hand, can be a puzzling. Recent advances in "molecular autopsies" have begun to uncover some of the sudden unex-plained natural deaths by identifying mutations that can predispose an apparently healthy individual to sudden death (Di Paolo et al. 2004).

18.2.1.2 Positive Autopsy

In the following, pathological conditions are described based on the frequency with which they are seen at the forensic autopsy material. More extensive reviews of a specific disease and the genes and loci linked to that disease are referenced and summarized in Table 18.1.

Table 18.1   **Genetics and Genomics in Sudden Natural Deaths**

Condition	Genes or Locus
Coronary artery atherosclerosis (familial)	
Coronary artery disease, autosomal dominant	MEF2A (Wang et al. 2003) and LRP6 (Mani et al. 2007)
Coronary artery atherosclerosis (sporadic)	
Ventricular fibrillation in acute myocardial infarction	21q21(CXADR) (Bezzina et al. 2010)
Coronary artery disease	9p21.3 (CDKN2A, CDKN2B) (Preuss et al. 2010)
Myocardial infarction, premature, susceptibility to	1p34–36 (LRP8) (Wang et al. 2004), 1q25.1(TNFSF4) (Wang et al. 2005; Ria et al. 2011), 6p12.1 (GCLC) (Koide 2003) 6q25.1(ESR1) (Aouizerat et al. 2011) 13q12 (ALOX5AP) (Helgadottir et al. 2004)
Hypertension	
Monogenic forms of hypertension	AGTR1 (Gribouval et al. 2005), PTGIS (Nakayama et al. 2003) ECE1 (Funke-Kaiser et al. 2003), RGS5 (Chang et al. 2007), ATP1B1 (Chang et al. 2007), SELE (Chang et al. 2007), AGT (Markovic et al. 2005), HYT3 (Angius et al. 2002), AGTR1 (Gribouval et al. 2005), ADD1 (Manunta et al. 1999), HYT6 (Joe et al. 2009), CYP3A5 (Thompson et al. 2004), NOS3 (Wang et al. 1996), HYT4 (Gong et al. 2003), HYT2 (Xu et al. 1999), HYT1 (Hilbert et al. 1991), NOS2A (Rutherford et al. 2001), HYT5 (Wallace et al. 2006), PTGIS (Nakayama et al. 2003)
Hypertension, genome-wide	1p36 (Newton-Cheh et al. 2009), 3p22.1 (Padmanabhan et al. 2010), 4q21 (Newton-Cheh et al. 2009), 10q21 (Newton-Cheh et al. 2009), 10q24 (Newton-Cheh et al. 2009; Padmanabhan et al. 2010), 10p12 (Padmanabhan et al. 2010), 11p15 (Padmanabhan et al. 2010), 12q21 (Padmanabhan et al. 2010), 12q24 (Newton-Cheh et al. 2009) (Padmanabhan et al. 2010), 12q24.21 (Padmanabhan et al. 2010), 15q24 (Newton-Cheh et al. 2009; Padmanabhan et al. 2010), 17q21 (Newton-Cheh et al. 2009), 17q21 (Newton-Cheh et al. 2009)
Pulmonary thromboembolism	Factor V Leiden (Tang et al. 2011), Prothrombin G20210A (Tang et al. 2011), protein S (Ikejiri et al. 2010), protein C (Vossen et al. 2005), antithrombin (Vossen et al. 2005)
Aortic aneurysm rupture	
Marfan syndrome	FBN1 (Dietz et al. 1991)
Ehlers–Danlos syndrome type IV	COL3A1 (Schwarze et al. 1997)
Genomic risks	9p21.3 (rs7025486 in DAB2IP) (Gretarsdottir et al. 2010)
Intracranial Berry aneurysm	2q33.1 (Bilguvar et al. 2008), 9p21 (Yasuno et al. 2010), 7q11.2 (Akagawa et al. 2006), 19q13 (van der Voet et al. 2004), 1p36.13–p34.3 (Nahed et al. 2005), 5p15.2–p14.3 (Verlaan et al. 2006), 14q23 (Mineharu et al. 2008), 2p15–q14 (Roos et al. 2004), 11q24–25 (Ozturk et al. 2006)
Cardiomyopathy	
Arrhythmogenic right ventricular cardiomyopathy (ARVC)	TGFB3 (Rampazzo et al. 2003), PKP2 (Gerull et al. 2004), DSC2 (Syrris et al. 2006), DSG2 (Syrris et al. 2007), JUP (Asimaki et al. 2007), DSP (Yang et al. 2006), TMEM43 (Merner et al. 2008), RyR 2 (Tiso et al. 2001)

(Continued)

Table 18.1 (Continued)   Genetics and Genomics in Sudden Natural Deaths

Condition	Genes or Locus
Hypertrophic cardiomyopathy (HCM)	MYH7 (Geisterfer-Lowrance et al. 1990), MYBPC3 (Watkins et al. 1995), TNNC1 (Hoffmann et al. 2001), TNNT2 (Thierfelder et al. 1994), TNNI3 (Kimura et al. 1997), TPM1 (Richard et al. 2003), ACTC (Richard et al. 2003), MYL2 (Richard et al. 2003), MYL3 (Richard et al. 2003), PLN (Landstrom et al. 2011), LAMP2 (Charron et al. 2004), PRKAG2 (Kelly et al. 2009), NEXN (Wang et al. 2010)
Dilated cardiomyopathy (DCM) reviewed by Dellefave and McNally (2010)	Sarcomere genes: MYH7, MYH6, MYL3, MYL2, TPM1, TNNT2, TNNI3, TNNC1, MYBPC3, ACTC
	Z-disk genes: TTN, LDB3, VCL, CSRP3, TCAP, TTID
	Nuclear genes: TAZ, LMNA, EMD, EYA4, SYNE1, TMPO, RBM20
	Mitochondrial genes: TTR, FRA, Cox15, HFE
	Cytoskeletal-membrane linkers genes: DMS, SGCG, SGCD, SGCB, SBCA, FKRP
	Others genes: ANKRD1, LAMP2, SCN5A, DES, ZASP, NEBL, DNM1L
Left ventricular noncompaction cardiomyopathy (LVNC)	DTNA (Ichida et al. 2001), LDB3 (Vatta et al. 2003), TAZ (Bleyl et al. 1997), ACTC (Klaassen et al. 2008), MYH7 (Klaassen et al. 2008), TNNT2 (Klaassen et al. 2008), MYBPC3 (Vatta et al. 2003)
Sudden cardiac death susceptible loci	2q24.2 (Arking et al. 2011)
Monogenic forms of diabetes	
Maturity-onset diabetes of the young (MODY)	HNF4A (Yamagata et al. 1996), GCK (Vionnet et al. 1992), HNF1A (Vaxillaire et al. 1997), IPF1 (Stoffers et al. 1997), HNF1B (Horikawa et al. 1997), NeuroD1 (Malecki et al. 1999), KLF11 (Neve et al. 2005), CEL (Raeder et al. 2006), PAX4 (Plengvidhya et al. 2007), BLK (Borowiec et al. 2009)
Neonatal diabetes mellitus (NDM)	GCK (Njolstad et al. 2001), INS (Edghill et al. 2008), KCNJ11 (Gloyn et al. 2004), SUR1 (Ellard et al. 2007)
Cardiac conduction system	
Long QT syndromes (George et al. 1995; Berthet et al. 1999; Priori et al. 1999; Splawski et al. 2000; Plaster et al. 2001; Mohler et al. 2003; Splawski et al. 2005; Cronk et al. 2007; Medeiros-Domingo et al. 2008; Chen et al. 2007; Ueda et al. 2008; Yang et al. 2010)	KCNQ1 (LQT1), KCNH2 (LQT2), SCN5A (LQT3), ANK2 (LQT4), KCNE1 (LQT5), KCNE2 (LQT6), KCNJ2 (LQT7), CACNA1C (LQT8), CAV3 (LQT9), SCN4B (LQT10), AKAP9 (LQT11), SNTA1 (LQT12), KCNJ5 (LQT13)
Brugada syndrome (Chen et al. 1998; London et al. 2007; Antzelevitch et al. 2007; Watanabe et al. 2008; Delpon et al. 2008; Hu et al. 2009; Ueda et al. 2009)	SCN5A, GPD1L, SCN1B, SCN2B, KCNE3, CACNA1C, SCN3B, CACNB2
Catecholaminergic polymorphic ventricular tachycardia (CPVT) (Priori et al. 2001; di Barletta et al. 2006)	RYR2 (Priori et al. 2001), CASQ2 (Lahat et al. 2004)
Short QT syndrome	KCNH2 (Hong et al. 2005), KCNQ1 (Bellocq et al. 2004), KCNJ2 (Priori et al. 2005)
Ventricular fibrillation	KCNJ8/KATP (Haissaguerre et al. 2009), DPP6 (Alders et al. 2009), CACNA1C (Aouizerat et al. 2011)

(Continued)

Table 18.1 (Continued)   Genetics and Genomics in Sudden Natural Deaths

Condition	Genes or Locus
Wolff–Parkinson–White syndrome	PRKAG2 (Zhang et al. 2011)
Sick sinus syndrome	HCN4 (Ueda et al. 2009), SCN5A (Gui et al. 2010)
Cardiac conduction defect	AKAP10 (Patel et al. 2010)
Progressive familial heart block, type IB	TRPM4 (Kruse et al. 2009)
Atrial fibrillation and sudden death	KCNA5 (Yang et al. 2009)
Progressive cardiac conduction disease, atrial arrhythmias, and sudden death	LMNA (Marsman et al. 2011)
Central neural systems	
Sudden explained death in epilepsy	KCNA1 (Tu et al. 2011), HCN1–4 (Tu et al. 2011), SCN5A (Aurlien et al. 2009; Tu et al. 2011), KCNH2 (Tu et al. 2011)

1. Cardiovascular system

- *Coronary artery atherosclerosis.* Cardiovascular disease is the most common (Barsheshet et al. 2011) cause of natural unexpected death seen by forensic pathologists. The stenosis or occlusion of a particular coronary artery results in ischemia of the myocardium, which may result in acute myocardial infarction and/or a lethal cardiac arrhythmia. Fatal coronary artery atherosclerosis tends to occur in the proximal aspects of the left main coronary artery, right main coronary artery, at branching points, and the proximal aspects of the left anterior artery and left circumflex arteries. Acute critical occlusion of a vessel may result in the sudden onset of symptoms. The acute change may result from complications within an atheromatous plaque including plaque rupture with subsequent thrombosis of the vessel and sudden hemorrhage into a plaque resulting in loss of luminal area and subsequent myocardial ischemia. As most cases of sudden natural death due to coronary artery atherosclerosis occur within hours of the onset of symptoms, the forensic pathologist does not usually find acute myocardial infarction upon macroscopic and microscopic examination of the heart (Little and Applegate 1996). In addition, there are gender differences in the severity and extent of coronary artery atherosclerosis in relationship to the fatal outcome. Women who die of fatal ischemic heart disease tend to have less severe and extensive coronary atherosclerosis (Smilowitz et al. 2011) at autopsy.

  Coronary artery atherosclerosis is often the result of multiple factors, include lifestyle (e.g., smoking, high-fat diet, lack of exercise), compounding disorders (e.g., hypertension, diabetes mellitus, hyperlipidemias), and genetic or genomic risks (Sivapalaratnam et al. 2011). A few studies of families where coronary artery atherosclerosis is clearly dominantly inherited with multiple affected individuals and often early onset have allowed the identification of the candidate genes, such as *MEF2A* (Wang et al. 2003) and *LRP6* (Mani et al. 2007), which are responsible for the autosomal dominant form of the coronary artery disease. Testing for these genes might be helpful for identifying the underlying CoD in young victims with family histories.

  In the general population, however, coronary artery atherosclerosis usually follows a multifactorial disease model where multiple genetic loci (genomic risks) are involved and may have synergistic effects on the development of disease. (Table 18.1).

- *Hypertensive cardiovascular disease.* Hypertensive cardiovascular disease may be reliably diagnosed from the pathological features of an enlarged heart showing concentric left ventricular hypertrophy in the absence of valve disease or cardiomyopathy. Characteristic hypertensive changes may be observed in the kidney and, in older individuals, the lenticulostriate vessels within the basal ganglia of the brain. Hypertensive cardiovascular disease can lead to sudden death through several mechanisms, including invoking a cardiac arrhythmia, precipitating rupture of an abdominal aortic aneurysm, dissection of the thoracic aorta, or causing an intracerebral hemorrhage (Burke et al. 1996).

  Hypertension is another multifactorial disease with lifestyle (e.g., high-salt diet, obesity), compounding common diseases or conditions (e.g., diabetes mellitus or aging) and multiple genes involved in regulating and maintaining the blood pressure (Franceschini et al. 2011). Numerous genes associated with essential conditions are shown in Table 18.1

  The monogenic forms of hypertension (autosomal dominant or recessive inheritance) usually manifested at an early age (children or young adult) in families with a history of moderate to severe hypertension, and may be associated with underlying clinical syndromes such as Bilginturan's disease (autosomal-dominant hypertension with brachydactyly) (Bahring et al. 2008). Hypertension can also be exacerbated during pregnancy (Rafestin-Oblin et al. 2003; Geller et al. 2000). A suppressed rennin–angiotensin–aldosterone system is believed to the common pathophysiological basis for all hereditary forms of hypertension (reviewed by Simonetti et al. 2011). Genes involved in the renin–angiotensin system or regulatory pathways, such as angiotensin receptor 1 (*AGTR1*) (Gribouval et al. 2005) and prostaglandin 12 synthase (*PTGIS*) (Nakayama et al. 2003), can be the candidate genes for the monogenic form of hypertension (Simonetti et al. 2011). Molecular testing of the monogenic forms of hypertension might be particularly useful in young victims or death during pregnancy when related to hypertension.

- *Deep venous thrombosis and pulmonary embolism* (*PE*). Fatal PE is typically seen as a massive and acute embolus blocking central vascular zones and results in hemodynamic compromise. Fatal PE often is associated with deep venous thromboembolism (VTE). VTE is also a multifactorial disease and accounts for approximately 100,000 deaths annually in the United States (Heit et al. 1999).

  Most deaths are associated with the three predisposing factors described by Virchow: stasis, hypercoagulability, and vascular injury (Kumar et al. 2010). Hypercoagulability risk factors include the thrombophilia, that is, diseases or conditions associated with an increased thrombotic risk. Heritable thrombophilias include mutations in factor V Leiden and prothrombin, deficiencies of antithrombin, and proteins C and S, among other rare factors in the coagulation cascade (reviewed by Favaloro and Lippi 2011).

  Ethnicity also plays a significant role in fatal PE. A study of a large number of PE incidents in New York City revealed that PE victims are predominately Blacks and died at an earlier age when compared to Whites. Consistent with other studies, Blacks were found less likely to be carriers of the common thrombophilia variants found in Whites (Tang et al. 2011). Future research is warranted to understand these unique ethnic aspects of fatal PE.

- *Aortic aneurysm rupture*. Typically, an elderly individual, often with common risk factors such as coronary artery atherosclerosis or hypertension, complains of severe abdominal pain culminating in collapse from catastrophic hemorrhage, which is easily observable at autopsy.

  Aortic dissection may also occur in individuals with genetic disorders of collagen formation such as Marfan syndrome (McKusick 1991) or Ehlers–Danlos syndrome type IV (vascular type; Beighton et al. 1997) (Table 18.1). Postmortem molecular testing for the genes causing these syndromes (*FBN1* [Dietz et al. 1991] and *COL3A1* [Schwarze et al. 1997]) can be useful in determining the underlying CoD.

Cardiomyopathy. Common types of the cardiomyopathies seen at autopsy, including dilated, hypertrophic, left ventricular noncompaction, as well as arrhythmogenic right ventricular cardiomyopathy, often have gross and/or microscopically distinct features at autopsy in adults (reviewed by Grant and Evans 2009). Cardiomyopathy is the well-known substrate of cardiac arrhythmia, which when is not reversed immediately may result in sudden unexplained death (SUD). Death can occur in an individual who has no previous clinical symptom or complaint, and the diagnosis of a specific type of cardiomyopathy is made only at autopsy. In addition, deaths may occur as a result of cardiac arrhythmias before the cardiac myocardium structural changes become apparent at autopsy. This is especially true in infants and children when the heart is still under development and structural changes are less apparent (Liu et al. 2008; Tester and Ackerman 2009). Therefore, cardiomyopathy should not be excluded in autopsy-negative infant or childhood deaths. The etiologies for dilated cardiomyopathy can be quite diverse: they may occur as a consequence of remote viral myocarditis, chronic exposure to cardiac toxins (derived from medications or from a metabolic disease), or defects of one of the sarcomeric, dystrophin complex or others genes. Hypertrophic cardiomyopathy, on the other hand, is more a genetic disorder due to defects in any of the genes encoding or processing sarcomere proteins. It is characterized by disarray of myocardial fibers on microscopic examination of heart muscle. Pathologically, arrhythmogenic right ventricular cardiomyopathy (ARVC) shows inflammatory and fibrofatty infiltration of the right ventricle. A few genes encoding the desmosomal proteins are responsible for ARVC. The pathogenesis of the common cardiomyopathies has been reviewed in detail by Watkins et al. (2011) (Table 18.1). At the molecular level, all types of cardiomyopathies are genetically heterogeneous; multiple genes (Table 18.1) are responsible for each type, and there are many different rare variants within each gene. Furthermore, different types of cardiomyopathies may have similar molecular bases. For instance, the *MYH7* gene is responsible for all three types of cardiomyopathies: hypertrophic, dilated, and left ventricular noncompaction (Klaassen et al. 2008). Cardiomyopathy can be a disease alone or part of a genetic syndrome, such as Emery–Dreifuss dystrophy (Bonne et al. 2000) and LEOPARD syndrome (multiple *l*entigines, *e*lectrocardiographic conduction abnormalities, *o*cular hypertelorism, *p*ulmonic stenosis, *a*bnormal genitalia, *r*etardation of growth, and sensorineural *d*eafness) (Carvajal-Vergara et al. 2010). In addition, genotype–phenotype correlation in cardiomyopathies is complicated by reduced penetrance, variable expressivity, genetic and allelic heterogeneity, as well as genetic and environmental modifier effects. Nevertheless, molecular testing of candidate genes responsible for cardiomyopathies are especially useful when the structure changes are subtle and do not meet the classic morphological diagnosis. Currently, DNA sequencing of coding regions and splice junctions of a large panel of OMIM genes associated with cardiomyopathy is available commercially and includes the following genes: *ABCC9, ACTC1, ACTN2, ALMS1, ALPK3, ANKRD1, BAG3, BRAF, CAV3, CHRM2, CRYAB, CSRP3, DES, DMD, DOLK, DSC2, DSG2, DSP, DTNA, EMD, FHL1, FKRP, FKTN, GATAD1, GLA, HCN4, HRAS, ILK, JPH2, JUP, KRAS, LAMA4, LAMP2, LDB3, LMNA, MAP2K1, MAP2K2, MIB1, MTND1, MTND5, MTND6, MTTD, MTTG, MTTH, MTTI, MTTK, MTTL1, MTTL2, MTTM, MTTQ, MTTS1, MTTS2, MURC, MYBPC3, MYH6, MYH7, MYL2, MYL3, MYLK2, MYOZ2, MYPN, NEBL, NEXN, NKX2-5, NRAS, PDLIM3, PKP2, PLN, PRDM16, PRKAG2, PTPN11, RAF1, RBM20, RIT1, RYR2, SCN5A, SGCD, SOS1, TAZ, TCAP, TGFB3, TMEM43, TMPO, TNNC1, TNNI3, TNNT2, TPM1, TTN, TTR, TXNRD2, VCL* (see GeneDx.com).

2. Other systems

- *Intracranial hemorrhage.* Sudden and unexpected death attributable to the central nervous system most commonly occurs as a consequence of intracranial hemorrhage, often in individuals with hypertension. Hemorrhage into the subarachnoid space most commonly occurs from rupture of an aneurysm within the circle of Willis at the base of the brain. Genetic and genomic risks for hypertension described earlier, therefore, can contribute to intracranial hemorrhage. In addition, genomic risks responsible for intracranial berry aneurysm were mapped to a handful of loci (Table 18.1). The significance of these loci in intracranial hemorrhage is to be determined.
- *Others.* Another less frequent yet broad group of diseases can also be the causes of SUD. These diseases may include metabolic diseases that are not on newborn screening panels (e.g., mitochondrial diseases) or chromosomal abnormalities that are not detected by the prenatal karyotyping. Diagnosis of these diseases can be very challenging during postmortem investigation because the autopsy findings are often subtle or nonspecific. Consider consulting a clinical geneticist, or molecular or cytogenetic specialist when needed.

18.2.1.3 Negative Autopsy

Sudden unexplained death is sudden death in apparently healthy individuals, which remains unexplained after a complete autopsy, extended laboratory testing, examination of the death scene, and review of clinical history. The genetic risks of SUDs have been studied from several different perspectives, with the majority of these studies focused on cardiac conduction and the central nervous system, which are discussed in this section.

1. *Cardiac conduction system.* Irregular heart rhythms that degenerate into ventricular fibrillation can be fatal if not reversed immediately. It is estimated that the 10%–15% of sudden unexplained deaths in infants (Arnestad et al. 2007; Ackerman et al. 2001; Millat et al. 2009; Chung et al. 2007) and up to 35% in children and adults (Tester and Ackerman 2009) can be attributable to a group of heritable cardiac arrhythmia syndromes, collectively known as channelopathies (i.e., cardiac ion channel diseases), including long QT syndrome, Brugada syndrome, short QT syndrome, catecholaminergic polymorphic ventricular tachycardia,

and familiar ventricular fibrillation (reviewed by Cerrone and Priori 2011; Table 18.1). Each of the syndromes is typically associated with a characteristic change in electrocardiograms (reviewed by Cerrone and Priori 2011). At the molecular level, channelopathies are caused either by abnormal functions of the cardiac ion channels (Na^+, K^+, and Ca^{2+}) or protein processing. Genes encoding for most of the cardiac ion channels have been mapped and cloned (Cerrone and Priori 2011). Postmortem molecular testing for these genes is the only way of making the diagnosis. Channelopathies are often undetected before sudden death in apparently healthy individuals who have structurally normal hearts. Sudden cardiac arrest (SCA) has been shown to be a significant cause of mortality in many parts of the world, as less than 10% of individuals in the US are successfully resuscitated after an out-of-hospital cardiac arrest (Prutkin et al. 2008; Haissaguerre et al. 2008; London et al 2018; Thiene 2018). The majority of sudden cardiac arrest or sudden cardiac death results from structural heart disease, while unexplained cardiac arrest in those under 35 should raise suspicion of the possibility of a heritable heart condition. For example, long QT syndrome (LQTS), Brugada syndrome (BrS), and catecholaminergic polymorphic ventricular tachycardia (CPVT) are heritable forms of arrhythmia often manifesting before age 45, even as early as infancy, often with sudden death as the first and only symptom. Next generation sequencing (NGS) of the coding regions and splice junctions of several genes is now available commercially and test panels include some of the following OMIM genes: *ANK2, CALM1, CALM2, CALM3, CASQ2, CAV3, KCNE1, KCNE2, KCNH2 (HERG), KCNJ2, KCNQ1, PPA2, RYR2, SCN5A* (see GeneDx.com).

2. *Central nervous system.* Forensic pathologists are aware of deaths in individuals with epilepsy in the absence of morphological abnormalities. Sudden unexpected death in epilepsy (SUDEP) has been extensively studied, yet the causes remained largely unknown. It is thought that various and overlapping pathophysiologic events may contribute to sudden death in epilepsy. Respiratory events (e.g., airway obstruction) or cardiac arrhythmia leading to cardiac arrest may play an important role (Tu et al. 2011). Recently, molecular links to several genes in SUDEP have been suggested, including hyperpolarization-activated cyclic-nucleotide gated cation (HCN1–4) channels (Tu et al. 2011), the Kv1.1 Shaker-like potassium channels gene *Kcna1* (Tu et al. 2011), and the cardiac ion channel genes *SCN5A* (Aurlien et al. 2009; Tu et al. 2011) and *KCNH2* (Tu et al. 2011).

18.3 Postmortem Pharmacogenetics

18.3.1 Pharmacogenetics and Medicolegal Death Investigation

With new research tools such as next generation sequencing and bioinformatics, the amount of information available from forensic sample types has multiplied. New directions in forensic research promise the near future of using previously unavailable biological samples, as well as data from crime scenes. Similarly, the same genomic advances are of utmost importance in the future for medicolegal autopsy practice, where these new directions lead to molecular autopsies (Tester et al. 2012) in a multidisciplinary CoD investigation utilizing knowledge from pathology, toxicology, and genetics.

18.3.2 Development of Pharmacogenetic Concept to Pharmacogenomics

The beginning of pharmacogenetics is generally considered to be based on observations from the 1950s. First, Hockwald et al. (1952) found that the antimalarial drug primaquine causes intravascular hemolysis, specifically among the African Americans but rarely among Caucasians. At about same time, the interindividual differences in responses to the tuberculosis drug isoniazid was studied in humans. After Carson et al. (1956) described the basis for the aforementioned primaquine-induced intravascular hemolysis (deficiency of the red cell enzyme glucose-6-phosphate dehydrogenase), and Lehman and Ryan (1956) and Kalow and Staron (1957) described inherited variation in the neuromuscular paralysis time after succinylcholine administration (deficiency in the metabolizing enzyme pseudocholinesterase), Motulsky (1957) summarized the concept of a genetic basis for the phenomenon of inherited variation in response to drug treatment. The term "pharmacogenetics" was then coined and referred at that time to monogenic traits. In the 1970s, the variability in the drug response was first uncovered using molecular genetics methods. The disadvantageous effects of debrisoquine (Mahgoub et al. 1977) and spartaine (Eichelbaum et al. 1979) to the patient were shown to be due to cytochrome P450 (CYP) monooxygenase variation, and CYP2D6 was characterized in 1988 as the first polymorphic gene affecting the drug response (Gonzalez et al. 1988).

It is now known that the drug-metabolizing enzymes are part of the xenobiotic metabolism system, which protects us against the harmful effects of foreign compounds introduced to our body. The metabolism of xenobiotics can be divided into three concerted phases with the aim of detoxifying and removing foreign substances from our cells: modification (phase I), conjugation (phase II), and excretion (phase III). In phase I, the functional groups of the foreign compounds are modified by a variety of enzymes. The most common and largely studied is the CYP family of enzymes (such as CYP2D6, CYP2C9, CYP2C19). In the subsequent phase II reaction, enzymes conjugate the modified xenobiotics with enzymes (e.g., acetate, sulfate, glutathione, and glucuronic acid) in a reaction that is catalyzed by transferases (e.g., glutathione S-transferases, N-acetyltransferases [NAT1 and NAT2], thiopurine methyltransferase). After phase II reactions, the xenobiotic conjugates and their metabolites are excreted from cells. This phase III metabolism includes the multidrug resistance protein (MPR) family, which is part of the ATP-binding cassette transporters, which can catalyze the ATP-dependent transporters and thus remove phase II products to the extracellular matrix to be excreted from the body (Figure 18.1).

The development of genomics and bioinformatics has progressed the straightforward ideas of PGt to a new complex level that is currently better described as pharmacogenomics (Wang et al. 2011).

Because of the great interest and importance of the emerging field(s), the US Department of Health and Human Services has released the "Definitions for Genomic Biomarkers, Pharmacogenomics, Pharmacogenetics, Genomic Data and Sample Coding Categories," defining *pharmacogenomics* (PGx) as the study of variations of DNA and RNA characteristics as related to drug response, and *pharmacogenetics* (PGt) as a subset of PGx defined as the study of variations in DNA sequence as related to drug response.

Essentially, PGt and PGx research focuses on understanding the underlying inherited molecular factors in individual variations in a drug and offer a molecular-level explanation to medical practitioners' notion of "responders" and "nonresponders." Juxtaposing pharmacology with genetics, PGt and PGx now form—together with pharmacokinetics and

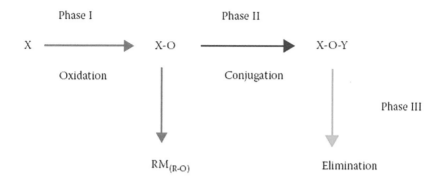

Figure 18.1 Xenobiotic metabolisms. Phase I, phase II, phase III. $RM_{(R-O)}$ is a usually minor reactive metabolite of the oxidation in the phase I reaction. See the text for details.

pharmacodynamics—the basis of clinical pharmacology. The ultimate objective of PGt and PGx is to aid clinicians to prescribe the most effective and safe, individually tailored drug treatment for the patient, and to minimize the adverse effects of those drugs.

18.3.3 Adverse Drug Reactions

According to the World Health Organization (WHO), an adverse drug reaction (ADR) is defined as an unintended response to a drug occurring at a conventional dose and used for disease in prophylaxis, diagnosis, therapy, or for modification of physiological functions (WHO, Factsheet No. 293). ADRs are divided into type A and B reactions (Pirmohamed et al. 1998), and further to type C–F reactions (Edwards and Aronson 2000). These definitions enable medical and pharmaceutical societies to compare the ADRs of different drugs and to collect them into a large database (see, e.g., www.who-umc.org/).

Although the efficacy of a drug is scrutinized in a clinical trial, more than 50% of ADRs are detected after marketing approval (Moore et al. 1998). In the clinical setting, ADRs are difficult to identify and report. The number of reported ADRs depends on several factors, such as type of ADR (mild, severe, fatal) and reporting institution (general practice, general hospital, university hospital), which can affect the data generated, and thus the data can remain underreported. Studies on pharmacovigilance, defined by the WHO as a science and activities relating to the detection, assessment, understanding and prevention of adverse effects or any other drug-related problem, and pharmacoepidemiology, defined as nonexperimental study of the use and effects of drugs in a large number of people (International Society of Pharmacoepidemiology, www.pharmacoepi.org; Launiainen et al. 2011), offer a means of detecting ARDs. Lazarou et al. (1998) found that about 100,000 deaths of US hospital inpatients occur per year because of ADRs, and calculated an overall incidence of fatal ADRs to be 0.32% among US hospital inpatients. This figure would make ADRs one of the leading causes of death in the United States. In a prospective Scandinavian study (Ebbesen et al. 2001) with 13,992 patients of internal medicine, the incidence of lethal ADR was estimated to be 0.95%, and Pirmohamed et al. (2004) observed a 6.5% ADR rate for hospital admissions in 18,820 patients, and 0.15%–2.3% ADR lethality, both of the figures being in gross agreement with the estimates from the United States. As a conclusion, it is clear that ADR-related deaths and the possibilities of using PGt and PGx methods in investigation of those are of great medicolegal interest (Musshoff et al. 2010; Sajantila et al. 2010).

18.3.4 Investigation of Death and Toxicology

"Autopsy negative" cases (Cohle and Sampson 2001; Corrado et al. 2001), where a traditional autopsy is not sufficient to determine the CoD and/or MoD, are considered of special nature in the postmortem field. In addition to a thorough description of macroscopical findings, further auxiliary tests (histology, toxicology, biochemistry, imaging, etc.) are requested. Examples of autopsy negative cases are the different types of SUDs such as sudden cardiac deaths, SUDEPs, sudden infant death syndrome, or untypical drug-related deaths. Series of negative cases are abundantly reported, and the incidence varies depending on the institution, population, and, particularly, among age groups.

One of the most widely used and important auxiliary analyses is forensic toxicology, which is derived from the toxicological study of the adverse effects of drugs and chemicals on humans, and bases its practice in published and widely accepted scientific methods. The objective of forensic toxicology is to analyze drugs, poisons, and any toxic agents in living and deceased individuals for legal purposes. An important part of forensic toxicology, and its most essential duty in CoD investigation, is interpreting the results in a scientific manner.

Academic research is an inherent part of forensic toxicology whose aims is to develop and evaluate new analytical methods and instrumentation (Ojanperä et al. 2012), to study postmortem effects (e.g., postmortem redistribution of drugs) (Pounder and Jones 1990) and postmortem storage and handling of samples, and to interpret the results in the medicolegal context (Koski 2005).

As in other forensic disciplines, forensic toxicology uses methods derived from innovations in clinical medicine and academic laboratories (e.g., analytical chemistry, pharmacology, and clinical chemistry). Forensic toxicology can be further divided into subdisciplines, such as human performance toxicology (e.g., effects of alcohol and drugs in traffic, alcohol- and drug-facilitated crimes), doping control, workplace testing, and postmortem toxicology (i.e., death investigation toxicology), each having its own specific topics and questions.

Numerous factors contribute to a successful drug therapy, and the same issues are also involved when considering interpretation of forensic toxicology results. These include a patient's compliance, adherence, external factors, diseases, and genetics. A patient's behavioral habits may be of surprising medicolegal importance; for example, borrowing and sharing of medication (Petersen et al. 2008), co-use of prescription drugs with over-the-counter drugs, abuse of prescribed drugs (Zaracostas 2007), and enhancement of the effect of illicit drugs with prescription drugs (Vuori et al. 2003) are common knowledge for experienced forensic pathologists and forensic toxicologists.

Forensic toxicology results are particularly important in the investigation of CoD and MoD in cases of intoxication, particularly in the evaluation of suicides by alcohol, drugs, or their combination (Koski et al. 2003). Data from drug screening is also important in assessment of clinical maltreatment and medical negligence (Madea et al. 2009).

It is important to note that polypharmacy is a common phenomenon in modern society, and this is also reflected in forensic autopsy material. Therefore, it is important that data on drug–drug interactions from clinical data are to be translated into the forensic toxicology case-work interpretation (Launiainen et al. 2011) and can be mirrored against the data obtained from large forensic toxicology databases (Launiainen et al. 2009). Conversely, new insights can be revealed from forensic data (Launiainen et al. 2010), which can be used as an information basis for drug safety.

18.3.4.1 Postmortem Pharmacogenetics and CoD Investigation

Although CoD and MoD of some ambiguous deaths cannot be unequivocally established via traditional autopsy, there are newly developed molecular approaches that may solve some of the SUDs. For application of genetic testing in mortuary work, the term "molecular autopsy" has been coined (Tester et al. 2012). The use of molecular autopsies for autopsy negative cases are obvious: first, genetic analyses are likely to reduce the number of undetermined CoDs and MoDs; second, the mechanisms underlying such deaths can be better understood; and third, genetic information on risks for sudden unexpected deaths or for fatal ADRs are of utmost importance to relatives.

Data from clinical PGt (i.e., individualize drug treatment by facilitating the choice of the proper drug and dose for each individual based on genetic information) can be expected to be translated to the postmortem setting to aid in the determination of CoD and MoD. The promise of PGx in the medicolegal context has also stimulated discussion of individualized rights (Wong et al. 2010).

The cost-effective use of toxicogenetics in the medicolegal investigation of death requires integrating research in forensic pathology, toxicology, and genetics. The data points to be collected in an autopsy for systematic pharmacogenetic studies include relevant medical history and other background information combined with detailed knowledge on the pathophysiological of the diseased, concentrations of all drugs and their relevant metabolites in the body at the time of death, and genotype data from the gene pathway related to the drugs found. Attempts for such studies have been published, particularly for drugs or case-control studies (cf. Levo et al. 2003; Koski et al. 2006).

All the data collected in this manner should be considered along with the fact that an individual's pathophysiological phenotype affecting drug efficacy (the ability of a drug to produce the desired therapeutic effect) depends to some degree on the genetic constitution of an individual and several other factors. These additional factors include developmental stage, physiological and environmental factors, and diseases or specific conditions (Table 18.2).

Deducing phenotype from the genotype is crucial for clinical PGx, and some guidelines already exist for the genotype–drug dosage relationship (Kirchheiner et al. 2004; Kirchheiner 2008). Before large-scale genotyping possibilities, phenotype from an individual was obtained by urine analysis one drug at a time, and calculating the "metabolic ratio" equaling to the metabolic capacity for the drugs, one by one (Dahl et al. 1995). The impact of genetic variation among patients to drug efficacy is now better known, and prediction of the phenotype from genotype is of a significant potential for individualized drug therapy. However, even with the new analytical methods and easily accessible genotype data, the probability of prediction of the phenotype from the genotype has challenges. The *CYP2D6* gene is the best characterized example, with four phenotypic classes, which can be deduced from genotypes: individuals who lack the functional enzyme are poor metabolizers (PMs); carriers of two decreased-function variants or a combination of one decreased-function variant and one nonfunctional variant are intermediate metabolizers; individuals who have at least one fully functional variant are extensive metabolizers; and finally, the carriers of active gene duplication or multiplication or another mutation increasing the enzyme activity inherited together with a functional variant are ultrarapid metabolizers (UMs).

Case examples. Medicolegal cases are often a source for problems and thus for new research. The following briefly describe a few selected cases that are of considerable

Table 18.2 Factors Affecting an Individual's Pathophysiological Phenotype Related to Drug Efficacy

Genetic variation
Gene deletions, gene duplications
Sequence variation
Developmental stage
Age
Gender
Physiological factors
Mental and physical stress
Hormonal changes
Periodical changes (seasonal and circadian factors)
Environmental factors
Personal environmental history
Lifestyle
Diet
Exposure to environmental toxins
Concomitant use of alcohol(s) and drug(s)
Potential, specific associations
Diabetes, obesity, gut microbiology

medicolegal interest. The examples are related to psychiatric disorders, use of analgesics, and suicides, which constitute a large proportion of medicolegal autopsies cases in the Western world.

18.3.4.1.1 Poor Metabolizing Genotypes and Intoxication
Sallee et al. (2000) showed in their report how a genetically determined poor drug metabolism led to fatal drug intoxication in the case a nine-year-old boy, who had behavioral problems and died of fluoxetine intoxication. A high concentration of fluoxetine and its major active metabolite norfluoxetine was found in forensic toxicological analysis, which led to a further investigation of the case. Toxicogenetic analysis of the case revealed that the child had a completely defective *CYP2D6* gene resulting in a poor ability to metabolize fluoxetine.

Druid et al. (1999) and Levo et al. (2003) showed that *CYP* genotyping was feasible in postmortem sample material, and that genetic variation correlated with the observed phenotype, that is, parent drug/metabolite ratio. Interestingly, in two studies on fatal drug intoxications, the PM subjects were found to be underrepresented when compared to the general population, but no clear explanation has been offered for this observation. *CYP* genotyping and PM has been used to aid interpretation of postmortem toxicology results in oxycodone (Jannetto et al. 2002), methadone (Wong et al. 2003), and fentanyl-related deaths (Jin et al. 2005).

18.3.4.1.2 Ultrarapid Metabolizing Genotypes and Intoxication
UM drug metabolism can also be associated with severe or fatal ADRs, if the enzyme catalyzes the conversion of a prodrug into an active compound. Three case reports involving CYP2D6 and codeine have recently been described. Koren et al. (2006) reported a case where a neonate was found dead at the age of 13 days. Postmortem analysis revealed that the CoD of the child was morphine intoxication. The newborn received the morphine

via breast milk from the mother who had been prescribed codeine for episiotomy pain. Codeine is O-demethylated to morphine in a reaction catalyzed by CYP2D6. After the child's death, it was discovered that the mother carried an active *CYP2D6* gene duplication associated with increased metabolism of codeine to morphine, which was lethal to the neonate. This case illustrates well that without pharmacogenetic testing and careful medicolegal consideration, the case could have been interpreted as an infanticide or medical misconduct/negligence. Although the CoD was morphine poisoning, the MoD was demonstrably accidental.

The CYP2D6 genotyping has been used interestingly to analyze cases with MoDs classified as suicide. From a previous study (Zackrisson et al. 2010), it was known that the frequency of *CYP2D6* gene duplication was higher than in normal populations, indicating UM metabolism, and Ahlner et al. (2010) analyzed two medicolegal autopsy groups: (violent) suicides and natural deaths. They found that the *CYP2D6* gene with more than two functional alleles occurred more than tenfold greater in the suicide group, when compared with the group of natural deaths.

In addition to CYP genes, cases possibly explained by transporter genes (Neuvonen et al. 2011) may prove to be interesting for medicolegal purposes.

The aforementioned cases demonstrate that applying pharmacogenetic concepts can have tremendous clinical and medicolegal interest. The toxicogenetic approach should also ultimately add important data for drug safety and translation of this knowledge to bedside medicine.

18.4 Investigation of Death Due to Neglect or Abuse

While somewhat beyond the scope of this chapter, it is crucial for the forensic autopsy to identify injuries that may raise suspicion of abuse but may be overlooked. These include bruises, oral injuries, or recent or healing fractures. Injuries in unusual locations, such as over the torso, ears or neck along with pattern injuries, injuries to multiple organ systems, or evidence of multiple injuries in different stages of healing. Any history of possible head trauma, including history of vomiting, lethargy, irritability, apnea, or seizures requires brain imaging and a skeletal survey.

It is crucial during the forensic autopsy to distinguish causes of injury leading to death from those that may in fact be due to genetic causes (Shur and Carey 2015). Examples include bone fractures assumed to be due to trauma with failure to recognize heritable bone fragility syndromes (Christian and States 2017), cases of sudden death in infants suspected to be the result of intentional poisoning but due instead to metabolic effects of rare heritable biochemical disorders (Shoemaker 1992), and CNS anomalies mistakenly thought to be the result of coup-contrecoup injuries (seen in "shaken-baby syndrome") but instead resulting from cerebral atrophy in rare metabolic disorders (Olpin 2004; Vester 2015).

18.5 Conclusion

Although molecular genetic testing is not yet commonly performed at autopsy, it is expected to become an integrated component in forensic investigations in the near future. A multidisciplinary team approach, involving forensic pathologists, scene investigators, epidemiologists, molecular geneticists, research scientists, and clinicians, has allowed us to gain an understanding of the genetic and genomic causes of sudden natural deaths. Such collaborative work will continue to be the theme of sudden natural death investigations and research. The recognition of rare genetic disorders that can mimic child abuse is crucial to avoid further tragedy to families after death of an infant or child.

Important issues need to be considered when performing genetic testing, including family consent, privacy, emotional stress, uncertainty of the clinical effects of a unreported or undercharacterized genetic variant, as well as the possibility of uncovering a previously unrecognized nonpaternity. Recognizing some of these important issues, the Genetic Information Nondiscrimination Act was signed into law in the United States in 2009 (Asmonga 2008), which protects individuals from the misuse of genetic information in health insurance and employment and protects the individual's privacy and rights. Furthermore, genetic counseling is expected to play an important role in postmortem molecular testing and family consultation.

References

Ackerman, M.J., B.L. Siu, W.Q. Sturner, et al. 2001. Postmortem molecular analysis of SCN5A defects in sudden infant death syndrome. *JAMA* 286(18):2264–9.

Ahlner, J., A.L. Zackrisson, B. Lindblom, and L. Bertilsson. 2010. CYP2D6, serotonin and suicide. *Pharmacogenomics* 11(7):903–5.

Akagawa, H., A. Tajima, Y. Sakamoto, et al. 2006. A haplotype spanning two genes, ELN and LIMK1, decreases their transcripts and confers susceptibility to intracranial aneurysms. *Hum Mol Genet* 15(10):1722–34.

Alders, M., T.T. Koopmann, I. Christiaans, et al. 2009. Haplotype-sharing analysis implicates chromosome 7q36 harboring DPP6 in familial idiopathic ventricular fibrillation. *Am J Hum Genet* 84(4):468–76.

Angius, A., E. Petretto, G.B. Maestrale, et al. 2002. A new essential hypertension susceptibility locus on chromosome 2p24–p25, detected by genomewide search. *Am J Hum Genet* 71(4):893–905.

Antzelevitch, C., G.D. Pollevick, J.M. Cordeiro, et al. 2007. Loss-of-function mutations in the cardiac calcium channel underlie a new clinical entity characterized by ST-segment elevation, short QT intervals, and sudden cardiac death. *Circulation* 115(4):442–9.

Aouizerat, B.E., E. Vittinghoff, S.L. Musone, et al. 2011. GWAS for discovery and replication of genetic loci associated with sudden cardiac arrest in patients with coronary artery disease. *BMC Cardiovasc Disord* 11:29.

Arking, D.E., M.J. Junttila, P. Goyette, et al. 2011. Identification of a sudden cardiac death susceptibility locus at 2q24.2 through genome-wide association in European ancestry individuals. *PLOS Genet* 7(6):e1002158.

Arnestad, M., L. Crotti, T.O. Rognum, et al. 2007. Prevalence of long-QT syndrome gene variants in sudden infant death syndrome. *Circulation* 115(3):361–7.

Asimaki, A., P. Syrris, T. Wichter, P. Matthias, J.E. Saffitz, and W.J. McKenna. 2007. A novel dominant mutation in plakoglobin causes arrhythmogenic right ventricular cardiomyopathy. *Am J Hum Genet* 81(5):964–73.

Asmonga, D. 2008. Getting to know GINA. An overview of the Genetic Information Nondiscrimination Act. *JAHIMA* 79(7):18, 20, 2.

Aurlien, D., T.P. Leren, E. Tauboll, and L. Gjerstad. 2009. New SCN5A mutation in a SUDEP victim with idiopathic epilepsy. *Seizure* 18(2):158–60.

Bahring, S., M. Kann, Y. Neuenfeld, et al. 2008. Inversion region for hypertension and brachydactyly on chromosome 12p features multiple splicing and noncoding RNA. *Hypertension* 51(2):426–31.

Barsheshet, A., A. Brenyo, A.J. Moss, and I. Goldenberg. 2011. Genetics of sudden cardiac death. *Curr Cardiol Rep* Oct; 13(5):364–76.

Beighton, P., A. De Paepe, B. Steinmann, P. Tsipouras, and R.J. Wenstrup. 1997. Ehlers–Danlos syndromes: Revised nosology, Villefranche, Ehlers–Danlos National Foundation (USA) and Ehlers–Danlos Support Group (U.K.). *Am J Med Genet* 77(1):31–7.

Bellocq, C., A.C. van Ginneken, C.R. Bezzina, et al. 2004. Mutation in the KCNQ1 gene leading to the short QT-interval syndrome. *Circulation* 109(20):2394–7.

Berthet, M., I. Denjoy, C. Donger, et al. 1999. C-terminal HERG mutations: The role of hypokalemia and a KCNQ1-associated mutation in cardiac event occurrence. *Circulation* 99(11):1464–70.

Bezzina, C.R., R. Pazoki, A. Bardai, et al. 2010. Genome-wide association study identifies a susceptibility locus at 21q21 for ventricular fibrillation in acute myocardial infarction. *Nat Genet* 42(8):688–91.

Bilguvar, K., K. Yasuno, M. Niemela, et al. 2008. Susceptibility loci for intracranial aneurysm in European and Japanese populations. *Nat Genet* 40(12):1472–7.

Bleyl, S.B., B.R. Mumford, M.C. Brown-Harrison, et al. 1997. Xq28-linked noncompaction of the left ventricular myocardium: Prenatal diagnosis and pathologic analysis of affected individuals. *Am J Med Genet* 31 72(3):257–65.

Bonne, G., E. Mercuri, A. Muchir, et al. 2000. Clinical and molecular genetic spectrum of autosomal dominant Emery–Dreifuss muscular dystrophy due to mutations of the lamin A/C gene. *Ann Neurol* 48(2):170–80.

Borowiec, M., C.W. Liew, R. Thompson, et al. 2009. Mutations at the BLK locus linked to maturity onset diabetes of the young and beta-cell dysfunction. *Proc Natl Acad Sci U S A* 106(34):14460–5.

Burke, A.P., A. Farb, Y.H. Liang, J. Smialek, and R. Virmani. 1996. Effect of hypertension and cardiac hypertrophy on coronary artery morphology in sudden cardiac death. *Circulation* 94(12):3138–45.

Carson, P.E., C.L. Flanagan, C.E. Ickes, and A.S. Alving. 1956. Enzymatic deficiency in primaquinesensitive erythrocytes. *Science* 124(3220):484–5.

Carvajal-Vergara, X., A. Sevilla, S.L. D'Souza, et al. 2010. Patient-specific induced pluripotent stem-cell-derived models of leopard syndrome. *Nature* 465(7299):808–12.

Cerrone, M., and S.G. Priori. 2011. Genetics of sudden death: Focus on inherited channelopathies. *Eur Heart J* 32(17):2109–18.

Chang, Y.P., X. Liu, J.D. Kim, et al. 2007. Multiple genes for essential-hypertension susceptibility on chromosome 1q. *Am J Hum Genet* 80(2):253–64.

Charron, P., E. Villard, P. Sebillon, et al. 2004. Danon's disease as a cause of hypertrophic cardiomyopathy: A systematic survey. *Heart* 90(8):842–6.

Chen, L., M.L. Marquardt, D.J. Tester, K.J. Sampson, M.J. Ackerman, and R.S. Kass. 2007. Mutation of an A-kinase-anchoring protein causes long-QT syndrome. *Proc Natl Acad Sci U S A* 104(52):20990–5.

Chen, Q., G.E. Kirsch, D. Zhang, et al. 1998. Genetic basis and molecular mechanism for idiopathic ventricular fibrillation. *Nature* 392(6673):293–6.

Chung, S.K., J.M. MacCormick, C.H. McCulley, et al. 2007. Long QT and Brugada syndrome gene mutations in New Zealand. *Heart Rhythm* 4(10):1306–14.

Christian, C.W., and L.J. States. 2017. Medical mimics of child abuse. *Amer J Roentgenol* 208(5):982–90.

Cohle, S.D., and B.A. Sampson. 2001. The negative autopsy: Sudden cardiac death or other? *Cardiovasc Pathol* 10(5):219–22.

Corrado, D., C. Basso, and G. Thiene. 2001. Sudden cardiac death in young people with apparently normal heart. *Cardiovasc Res* 50(2):399–408.

Cronk, L.B., B. Ye, T. Kaku, et al. 2007. Novel mechanism for sudden infant death syndrome: Persistent late sodium current secondary to mutations in caveolin-3. *Heart Rhythm* 4(2):161–6.

Dahl, M.L., I. Johansson, L. Bertilsson, M. Ingelman-Sundberg, and F. Sjöqvist. 1995. Ultrarapid hydroxylation of debrisoquine in a Swedish population. Analysis of the molecular genetic basis. *J Pharmacol Exp Ther* 274(1):516–20.

Dellefave, L., and E.M. McNally. 2010. The genetics of dilated cardiomyopathy. *Curr Opin Cardiol* 25(3):198-204.

Delpon, E., J.M. Cordeiro, L. Nunez, et al. 2008. Functional effects of KCNE3 mutation and its role in the development of Brugada syndrome. *Circ Arrhythm Electrophysiol* 1(3):209–18.

di Barletta, M.R., S. Viatchenko-Karpinski, A. Nori, et al. 2006. Clinical phenotype and functional characterization of CASQ2 mutations associated with catecholaminergic polymorphic ventricular tachycardia. *Circulation* 114(10):1012–9.

Di Paolo, M., D. Luchini, R. Bloise, and S.G. Priori. 2004. Postmortem molecular analysis in victims of sudden unexplained death. *Am J Forensic Med Pathol* 25(2):182–4.

Dietz, H.C., G.R. Cutting, R.E. Pyeritz, et al. 1991. Marfan syndrome caused by a recurrent de novo missense mutation in the fibrillin gene. *Nature* 352(6333):337–9.

Druid, H., P. Holmgren, B. Carlsson, and J. Ahlner. 1999. Cytochrome P450 2D6 (CYP2D6) genotyping on postmortem blood as a supplementary tool for interpretation of forensic toxicological results. *Forensic Sci Int* 99(1):25–34.

Ebbesen, J., I. Buajordet, J. Erikssen, et al. 2001. Drug-related deaths in a department of internal medicine. *Arch Intern Med* 161(19):2317–23.

Edghill, E.L., S.E. Flanagan, A.M. Patch, et al. 2008. Insulin mutation screening in 1,044 patients with diabetes: Mutations in the INS gene are a common cause of neonatal diabetes but a rare cause of diabetes diagnosed in childhood or adulthood. *Diabetes* 57(4):1034–42.

Edwards, I.R., and J.K. Aronson. 2000. Adverse drug reactions: Definitions, diagnosis, and management. *Lancet* 356(9237):1255–9.

Eichelbaum, M., N. Spannbrucker, B. Steincke, and H.J. Dengler. 1979. Defective N-oxidation of sparteine in man: A new pharmacogenetic defect. *Eur J Clin Pharmacol* 16(3):183–7.

Ellard, S., S.E. Flanagan, C.A. Girard, et al. 2007. Permanent neonatal diabetes caused by dominant, recessive, or compound heterozygous SUR1 mutations with opposite functional effects. *Am J Hum Genet* 81(2):375–82.

Favaloro, E.J., and G. Lipp. 2011. Coagulation update: What's new in hemostasis testing? *Thromb Res* 127(Suppl 2):S13–S6.

Franceschini, N., A.P. Reiner, and G. Heiss. 2011. Recent findings in the genetics of blood pressure and hypertension traits. *Am J Hypertens* 24(4):392–400.

Funke-Kaiser, H., F. Reichenberger, K. Kopke, et al. 2003. Differential binding of transcription factor E2F-2 to the endothelin-converting enzyme-1b promoter affects blood pressure regulation. *Hum Mol Genet* 12(4):423–33.

Geisterfer-Lowrance, A.A., S. Kass, G. Tanigawa, et al. 1990. A molecular basis for familial hypertrophic cardiomyopathy: A beta cardiac myosin heavy chain gene missense mutation. *Cell* 62(5):999–1006.

Geller, D.S., A. Farhi, N. Pinkerton, et al. 2000. Activating mineralocorticoid receptor mutation in hypertension exacerbated by pregnancy. *Science* 289(5476):119–23.

George Jr., A.L., T.A. Varkony, H.A. Drabkin, et al. 1995. Assignment of the human heart tetrodotoxin-resistant voltage-gated Na^+ channel alpha-subunit gene (SCN5A) to band 3p21. *Cytogenet Cell Genet* 68(1–2):67–70.

Gerull, B., A. Heuser, T. Wichter, et al. 2004. Mutations in the desmosomal protein plakophilin-2 are common in arrhythmogenic right ventricular cardiomyopathy. *Nat Genet* 36(11):1162–4.

Gill, J.R., and I.A. Scordi-Bello. 2010. Natural, unexpected deaths: Reliability of a presumptive diagnosis. *J Forensic Sci* 55(1):77–81.

Gloyn, A.L., E.R. Pearson, J.F. Antcliff, et al. 2004. Activating mutations in the gene encoding the ATP-sensitive potassium-channel subunit Kir6.2 and permanent neonatal diabetes. *N Engl J Med* 350(18):1838–49.

Gong, M., H. Zhang, H. Schulz, et al. 2003. Genome-wide linkage reveals a locus for human essential (primary) hypertension on chromosome 12p. *Hum Mol Genet* 12(11):1273–7.

Gonzalez, F.J., R.C. Skoda, S. Kimura, et al. 1988. Characterization of the common genetic defect in humans deficient in debrisoquine metabolism. *Nature* 331(6155):442–6.

Grant, E.K., and M.J. Evans. 2009. Cardiac findings in fetal and pediatric autopsies: A five-year retrospective review. *Pediatr Dev Pathol* 12(2):103–10.

Gretarsdottir, S., A.F. Baas, G. Thorleifsson, et al. 2010. Genome-wide association study identifies a sequence variant within the DAB2IP gene conferring susceptibility to abdominal aortic aneurysm. *Nat Genet* 42(8):692–7.

Gribouval, O., M. Gonzales, T. Neuhaus, et al. 2005. Mutations in genes in the renin–angiotensin system are associated with autosomal recessive renal tubular dysgenesis. *Nat Genet* 37(9):964–8.

Gui, J., T. Wang, R.P. Jones, D. Trump, T. Zimmer, and M. Lei. 2010. Multiple loss-of-function mechanisms contribute to SCN5A-related familial sick sinus syndrome. *PLOS ONE* 5(6):e10985.

Haissaguerre, N. Derval, F. Sacher, et al. 2008. Sudden cardiac arrest associated with early repolarization. *N Engl J Med* 358(19):2016–23.

Haissaguerre, M., S. Chatel, F. Sacher, et al. 2009. Ventricular fibrillation with prominent early repolarization associated with a rare variant of KCNJ8/KATP channel. *J Cardiovasc Electrophysiol* 20(1):93–8.

Heit, J.A., M.D. Silverstein, D.N. Mohr, T.M. Petterson, W.M. O'Fallon, and L.J. Melton 3rd. 1999. Predictors of survival after deep vein thrombosis and pulmonary embolism: A population-based, cohort study. *Arch Intern Med* 159(5):445–53.

Helgadottir, A., A. Manolescu, G. Thorleifsson, et al. 2004. The gene encoding 5-lipoxygenase activating protein confers risk of myocardial infarction and stroke. *Nat Genet* 36(3):233–9.

Hilbert, P., K. Lindpaintner, J.S. Beckmann, et al. 1991. Chromosomal mapping of two genetic loci associated with blood-pressure regulation in hereditary hypertensive rats. *Nature* 353(6344):521–9.

Hockwald, R.S., J. Arnold, C.B. Clayman, and A.S. Alving. 1952. Toxicity of primaquine in Negroes. *J Am Med Assoc* 149(17):1568–70.

Hoffmann, B., H. Schmidt-Traub, A. Perrot, K.J. Osterziel, and R. Gessner. 2001. First mutation in cardiac troponin C, L29Q, in a patient with hypertrophic cardiomyopathy. *Hum Mutat* 17(6):524.

Hong, K., P. Bjerregaard, I. Gussak, and R. Brugada. 2005. Short QT syndrome and atrial fibrillation caused by mutation in KCNH2. *J Cardiovasc Electrophysiol* 16(4):394–6.

Horikawa, Y., N. Iwasaki, M. Hara, et al. 1997. Mutation in hepatocyte nuclear factor-1 beta gene (TCF2) associated with MODY. *Nat Genet* 17(4):384–5.

Hu, D., H. Barajas-Martinez, E. Burashnikov, et al. 2009. A mutation in the beta 3 subunit of the cardiac sodium channel associated with Brugada ECG phenotype. *Circ Cardiovasc Genet* 2(3):270–8.

Ichida, F., S. Tsubata, K.R. Bowles, et al. 2001. Novel gene mutations in patients with left ventricular noncompaction or Barth syndrome. *Circulation* 103(9):1256–63.

Ikejiri, M., A. Tsuji, H. Wada, et al. 2010. Analysis three abnormal protein S genes in a patient with pulmonary embolism. *Thromb Res* 125(6):529–32.

Ilina, M.V., C.A. Kepron, G.P. Taylor, D.G. Perrin, P.F. Kantor, and G.R. Somers. 2011. Undiagnosed heart disease leading to sudden unexpected death in childhood: A retrospective study. *Pediatrics* 2011 128(3):e513-20.

International Statistical Classification of Diseases and Related Health Problems. 10th revision (ICD-10). 1993. Vol 2: Instruction Manual, World Health Organisation, Geneva.

Jannetto, P.J., S.H. Wong, S.B. Gock, E. Laleli-Sahin, B.C. Schur, and J.M. Jentzen. 2002. Pharmacogenomics as molecular autopsy for postmortem forensic toxicology: Genotyping cytochrome P450 2D6 for oxycodone cases. *J Anal Toxicol* 26(7):438–47.

Jin, M., S.B. Gock, P.J. Jannetto, J.M. Jentzen, and S.H. Wong. 2005. Pharmacogenomics as molecular autopsy for forensic toxicology: Genotyping cytochrome P4503A4*1B and3A5*3 for 25 fentanyl cases. *J Anal Toxicol* 29(7):590–8.

Joe, B., Y. Saad, N.H. Lee, et al. 2009. Positional identification of variants of Adamts16 linked to inherited hypertension. *Hum Mol Genet* 18(15):2825–38.

Kalow, W., and N. Staron. 1957. On distribution and inheritance of atypical forms of human serum cholinesterase, as indicated by dibucaine numbers. *Can J Biochem Physiol* 35(12):1305–20.

Kelly, B.P., M.W. Russell, J.R. Hennessy, and G.J. Ensing. 2009. Severe hypertrophic cardiomyopathy in an infant with a novel PRKAG2 gene mutation: Potential differences between infantile and adult onset presentation. *Pediatr Cardiol* 30(8):1176–9.

Kimura, A., H. Harada, J.E. Park, et al. 1997. Mutations in the cardiac troponin I gene associated with hypertrophic cardiomyopathy. *Nat Genet* 16(4):379–82.

Kirchheiner, J. 2008. CYP2D6 phenotype prediction from genotype: Which system is the best? *Clin Pharmacol Ther* 83(2):225–27.

Kirchheiner, J., K. Nickchen, M. Bauer, et al. 2004. Pharmacogenetics of antidepressants and antipsychotics: The contribution of allelic variations to the phenotype of drug response. *Mol Psychiatry* 9(5):442–73.

Klaassen, S., S. Probst, E. Oechslin, et al. 2008. Mutations in sarcomere protein genes in left ventricular noncompaction. *Circulation* 117(22):2893–901.

Koide, S., K. Kugiyama, S. Sugiyama, et al. 2003. Association of polymorphism in glutamate–cysteine ligase catalytic subunit gene with coronary vasomotor dysfunction and myocardial infarction. *J Am Coll Cardiol* 41(4):539–45.

Koren, G., J. Cairns, D. Chitayat, A. Gaedigk, and S.J. Leeder. 2006. Pharmacogenetics of morphine poisoning in a breastfed neonate of a codeine-prescribed mother. *Lancet* 368(9536):704.

Koski, A. 2005 Interpretation of Postmortem Toxicology Results. Academic dissertation. Helsinki University Printing House, Helsinki. http://e-thesis.helsinki.fi.

Koski, A., I. Ojanperä, and E. Vuori. 2003. Interaction of alcohol and drugs in fatal poisonings. *Hum Exp Toxicol* 22(5):281–7.

Koski, A., J. Sistonen, I. Ojanpera, M. Gergov, E. Vuori, and A. Sajantila. 2006. CYP2D6 and CYP2C19 genotypes and amitriptyline metabolite ratios in a series of medicolegal autopsies. *Forensic Sci Int* 158(2–3):177–83.

Kruse, M., E. Schulze-Bahr, V. Corfield, et al. 2009. Impaired endocytosis of the ion channel TRPM4 is associated with human progressive familial heart block type I. *J Clin Invest* 119(9):2737–44.

Kumar, D.R., E. Hanlin, I. Glurich, J.J. Mazza, and S.H. Yale. 2010. Virchow's contribution to the understanding of thrombosis and cellular biology. *Clin Med Res* 8(3–4):168–72.

Lahat, H., E. Pras, and M. Eldar. 2004. A missense mutation in CASQ2 is associated with autosomal recessive catecholamine-induced polymorphic ventricular tachycardia in Bedouin families from Israel. *Ann Med* 36(Suppl 1):87–91.

Landstrom, A.P., B.A. Adekola, J.M. Bos, S.R. Ommen, and M.J. Ackerman. 2011. PLN-encoded phospholamban mutation in a large cohort of hypertrophic cardiomyopathy cases: Summary of the literature and implications for genetic testing. *Am Heart J* 161(1):165–71.

Launiainen, T., A. Sajantila, I. Rasanen, E. Vuori, and I. Ojanperä. 2010. Adverse interaction of warfarin and paracetamol: Evidence from a post-mortem study. *Eur J Clin Pharmacol* 66(1):97–103.

Launiainen, T., E. Vuori, and I. Ojanperä. 2009. Prevalence of adverse drug combinations in a large post-mortem toxicology database. *Int J Legal Med* 123(2):109–15.

Launiainen, T., I. Rasanen, E. Vuori, and I. Ojanperä. 2011. Fatal venlafaxine poisonings are associated with a high prevalence of drug interactions. *Int J Legal Med* 125(3):349–58.

Lazarou, J., B.H. Pomeranz, and P.N. Corey. 1998. Incidence of adverse drug reactions in hospitalized patients: A meta-analysis of prospective studies. *JAMA* 279(15):1200–05.

Lehmann, H., and E. Ryan. 1956. The familial incidence of low pseudocholinesterase level. *Lancet* 271(6934):124.

Levo, A., A. Koski, I. Ojanperä, E. Vuori, and A. Sajantila. 2003. Post-mortem SNP analysis of *CYP2D6* gene reveals correlation between genotype and opioid drug (tramadol) metabolite ratios in blood. *Forensic Sci Int* 135(1):9–15.

Little, W.C., and R.J. Applegate. 1996. Role of plaque size and degree of stenosis in acute myocardial infarction. *Cardiol Clin* 14(2):221–8.

Liu, N., Y. Ruan, and S.G. Priori. 2008. Catecholaminergic polymorphic ventricular tachycardia. *Prog Cardiovasc Dis* 51(1):23–30.

London, B., M. Michalec, H. Mehdi, et al. 2007. Mutation in glycerol-3-phosphate dehydrogenase 1 like gene (*GPD1-L*) decreases cardiac Na$^+$ current and causes inherited arrhythmias. *Circulation* 13 116(20):2260–8.

London, B., A.M. Greiner, H. Mehdi, and R. Gutmann. 2018. Identifying new sudden death genes. *Trans Am Clin Climatol Assoc* 129:183–4.

Madea, B., F. Musshoff, and J. Preuss. 2009. Medical negligence in drug associated deaths. *Forensic Sci Int* 190(1–3):67–73.

Mahgoub, A., J.R. Idle, L.G. Dring, R. Lancaster, and R.L. Smith. 1977. Polymorphic hydroxylation of debrisoquine in man. *Lancet* 2(8038):584–86.

Malecki, M.T., U.S. Jhala, A. Antonellis, et al. 1999. Mutations in NEUROD1 are associated with the development of type 2 diabetes mellitus. *Nat Genet* 23(3):323–8.

Mani, A., J. Radhakrishnan, H. Wang, et al. 2007. LRP6 mutation in a family with early coronary disease and metabolic risk factors. *Science* 315(5816):1278–82.

Manunta, P., M. Burnier, M. D'Amico, et al. 1999. Adducin polymorphism affects renal proximal tubule reabsorption in hypertension. *Hypertension* 33(2):694–7.

Markovic, D., X. Tang, M. Guruju, M.A. Levenstien, J. Hoh, A. Kumar, and J. Ott. 2005. Association of angiotensinogen gene polymorphisms with essential hypertension in African-Americans and Caucasians. *Hum Hered* 60(2):89–96.

Marsman, R.F., A. Bardai, A.V. Postma, et al. 2011. A complex double deletion in LMNA underlies progressive cardiac conduction disease, atrial arrhythmias, and sudden death. *Circ Cardiovasc Genet* 4(3):280–7.

McKusick, V.A. 1991. The defect in Marfan syndrome. *Nature* 352(6333):279–81.

Medeiros-Domingo, A., T. Kaku, and D.J. Tester. 2008. SCN4B-encoded sodium channel beta4 subunit in congenital long-QT syndrome. *Circulation* 116(2):134–42.

Merner, N.D., K.A. Hodgkinson, A.F. Haywood, et al. 2008. Arrhythmogenic right ventricular cardiomyopathy type 5 is a fully penetrant, lethal arrhythmic disorder caused by a missense mutation in the *TMEM43* gene. *Am J Hum Genet* 82(4):809–21.

Millat, G., B. Kugener, P. Chevalier, et al. 2009. Contribution of long-QT syndrome genetic variants in sudden infant death syndrome. *Pediatr Cardiol* 30(4):502–9.

Mineharu, Y., K. Inoue, S. Inoue, et al. 2008. Association analyses confirming a susceptibility locus for intracranial aneurysm at chromosome 14q23. *J Hum Genet* 53(4):325–32.

Mohler, P.J., J.J. Schott, A.O. Gramolini, et al. 2003. Ankyrin-B mutation causes type 4 long-QT cardiac arrhythmia and sudden cardiac death. *Nature* 421(6923):634–9.

Moore, T.J., B.M. Psaty, and C.D. Furberg. 1998. Time to act on drug safety. *JAMA* 279(19): 1571–3.

Motulsky, A.G. 1957. Drug reactions enzymes, and biochemical genetics. *JAMA* 165(7):835–7.

Musshoff, F., U.M. Stamer, and B. Madea. 2010. Pharmacogenetics and forensic toxicology. *Forensic Sci Int* 203(1–3):53–62.

Nahed, B.V., A. Seker, B. Guclu, et al. 2005. Mapping a Mendelian form of intracranial aneurysm to 1p34.3–p36.13. *Am J Hum Genet* 76(1):172–9.

Nakayama, T., M. Soma, Y. Watanabe, et al. 2003. Splicing mutation of the prostacyclin synthase gene in a family associated with hypertension. *Adv Exp Med* 525:165–8.

Neuvonen, A.M., J.U. Palo, and A. Sajantila. 2011. Post-mortem ABCB1 genotyping reveals an elevated toxicity for female digoxin users. *Int J Legal Med* 125(2):265–9.

Neve, B., M.E. Fernandez-Zapico, V. Ashkenazi-Katalan, et al. 2005. Role of transcription factor KLF11 and its diabetes-associated gene variants in pancreatic beta cell function. *Proc Natl Acad Sci U S A* 102(13):4807–12.

Newton-Cheh, C., T. Johnson, V. Gateva, et al. 2009. Genome-wide association study identifies eight loci associated with blood pressure. *Nat Genet* 41(6):666–76.

Njolstad, P.R., O. Sovik, A. Cuesta-Munoz, et al. 2001. Neonatal diabetes mellitus due to complete glucokinase deficiency. *N Engl J Med* 344(21):1588–92.

Ojanperä, I., M. Kolmonen, and A. Pelander. 2012. Current use of high-resolution mass spectrometry in drug screening relevant to clinical and forensic toxicology and doping control. *Anal Biochem* 403(5):1203–20.

Olpin, S.E. 2004. The metabolic investigation of sudden infant death. *Ann Clin Biochem* 41(4):282–93.

Ozturk, A.K., B.V. Nahed, M. Bydon, et al. 2006. Molecular genetic analysis of two large kindreds with intracranial aneurysms demonstrates linkage to 11q24–25 and 14q23–31. *Stroke* 37(4):1021–7.

Padmanabhan, S., O. Melander, T. Johnson, et al. 2010. Genome-wide association study of blood pressure extremes identifies variant near UMOD associated with hypertension. *PLOS Genet* 6(10):e1001177.

Patel, H.H., L.L. Hamuro, B.J. Chun, et al. 2010. Disruption of protein kinase A localization using a trans-activator of transcription (TAT)-conjugated A-kinase-anchoring peptide reduces cardiac function. *J Biol Chem* 285(36):27632–40.

Petersen, E.E., S.A. Rasmussen, K.L. Daniel, M.M. Yazdy, and M.A. Honein. 2008. Prescription medication borrowing and sharing among women of reproductive age. *J Womens Health (Larchmt)* 17(7):1073–80.

Pirmohamed, M., A.M. Breckenridge, N.R. Kitteringham, and B.K. Park. 1998. Adverse drug reactions. *BMJ* 316(7140):1295–8.

Pirmohamed, M., S. James, S. Meakin, et al. 2004. Adverse drug reactions as cause of admission to hospital: Prospective analysis of 18 820 patients. *Br Med J* 329(7456):15–9.

Plaster, N.M., R. Tawil, M. Tristani-Firouzi, et al. 2001. Mutations in Kir2.1 cause the developmental and episodic electrical phenotypes of Andersen's syndrome. *Cell* 105(4):511–9.

Plengvidhya, N., S. Kooptiwut, N. Songtawee, et al. 2007. PAX4 mutations in Thais with maturity onset diabetes of the young. *J Clin Endocrinol Metab* 92(7):2821–6.

Podgoreanu, M.V., and D.A. Schwinn. 2005. New paradigms in cardiovascular medicine: Emerging technologies and practices: Perioperative genomics. *J Am Coll Cardiol* 46(11):1965–77.

Pounder, D., and G.R. Jones. 1990. Post-mortem drug redistribution—A toxicological nightmare. *Forensic Sci Int* 45(3):253–63.

Preuss, M., I.R. Konig, J.R. Thompson, et al. 2010. Design of the coronary artery disease genome-wide replication and meta-analysis (CARDIoGRAM) study: A genome-wide association meta-analysis involving more than 22 000 cases and 60 000 controls. *Circ Cardiovasc Genet* 3(5):475–83.

Priori, S.G., C. Napolitano, and P.J. Schwartz. 1999. Low penetrance in the long-QT syndrome: Clinical impact. *Circulation* 99(4):529–33.

Priori, S.G., C. Napolitano, N. Tiso, et al. 2001. Mutations in the cardiac ryanodine receptor gene (*hRyR2*) underlie catecholaminergic polymorphic ventricular tachycardia. *Circulation* 2(2):196–200.

Priori, S.G., S.V. Pandit, I. Rivolta, et al. 2005. A novel form of short QT syndrome (SQT3) is caused by a mutation in the *KCNJ2* gene. *Circ Res* 96(7):800–7.

Prutkin, J.M., J.A. Strote, and K.K. Stout. 2008. Percutaneous right ventricular assist device as support for cardiogenic shock due to right ventricular infarction. *J Invasive Cardiol.* 20(7):E215-6.

Raeder, H., S. Johansson, P.I. Holm, et al. 2006. Mutations in the CEL VNTR cause a syndrome of diabetes and pancreatic exocrine dysfunction. *Nat Genet* 38(1):54–62.

Rafestin-Oblin, M.E., A. Souque, B. Bocchi, G. Pinon, J. Fagart, and A. Vandewalle. 2003. The severe form of hypertension caused by the activating S810L mutation in the mineralocorticoid receptor is cortisone related. *Endocrinology* 144(2):528–33.

Rampazzo, A., G. Beffagna, A. Nava, et al. 2003. Arrhythmogenic right ventricular cardiomyopathy type 1 (ARVD1): Confirmation of locus assignment and mutation screening of four candidate genes. *Eur J Hum Genet* 11(1):69–76.

Ria, M., J. Lagercrantz, A. Samnegard, S. Boquist, A. Hamsten, and P. Eriksson. 2011. A common polymorphism in the promoter region of the *TNFSF4* gene is associated with lower allele-specific expression and risk of myocardial infarction. *PLOS ONE* 6(3):e17652.

Richard, P., P. Charron, L. Carrier, et al. 2003. Hypertrophic cardiomyopathy: Distribution of disease genes, spectrum of mutations, and implications for a molecular diagnosis strategy. *Circulation* 107(17):2227–32.

Roos, Y.B., G. Pals, P.M. Struycken, et al. 2004. Genome-wide linkage in a large Dutch consanguineous family maps a locus for intracranial aneurysms to chromosome 2p13. *Stroke* 35(10):2276–81.

Rutherford, S., M.P. Johnson, R.P. Curtain, and L.R. Griffiths. 2001. Chromosome 17 and the inducible nitric oxide synthase gene in human essential hypertension. *Hum Genet* 109(4): 408–15.

Sajantila, A., J.U. Palo, I. Ojanperä, C. Davis, and B. Budowle. 2010. Pharmacogenetics in medico-legal context. *Forensic Sci Int* 203(1–3):44–52.

Sallee, F.R., C.L. DeVane, and R.E. Ferrell. 2000. Fluoxetine-related death in a child with cytochrome P-450 2D6 genetic deficiency. *J Child Adolesc Psychopharmacol* 10(1):27–34.

Schwarze, U., J.A. Goldstein, and P.H. Byers. 1997. Splicing defects in the *COL3A1* gene: Marked preference for 5' (donor) spice-site mutations in patients with exon-skipping mutations and Ehlers–Danlos syndrome type IV. *Am J Hum Genet* 61(6):1276–86.

Shoemaker, J.D., R.E. Lynch, J.W. Hoffmann, and W.S. Sly. 1992. Misidentification of propionic acid as ethylene glycol in a patient with methylmalonic acidemia. *J Pediatr* 120(3):417–21.

Shur, N., and J.C. Carey. 2015. Genetic differentials of child abuse: Is your case rare or real? Am J Med Genet C Semin Med Genet 169(4):281–8.

Simonetti, G.D., M.G. Mohaupt, and M.G. Bianchetti. 2011. Monogenic forms of hypertension. *Eur J Pediatr* Oct; 171(10):1433–9.

Sivapalaratnam, S., M.M. Motazacker, S. Maiwald, et al. 2011. Genome-wide association studies in atherosclerosis. *Curr Atheroscler Rep* 13(3):225–32.

Smilowitz, N.R., B.A. Sampson, C.R. Abrecht, J.S. Siegfried, J.S. Hochman, and H.R. Reynolds. 2011. Women have less severe and extensive coronary atherosclerosis in fatal cases of ischemic heart disease: An autopsy study. *Am Heart J* 161(4):681–8.

Splawski, I., J. Shen, K.W. Timothy, et al. 2000. Spectrum of mutations in long-QT syndrome genes. KVLQT1, HERG, SCN5A, KCNE1, and KCNE2. *Circulation* 102(10):1178–85.

Splawski, I., K.W. Timothy, N. Decher, et al. 2005. Severe arrhythmia disorder caused by cardiac L-type calcium channel mutations. *Proc Natl Acad Sci U S A* 102(23):8089–96; discussion 6–8.

Stoffers, D.A., J. Ferrer, W.L. Clarke, and J.F. Habener. 1997. Early-onset type-II diabetes mellitus (MODY4) linked to IPF1. *Nat Genet* 17(2):138–9.

Syrris, P., D. Ward, A. Asimaki, et al. 2007. Desmoglein-2 mutations in arrhythmogenic right ventricular cardiomyopathy: A genotype–phenotype characterization of familial disease. *Eur Heart J* 28(5):581–8.

Syrris, P., D. Ward, A. Evans, et al. 2006. Arrhythmogenic right ventricular dysplasia/cardiomyopathy associated with mutations in the desmosomal gene desmocollin-2. *Am J Hum Genet* 79(5):978–84.

Tang, Y., B. Sampson, S. Pack, et al. 2011. Ethnic differences in out-of-hospital fatal pulmonary embolism. *Circulation* 123(20):2219–25.

Tester, D.J., A. Medeiros-Domingo, M.L. Will, C.M. Haglund, and M.J. Ackerman. 2012. Cardiac channel molecular autopsy: Insights from 173 consecutive cases of autopsy-negative sudden unexplained death referred for postmortem genetic testing. *Mayo Clin Proc* 87(6):524–39.

Tester, D.J., and M.J. Ackerman. 2009. Cardiomyopathic and channelopathic causes of sudden unexplained death in infants and children. *Annu Rev Med* 60:69–84.

Thiene, G. 2018. Sudden cardiac death in the young: A genetic destiny? *Clin Med (Lond)* 18(Suppl 2):s17–s23.

Thierfelder, L., H. Watkins, C. MacRae, et al. 1994. Alpha-tropomyosin and cardiac troponin T mutations cause familial hypertrophic cardiomyopathy: A disease of the sarcomere. *Cell* 77(5):701–12.

Thompson, E.E., H. Kuttab-Boulos, D. Witonsky, L. Yang, B.A. Roe, and A. Di Rienzo. 2004. CYP3A variation and the evolution of salt-sensitivity variants. *Am J Hum Genet* 75(6):1059–69.

Tiso, N., D.A. Stephan, A. Nava, et al. 2001. Identification of mutations in the cardiac ryanodine receptor gene in families affected with arrhythmogenic right ventricular cardiomyopathy type 2 (ARVD2). *Hum Mol Genet* 10(3):189–94.

Tu, E., R.D. Bagnall, J. Duflou, and C. Semsarian. 2011. Post-mortem review and genetic analysis of sudden unexpected death in epilepsy (SUDEP) cases. *Brain Pathol* 21(2):201–8.

Tu, E., R.D. Bagnall, L. Waterhouse, J. Duflou, and C. Semsarian. 2011. Genetic analysis of hyper-polarization-activated cyclic nucleotide-gated cation channels in sudden unexpected death in epilepsy cases. *Brain Pathol* 21(6):692-8.

Ueda, K., Y. Hirano, Y. Higashiuesato, et al. 2009. Role of HCN4 channel in preventing ventricular arrhythmia. *J Hum Genet* 54(2):115–21.

Ueda, K., C. Valdivia, A. Medeiros-Domingo, et al. 2008. Syntrophin mutation associated with long QT syndrome through activation of the nNOS–SCN5A macromolecular complex. *Proc Natl Acad Sci U S A* 105(27):9355–60.

van der Voet, M., J.M. Olson, H. Kuivaniemi, et al. 2004. Intracranial aneurysms in Finnish families: Confirmation of linkage and refinement of the interval to chromosome 19q13.3. *Am J Hum Genet* 74(3):564–71.

Vatta, M., B. Mohapatra, S. Jimenez, et al. 2003. Mutations in Cypher/ZASP in patients with dilated cardiomyopathy and left ventricular non-compaction. *J Am Coll Cardiol* 42(11):2014–27.

Vaxillaire, M., M. Rouard, K. Yamagata, et al. 1997. Identification of nine novel mutations in the hepatocyte nuclear factor 1 alpha gene associated with maturity-onset diabetes of the young (MODY3). *Hum Mol Genet* 6(4):583–6.

Verlaan, D.J., M.P. Dube, J. St-Onge, et al. 2006. A new locus for autosomal dominant intracranial aneurysm, ANIB4, maps to chromosome 5p15.2–14.3. *J Med Genet* 43(6):e31.

Vionnet, N., M. Stoffel, J. Takeda, et al. 1992. Nonsense mutation in the glucokinase gene causes early-onset non-insulin-dependent diabetes mellitus. *Nature* 356(6371):721–2.

Vester, M.E.M., R.A.C. Bilo, W.A. Karst, J.G. Daams, W.L.J.M. Duijst, and R.R. van Rijn. 2015. Subdural hematomas: Glutaric aciduria type 1 or abusive head trauma? A systematic review. Forensic Sci Med Pathol 11(3):405–15.

Vossen, C.Y., J. Conard, J. Fontcuberta, et al. 2005. Risk of a first venous thrombotic event in carriers of a familial thrombophilic defect. The European prospective cohort on thrombophilia (EPCOT). *J Thromb Haemost* 3(3):459–64.

Vuori, E., J.A. Henry, I. Ojanpera, et al. 2003. Death following ingestion of MDMA (ecstasy) and moclobemide. *Addiction* 98(3):365–8.

Wallace, C., M.Z. Xue, S.J. Newhouse, et al. 2006. Linkage analysis using co-phenotypes in the BRIGHT study reveals novel potential susceptibility loci for hypertension. *Am J Hum Genet* 79(2):323–31.

Wang, Q., S. Rao, G.Q. Shen, et al. 2004. Premature myocardial infarction novel susceptibility locus on chromosome 1P34–36 identified by genomewide linkage analysis. *Am J Hum Genet* 74(2):262–71.

Wang, H., Z. Li, J. Wang, et al. 2010. Mutations in *NEXN*, a Z-disc gene, are associated with hypertrophic cardiomyopathy. *Am J Hum Genet* 87(5):687–93.

Wang, L., C. Fan, S.E. Topol, E.J. Topol, and Q. Wang. 2003. Mutation of MEF2A in an inherited disorder with features of coronary artery disease. *Science* 302(5650):1578–81.

Wang, L., H.L. McLeod, and R.M. Weinshilboum. 2011. Genomics and drug response. *N Engl J Med* 364(12):1144–53.

Wang, X., M. Ria, P.M. Kelmenson, et al. 2005. Positional identification of TNFSF4, encoding OX40 ligand, as a gene that influences atherosclerosis susceptibility. *Nat Genet* 37(4):365–72.

Wang, X.L., A.S. Sim, R.F. Badenhop, R.M. McCredie, and D.E. Wilcken. 1996. A smoking-dependent risk of coronary artery disease associated with a polymorphism of the endothelial nitric oxide synthase gene. *Nat Med* 2(1):41–5.

Watanabe, H., T.T. Koopmann, S. Le Scouarnec, et al. 2008. Sodium channel beta1 subunit mutations associated with Brugada syndrome and cardiac conduction disease in humans. *J Clin Invest* 118(6):2260–8.

Watkins, H., D. Conner, L. Thierfelder, et al. 1995. Mutations in the cardiac myosin binding protein-C gene on chromosome 11 cause familial hypertrophic cardiomyopathy. *Nat Genet* 11(4):434–7.

Watkins, H., H. Ashrafian, and C. Redwood. 2011. Inherited cardiomyopathies. *N Engl J Med* 364(17):1643–56.

Wong, S.H., C. Happy, D. Blinka, et al. 2010. From personalized medicine to personalized justice: The promises of translational pharmacogenomics in the justice system. *Pharmacogenomics* 11(6):731–7.

Wong, S.H., M.A. Wagner, J.M. Jentzen, et al. 2003. Pharmacogenomics as an aspect of molecular autopsy for forensic pathology/toxicology: Does genotyping CYP2D6 serve as an adjunct for certifying methadone toxicity? *J Forensic Sci* 48(6):1406–15.

Xu, X., J. Yang, J. Rogus, C. Chen, and N. Schork. 1999. Mapping of a blood pressure quantitative trait locus to chromosome 15q in a Chinese population. *Hum Mol Genet* 8(13):2551–5.

Yamagata, K., H. Furuta, N. Oda, et al. 1996. Mutations in the hepatocyte nuclear factor-4alpha gene in maturity-onset diabetes of the young (MODY1). *Nature* 384(6608):458–60.

Yang, Y., B. Liang, J. Liu, et al. 2010. Identification of a Kir3.4 Mutation in congenital long QT syndrome. *Am J Hum Genet* 86(6):872-80.

Yang, Y., J. Li, X. Lin, et al. 2009. Novel KCNA5 loss-of-function mutations responsible for atrial fibrillation. *J Hum Genet* 54(5):277–83.

Yang, Z., N.E. Bowles, S.E. Scherer, et al. 2006. Desmosomal dysfunction due to mutations in desmoplakin causes arrhythmogenic right ventricular dysplasia/cardiomyopathy. *Circ Res* 99(6):646–55.

Yasuno, K., K. Bilguvar, P. Bijlenga, et al. 2010. Genome-wide association study of intracranial aneurysm identifies three new risk loci. *Nat Genet* 42(5):420–5.

Zackrisson, A.L., B. Lindblom, and J. Ahlner. 2010. High frequency of occurrence of *CYP2D6* gene duplication/multiduplication indicating ultrarapid metabolism among suicide cases. *Clin Pharmacol Ther* 88(3):354–9.

Zaracostas, J. 2007. Misuse of prescription drugs could soon exceed that of illicit narcotics, UN panel warns. *BMJ* 334(7591):444.

Zhang, L.P., B. Hui, and B.R. Gao. 2011. High risk of sudden death associated with a PRKAG2-related familial Wolff–Parkinson–White syndrome. *J Electrocardiol* 44(4):483–6.

Genetic Genealogy in the Genomic Era

19

JAKE K. BYRNES AND PETER
A. UNDERHILL

Contents

19.1 Introduction

Genealogy is the study of family history, which involves tracing one's family lineages back through time by documenting vital statistics for each ancestor, such as dates and places of births, deaths, and marriages. Genealogy has grown into a popular pursuit for millions of people worldwide, facilitated in recent decades by the digitization of records made accessible via the internet. Recently, genealogy ranked as the second most popular search topic on the internet (Falconer 2012), and there are currently a substantial number of companies offering genealogical products, many of which include genetic analysis (Shriver and Kittles 2004). The primary factors that motivate individuals to engage in genealogy research involve the combined thrill of discovering new family information while also experiencing an enhanced sense of personal identity. Serious genealogists and hobbyists alike describe the research as "addictive" because there is always a new ancestor to discover and learn about.

Genealogical research depends on written historical records for determining family relationships, and reliance on a chain of documents back through time inevitably leads to

DOI: 10.4324/9780429019944-22

brick walls for researchers. While there are numerous reasons for such barriers, two common causes are that records have been destroyed or they were never kept in the first place. For example, during the American Civil War the Southern states experienced extensive record loss resulting from a scorched earth policy that was implemented by Union General William T. Sherman. As a result, many genealogy researchers with ancestry from Southern states cannot extend these lineages beyond the late 1800s. Similarly, African Americans have a particularly difficult task when tracing ancestors beyond the late 1800s since many of their African ancestors before this time were enslaved, and only limited records, if any, were kept on slaves. Prior to emancipation, slaves were considered property and therefore they could not legally marry, vote, own land, or participate in other activities recorded in vital records. Furthermore, the 1870 census was the first time African Americans were listed by name as opposed to simply being listed by quantity, age, and gender.

The idea of applying molecular methods to overcome otherwise impenetrable brick walls in genealogical research emerged in the 1990s with several highly publicized reports of long-standing family mysteries being solved with genetic evidence in cases where traditional methods were unsuccessful. Examples include analysis conclusively disproving the claims by Anna Anderson that she was actually Anastasia Romanov, daughter of Tsar Nicholas II of Russia (Coble 2011), and Y chromosome analysis showing that Thomas Jefferson, or a close paternal relative, fathered the children of one of Jefferson's slaves, Sally Hemmings (Foster et al. 1998). From these and other studies, it became readily apparent that genetic testing was a powerful tool to discover unknown relationships beyond the point where traditional records end.

During the same time, the cost of genetic testing dropped precipitously, enabling numerous companies to emerge offering genetic testing for genealogical purposes. Companies initially focused on Y chromosome short tandem repeat (STR) testing for surname-based patrilineal studies. Soon however, product lines expanded to include mitochondrial genome sequence analysis for maternal line research. The popularity of these haploid tests has resulted in large and growing haplotype databases comprising hundreds of thousands of genetic profiles. Genetic genealogy companies are now offering autosomal microarray-based testing options, in some cases providing medical risk assessments in addition to genealogical matching services. As the genetic genealogy market continues to expand and its DNA databases continue to grow in size and diversity, significant repositories of human genetic variation are being created that may serve as a valuable resource for numerous disciplines focused on the study of human genetics.

19.2 DNA Testing for Genealogy

Genealogy researchers typically turn to DNA testing for two primary reasons. First, they want to confirm a particular family connection or piece of family lore that is unsubstantiated by written records. Second, they are searching for new leads after getting stuck. To address these challenges, DNA tests can be purchased over the internet from any one of several DNA testing companies. Typically, a company will mail a DNA collection kit to the client/genealogist that will contain cheek swabs, mouth rinse, or, more recently, saliva tubes to collect a biological sample from which to isolate DNA. The recipient returns the collection kit to the testing company and receives results several weeks later. The test results that the client receives typically include a personal DNA profile and information about deep (ancient) ancestry in the form of haplogroup membership or ethnicity estimation. The DNA profile is usually added to the company's DNA database and the client obtains access to online tools for comparing his/her genetic profile with others. In general, the primary objective of taking a DNA test is to compare one's DNA profile with those of other individuals to find close matches, often referred to as "genetic cousins," to identify a shared ancestor within the recent past (within the past 1000 years). In the following we introduce the tools and techniques applied to genetic data to extract the most useful genealogical information.

19.3 Haploid Chromosome Testing

Until the recent introduction of autosomal testing the primary options available to genealogists were Y chromosome and mtDNA (mitochondrial DNA) analyses. The haploid nature of these chromosomes makes them powerful systems for identifying common ancestors shared between individuals along specific ancestral lineages. Obvious differences exist between the Y chromosome and mtDNA genomes, such as the size of their informative sequence, 10 million versus 16,000 nucleotides, respectively, plus a higher overall mutation rate for the mtDNA resulting in more homoplasy (recurrent/revertant mutational events) within the mtDNA phylogenetic tree. Nonetheless, because these loci follow strict lineal inheritance patterns and are not subject to recombination and assortment, extant variation is due solely to mutations that have accumulated sequentially over time, thereby providing a molecular clock for inferring the number of generations separating two haploid genetic profiles or haplotypes. Additionally, because haploid chromosomes are not subjected to loss due to independent assortment, these chromosomes trace lineages both through recent history as well as into the very distant past. We discuss both haploid systems next, beginning with the Y chromosome.

19.3.1 Y Chromosome Testing

19.3.1.1 Discovering Paternal Lineages

In addition to being the first type of DNA test provided for genealogical purposes, Y chromosome testing remains a favorite tool among genealogists for researching paternal line ancestors. The Y chromosome is carried only by males and follows a strict paternal inheritance pattern, being passed from father to son with each successive generation (Figure 19.1a). This mode of inheritance mirrors the transmission of surnames in most Western societies making the Y chromosome a particularly useful tool for genealogists (King and Jobling 2009).

The popularity of surname studies and Y chromosome databases demonstrates the utility of the Y chromosome for identifying relatedness along paternal lineages. There are currently thousands of surname studies underway which involve recruiting males who share the same surname or similar surname variants to undergo Y chromosome testing in order to identify the number of unique origins of their surname and to determine the individual lineages that converge within recent history. This enables related individuals to reconstruct the details of how they are connected to each other along their paternal lines and to collaborate on extending their shared lineage further into the past. At the same time false leads are eliminated that may have been previously pursued, saving both time and money.

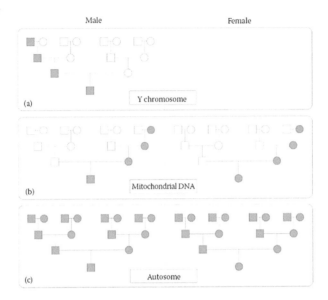

Figure 19.1 Utility of different genetic sources for genealogical research. This figure shows the familial inheritance patterns of different genetic material. Each panel (a–c) contains two family trees, one for a male descendant (left) and one for a female descendant (right) with contributing lineages highlighted in green. Male ancestors are represented as squares and female ancestors as circles. (a) The Y chromosome is inherited paternally and is only passed to male children. Therefore, it only reveals information about your paternal lineage and is only present in men. (b) Mitochondrial DNA is inherited maternally but is passed on to both sexes. It reveals information about your maternal lineage and is useful for both men and women. (c) Autosomal DNA is inherited from both parents. Due to recombination and assortment, an individual may inherit autosomal DNA from any ancestor in the tree. It is also passed on to males and females, making it useful for both males and females.

In addition to surname studies, Y chromosome databases are the largest and among the most frequently searched DNA databases. There are numerous public and private Y chromosome databases comprising tens to hundreds of thousands of Y chromosome haplotypes associated with various other types of data such as surnames, geographic locations, and pedigrees. Genealogists often search Y chromosome databases when they are looking for new leads or attempting to solve problems caused by surname changes along paternal lineages. For example, surnames in Scandinavian countries change at each generation making paternal ancestry within these regions more difficult to trace. Similarly, many European American immigrants experienced surname changes when transiting through Ellis Island. Y chromosome analysis can circumvent confusing surname changes by identifying individuals within Y databases who have different surnames but who share closely related haplotypes.

19.3.1.2 Future of Y Chromosome Testing

The field of genetic genealogy is poised on the threshold of having the ability to more comprehensively profile a male individual's Y chromosome at the genome scale. Recently, a pioneering study was published (Wei et al. 2013) that provides a possible didactic blueprint for how, in the not too distant future, genetic genealogists may be able to extract a much more detailed profile of an individual's constellation of single nucleotide variation on his Y chromosome. The research team reported the patterns of sequence variation detected across the same identical territory of approximately 9 million bp for each of 36 men of diverse continental ancestry in which 8 of the 20 major haplogroups were represented. They listed the more than 6600 SNPs they detected in the sample set, which ranged from ancient haplogroup defining SNPs located deep in the tree to much younger (i.e., rarer) and more "private-like" SNPs located within some of haplogroups, especially those which had a larger membership sample size. Obviously, reliable knowledge of these rare recently evolved SNPs would be particularly valuable in a genealogical context. A representation of their phylogeny with the number of SNPs discovered within 3.2 Mb of sequence subjected to high resequencing coverage on each branch is shown in Figure 19.2. While the calibration of the rate at which SNPs accrue on the line of descent of a lineage is yet to be finalized, it is data like these, namely, the number of accumulated SNPs on lineages, which provide the opportunity to estimate both the entire time depth of the tree as well as the individual branches. Noteworthy are the comblike branching patterns associated with the more numerous E and R haplogroup affiliated samples. Generally, in flawless data, all samples should have approximately the same overall number of mutations tracing back to the root of the phylogeny under a model of neutrality. The variability of overall lengths observed

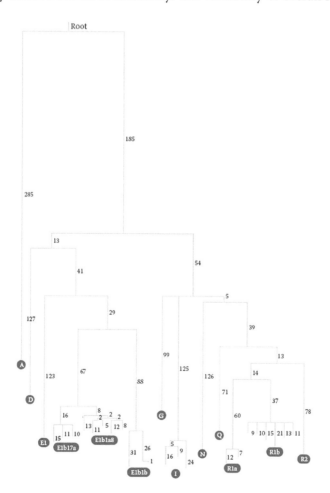

Figure 19.2 Gene tree of variants detected while resequencing identical 3.2 Mb regions in 36 Y chromosomes: Branch length is proportional to the number of mutations indicated along each branch. (Adapted from Wei, W., Ayub, Q., Chen, Y. et al., *Genet Res*, 23 (92), 388–395, 2013.)

in this dataset can, to a large extent, be attributed to sequencing errors. While challenges associated with sequence errors, especially false negatives, must be resolved in the future, sequence data similar to these demonstrate a path forward to more fully reveal the recent genealogical relationships between Y chromosome genomes.

19.3.2 mtDNA Testing

19.3.2.1 Discovering Maternal Lineages

Like the Y chromosome, mtDNA follows a strict lineal inheritance pattern (Figure 19.1b). MtDNA is passed along the direct maternal line from a mother to all her children. Unlike the Y chromosome, both males and females carry mtDNA, but only females pass it to the next generation. Genealogists primarily use mtDNA testing as evidence to support or disprove a suspected maternal relationship

19.3.2.2 mtDNA Testing Options

The most common type of mtDNA testing used for genealogical purposes involves sequencing one or more of the three hypervariable segments (HVS1, HVS2, HVS3) located within the mtDNA control region. With a mutation rate of 10^{-6} mutations/site/ generation (Howell et al. 2003), the hypervariable segments display a significantly higher mutation rate than that of the coding region (Soares et al. 2009). By targeting the regions with the highest mutation rate, HVS testing captures the greatest amount of variability for the lowest cost, making the test affordable for consumers. Similar to Y chromosome testing, mtDNA haplotypes are compiled into databases and often linked to genealogical information such as geographic locations and pedigree charts. Additionally, mtDNA haplogroup analysis is also included as part of most mtDNA tests to shed light on ancient maternal ancestry.

19.3.3 Future of Haploid Testing

The general trend in both Y and mtDNA testing is to continually test more markers to achieve greater resolution for matching purposes. By increasing the number of markers tested, rare alleles can be identified that connect populations and family groups on an ever-finer scale. However, both the Y chromosome and mtDNA are limited in this capacity due their small sizes. Moving to autosomal testing offers a much greater opportunity to identify those rare alleles that have arisen over the recent past and are therefore of most use for genealogical purposes.

Numerous but rare sequence variants have arisen in the recent past as populations have experienced recent rapid growth (Keinan and Clark 2012). Furthermore, the probability that these newly arising mutations are passed on increases because they have not yet been lost by drift, purifying selection. Analyses of a relatively small number of common variants can typically reveal continental (Lao et al. 2006) and subcontinental patterning, especially as additional possible source populations are added to genealogy databases. While currently much more challenging from both an allele calling accuracy and database size perspective, the large reservoir of rare variants provides a potentially salient and powerful tool for detecting the distinguishing shared ancestry sought by genealogists ultimately at the familial level since the most recent and most heavily populated generations contribute the most recent de novo variation.

19.4 Autosomal Testing

19.4.1 Why Assay the Autosomes?

Both the Y chromosome and mtDNA reveal deep, detailed genealogical information that may provide important insights about personal ancestry. Uniparental inheritance patterns and a lack of recombination also make the interpretation of results in these haploid systems more straightforward. However, inferences based on Y and mtDNA will only be relevant to the paternal and maternal lineages, respectively, and these lineages only represent a small proportion of a complete family pedigree (Figure 19.1). More important, the history associated with the purely paternal and maternal lineages may be biased. In cases of sex-biased migration, the Y chromosome and mitochondrial DNA may reveal different details about an individual's genealogical history. For example, the initial European colonization of the Americas in the 1500s primarily involved the migration of Southern European men to South and Central America (Mesa et al. 2000; Bryc et al. 2010). These men frequently fathered children with Native American women and living descendants of these couples retain admixed genetic ancestry having inherited genetic material from at least two distinct continental populations. Thus, many Y chromosome lineages present among Latinos in the Americas will be indicative of European ancestry. On the other hand, mtDNA analysis is more likely to suggest Native American ancestry. In any case, exclusively analyzing Y and mtDNA is unlikely to tell a complete story.

Fortunately, technological developments in the last decade have led to the opportunity to interrogate the entire human genome at greatly increased breadth. Although the rapid pace of next-generation sequencing innovation has made forensic genetics and genetic genealogy based on full genome sequences a possibility, this section will focus on what is presently possible with single-nucleotide polymorphism genotyping microarrays, often called SNP chips, as this is currently the preferred technology in the industry. With genotyping microarrays, it is possible to assay millions of positions across not only the Y and mtDNA systems, but also within the autosomes. SNP chips assay alleles at a prescribed set of genomic sites known to be variable across multiple human populations. Since the Y and mtDNA make up less than 2% of the genome, a significantly more complete picture of genetic inheritance across the entire genome is gained by surveying autosomal sites as well. More important, as explained in the following section, the diploid nature of the autosomes makes it possible to inherit genomic segments from any ancestor in a pedigree.

19.4.2 Meiosis: A Genealogical Double-Edged Sword

Each human carries 22 pairs of autosomal chromosomes. One copy of each pair has been inherited maternally and the other paternally. During sexual reproduction, each parent transmits one copy of each autosome plus one sex chromosome (X or Y) through a haploid gamete to his or her child. Gametes are created by meiotic cell division, a process by which a single diploid cell replicates its DNA and then divides twice into four haploid cells. During meiosis I, autosomal DNA is replicated, undergoes recombination, and assorts independently into two haploid cells. The process of recombination allows genetic exchange between the non-sister chromatid copies of a chromosome. The result is the creation of mixed chromatids containing segments from both the paternal and maternal chromosome copies (Figure 19.3a). Following recombination, independent assortment partitions the chromosome copies so that each daughter cell gets only one copy of each chromosome. This partitioning occurs independently for each chromosome pair. After meiosis I has completed, each daughter cell will complete a further division to separate sister chromatids into four haploid cells, and at least one of these cells will become a gamete. Both the process of recombination and the process of assortment have probabilistic aspects, and this leads to the following facts that are specifically relevant for genetic genealogy:

1. An inherited autosome typically includes a mixture of genetic material from both copies of the autosome contained in the parent.
2. Autosomal inheritance is semiconservative, meaning that a child's genome will only contain 50% of the genetic material present in each parent.
3. The mixture present in a transmitted autosome will likely be different for each inheritance event (i.e., two children will inherit different mixtures of each parent's autosomal chromosome copies unless they are monozygotic twins).

The first point implies that it is possible for an individual to inherit autosomal DNA from any ancestor in their pedigree (Figure 19.3b). While this suggests that the autosomes will provide much broader genealogical information, an additional consequence of recombination and independent assortment is genetic loss, described in point 2. During meiotic division, only one of the two mixed chromosome copies is transmitted into the haploid gamete. Because transmission is random, there is a 50% chance that a particular autosomal segment will not be transmitted to a child (Figure 19.3b). Since the process repeats each generation, the likelihood that a DNA segment is transmitted successfully from an ancestor to a descendant decrease with the number of generations between them. Point 3 highlights that each transmission event results in a unique mixture of parental autosomal DNA. This implies that while two siblings may each inherit 50% of their autosomal DNA from their mother, they are only expected to share 25% of that material with each other. Since they are also expected to share 25% of their paternally transmitted autosomal DNA, the total expected DNA sharing for full siblings is 50%. However, due to the stochastic nature of recombination this percentage can vary widely. Thus, each autosome is a mosaic, made up of contiguous blocks of DNA each inherited from a single lineage in the pedigree (Figure 19.1c). The maternal copy contains segments inherited from lineages on the maternal side, while the paternal copy will contain segments inherited from lineages on the paternal side.

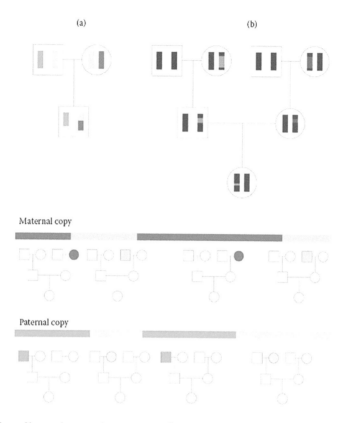

Figure 19.3 The effect of recombination and assortment on autosomal inheritance: (a) For every autosomal chromosome, a child gets one copy from each parent. The inherited copy from each parent is a mixture of each parents' two copies of that chromosome. On the scale of whole chromosomes, the breakpoints for these mixtures are distributed uniformly. (b) Through independent assortment following recombination, it is possible to inherit autosomal DNA from any ancestor (e.g., a paternal grandmother shown in blue). However, this assortment process also leads to the fact that any particular piece of DNA has a chance of being "lost" at each generation (red). (c) Contiguous blocks of the genome are inherited from the same ancestor, but each block may come from a different source, leading to an inheritance mosaic along each chromosome.

19.4.3 Genetic Genealogy with Autosomal DNA

There are two primary avenues by which autosomal genetic analysis enhances genealogical research. First, it is possible to determine the historical populations that contributed significant amounts of genetic material to an individual. This is analogous to Y haplogroup analysis and mtDNA maternal lineage identification. By inferring ancestral genetic origins, an individual may gain insight into deep family history. Second, autosomal DNA can be used to identify relatives. By comparing the genetic content of two individuals, it is possible to infer not only that a relationship exists, but also to estimate how closely related two people are to one another. Identifying relatives creates opportunities to add these individuals to a family tree. Moreover, if an identified relative has recorded his or her family history, it can be used to help grow one's own family tree if a common ancestor can be identified.

19.4.4 Ancestral Origin Estimation

Estimating ancestral origins using autosomal DNA has recently become a very active area of research. Not only has this pursuit been successful for individuals alive today (Li et al. 2008; Nelson et al. 2008; Novembre et al. 2008; Winney et al. 2011; Reich et al. 2012), a surprising amount of knowledge has been gained from estimating ancestry in studies involving ancient DNA (Green et al. 2010; Reich et al. 2010; Rasmussen et al. 2010; Keller et al. 2012). As more genetic data is amassed across an ever-broadening range of modern human populations a number of powerful methods have been developed to use these data to answer questions regarding population structure and migration (Patterson, Price, and Reich 2006; Pritchard, Stephens, and Donnelly 2000; Alexander, Novembre, and Lang 2009; Lawson et al. 2012). While it should be noted that these methods have primarily been developed for demographic inference, they can often be easily adapted to infer ancestral origins using reference panels consisting of individuals with known ancestry. Though each approach may apply a unique algorithm, they all rely on genetic variation data, created primarily by population bottlenecks, mutation, genetic drift, and selection. The methods developed for ancestral origin estimation can be broadly grouped into two classes based on whether they are designed to provide a single genome-wide ancestry estimate or localized ancestry estimate for each segment of the genome. Here we will provide a discussion of genome-wide approaches.

Though the first method for quantifying genetic variation within and between populations is probably due to Sewall Wright (Wright 1949), genome-wide ancestry estimation began in earnest with early work from Luca Cavalli-Sforza and colleagues, in which they demonstrated that patterns in human genetic variation are indicative of population history and can thus be used to identify ancestral origins (Cavalli-Sforza and Bodmer 1971). Cavalli-Sforza went on to be the first to apply principal components analysis (PCA) to genetic data, and this is the inference method we consider first (Menozzi, Piazza, and Cavalli-Sforza 1978).

PCA is a mathematical method for performing an orthogonal transformation of a data matrix (Pearson 1901). The result is a matrix containing a set of linearly orthogonal vectors (principal components) ordered such that the first has maximal variance and each successive vector has the next largest variance. When using genotyping array data, PCA is applied to a data matrix containing I rows, each row representing an individual, and J columns, each column representing an SNP. Entry $g_{i,j}$ in the matrix represents the genotype of SNP j for individual i. Genotypes are codified as 0, 1, or 2 representing the count of a specific allele at the given site. Since PCA is an unsupervised method that simply returns a set of orthogonal vectors decreasingly ordered by variance, care must be taken in preparing data for PCA if the goal is to reveal population structure and infer origins. It is first important to "thin" the data set to reduce genotype correlation between alleles observed at neighboring sites known as linkage disequilibrium (LD). If the data are not thinned, it is possible that the components containing maximal variance will represent the information in a highly correlated set of variants in a single localized region of the genome. This type of variation is typically due to a genomic event that is specific to this region, such as a selective sweep, rather than capturing variation that is attributable to demographic forces like migration and differentiation that have a genome-wide effect on patterns of variation. It is also necessary to remove closely related individuals, as these too may skew the transformation process. Finally, PCA is sensitive to the relative scaling of each variable, so each SNP column must be normalized to have mean zero and standard deviation of one. Following thinning, removal of related samples, and normalization we are typically guaranteed that the first few principal components will represent large-scale population structure. We can then make a scatter plot of the first two principal components in which each individual is given an x-coordinate based on PC1 and a y-coordinate based on PC2. Typically, we will see that the first two axes of the transformed matrix, the two accounting for maximal amounts of variance in the original genotype data, recapitulate the geographic origin of each sample (Figure 19.4a). Individuals from the same population will cluster together, and population clusters themselves will tend to be in the proximity of their nearest geographic neighbors. Moreover, the Euclidean distance between pairs of points in this scatter plot is proportional to the genetic differentiation between them. For example, in Figure 19.4a, Greek samples ("G's" lower right) are more genetically similar to Italian samples ("I's" lower right) than Estonian samples ("E's" top center).

If we wish to infer ancestry for an individual of unknown origin, we first require a collection of reference samples, each with deep ancestry from a single point of origin, such as those in Figure 19.4a. Given this reference PCA vector space, we simply project the new individual into this space and infer the likely ancestral origin to be the population in which the new individual clusters. This approach works well for individuals with a single point of origin. However, for admixed individuals with recent ancestry from multiple source populations, the problem is more complex. If an individual has a mixture of two ancestries, they will be projected along the axis between these two population clusters at a position that is proportional to the relative amounts of each ancestry the individual's genome contains. For example, an individual with 50% Spanish ancestry and 50% Swedish ancestry would fall at the midpoint between these two clusters. In this case they would appear to fall within the French cluster. Thus, without further information, PCA-based inference can be misleading for admixed individuals. The software package EIGENSTRAT represents the most developed application of PCA to ancestry inference (Price et al. 2006).

Another popular approach for genome-wide ancestral origin estimation is based on work by Jonathan Pritchard and colleagues (Pritchard et al. 2000). They developed a Bayesian model-based clustering algorithm called STRUCTURE that uses genome-wide genotyping data to infer ancestry. Given the number of expected population clusters in the data K and a genotype data matrix G, where $g_{i,j}$ is the genotype for individual i at SNP j, the model consists of a $K \times J$ matrix P, where $p_{k,j}$ is the frequency of one allele for population k at SNP j and an $I \times K$ matrix Q, where q_{ik} is the proportion of ancestry individual i contains from population k. In the initial implementation the authors use a Markov chain Monte Carlo approach to simultaneously estimate both P and Q. While the P matrix can be informative with regard to the architecture of populations contained in the sample, it is the Q matrix that provides an ancestral origin estimate for each sample. One obvious advantage with this approach is that q_i is a numerical estimate of ancestry proportions, whereas quantifying ancestry from PCA requires further extrapolation. More important, this method does not have difficulties distinguishing individuals of admixed ancestry from those with a single ancestral origin. It does, however, require both data thinning and removal of related samples much like PCA. Recently, a significantly faster model-based method called ADMIXTURE has become a standard clustering method for ancestry inference (Alexander, Novembre, and Lang 2009). ADMIXTURE is based on the likelihood model from the program *structure*, but uses a novel optimization approach to reduce computation time while returning comparable estimates of *P* and *Q*. An example of the estimated *Q* matrix for ~700 Europeans is provided in Figure 19.4b. In this case, *K* was set to five. As you can see, the majority of individuals are estimated to be composed of multiple ancestries. However, much like we observed in Figure 19.4a individuals from neighboring countries tend to show similar ancestry proportions. It is also worth noting that individuals from the same country often show a gradient in the proportion of a given ancestry. For example, the "red" cluster in English samples varies from 0 to nearly 50%. This gradient may be useful in further identifying ancestral origins. In the preceding example, the "red" cluster seems to represent Scandinavian ancestry and this suggests that English individuals with significant amounts of this ancestry may have ancestors from along the Eastern coast of England where Scandinavian admixture was historically more prevalent. ADMIXTURE also provides the opportunity to perform a semisupervised clustering analysis in which a reference panel of single-origin individuals is provided and ancestry is estimated for only a subset of individuals of unknown origin.

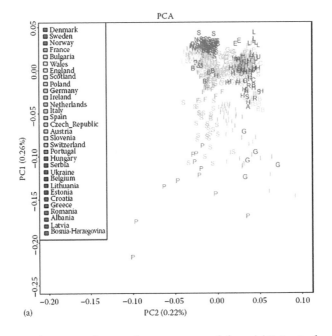

Figure 19.4a Ancestral origin inference from autosomal data. (a) Principal components analysis is used to project high-dimensional data to a low-dimensional space while still capturing the maximum amount of variation possible in the original data. In this case, we project a matrix of ~135,000 SNPs across ~1000 individuals of European ancestry down to a two-dimensional space. The first two components of variation in this data reveal that individuals from the same geographic location tend to be more genetically similar. This is evidence of population structure and can be used to identify the ancestral sources of genetic information. (b) Results of running the clustering method admixture with five clusters on ~100,000 SNPs for ~700 individuals. Each vertical bar represents an individual colored by proportions of ancestry belonging to each population cluster.

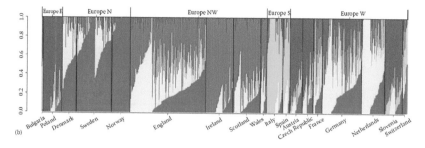

(b)

Figure 19.4b

19.4.5 Relative Identification

Ancestral origin inference can provide exciting clues to one's history, but the primary use of genetic data in genealogical research is in the discovery of relatives. Physical record databases, including census data, military records, and parish archives, are typically the primary sources of information used to research and build one's family tree. However, the addition of genetic data provides a powerful independent source of information with which to confirm suspected relationships and discover new, previously unknown relatives.

In genetics, relative identification is accomplished by identifying pairs of individuals who share one or more common segments of DNA that are identical by descent (IBD). This implies that this pair of individuals is related and has inherited this segment of DNA from a common ancestor. More important, the amount of shared DNA and the number and length of shared segments varies with the number of generations back to the common ancestor. By deriving some simple quantities relating to expected IBD-sharing proportions, we can demonstrate the power of this approach for relative identification. The process of independent assortment in sexual reproduction implies that the probability that a particular position in the genome is shared IBD for a pair of individuals that are each g generations descended from a single common ancestor is $2^{-(2g-1)}$. This is also the expected proportion of the genome that will be shared IBD. If we wish to calculate the expected number of IBD segments, we simply need to multiply the probability of IBD by the expected number of independently inherited contiguous segments. Under simplifying assumptions, recombination can be thought to occur following a Poisson process each generation with a rate of one event per morgan of sequence. A morgan is a unit of genetic length, measuring the distance between two chromosome positions in which the expected number of crossover recombination events per meiosis is one. Thus, the expected number of recombination events per generation is one multiplied by the length in morgans, L, of the autosomal portion of the genome (for humans this is approximately 35.3 morgans [McVean et al. 2004]). Since this process occurs independently in each generation, we can use the fact that the sum of n independent Poisson processes each with a rate of r can be represented as a single Poisson process of rate nr. In our case, this implies that the expected number of recombinations between two individuals each g generations descended from a single common ancestor is $2Lg$. It is important to remember that the autosomal portion of the genome is already separated into 22 independently inherited chromosomes, thus the final number of independently inherited autosomal segments is expected to be $(2Lg + 22)$ (Thomas et al. 1994; Huff et al. 2011). By multiplying this quantity by the chance that each segment has been inherited IBD, we see that the expected number of IBD segments, $N_s(g)$, in this case is $(2Lg + 22)2^{-(2g-1)}$. As g increases, this expectation is dominated by the $2^{-(2g-1)}$ term and thus for $g > 4$ the expected number of autosomal IBD segments is less than one (Table 19.1). This means that for distantly

Table 19.1 Expected Identity by Descent and Numbers of Relatives by Degree of Relatedness

Relationship	Generations to Common Ancestor (g)	Expected Number of Shared Autosomal Segments per Ancestor ($N_s(g)$) $(L \cdot (2g) + 22)^* \, 2^{-(2g-1)}$	Expected Number of Relatives of This Degree ($N_r(g)$) $(2^{g-1})^* \, o^{(g-1)*} (o-1)$	Expected Number of IBD Segments from All Relatives of This Degree 2^* $N_s(g)^* \, N_r(g)$
1st cousins	2	20.40	7.5	306.0
2nd cousins	3	7.31	37.5	548.0
3rd cousins	4	2.38	187.5	891.8
4th cousins	5	0.73	937.5	1373.3
5th cousins	6	0.22	4687.5	2039.8
6th cousins	7	0.06	23,437.5	2953.7
7th cousins	8	0.02	117,187.5	4197.1

Note: This table shows how the expected number of IBD segments (column 3) and relatives (column 4) vary by the degree of relationship (column 1). Column 4 represents the expected number of IBD segments inherited from all full relatives of a given degree. In this example, $L = 35.3$ and $o = 2.5$.

related individuals there is often a large probability that no IBD segments exist. The expected length of IBD segments is also related to the number of generations since common ancestry is due to the effect of recombination. Segment lengths, again measured in morgans, tend to be exponentially distributed with a mean of length of $1/2g$, where g is again the number of generations back to the common ancestor for each individual. For example, for two individuals that are half first cousins (sharing one grandparental ancestor two generations in the past) the expected length of a shared segment is 0.25 morgans, or 25 centimorgans (cM). Similarly, to the expected number of segments, the expected length of segments decreases with increasing g. Beyond $g > 10$, the expected segment length is only 5 cM and without very high-quality data it becomes very difficult to identify these segments.

While it may seem that IBD analysis is doomed to identify only recent shared ancestry, and only a small portion at that, this fact is alleviated by the sheer numbers of relatives each individual has for a given degree of relatedness. To calculate the expected number of relatives $N_r(g)$ an individual has due to common ancestry from g generations in the past we simply multiply the number of ancestor couples from generation g by the expected number of descendants each of these ancestral couples will produce in the current generation. This is equal to $2^{g-1} o^{g-1}$ $(o - 1)$, where o is equal to the average number of offspring per generation (Henn et al. 2012). If we assume $o = 2.5$, which is not unreasonable for many human populations, we can see how this quantity grows (Table 19.1). What is surprising is the sheer number of relatives one expects at each level. Thus, although the expected number of IBD segments for a given pair of relatives is often far less than one, the number of relatives at a given generation of descent g implies that the chance of sharing an IBD segment with at least one individual who is related through an ancestor g generation actually increases with g (Table 19.1). While this demonstrates the power of using genetic data to identify relatives, the ability to successfully identify IBD segments is heavily dependent on data quality and the methods applied.

As with the ancestral origin inference, the identification of relatives using IBD analysis has been an active area of research for many years (Purcell et al. 2007; Browning and Browning 2010; Gusev et al. 2009; Browning and Browning 2011). While most methods rely on phased data as input, a few have been developed to identify long sequences of IBD-consistent genotypes from unphased data. In this case, as long as two individuals share at least one allele at each position they are said to be IBD consistent. Methods that do not require phasing are fast, however, minimizing false positives using only genotype data requires setting a fairly conservative threshold on the minimum number of consecutive IBD-consistent genotypes one must observe before calling a segment IBD. This problem is worsened by the presence of linkage disequilibrium (LD), which may appear to increase the length of an IBD-consistent segment due to the fact that consistency at one site will increase the probability of consistency at correlated neighboring sites. This implies that such methods will only have power to identify fairly large IBD segments and thus fairly recent common ancestry. One such example is the IBD method included as part of the PLINK analysis package (Purcell et al. 2007).

In contrast, significantly more sensitive methods can be developed that either take as input phased data or estimate IBD in conjunction with phasing. Many of these model-based methods are computationally expensive and thus only suitable for small datasets. However, two more recently developed methods of this type, GERMLINE and fastIBD, use hashing to regain some speed while maintaining improved sensitivity (Gusev et al. 2009; Browning and Browning 2011). Beginning with phased data, GERMLINE first identifies short, exact matching segments as anchor matches. From each anchor a matching segment may be extended using a more permissive "fuzzy" matching algorithm. This procedure allows for both phasing mistakes and genotyping error (Figure 19.5). Both are major sources of error in IBD analysis. A minimum segment length threshold is still applied, though it can be significantly shorter than that used for the unphased methods. The fastIBD method works in a similar fashion maintaining a probabilistic model of each individual's haplotypes and identifying candidate IBD segments as shared "rare" haplotypes. While a frequency threshold may appear to be distinct from the

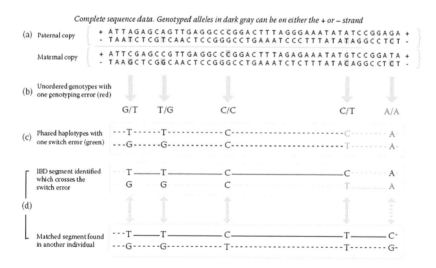

Figure 19.5 Identifying relatives through identity-by-descent (IBD) analysis. (a and b) Genotyping arrays provide unordered pairs of alleles observed from both copies of a chromosome assayed on one strand. Occasionally the assay returns erroneous genotypes (red). (c) These allele pairs must be phased to determine the underlying relationship between neighboring sites and to separate the chromosome copies. This process is imperfect and some switch errors (phasing continuity breaks) will remain (green). (d) Once phased, an individual's haplotypes can be compared to haplotypes from other individuals to identify long-shared stretches that have been inherited identical by descent. IBD algorithms have been designed to allow for both genotyping and switch errors.

length threshold, they are actually much the same. As the length of the segment in question increases, the observed haplotype tends to become less common as more and more unique haplotypes are possible. Both of these methods can be run on datasets consisting of thousands of samples and provide accurate IBD segment identification of segments as small as 2–5 cM. Given this information, it is possible to identify even fifth cousins (six generations of separation) with some degree of certainty.

19.4.6 Conclusions

It is clear that autosomal analysis can add a tremendous amount of information to any genetic genealogical research project. Whereas the Y chromosome and mtDNA provide detailed information on single lineages, autosomal chromosomes undergo recombination and independent assortment during transmission, and thus they may provide information about all ancestral lineages simultaneously. While this is a significant benefit, it comes at a cost, as genetic material is also lost through the same process each generation. Recombination and assortment also make the process of inference more difficult. A lack of recombination in haploid systems means that all samples can be placed on a single phylogenetic tree, and the exact sequence of alleles observed implies only one possible history for each sample. On the other hand, the combination of alleles observed along an autosomal chromosome can be inherited from one or more ancestors and each segment may have a history that is effectively independent from neighboring regions, and the effect of this fact must be accounted for in any analysis.

19.5 Discussion

The previous sections introduced and discussed the two primary genealogical questions on which we can gain insight via genetic analysis: haplogroup analysis or ancestral origin inference, and haplotype matching or relative identification. From a broader perspective, these are in fact the same question but on different timescales. In the context of ancestral origin inference, a population is a group of individuals with a somewhat distinct genetic signature due to mating within the group and isolation from other neighboring groups. In some sense this is also a family. On the other hand, a family can be thought of as a narrowly defined population, since members of a family share genetic material due to common ancestry and this material can be used to assign individuals to this group. Currently, the majority of methods for identifying ancestral origins are effective at distinguishing large-scale differences in populations that may be separated by many hundreds or thousands of years of evolution. On the other hand, our best IBD methods can identify relationships at a depth of roughly 5–15 generations, which is approximately 150–450 years ago. Improvements in the methods for both of these types of analyses continue to close this gap, with the end goal that we can trace single DNA segments through space and time capturing the entire inheritance history from a large-scale continental population in the distant past to a specific family in the last few generations. While this is obviously an extremely ambitious goal, active research in these areas is leading to progress. For more detail on these topics, please see the recent review by Novembre and Ramachandran (2011) on ancestry inference and the recent review by Browning and Browning (2012) on IBD inference. Progress has not been limited to methods development. Most notably next-generation sequencing (NGS) technologies are rapidly improving our catalog of global human variation.

It should also be noted that the growth in this field is not limited to academic and casual users. Companies offering genetic testing can boast of delivering customers some truly amazing discoveries. Demand for these tests is due in part by an assumption that genetics can be used to "prove" the existence of a particular ethnic background in one's past, or "confirm" a relationship based on tenuous records. Similar to DNA forensics, the allure of using genetic analysis for genealogy is in part due to the perceived infallibility of the information. However, it is important to remember that although genetic analysis offers a powerful independent source of information for genealogical research, it too provides an incomplete, imperfect record of the past and is best used in conjunction with record-based research.

Acknowledgments

The authors would like to acknowledge Natalie Myers for her authorship on an earlier edition of this chapter, her extensive contributions to the field of genetic genealogy, and her friendship.

We would also like to thank Jared Lewandowski and Josh Callaway for their graphic design expertise in creating the figures found in this chapter.

References

Alexander, D.H., J. Novembre, and K. Lange. 2009. Fast model-based estimation of ancestry in unrelated individuals. *Gen Res* 19(9): 1655–1664.

Browning, B.L., and S.R. Browning. 2011. A fast, powerful method for detecting identity by descent. *Am J Hum Genet* 88(2): 173–182.

Browning, S.R., and B.L. Browning. 2010. High-resolution detection of identity by descent in unrelated individuals. *Am J Hum Genet* 86(4): 526–539.

Browning, S.R., and B.L. Browning. 2012. Identity methods in human gene mapping. *Annu Rev Genet* 46(1):617–33.

Bryc, K., C. Velez, T. Karafet, et al. 2010. Genome-wide patterns of population structure and admixture among Hispanic/Latino populations. *Proc Natl Acad Sci U S A* 107(Supplement 2): 8954–8961.

Cavalli-Sforza, L.L., and W.F. Bodmer. 1971. *The genetics of human populations*. W.H. Freeman, San Francisco (reprinted 1999 by Dover Publications).

Coble, M.D. 2011. The identification of the Romanovs: Can we (finally) put the controversies to rest? *Investig Genet* 2(1): 1–7.

Falconer, B. 2012. We are family? *Bloomberg Businessweek* 30: 81–93.

Foster, E.A., M.A. Jobling, P.G. Taylor, et al. 1998. Jefferson fathered slave's last child. *Nature* 396(6706): 27–28.

Green, R.E., J. Krause, A.W. Briggs, et al. 2010. A draft sequence of the Neandertal genome. *Science* 328(5979): 710–722.

Gusev, A., J.K. Lowe, M. Stoffel, et al. 2009. Whole population, genome-wide mapping of hidden relatedness. *Gen Res* 19(2): 318–326.

Henn, B.M., L. Hon, J.M. Macpherson, et al. 2012. Cryptic distant relatives are common in both isolated and cosmopolitan genetic samples. *PLOS ONE* 7(4): e34267.

Howell, N., C.B. Smejkal, D.A. Mackey, P.F. Chinnery, D.M. Turnbull, and C. Herrnstadt. 2003. The pedigree rate of sequence divergence in the human mitochondrial genome: There is a difference between phylogenetic and pedigree rates. *Am J Hum Genet* 72(3): 659–670.

Huff, C.D., D.J. Witherspoon, T.S. Simonson, et al. 2011. Maximum-likelihood estimation of recent shared ancestry (ERSA). *Gen Res* 21(5): 768–774.

Keinan, A., and A.G. Clark. 2012. Recent explosive human population growth has resulted in an excess of rare genetic variants. *Science* 336(740): 740–743.

Keller, A., A. Graefen, M. Ball, et al. 2012. New insights into the Tyrolean Iceman's origin and phenotype as inferred by whole-genome sequencing. *Nat Commun* 3: 698.

King, T.E., and M.A. Jobling. 2009. What's in a name? Y chromosomes, surnames and the genetic genealogy revolution. *Trends Genet* 25(8): 351–360.

Lao, O., K. vanDuijn, P. Kersbergen, P. deKnijff, and M. Kayser. 2006. Proportioning whole-genome single-nucleotide-polymorphism diversity for the identification of geographic population structure and genetic ancestry. *Am J Hum Genet* 78(4): 680–690.

Lawson, D.J., G. Hellenthal, S. Myers, and D. Falush. 2012. Inference of population structure using dense haplotype data. *PLOS Genet* 8(1): e1002453.

Li, J.Z., D.M. Absher, H. Tang, et al. 2008. Worldwide human relationships inferred from genomewide patterns of variation. *Science* 319(5866): 1100–1104.

McVean, G.A.T., S.R. Myers, S. Hunt, P. Deloukas, D.R. Bentley, and P. Donnelly. 2004. The fine-scale structure of recombination rate variation in the human genome. *Science* 304(5670): 581–584.

Menozzi, P., A. Piazza, and L.L. Cavalli-Sforza. 1978. Synthetic maps of human gene frequencies in Europeans. *Science* 201(4358): 786.

Mesa, N.R., M.C. Mondragon, I.D. Soto, et al. 2000. Autosomal, mtDNA, and Y-chromosome diversity in Amerinds: Pre-and post-Columbian patterns of gene flow in South America. *Am J Hum Genet* 67(5): 1277–1286.

Nelson, M.R., K. Bryc, K.S. King, et al. 2008. The population reference sample, POPRES: A resource for population, disease, and pharmacological genetics research. *Am J Hum Genet* 83(3): 347–358.

Novembre, J., T. Johnson, K. Bryc, et al. 2008. Genes mirror geography within Europe. *Nature* 456(7218): 98–101.

Novembre, J., and S. Ramachandran. 2011. Perspectives on human population structure at the cusp of the sequencing era. *Annu Rev Genomics Hum Genet* 12: 245–274.

Patterson, N., N. Hattangadi, B. Lane, et al. 2004. Methods for high-density admixture mapping of disease genes. *Am J Hum Genet* 74(5): 979.

Pearson, K. 1901. LIII. On lines and planes of closest fit to systems of points in space. *Lond Edinb Dublin Philos Mag J Sci* 2(11): 559–572.

Price, A.L., N.J. Patterson, R.M. Plenge, M.E. Weinblatt, N.A. Shadick, and D. Reich. 2006. Principal components analysis corrects for stratification in genome-wide association studies. *Nat Genet* 38(8): 904–909.

Pritchard, J.K., M. Stephens, and P. Donnelly. 2000. Inference of population structure using multilocus genotype data. *Genetics* 155(2): 945–959.

Purcell, S., B. Neale, K. Todd-Brown, et al. 2007. PLINK: A tool set for whole-genome association and population-based linkage analyses. *Am J Hum Genet* 81(3): 559–575.

Rasmussen, M., Y. Li, S. Lindgreen, et al. 2010. Ancient human genome sequence of an extinct Palaeo-Eskimo. *Nature* 463(7282): 757–762.

Reich, D., R.E. Green, M. Kircher, et al. 2010. Genetic history of an archaic hominin group from Denisova Cave in Siberia. *Nature* 468(7327): 1053–1060.

Reich, D., N. Patterson, D. Campbell, et al. 2012. Reconstructing Native American population history. *Nature* 488(7411): 370–374.

Shriver, M.D., and R.A. Kittles. 2004. Genetic ancestry and the search for personalized genetic histories. *Nat Rev Genet* 5(8): 611–618.

Soares, P., L. Ermini, N. Thomson, et al. 2009. Correcting for purifying selection: An improved human mitochondrial molecular clock. *Am J Hum Genet* 84(6): 740–759.

Thomas, A., M.H. Skolnick, and C.M. Lewis. 1994. Genomic mismatch scanning in pedigrees. *Math Med Biol* 11(1): 1–16.

Wei, W., Q. Ayub, Y. Chen, et al. 2013. A calibrated human Y-chromosomal phylogeny based on resequencing. *Genet Res* 23(2): 388–395.

Winney, B., A. Boumertit, T. Day, et al. 2011. People of the British Isles: Preliminary analysis of genotypes and surnames in a U.K.-control population. *Eur J Hum Genet* 20(2): 203–210.

Wright, S. 1949. The genetical structure of populations. *Ann Hum Genet* 15(1): 323–354.

Law, Ethics, and Policy IV

DNA as Evidence in the Courtroom

20

STEVEN W. BECKER, DAVOR
DERENČINOVIĆ, AND DAMIR PRIMORAC

Contents

20.1 The American Experience

20.1.1 Introduction

As acknowledged by the United States Supreme Court:

> Modern DNA testing can provide powerful new evidence unlike anything known before. Since its first use in criminal investigations in the mid-1980s, there have been several major advances in DNA technology, culminating in STR technology. It is now often possible to determine whether a biological tissue matches a suspect with near certainty. DNA testing has exonerated wrongly convicted persons, and has confirmed the convictions of many others.
>
> *District Attorney's Office v. Osborne*, **557 U.S. 52, 62 (2009)**

Yet, especially in the post-conviction context, in which DNA testing has gained the greatest public notoriety, there is an inherent tension between finality and justice. How does one balance the need for juridical closure with countervailing claims of actual innocence? As explained by America's highest court:

> The availability of technologies not available at trial cannot mean that every criminal conviction, or even every criminal conviction involving biological evidence, is suddenly in doubt.

DOI: 10.4324/9780429019944-24

The dilemma is how to harness DNA's power to prove innocence without unnecessarily over-throwing the established system of criminal justice.

District Attorney's Office v. Osborne, **557 U.S. 52, 62 (2009)**

Ultimately, the United States Supreme Court ruled that an inmate does not have a free-standing, constitutional due process right to seek post-conviction DNA testing. *Id.* at 73–75. Yet, the court subsequently held that a state-court inmate does have the right to seek such testing in a civil lawsuit for deprivation of constitutional rights. *Skinner v. Switzer*, 562 U.S. 521, 534 (2011).

Although there is inevitable overlap, issues involving DNA and its application in prac-tice vary between trial and post-conviction proceedings, given the greater procedural chal-lenges a convicted defendant faces in seeking collateral relief.

In the pretrial setting, there was much litigation surrounding whether law enforce-ment personnel could extract DNA from arrestees—as opposed to defendants already convicted of serious offenses. The objection from pretrial detainees, who are presumed innocent, was that such collection of biologic material constituted an unlawful search and seizure of their private information without probable cause, which could result in them being implicated in previous crimes through a match made by means of a DNA database search. This is precisely what occurred in the case of *Maryland v. King*, 569 U.S. 435 (2013), where the pretrial collection of an arrestee's DNA led to his being identified in the Federal Bureau of Investigation's (FBI's) Combined DNA Index System (CODIS) database as being the perpetrator of an earlier rape. *Id.* at 441. In *King*, however, the United States Supreme Court ruled that the collection of one's skin cells by means of a buccal swab applied inside a detainee's cheek was minimally invasive, did not unconstitutionally interfere with the arrestee's privacy, and was a permissible identification procedure during the booking pro-cess akin to photographing or fingerprinting. *Id.* at 464–66.

20.1.2 Standards for Admissibility

At the trial level, one of the most important issues practitioners must assess at the outset of a case is the legal admissibility of DNA or other scientific evidence, which could ultimately prove to be outcome-determinative. There are two recognized standards that are utilized for this appraisal.

The first of these standards is known as the *Frye* test, which originated from the case of *Frye v. United States*, 293 F. 1013 (D.C. Cir. 1923). Under the *Frye* standard, the focus is on the underlying methodology employed. Thus, forensic evidence is admissible at trial only if the methodology or scientific principle upon which the opinion is based is "suf-ficiently established to have gained general acceptance in the particular field in which it belongs," i.e., in the relevant scientific community. *Id.* at 1014. It is important to note that this test applies not only to scientific methodology that is "new" or "novel" in the sense that it is may be in the transitional phase between experimentation and general accep-tance but also to techniques that have been utilized for decades but have never previously been challenged or tested in court for general acceptance. See *People v. McKown*, 226 Ill. 2d 245, 257–58 (2007).

The second principal test is known as the *Daubert* standard, which takes its name from the landmark United States Supreme Court decision in *Daubert v. Merrell Dow*

Pharmaceuticals, Inc., 509 U.S. 579 (1993). This standard, which is utilized in federal court proceedings, assigns to the judge the role of a gatekeeper, who must make a preliminary determination as to whether "the reasoning or methodology underlying the [expert] testimony is scientifically valid and of whether that reasoning or methodology properly can be applied to the facts of the case." *Id.* at 592–93. Said another way, the overarching subject of the judge's inquiry, which is flexible in nature, is "the scientific validity—and thus the evidentiary relevance and reliability—of the principles that underlie a proposed submission. The focus, of course, must be solely on principles and methodology, not on the conclusions that they generate." *Id.* at 594–95.

In order to give guidance to judges on how to make this "relevance and reliability" determination, the court set forth the following factors as guideposts: (1) whether a theory or technique "can be (and has been) tested"; (2) "whether the theory or technique has been subjected to peer review and publication"; (3) "the known or potential rate of error"; (4) "the existence and maintenance of standards controlling the technique's operation"; and (5) the degree of general acceptance within the relevant scientific community. *Id.* at 593–94.

20.1.3 Contemporary Issues in the Courtroom

DNA evidence has dramatically changed the legal landscape for both the prosecution and the defense, especially in the context of jury trials in the common law system. Because of jurors' exposure to popular television programs involving forensic scientists solving seemingly "perfect" crimes, many jurors now expect the prosecution to put on forensic evidence in cases in which such scientific evidence was never traditionally introduced, making cases more difficult to bring to a successful verdict. This has become known as the "*CSI* effect," named after the television show, *Crime Scene Investigation* and its various spinoffs.

On the other side of the coin, because of the very persuasive, almost magical, quality that DNA evidence holds in the minds of jurors, the defense has had to contend with claims by the prosecution of astronomical probability figures associated with a DNA "match" that seem to infallibly indicate guilt. These figures become especially problematic in cases of a "partial" or incomplete DNA profile, i.e., a profile for which results are not available at all tested loci.

The threshold evidentiary question in cases of a partial DNA profile is whether the prejudicial effect of such evidence substantially outweighs its probative value. Fed. R. Evid. 403. Of course, as the prosecution will argue, such evidence is relevant to the question of the defendant's guilt, as it constitutes circumstantial evidence that the defendant could be the perpetrator. On the other hand, when there is not a definitive "match," the trial court must determine whether the introduction of such evidence unfairly prejudices a defendant by misleading the jury; in other words, whether the evidence might overpersuade a fact finder of the defendant's guilt even though the evidence, if it were capable of being fully tested, might exclude the defendant entirely as the offender.

The FBI's CODIS database is based on 13 loci at which the STR alleles are noted and compared; making possible "extreme accuracy in matching individual samples." *King*, 569 U.S. at 445. How few loci, however, can the prosecution's expert test before the evidence should be rendered inadmissible? For example, in the case of *People v. Mitchell*, 955 N.E.2d 1180 (Ill. App. Ct. 2011), the expert extracted DNA from swabs taken from a pair of gloves found in a garage in which a rifle was recovered in a murder case. *Id.* at 1183. The expert testified that he was only able to obtain a profile at four loci and that the swabs contained a mixture of human DNA profiles from at least three individuals. *Id.* He further opined that "1 in 71 black, 1 in 71 white, and 1 in 82 Hispanic unrelated individuals cannot be excluded from having contributed to this mixture of human profiles." *Id.* The expert concluded that the defendant could not be excluded from the profile. *Id.* The defense argued that the evidence should be barred as being "too inconclusive," yet, the appellate panel affirmed the trial court's admission of the DNA evidence on the gloves, holding that although "the circuit court certainly had discretion to disallow the expert's testimony if it determined its probative value was minimal, the court had the discretion to allow the triers of fact to assess that testimony and assign the weight, if any, the testimony warranted." *Id.* at 1188. With any like forensic evidence, however, whether it is the number of loci examined along the chromosome or the points of comparison in a fingerprint analysis, as the number gets closer to zero the question should become one not of weight, but of admissibility. *Id.* at 1186 n. 1.

Cases involving "cold hits" and partial DNA profiles have revealed one of the most substantial challenges to date of the accuracy of probability figures used against defendants in connection with DNA prosecutions. In *People v. Wright*, 971 N.E.2d 549 (Ill. App. Ct. 2012), a case of first impression, the only evidence linking the defendant to a criminal sexual assault was a nine-loci DNA profile found on a rectal swab, which the DNA expert incorrectly described as a "match." *Id.* at 559. The victim was unable to identify the defendant as her assailant. *Id.* at 557. The expert further testified that the frequency with which he would expect to see this nine-loci male profile occurring in the general population was approximately "one in 420 trillion black[s], one in 670 trillion white[s], and one in 2 point 9 quadrillion Hispanic unrelated individuals." *Id.* at 559.

In light of the incomplete DNA profile obtained from the rectal swab, the defense filed a motion under a unique state statute allowing defendants to obtain a DNA database search upon court order. The defendant's request was predicated upon what has become known as the "Arizona Study," in which a search of the Arizona convicted offender DNA database startlingly revealed 120 pairs between inmates at nine loci in a database of 65,493 offenders. *Id.* at 555. Thus, "in Arizona, there is a 1 in 700 chance that two individuals will match up at nine locations." *Id.* Other database studies resulted in similar unexpected findings; for example, a study of the Illinois database revealed 903 pairs at nine loci in a database of 220,456 offenders, while a study of the Maryland database, which was markedly smaller, yielded 32 pairs at nine loci in a database containing fewer than 30,000 inmate profiles. *Id.* at 561, 567.

Such results have led commentators, scientists, and mathematicians to decry the astronomical random match probability (RMP) figures calculated by use of the product rule and introduced at trial by prosecutors as "no better than alchemy" (Kaye 2009, 145). For example, with reference to the Arizona Study: "If the RMP for a nine-locus match is anything like 'one in 754 million for whites, and one in 561 million for blacks,' how can it be that a database as small as 'a mere 65,493 entries' produces even one such match?" (Kaye 2009).

In *Wright*, the appellate court concluded that the trial judge abused his discretion in failing to order a database search at nine loci, holding that "the error challenged the integrity of the judicial process by barring defendant's access to evidence that could have assisted him in establishing his innocence, by casting serious doubt on the State's identification evidence." *Wright*, 971 N.E.2d at 571. Accordingly, the reviewing court reversed the defendant's conviction and remanded the case for a new trial. *Id.* at 576.

Another issue that has arisen in the trial context is whether the prosecution's forensic expert can testify about another laboratory's processing of a DNA sample without violating a defendant's right to cross-examine the other laboratory's personnel. In *Williams v. Illinois*, 567 U.S. 50 (2012), in a sharply divided decision, the plurality ruled that an expert, at least in the context of a bench trial, can introduce the results of a private laboratory's testing without running afoul of the Confrontation Clause, even though the defendant is not afforded the opportunity to examine the other laboratory's analyst who generated the results. *Id.* at 70-79.

20.1.4 Elements for Statutory DNA Testing

With respect to post-conviction DNA testing, in addition to a federal statute, all 50 states in the United States now have statutes allowing a defendant to seek DNA testing in collateral proceedings (Becker 2016).

Although the statutes vary greatly, the most common elements necessary to obtain post-conviction DNA testing under these provisions are as follows:

1. The evidence must have been secured in relation to the trial;
2. (a) The evidence was not previously subjected to testing; (b) the evidence was not previously subjected to testing because the technology was not available at the time of trial; and/or (c) although previously tested, the evidence can be subjected to additional testing using a method not available at the time that provides a reasonable likelihood of more probative results;
3. Identity was an issue at trial;
4. There must be a chain of custody of the evidence;
5. The evidence must have the potential to produce new, non-cumulative evidence materially relevant to actual innocence; and
6. The testing must employ a scientific method generally accepted in the relevant scientific community.

(Becker 2016)

Some of these statutes permit DNA testing for only the most serious felony offenses, many limit the time period in which a defendant can file a motion for post-conviction testing, and still others provide for comparison analysis of genetic marker groupings. One of the most controversial areas, however, surrounds guilty pleas, which are so prevalent in the United States. In some jurisdictions, a defendant who pled guilty is not permitted to file a collateral motion for DNA testing. Other jurisdictions, however, insist that, by entering a guilty plea, a defendant does not waive the right to subsequently request DNA testing (Becker 2016).

Yet, not all post-conviction DNA testing statutes are necessarily favorable to the defense. Certain states require that, in order to obtain collateral DNA testing, the defendant must consent to have his DNA run in databases for the purpose of investigating unsolved offenses, which could potentially link the inmate to a previous crime—thus leaving a defendant who committed an earlier unsolved violent crime but is innocent of the current offense with the agonizingly difficult decision of whether or not to roll the die (Becker 2016).

As of this writing, more than 350 individuals in the United States have been exonerated based upon DNA (Innocence Project 2019). The most significant factor in these wrongful convictions, by far, is eyewitness misidentification. Other identifiable factors include the misapplication of forensic science, false confessions, and the use of informants. Moreover, it is important to note that DNA exonerations are also beneficial to police and prosecutors, as in slightly less than half of these cases the actual perpetrators or true suspects are identified (Innocence Project 2019).

20.2 The European Experience

20.2.1 Introduction

In the context of the use of modern technologies for the purpose of criminal proceedings, a special place belongs to DNA analysis. Data obtained using DNA analysis can be used as evidence in criminal proceedings under the conditions prescribed by national legislation. This increases the efficiency of the prosecution in terms of both positive and negative identification of potential suspects. In other words, the use of DNA analysis in criminal proceedings represents a significant development as it enables identification of the culprit based on the evidence gathered at the crime scene (positive identification). In addition, DNA analysis serves as a method of excluding potential suspects and thus avoiding the situation where persons who did not commit the offense are held liable (negative identification).

However, despite all the advantages and relatively high degree of reliability, the use of DNA analysis in criminal proceedings raises a series of issues about potential violations of some human rights (right to privacy, presumption of innocence, protection from unlawful search and seizure, etc.). As one commentator observed, "if a person gives a genetic sample during the course of a criminal investigation for the limited purpose of eliminating himself as a suspect, can a state then include that sample in a database for use and comparison in subsequent investigations?" (Taylor 2003). Furthermore, a question arises as to the legal justification of DNA retention in cases of acquittal, as well as indefinite retention of DNA of persons convicted for some less serious criminal offences. Finally, should DNA databases contain profiles of some categories of perpetrators who enjoy special status (e.g., juveniles)?

Debate on these issues in the United States has been focused on the collection and processing of DNA profiles, while in Europe the focus is primarily on the removal of DNA profiles from databases (local, national, and international). Commentators from the United States approach this issue by analyzing the constitutional protection against unlawful searches prohibited by the Fourth Amendment, while the European emphasis is somewhat different and largely focuses on privacy protection in terms of the provisions of the European Convention on Protection of Human Rights and Fundamental Freedoms (hereinafter: European Convention) and constitutions of European countries. The focus of this article is on standards for DNA retention established in jurisprudence of the European Court of Human Rights (hereinafter: ECHR), which has been duly reflected in decisions of constitutional courts in certain European countries.

20.2.2 Jurisprudence of the European Court of Human Rights

The ECHR has discussed issues concerning DNA analysis and retention of DNA primarily in the context of potential violations of Article 8 of the European Convention (right to personal and family life). The court found DNA analysis "necessary in a democratic society" as a measure that substantially contributed to law enforcement in recent years (*Van der Velden v. Netherlands* 2006). Likewise, the ECHR came to the conclusion that setting up and operating national DNA databases, if subjected to stringent legal regulation, are not inconsistent with privacy rights enshrined in Article 8 of the European Convention. This particularly goes for perpetrators of serious criminal offences. In *Peruzzo and Martens v. Germany*, the applicants, who were convicted for serious criminal offences (related to narcotic drugs), claimed that national authorities violated the European Convention by implementing legislation governing the collection and storage of their DNA in the national database. The ECHR found the application inadmissible and manifestly ill-founded. According to the court, domestic legislation on the collection and storage of DNA from persons convicted for serious criminal offences is permissible under Article 8 of the European Convention. Therefore, the government had struck a fair balance between the competing public and private interests, and fell within the respondent state's acceptable margin of appreciation (*Peruzzo and Martens v. Germany* 2013).

Notwithstanding general acceptance of using DNA technology in the course of criminal proceedings, as well as of establishing and running of DNA databases, in some of its landmark decisions the ECHR established standards where certain measures taken by the authorities in this field fell outside their margin of appreciation, thus violating privacy rights of the applicants. Probably the best known and extensively quoted decision is *S and Marper v. UK* (Asplen 2014). The case concerned two applicants prosecuted by national authorities but then released on all charges. According to the legislation then in force (UK Police and Crime Act), their fingerprints and cell samples were stored in a database for an unlimited period of time despite the fact that criminal proceedings against them had been terminated. Their motion for expungement was denied and the House of Lords ruled that their privacy rights were not violated. In its preliminary findings, the ECHR reiterated that there is nothing in the European Convention prohibiting establishment and operation of DNA databases. However, this requires very stringent legal regulation that should be in accordance with principles of proportionality and necessity. In other words, restrictions to privacy rights of individuals will be permitted if they are prescribed by law, proportionate, and necessary in a democratic society. The court also pointed out that each state claiming a pioneer role in the development of new technologies bore special responsibility for "striking the right balance." Deciding on the merits, the court ruled that indefinite fingerprint DNA retention of all convicted persons is inconsistent with the principle of proportionality, particularly in the context of the protection of particular groups of delinquents (e.g., juveniles). In conclusion, the court was struck by the blanket and indiscriminate nature of the power of retention in England and Wales. In particular, the data in question could be retained irrespective of the nature or gravity of the offence with which the individual was originally suspected or of the age of the suspected offender; the retention was not time-restricted; and there existed only limited possibilities for an acquitted individual to have the data removed from the nationwide database or to have the materials destroyed (*S and Marper v. UK* 2008).

A similar path of reasoning was followed in another case in which the court also found a violation of Article 8 of the European Convention. The applicant, a French national, was convicted by the domestic court for refusing to give his biological sample to the national DNA database (FNAEG). The court referred to the French Conseil Constitutionnel decision on the constitutionality of legislation regulating FNAEG, particularly the conclusion reached by the highest constitutional authority in France that a one-size-fits-all retention period (40 years irrespective of the gravity of the criminal offense) is not consistent with the principle of proportionality (see earlier). The Conseil Constitutionnel ruled that provisions on the FNAEG were in conformity with the Constitution, subject, inter alia, to "determining the duration of storage of such personal data depending on the purpose of the file stored and the nature and/or seriousness of the offences in question" (*compte tenu de l'objet du fichier, à la nature ou à la gravité des infractions concernées*) (*La décision du Conseil Constitutionnel nᵒ 2010-25 QPC, 706-54 à 706-56 du CPP*). The ECHR concluded that "the regulations therefore failed to strike a fair balance between the competing public and private interests" (*Aycaguer v. France* 2017).

20.2.3 Croatian Constitutional Court Decision (2012)

The principle of proportionality was extensively dealt with by the Croatian Constitutional Court in its decision issued in 2012. The court found that Article 186 of the Criminal Procedure Act contradicts the constitution because it gives the police, the state prosecution and the court the right to "collect, store and process personal data of citizens relevant to the purpose of criminal proceedings, taking into account the appropriateness of the need for such data in the specific case." The Constitutional Court ruled that the provision was "too broad and vague, as it allows the collection of personal data of all citizens irrespective of the gravity of the criminal offence that was committed … that provision also does not distinguish between the various types of personal data being collected." The general condition that reads "important for the purpose of criminal proceedings" is general and ambiguous leaving room for various interpretation and thus leading to legal uncertainty.

The Constitutional Court also found that the aforementioned provision of the Criminal Procedure Act departs from the EU Council Framework Decision 2008/977/JHA of 27 November 2008 on the protection of personal data processed in the framework of police and judicial cooperation in criminal matters (*Official Journal L350*, 30. 12. 2008, 60–71). This is because the purpose of personal data collection (including DNA) has not been specified from the perspective of a legitimate aim, and it is not clear for what purposes personal data could be used and processed. Furthermore, there are no rules on periodic assessment of the data, and no rules on restricted access to this information in cases of data concerning the racial or ethnic origin of the person, his political, religious and other beliefs, sexual life, etc. (Constitutional Court of the Republic of Croatia, decision UI-448/2009 of 19 July 2012).

20.2.4 Implications of *S and Marper* in the UK

Following the ECHR judgment in *S and Marper*, the government of the United Kingdom adopted new legislation on DNA retention but, due to the change of government in May 2010, the legislation never entered into force (https://ukhumanrightsblog.com/2011/ 05/18 /retention-of-dna-breaches-article-8-says-supreme-court/, August 6, 2018). However, a change of attitude in domestic jurisprudence and the overruling of the House of Lords' decision in *S and Marper* took place in 2011 in a UK Supreme Court decision, *R (on the Application of GC) (FC) (Appellant) v. The Commissioner of Police of the Metropolis*. The case concerned two appellants who, after being acquitted on criminal charges, requested destruction of their DNA profiles from the database. The police refused the motion referring to the guidelines issued by the Association of Chief Police Officers (ACPO). According to the ACPO Guidelines, stored DNA profiles should be destroyed only in exceptional cases. The UK Supreme Court ruled that such a policy is not in accordance with the ECHR decision in *S and Marper* and consequently not compatible with Article 8 of the European Convention. In other words, it was not the legislation itself that was incompatible with the convention, but the policy disregarding the ECHR reasoning that domestic legislation (Police and Criminal Evidence Act) should be interpreted in accordance with privacy rights protection under the convention (https://ukhumanrightsblog. com/ 2011/ 05/18/rete ntion-of-dna-breaches-article-8-says-supreme-court/, August 6, 2018). The most recent implication of the *S and Marper* judgment in the UK is the adoption and entering into force of the Protection of Freedoms Act (adopted in 2012, in force since October 2013). In a nutshell, it provides that DNA profiles of those persons arrested or charged but not convicted may now be destroyed. For those convicted, the police are authorized to indefinitely retain their DNA profiles. Oversight on retention is the responsibility of the newly created Commissioner for the Retention and Use of Biometric Material (https://www.libertyhumanrights.org.uk/human-rights/privacy/dna-retention, August 6, 2018).

20.2.5 Standards for DNA Retention and Use in Criminal Proceedings

In sum, as a genetic material, DNA is considered as specific personal data. But it is even more than just personal data. DNA is also personally identifiable information. As one of the biometric identifiers, DNA requires more stringent legal protection than other personal data. This is because "DNA contains far more intimate information about a donor than other identification technologies, such as fingerprints" (Taylor 2003). Therefore, when prescribing and implementing legislation related to retention of DNA profiles, it is important to adhere to the proportionality standards established in ECHR jurisprudence.

The first standard concerns validity (legality) of retention of DNA profiles for various categories of persons targeted by criminal investigation and prosecution. The European Convention allows states a broader margin of appreciation to restrict the right to privacy of convicted persons in comparison to those who have just been arrested or acquitted. Apart from the European Court, this principle has been confirmed by US courts. For example, in *Rise v. Oregon*, the court expressly differentiated between the privacy rights of convicted felons and those of "free persons or even mere arrestee—certain classes of convicted felons do not have the same expectations of privacy in their identifying genetic information that 'free persons' have" (59 F.3d 1556, 9th Cir. 1995). Along these same lines, the Minnesota appellate court ruled that a statute requiring officers to obtain a sample of DNA from arrestees was unconstitutional because a person who is merely charged with a crime has a greater expectation of privacy than someone who was convicted (*In re C.T.L.*, 722 N.W.2d 484, 492, Minn. Ct. App. 2006).

The second standard requires that the period of DNA retention should depend on the gravity of the crime that was committed. A more serious crime should, in principle, mean a longer retention period. A good example of such a model is Croatian legislation that provides for a 20-year general retention period (from the moment criminal proceedings are ended) and a 40-year period for more serious criminal offences punishable with imprisonment of at least ten years or for sexual crimes punishable with imprisonment of more than five years. Indefinite retention would not per se be considered as contrary to the proportionality principle. However, it may not entirely be in accordance with the necessity principle. Namely, research on recidivism suggests that the risk of reoffending significantly decreases over time, which means that the usefulness of indefinite retention in practice remains quite questionable. The exception is sexual offenders. Analysts and legislatures can use available statistics to determine the types of crimes (e.g., sexual assaults) that are typically committed by repeat offenders, and only in those situations is it appropriate to permanently maintain DNA samples in the database (Deray 2011).

Another standard concerns the procedure of removing DNA samples from the database. The burden of proof for removal should never be on the applicant. An example of improperly shifting the burden of proof to the applicant is a Florida statute that provides that individuals from whom DNA is collected upon arrest can only have the sample and information destroyed if they provide official documentation demonstrating that the charges were dismissed or the conviction overturned. *Id.* A more favorable solution would be through automatic expungement from the database by the authorities (ex officio) under the conditions prescribed by law.

20.3 The Croatian Experience

20.3.1 Introduction

In the Republic of Croatia, molecular genetic analysis in criminal procedure is regulated by the Criminal Procedure Act; the Ordinance on Sampling Biological Materials and Performing Molecular Genetic Analysis; the Ordinance on Organization and Managing of Collections with Automatic Data Processing Related to Identification of the Suspects; the Act on Simplification of Exchange of Information Between the European Union Member State Bodies in Charge of Law Enforcement; and the Act on Protection of Individuals Related to Personal Data Processing and Exchange Aimed at Preventing, Investigating, Discovering or Prosecuting Felonies or the Implementation of Penal Sanctions, whereas the procedure related to imprisoned persons is regulated by the Imprisonment Act.

20.3.2 European Law and Application of DNA Analysis in Criminal Procedure

Of exceptional importance in using DNA analysis for the purpose of identification of the suspect or other persons in criminal proceedings is Recommendation No. R (92) 1 of the Committee of Ministers to Member States on the use of analysis of deoxyribonucleic acid (DNA) within the framework of the Criminal Justice System (adopted by the Committee of Ministers on February 10, 1992). Undoubtedly, fighting crime requires using the latest and most efficient methods, and in this fight, DNA analysis can offer advantages to the criminal law system. The recommendation consists of 12 chapters, each of them treating the most important principles and recommendations concerning the use of DNA analysis in the penal system. The chapters are definition, scope and limits, using samples and the information obtained from them, sampling for DNA analysis, deciding to analyze DNA, accreditation of laboratories and institutions and controlling DNA analysis, data protection, string samples and data, equality of arms, technical standards, intellectual property, and cross-border exchange of information.

Aimed at efficient police and judicial international cooperation in criminal matters in combating cross-border crime and terrorism, the convention between the Kingdom of

Belgium, the Federal Republic of Germany, the Kingdom of Spain, the French Republic, the Grand Duchy of Luxembourg, the Kingdom of the Netherlands, and the Republic of Austria, signed in Prüm, Germany, on May 27, 2005 (the Prüm Convention), plays a very important role in fostering cross-border cooperation, particularly in combating terrorism, cross-border crime, and illegal migration. This convention enables DNA profiles and fingerprints to be shared between the signatory states, as well as the availability of this data to other signatory states for their automatic search and comparison with the data in their national databases. Since the convention was signed by a significant number of the European Union member states, and other European Union member states subsequently expressed their interest in such a form of cooperation, Council Decision 2008/615/JHA of 23 June 2008 on the stepping up of cross-border cooperation, particularly in combating terrorism and cross-border crime was adopted, and, on the same day, Council Decision 2008/616/JHA on the implementation of Council Decision 2008/615/JHA of 23 June 2008 on the stepping up of cross-border cooperation, particularly in combating terrorism and cross-border crime, was approved. In this way, the so-called Prüm Convention of 2005 was adopted at the level of the entire European Union, and cross-border cooperation between the European Union member states about exchanging DNA profiles and fingerprinting data between the authorities in charge for prevention and investigating felonies has been further enhanced.

Also aimed at a more efficient and expeditious exchange of existing information and other relevant data, including DNA profile data, required in criminal investigations (crime prevention, discovering and gathering data on crimes, performing inspections, research, investigations, etc.) between the competent bodies of the European Union member states, the European Union Council adopted Council Framework Decision 2006/960/JHA of 18 December 2006 on simplifying the exchange of information and intelligence between law enforcement authorities of the member states of the European Union.

Furthermore, having in mind the fundamental principles and the existing norms related to the exchange of DNA data, the rules on applications and responses related to DNA data, the process of transferring DNA profiles, and establishing efficient police and judicial cooperation in criminal matters, the Council of the European Union adopted Resolution 2009/C of 30 November 2009 on exchanging DNA analysis results.

With regard to delivering and exchanging DNA analysis results between the European Union member states, Article 25 of Council Decision 2008/615/JHA of 23 June 2008 on the stepping up of cross-border cooperation, particularly in combating terrorism and cross-border crime, is very important and relates to personal data protection. Namely, paragraph 2 of said article stipulates that submitting personal data described in Council Decision 2008/615/JHA may not be delivered before the stipulations of said decision related to personal data protection are incorporated into the national legislation of the member states involved in the delivery. It is to be emphasized that this stipulation does not apply to the member states that initiated their personal data deliveries under the Prüm Convention of 2005.

The Directive (EU) 2016/680 of the European Parliament and Council of 27 April 2016 on the protection of natural persons with regard to the processing of personal data by competent authorities for the purposes of the prevention, investigation, detection or prosecution of criminal offences or the execution of criminal penalties, and on the free movement of such data, and repealing the Council Framework Decision 2008/977/JHA, stipulates the rules regulating protection of natural persons with regard to the processing of personal data by competent authorities for the purposes of the prevention, investigation, detection, or prosecution of criminal offences or the execution of criminal penalties, including protection against threats to the public safety and their prevention. Accordingly, this is a new legal framework in the field of personal data protection at the level of the European Union. It is important to emphasize that Article 3, point 12, of said directive stipulates that genetic data constitutes personal data related to an individual's inherited or acquired genetic characteristics that provide unique information on that individual's physiology and health and that are acquired by analyzing that individual's biological sample. Thus, this provision covers data obtained by analyzing biological samples of chromosomes, deoxyribonucleic acid (DNA), or ribonucleic acid (RNA), or by analyzing another element that enables one to obtain any other equally valid information belonging to the personal data category. On the other hand, if the data processing is not directed toward the prevention, investigation, discovery, or prosecution of felonies or implementation of penal sanctions, including the protection against threats to the public safety and their prevention, Regulation (EU) 2016/679 of the European Parliament and of the Council of 27 April 2016 on the protection of natural persons with regard to the processing of personal data and on the free movement of such data, and repealing Directive 95/46/EC (General Data Protection Regulation), is instead applied. The difference between the regulation and the directive is that the regulation is a compulsory legislative act that is applicable directly in all the European Union member states, not requiring its incorporation into the national legislation, whereas the directive is a legislative act establishing the goal that all the European Union member states are to achieve and binding them to harmonize their national legislation with the directive.

It has already been pointed out herein that DNA data may not be delivered before the stipulations on data protection in Council Decision 2008/615/JHA of 23 June 2008 on the stepping up of cross-border cooperation, particularly in combating terrorism and cross-border crime, are incorporated in the national legislation of the European Union member states involved in the delivery. The decision on whether a particular European Union member state fully implemented the general regulations on data protection is to be made by the European Union Council unanimously. Meeting this condition is verified against the assessment report based on the questionnaire (the questionnaire relates to data protection and the DNA data exchange), testing, and the assessment visit. The questionnaire is prepared by the adequate work team of the European Union Council and is completed by each European Union member state deeming to have met the conditions required for delivering the DNA data. Besides completing the questionnaire, the European Union member state concerned is to perform a test of the DNA data distribution to one or more member states already delivering data under Council Decision 2008/615/JHA of 23 June 2008. After that, the European Union member state that completed the questionnaire is paid an assessment visit by the assessment team made up of members from one or several European Union member states with which the European Union member state wishes to begin delivering DNA data. The assessment team is to make its report on the assessment visit and to submit it to the European Union Council work team, whereafter the work team is to present to the European Union Council its comprehensive assessment report containing results of the questionnaire, the assessment visit, and the test. The European Union Council, after establishing that all European Union member states agree that the conditions have been met (an agreement of all the European Union member states bound by the Council Decision 2008/615/JHA of 23 June 2008 is required), will decide that the European Union member state concerned fully implemented the general stipulations of Council Decision 2008/615/

JHA of 23 June 2008 with regard to automated exchange of DNA data. Once such a decision is made, the European Union member state concerned is authorized to receive and send DNA data. It is to be particularly emphasized here that Croatia has passed the entire aforementioned procedure and adopted the Council Implementing Decision (EU) 2018/1035 of 16 July 2018 on the launch of automated data exchange with regard to DNA data in Croatia, meaning that Croatia has fully implemented the general stipulations of Council Decision 2008/615/JHA of 23 June 2008 on the stepping up of cross-border cooperation, particularly in combating terrorism and cross-border crime.

20.3.3 Application of DNA Analysis in Croatian Criminal Procedure

The matters related to expert reporting in Croatian criminal procedure are regulated in Articles 108-328 of the Criminal Procedure Act. Pursuant to this act, an expert report is to be requested by a written order when establishing or evaluating an important fact requires obtaining findings and an opinion from a person (a certified court expert) having the necessary professional knowledge or skill. A court expert is an unbiased person, expert in a nonlegal field, commissioned by the authority conducting a criminal proceeding to help in establishing crucial facts. Every expert report order is to state the facts related to which the reporting is requested and the name of the expert commissioned. An expert report is ordered by the authority conducting a criminal proceeding, that is, the public attorney's office or the court, and which of them is to order the report in a particular case depends on the stage of the criminal proceeding. Hence, an expert report prepared at the request of the defense does not constitute evidence in the criminal proceeding since an expert report may be ordered only by the authority conducting the criminal proceeding, whereas the force of a report ordered by any other person is limited to information and, possibly, an indication that an additional report or ordering a new expert report is required.

Molecular genetic analysis is determined in Article 202, paragraph 2, part 23, of the Criminal Procedure Act, whereas the molecular genetic analysis procedure is provided for in Articles 327 and 327a of the act. Molecular genetic analysis is a procedure of analyzing the DNA that makes the basic genetic material of man and other living beings. The authority conducting a criminal proceeding may order a molecular genetic analysis in order to have the biological traces collected at the crime scene or another place where traces of the felony exist compared with the biological samples taken from the accused, the victim, or a third person but not making biological samples of that person, or in order to have these traces or samples compared with molecular genetic analysis results. To this end, the competent authority will order taking biological material samples of the accused, the victim, or a third person but not making biological samples of that person. The competent authority ordering the sampling is to advise the person whose samples are being taken that the data will be subjected to molecular genetic analysis and stored.

Which competent authority will take or order the taking of biological material samples depends on whether the samples are to be taken from the crime scene, from the accused, the victim, or a third person but not making biological samples of that person. If a search, temporary seizure of objects, inspection, or other evidence taking is performed before commencing the criminal proceeding, the authority conducting these activities may only take samples from the crime scene or another place where traces of the felony exist. Pursuant to the Criminal Procedure Act, said activities may be performed by the public attorney or the police.

If a biological material sample is being taken from the accused, the victim, or a third person but not making biological samples of that person (taking samples from the victim or another person requires their written consent), taking samples and their molecular genetic analysis is to be ordered by the public attorney. If the victim or a third person did not consent in writing to taking samples from them, such sampling may be ordered by the court at the public attorney's request. A molecular genetic analysis of biological trace samples is entrusted to a court expert, and the material to be analyzed is presented to him or her anonymously.

Besides providing for the possibility of taking and analyzing biological materials, the periods of storing the data obtained from molecular genetic analyses are also provided for. The fundamental rule is that the tested biological material of the accused, the victim, or a third person but not making biological samples of that person may be used and processed only until linking to the trace or identity or origin is eliminated, whereupon it is to be destroyed. For how long data gathered in the molecular genetic analysis is to be stored depends on whether the accused on trial is finally found guilty or exonerated, or the trial is stopped, or the criminal complaint is dismissed.

Where the accused is finally found guilty in the trial, the data gathered in the molecular genetic analysis is to be stored for 20 years after the completion of the trial. If the case involves a felony punishable with ten or more years' imprisonment, or a felony against sexual freedom punishable with five or more years' imprisonment, such data is to be stored for up to 40 years after the completion of the trial. However, if the accused is finally exonerated in the trial, or the trial is stopped, or the criminal complaint dismissed, the data gathered from the molecular genetic analysis of the materials sampled from the accused, as well as the victim, is stored for ten years from the completion of the trial. It is very important to emphasize that the authority competent ex officio is to delete said data, regardless of whether the accused is finally found guilty or exonerated, or the trial is stopped, or the criminal complaint is dismissed, upon the expiration of the prescribed periods of time. Only the data obtained from the molecular genetic analysis of the materials found at the crime scene and other samples not related to any particular person are to be stored indefinitely.

As stated earlier, the Criminal Procedure Act provides for the possibility of sampling biological material related to suspects, the accused, and defendants, but the question is what about the persons serving time penalties (prisoners) from whom no biological material sample has been taken. Pursuant to Article 174a of the Imprisonment Act, prisoners sentenced for felonies punishable by at least six months' imprisonment, and who, on the date of the act coming into force (11 July 2009), were already serving time penalties are to be sampled for biological material for molecular genetic analysis within 180 days from the date when the act came into force. Pursuant to Article 51 of the Imprisonment Act, the penal sanction enforcement judge is to submit to the penitentiary or the jail copies of the penal sanction enforcement orders, sentences, social welfare examinations and criminal record extracts, and the information on whether any biological material was sampled in the trial for molecular genetic analysis purposes, and other information on file, and this must be done at least three days before the defendant is to report to commence serving the penalty.

Furthermore, the Ordinance on Sampling Biological Materials and Performing Molecular Genetic Analysis stipulates the conditions for storing and safekeeping biological materials, storing conditions, processing and storing the data obtained in the molecular genetic analysis, as well as supervision over storing, processing, and safekeeping of such obtained data. The biological material samples are to be processed, that is, subjected to molecular genetic analysis in line with the established standards and rules of the trade, meaning that new methods and technologies are to be introduced, and the professional staff is to be educated regularly. Under the ordinance, the data obtained from the molecular genetic analysis is stored and kept in the database of the Ivan Vučetić Forensic Science Centre, and the biological material samples and the obtained data may be delivered to the competent institutions in Croatia and abroad. The Ivan Vučetić Forensic Science Centre is an organizational unit within the Ministry of Interior of the Republic of Croatia. If biological samples have been taken from the accused, the victim, or a third person but not making biological samples of that person, the court that made the sentence of the first instance or a procedure terminating the process is to notify the Ivan Vučetić Forensic Science Centre on the finality of such ruling and order the storage of the data obtained from the molecular genetic analysis pursuant to Article 327a of the Criminal Procedure Act. Also, if the competent public attorney decided to stop the investigation, they are to notify the Ivan Vučetić Forensic Science Centre accordingly.

Aimed at preventing possible illegal practices about storing, processing, and safekeeping data obtained from molecular genetic analysis, the aforementioned ordinance stipulates that this is overseen by the Commission for Supervision of Storing, Processing and Safekeeping of the Data Obtained from Molecular Genetic Analysis. The commission is made up of five members appointed by the Minister of Interior in cooperation with the Minister of Justice, for a four-year period of office, and from among the persons of adequate professional expertise as stipulated for membership in the commission.

Organization, content, and manner of managing the collections with the automatic processing of data on the identification of accused persons and the ways of using the data in the collections are provided for in the Ordinance on Organization and Managing of Collections with Automatic Data Processing Related to Identification of the Suspects. Pursuant to said ordinance, two collections of data on the identification of accused persons are established within the Ministry of Interior: the Papillary Line Collection and the Collection of Data Obtained from Molecular Genetic Analysis. The collections are managed by the Ivan Vučetić Forensic Science Centre.

The Collection of Data Obtained from Molecular Genetic Analysis is maintained digitally by the CODIS system (Combined DNA Index System), and contains the DNA profile of the suspect, information on the type of the undisputed biological sample (blood or oral mucous membrane swab), and the information on the person from whom such undisputed sample is taken for the molecular genetic analysis (full name, birth date, register number, and national identification number). The head of the Ivan Vučetić Forensic Science Centre is to appoint one police officer as the only person authorized to enter, delete, or update the data, and all this data is to be entered immediately after the molecular genetic analysis. Particular attention is paid to protecting the data in the Papillary Line Collection and the Collection of Data Obtained from Molecular Genetic Analysis; wherefore, every access to the collections is to be registered in the form of logs, which are used in subsequent verification of the legality of data usage.

The Act on Simplification of Exchange of Information Between the European Union Member State Bodies in Charge of Law Enforcement is enabling efficient, that is, simplified and accelerated, international cooperation in exchanging information between the Croatian authorities in charge of prevention and discovering crimes, collection of information on crimes, performing inspections, research, and investigation of crimes, as well as searching and establishing the illegal gains resulting from crimes, and implementation of the legal powers within the aforementioned activities, on the one side, and the adequate authorities of the European Union member states on the other side. However, it is to be pointed out that this act provides only for submitting the information that the domestic authorities already had collected and stored at the moment they received the request for submitting the information. This means that if the authorities did not have the requested information already collected and stored, they are not obliged to collect the requested information, regardless of the authority they have. The act rules the area of its application, the way of submitting the data and the prescribed submission deadlines, national points of contact, the ways of spontaneous information exchange, reasons for denying the information, and exchange of information in the field of searching for and establishing illegal gains resulting from crimes.

The Act on Protection of Individuals Related to Personal Data Processing and Exchange Aimed at Preventing, Investigating, Discovering or Prosecuting Felonies or the Implementation of Penal Sanctions stipulates the rules of protection of physical persons related to processing and exchanging of personal data by competent authorities for the purpose of prevention, investigation, discovering, or prosecution of felonies or implementation of penal sanctions, including threats to public safety and prevention of such threats. The act, in its general stipulations, defines personal data as every piece of information related to a physical person whose identity is established or may be established, whereas genetic data is defined as all personal data, especially that obtained by analyzing a biological sample, related to inherited or acquired genetic characteristics that provide unique information on that individual's physiology and health. Therefore, and having in mind that according to the very Constitution of the Republic of Croatia, protection and privacy of personal data are among the personal freedoms and rights of each individual, the act stipulates that every processing of personal data is to be legal and fair, and aimed exclusively to the purposes set forth in the act.

References

Asplen, C. 2014. DNA databases. In: *Forensic DNA Applications: An Interdisciplinary Perspective*, eds. D. Primorac and M. Schanfield, 557–569. CRC Press.

Becker, S.W. 2016. Post-conviction DNA testing, actual innocence, and cold cases: A practitioner's guide to freeing the innocent, exhuming the past, and resurrecting the truth—Making a case for seeking justice over finality. *Glob. Comm. B. L. Juris* 2016:15–45.

Deray, E.S. 2011. The double-helix double-edged sword: Comparing DNA retention policies of the United States and the United Kingdom. *Vand. J. Transnat'l L.* 44:745–775.

Innocence Project, https://www.innocenceproject.org (last accessed March 30, 2019).

Kaye, D.H. 2009. Trawling DNA databases for partial matches: What is the FBI afraid of? *Cornell J. L. Pub. Pol.* 19:145–171.

Taylor, B.L. 2003. Storing DNA samples of non-convicted persons & the debate over DNA database expansion. *T.M. Cooley L. Rev.* 20:509–545.

Some Ethical Issues in Forensic Genetics

21

ERIN D. WILLIAMS

Contents

21.1 Introduction

Identifying individuals, living or dead, from their genotypes can help to convict offenders or exonerate innocent suspects. It can confirm one's presence at a crime scene or one's place in a family tree. However, the extensive and sensitive nature of the genetic information locked in the coils of the DNA molecule also gives rise to ethical dilemmas.

This chapter surveys such issues. Section 21.2 offers an overview of some relevant concepts in bioethics. Section 21.3 addresses the ethics of acquiring DNA samples—from crime scenes and other locations and from the bodies of individuals. Section 21.4 outlines a range of ethical concerns with DNA databanks and databases. Section 21.5 concerns efforts to infer phenotypes from genotypes to assist in criminal investigations. Section 21.6 identifies ethical concerns that arise in identifying human remains, particularly in mass disasters or conflicts. Finally, Section 21.7 discusses the professional and ethical standards for reporting forensic laboratory results and testifying about them.

21.2 General Concepts in Bioethics

The bioethical concepts explored in this chapter include justice, privacy and confidentiality, autonomy and informed consent, beneficence, and utility.

DOI: 10.4324/9780429019944-25

21.2.1 Justice

The two types of justice most relevant to forensic genetics are distributive and retributive. Distributive justice means the fair, equitable, and appropriate distribution of benefits and burdens (Beauchamp and Childress 2019). Retributive justice is concerned with the righting of wrongs, often through criminal justice (Williams and Wienroth 2017). When the criminal justice system burdens certain groups disproportionately, questions of comparative justice arise (Feinberg 1973). Societies and individuals seek to balance claims of retributive and comparative injustice.

21.2.2 Privacy and Confidentiality

Privacy, broadly speaking, means freedom from unwanted observation, intrusion, or distribution of information about an individual. Confidentiality refers to information or items revealed in a protected relationship, such as that between lawyer and client. Both privacy and confidentiality may involve control over how and when information is revealed, but confidentiality typically arises in a relationship of trust and refers to the duty of an individual entrusted with the information. In the medical arena, maintaining the confidentiality of a patient's medical records is an important professional norm and the subject of a complex set of legal rules, but confidentiality typically is legally overridden in the pursuit of justice.

21.2.3 Autonomy and Informed Consent

The Kantian principle of autonomy respects persons and their choices. In the context of patient care and human experimentation, it means that every human being "of adult years and sound mind has a right to determine what shall be done with his body" (*Schloendorff v. Society of New York Hospital*, 105 N.E. 92, 93 [N.Y. 1914]). It is "consistent with the individualistic temper of American life, which emphasizes privacy and self-determination" (Pellegrino 1993, 10).

Informed consent protects autonomy by disclosing the risks and benefits of, and alternatives to, a particular course of action, enabling the individual to make a reasoned, personal choice. A surgeon, for example, generally may not perform a procedure—even when it clearly would benefit a patient—without the patient's informed content (*Schloendorff v. Society of New York Hospital*, 105 N.E. 92 [N.Y. 1914]).

Informed consent has a reduced bearing on actions in the law enforcement context. A suspect who refuses to submit to dental impressions authorized by a search warrant may be forced to undergo general anesthesia and nasal intubation (*Carr v. State*, 728 N.E.2d 125, 128 [Ind. 2000]; cf. *Winston v. Lee*, 470 U.S. 753, 760 [1985]: "The reasonableness of surgical intrusions beneath the skin depends on a case-by-case approach, in which the individual's interests in privacy and security are weighed against society's interests in conducting the procedure."). The law routinely requires individuals to submit to much less invasive DNA sampling for investigative purposes.

21.2.4 Utility

Utilitarianism is a moral theory that evaluates acts or rules by their consequences. Roughly stated, it requires that we act to secure the greatest good for the greatest number of people.

However, different utilitarians have different conceptions of what counts as "good" (Beauchamp and Childress 2019). A common criticism of utilitarianism is that it permits (at least in principle) an individual's rights to be sacrificed for the general welfare. When indefeasible rights are not involved, there is general agreement that a utilitarian weighing of costs and benefits is crucial to developing good social policies.

With this catalog of general concepts and principles that should influence ethical judgments, we turn to a set of ethical issues that are apparent in forensic genetics. We start with the collection of DNA samples—from people and places.

21.3 Ethical Issues in Acquiring DNA Samples

21.3.1 Crime Scene Samples, and Shed or Abandoned DNA

Collecting DNA from crime scenes and cadavers can raise ethical issues. In some cases, DNA profiles from cadavers might reveal genetic conditions or relationships (or their absence, as in situations of "misattributed paternity") of which family members were not aware. This is a common occurrence in genetic research using pedigrees as well as in genetic counseling, where the obligations to individuals involved in and consenting to genetic tests as well as those affected by the testing have been debated for some time (Juengst and Goldenberg 2008; Lucast 2007; Tozzo, Caenazzo, and Parker 2013). The genetic research as well as genetic counseling does not typically infringe informational privacy, however, because the STR loci used in forensic identification are not known to cause disease or to have substantial correlations to sequences that affect health status (Kaye 2007).

DNA left at a crime scene is an example of abandoned DNA. An individual has some privacy interest in the secrecy of his travels, but it is implausible to contend that people have any right to prevent others from discovering, through trace evidence, that they were present in a particular place at some time in the past. A more appealing claim is that "genetic privacy" should encompass the genetic information in biological materials inadvertently left at ordinary places in the course of everyday life. Should a nosy neighbor be permitted to retrieve and analyze DNA from dental floss left in the trash to obtain possibly embarrassing information about genetic weaknesses? Whatever one thinks of a genetic privacy law that would protect individuals from such snooping (Joh 2011), the extraction of purely identifying information from the DNA that we routinely shed or abandon is less problematic.

21.3.2 Sampling with Consent

Police often search individuals or their possessions with the ostensible permission of the individuals. The same is true of DNA sampling. In a rape case, for example, police may request "elimination samples" from spouses or sexual partners of the victim. They might ask a suspect to give a sample, telling him that he can establish his innocence by cooperating. Indeed, in exceptional cases, police have systematically collected samples from entire communities. The earliest "DNA dragnet" occurred in England in 1987 in Narborough (Wambaugh 1987). Investigating the rape and strangulation of two teenage girls, police requested blood samples from more than 4000 local men. Eventually, the killer was identified after he asked a coworker to give a sample in his place. Since then, DNA dragnets have been used with some success in Europe and the United States (Ripley et al. 2005; Maschke 2008).

DNA dragnets raise two sets of ethical and legal questions. The first is whether permission obtained in face-to-face encounters with the police, and under the pressure of becoming a suspect if one refuses to cooperate, is truly voluntary (Maclin 2008). The medical model of informed consent does not apply to encounters between the police and the public (*Schneckloth v. Bustamonte*, 412 U.S. 218 [1973]), but the threat of force can render consent to a search legally ineffective (cf. *Bumper v. North Carolina*, 391 U.S. 543 [1968]).

The second ethical issue is whether the police may keep the samples or profiles for later investigations of unrelated crimes. The determination turns on the scope of the consent provided in the first instance (or whether the sample donor later agrees to the further use) and whether law enforcement practices require retention of all evidence collected as part of official investigations.

Whether the benefits of the practice outweigh the costs also has been questioned. Large-scale screening consumes considerable police resources and is an imposition on the many innocent people who are stopped and questioned. As a result, the practice is used only in the most serious cases, such as those of serial killings and rapes.

21.3.3 Acquiring DNA Samples from Medical Providers or Researchers

Ethical issues arise from the potential to use genetic information or samples collected for medical and other purposes for criminal investigations. DNA can be collected by medical providers either via genetic testing or from tissue abandoned after surgery. It can also be collected by companies selling direct-to-consumer genetic testing, genealogy and ancestry services, and nutrigenomics. Some employers also collect DNA to monitoring exposure to toxins.

In the United States, the medical community may only disclose health and other genetic information in certain circumstances (Privacy Rule, 42 C.F.R. §§ 160.101-552, 164.102-106, 164.500-534 [2002]; Genetic Information Nondiscrimination Act of 2008, Pub. L. No. 110-233, 122 Stat. 881). However, these restrictions only attach to specified entities involved in healthcare, and even then, protected health information may be disclosed when required by other laws or for law enforcement purposes (45 C.F.R. § 164.512(a)). Likewise, in some jurisdictions, the physician–patient privilege for communications to one's doctor does not apply to defendants in criminal trials and certain other proceedings (Broun et al. 2013, sec. 104).

Several examples illustrate the bounds of when genetic information collected for another purpose may be released for law enforcement uses. One involves a special act of the Swedish Parliament, which authorized access to a "neonatal database" to identify tsunami victims (Mund 2005). This humanitarian use was not contemplated originally, however, access was granted for the benefit of their families. In this situation, the unanticipated use was benign to the DNA donors and likely to be consistent with their wishes if they had been informed of the situation.

A second case in Texas also involved neonatal samples. These were collected in the form of dried blood spots for a compulsory newborn screening program. Over the years, the Texas State Department of Health Services supplied some of these cards to medical researchers studying club foot, childhood cancer, and lead exposure. Five parents sued

(Lewis et al. 2011), and the state promptly settled the case by agreeing to destroy millions of cards, to give parents clearer procedures to opt out of the storage of the cards, and to pay $26,000 in attorney's fees and costs. A journalist who read the public reports of the health department, then reported, confusingly, that nearly 800 of the samples had been provided to "the federal government to create a vast DNA database, one that could help crack cold cases and identify missing persons" (Ramshaw 2010). In reality, the cards—with personally identifying information removed—had gone to the United States Armed Forces DNA Identification Laboratory for research into the population frequencies of mitochondrial DNA types. In this case, then, there was even less of a risk to personal privacy than in the Swedish disclosure, but the use of the samples did not specifically benefit the individuals or families whose mitochondrial DNA was typed anonymously for statistical purposes.

The third example pertains to the apprehension of a notorious serial killer in the United States. Over a 30-year period, he killed at least ten people, taunting the police in Wichita, Kansas, with letters signed BTK (for Bind, Torture, Kill). As police developed a case against Dennis Rader, they obtained a court order for a Pap smear his daughter had given years earlier at a university medical clinic. Investigators compared an STR (short tandem repeat) profile of the clinic specimen to that of DNA taken from several BTK crime scenes. A reverse paternity (statistical) analysis led detectives to conclude that the daughter, who neither knew of nor consented to the testing, was most likely the biological child of the killer. Rader confessed and pled guilty to ten counts of murder (Nakashima 2008). Here, the innocent daughter's interest in the confidentiality of her Pap smear yielded to society's interest in the justice and utility of capturing the serial killer. However, if medical information is commonly used for familial criminal proceedings, patients may be deterred from seeking care, calling into question the utility of the tactic.

21.4 Law Enforcement DNA Databanks

Chapter 23 describes the use and expansion of law enforcement DNA databanks across the world. The proliferation of these databanks raise issues of distributive justice, privacy, and confidentiality. Complaints grounded in distributive justice have been raised with law enforcement's use of DNA databases in the United States and the United Kingdom. Profiles in these databases come disproportionately (compared to population percentages) from racial or ethnic minorities (Kaye and Smith 2003; Krimsky and Simoncelli 2011; Williams and Wienroth 2017).

Good governance can diminish privacy and confidentiality concerns related to databases. Using profiles that are not associated with diseases, and destroying samples once they no longer are needed can mitigate these issues. In addition, expunging profiles after certain time periods, and forbidding the release of samples for biological research (other than that related to the operation of the database or the interpretation of DNA matches) have a similar effect.

Biobanking raises additional privacy and confidentiality issues with research into human genetics and genomics, and the application of the resulting knowledge in the healthcare system (Budimir et al. 2011). Traditionally, pathology tissue repositories have used samples collected from patients for a variety of purposes without securing the informed consent of the donors for each and every use (Korn 1999). Some physicians and ethicists have questioned this tradition (Clayton et al. 1995), and the liberal use and sharing of samples from repositories created for research into the genetics of diseases also has been challenged (Greely 1999). Even with consent, the rights and interests of connected individuals may not be protected (Widdows and Cordell 2011).

The norms that govern research biobanks do not necessarily apply to law enforcement. For example, military and law enforcement repositories collect biological material maintained for the limited purpose of identification of remains or crime scene samples. Law enforcement personnel do not have a fiduciary duty to protect the interests of the suspects in criminal investigations, although they do have other ethical and legal duties toward suspects and witnesses (Kaye 2000).

DNA collection for law enforcement sparks pragmatic issues. When and why may samples be collected? Who should have access to the samples and how may they be used? How long may samples be retained? As with databanking, good biobanking governance practices can effectively balance the public good (utility) with distributive justice and privacy concerns.

21.5 Phenotypes and Racial Identifications from Genotypes

Although the STR loci used in forensic identification are not well suited for drawing inferences about physical traits or ancestral groups, geneticists can use other loci that provide clearer "ancestry and morphology information to infer a suspect's race and general appearance" (Ossorio 2006). Chapter 17 describes research into the genetics of eye, hair, and skin color; height; and other physical features. Once inferences of specific physical traits become available (with known limitations), it will be tempting to use them to guide investigations. Such publicly exposed traits cannot be considered private information.

The use of ancestry informative markers has been opposed, however, on the ground that it will reify race as a biologically meaningful category instead of the social construct and reflection of ancient migrations that it is (Ossorio 2006). There is also some concern that police authorities will not appreciate the limitations in inferring ancestry or phenotypes from crime scene samples. Whether the risks warrant the suppression of characterizations of ancestry or race from DNA analysis depends on how much the practice would contribute to public misunderstanding and the value the descriptions would have in criminal investigations. Some commentators argue that "[p]olicymakers, judges, law enforcement officers, and civil rights activists alike must develop a more nuanced understanding of genetics" and note that "DNA ancestry tests and indirect molecular photofitting [could] ... help remove the prejudicial blinders that investigators may have toward minorities" (Wagner 2009).

A very different issue is whether, in the courtroom or in reports to the police, experts should report population frequencies or likelihood ratios according to ethnicity or race. This practice is not an attempt to establish the race of an individual, but an effort to show that regardless of the subpopulation to which an accused offender might belong, it is improbable that the DNA match is coincidental. However, the same argument about reifying race has been made. According to one law professor, "the persuasive authority of DNA evidence and the reality of implicit prejudice call into question the legitimacy of using racialized DNA evidence" (Kahn 2009).

21.6 Identification of Remains

This chapter has focused on ethical issues in forensic genetics as it has been applied in the criminal justice systems of individual countries. Ethical issues also arise in the identification of human remains (Kelly 2010; 43 C.F.R. § 10.11 [2010]). Many of them are well illustrated by the exhumation and analysis of remains from mass graves. Some of the largest of these in recent history were located in the former Yugoslavia. Exhumation and examination helped to give closure to loved ones and assisted retributive justice efforts for war crimes in the International Criminal Tribunal for the Former Yugoslavia (Williams and Crews 2003; Neuffer 2007).

How remains should be treated is complicated in the case of mass graves. To confound criminal prosecution, Bosnian Serbs used earth-moving equipment to exhume and redeposit remains from mass graves into secondary grave sites, and then mixed secondary sites into tertiary sites, disarticulating, commingling, mangling, and crushing the remains. Bodies needed to be reassembled. Protocols for the return of a body to family members and for making declarations of death could depend in part on the amount and type of remains located. For example, a body with a missing finger might merit one outcome, and a finger missing a body might merit another.

This commingling and reassembly of remains affects families and states that wish to have remains returned and repatriated, or at least buried in accordance with their religious or cultural traditions. Some burial traditions maintain that rituals involving the dead body are necessary to ensure peace in the afterlife. Jewish and Muslim faiths require the burial of bodies including any body parts that may have been removed. The Hindu faith requires cremation.

When only a sampling of remains from a grave can be identified, and when it may not be certain that all, or even the majority of bodies in a particular grave come from a particular state or tradition, the appropriate way to dispose of the commingled remains is not obvious. Ideally, the treatment of remains will respect the wishes of the deceased or their family members. Survivor groups can guide the establishment of protocols that are respectful and fair to all involved.

21.7 Ethics of Forensic Laboratory Reporting and Expert Testimony

Expert reports convey useful information to the prosecution, the defense, and the court before trial. In the United States, due process principles require the prosecution to disclose information that is "favorable to [the] accused" and "material either to guilt or to punishment" (*Brady v. Maryland*, 373 U.S. 83, 87 [1963]), as well as "evidence that the defense might have used to impeach the government's witnesses by showing bias or interest" (*United States v. Bagley*, 473 U.S. 667, 676 [1985]). Although the prosecution is responsible for disclosing all such information, even if it is in the hands of the law enforcement agency (*Kyles v. Whitley*, 514 U.S. 419, 437 [1995]), and the forensic scientist cannot control the actions of police or prosecutors, the expert may have an ethical obligation to avoid practices such as "preparation of reports containing minimal information in order not to give the 'other side' ammunition for cross-examination," "reporting of findings without an interpretation on the assumption that if an interpretation is required it can be provided from the witness box," and "omitting some significant point from a report to trap an unsuspecting cross-examiner" (Lucas 1989; NIST 2012).

At trial, expert witnesses use or impart specialized knowledge and information to assist the trier of fact to understand the evidence. Although an expert can control the content of laboratory notes and subsequent pretrial reports, during a trial, questioning by counsel and judges frames the expert's presentation. The expert may not simply decide what information to discuss but must answer the questions. This can create some tension between the goal of being complete and the need to be responsive (NIST 2012). In resolving this tension, "ethical considerations and professional standards properly place a number of constraints on the expert's behavior" (Feinberg 1989). One such constraint is "a requirement of candor. While an expert is ordinarily under no legal obligation to volunteer information, professional ethics may compel this ... when the expert believes that withholding information will change dramatically the picture that his ... analyses, properly understood, convey" (Feinberg 1989). Thus, the "guiding principles" proposed by the American Society of Crime Laboratory Directors/Laboratory Accreditation Board advise forensic scientists to "attempt to qualify their responses while testifying when asked a question with the requirement that a simple 'yes' or 'no' answer be given, if answering 'yes' or 'no' would be misleading to the judge or the jury" (ANAB 2018).

These precepts may seem more significant in the Anglo-American system of adversarial trials than in the Continental system, in which court-appointed experts are the rule rather than the exception, but the basic principles apply in all jurisdictions. Thus, the European Network of Forensic Science Institutes (2004, Standard I3) urges analysts to "deal with questions truthfully, impartially and flexibly in a language which is concise, unambiguous and admissible"; to "give explanations to specific questions in a manner that facilitates understanding by nonscientists"; to "consider additional information and alternative hypotheses that are presented to you"; to "consider and evaluate these and express relevant opinions taking into account the limitations on opinions which cannot be given without further examination and investigation"; and to "clearly differentiate between fact and opinion and ensure that the opinions you express are within your area of expertise." It is only natural that a forensic scientist who is convinced that his work should guide the court would wish to be as persuasive as possible, but persuasion should come from the clarity of the presentation and the quality and completeness of the scientific investigation. The expert is not the decision maker; the forensic scientist's task is to educate the other participants in the proceedings so that the adjudication will use the scientific findings for what they are worth—no more and no less.

Acknowledgments

The author thanks David Kaye, Nanette Elster, and Joseph Montenegro for their contributions.

References

ANAB. 2018. Guiding Principles of Professional Responsibility for Forensic Service Providers and Forensic Personnel. GD 3150. https://anab.qualtraxcloud.com/ShowDocument.aspx?ID =6732 (accessed February 22, 2020).

Beauchamp, T. L., and J. F. Childress. 2019. *Principles of Biomedical Ethics*. 8th ed. Oxford: Oxford University Press.

Broun, K. S., G. Dix, E. Imwinkelried, et al. 2013. *McCormick on Evidence.* 7th ed. St. Paul, MN: West.

Budimir, D., O. Polašek, A. Marušić, et al. 2011. Ethical aspects of human biobanks: A systematic review. *Croat Med J* 52(3):262–279.

Clayton, E. W., K. K. Steinberg, M. J. Khoury, et al. 1995. Informed consent for genetic research on stored tissue samples. *JAMA* 274(22):1786–1792.

European Network of Forensic Science Institutes. 2004. Standing Committee for Quality and Competence. Performance Based Standards for Forensic Science Practitioners. http://enfsi.eu/wp-content/uploads/2016/09/performance_based_standards_for_forensic_science_practitioners_0.pdf (accessed February 22, 2020).

Feinberg, J. 1973. *Social Philosophy.* Upper Saddle River, NJ: Pearson.

Feinberg, S., ed. 1989. *The Evolving Role of Statistical Assessments as Evidence in the Courts.* New York: Springer-Verlag.

Greely, H. T. 1999. Breaking the stalemate: A prospective regulatory framework for unforeseen research uses of human tissue samples and health information. *Wake Forest Law Rev* 34(3):737–766.

Joh, E. E. 2011. DNA theft: Recognizing the crime of nonconsensual genetic collection and testing. *Boston Univ Law Rev* 91:665–700.

Juengst, E. T., and A. Goldenberg. 2008. Genetic diagnostic, pedigree, and screening research. In: *The Oxford Textbook of Clinical Research Ethics*, eds. E. J. Emanuel, C. Grady, R. A. Crouch, et al., 298–314. Oxford: Oxford University Press.

Kahn, J. 2009. Race, genes, and justice: A call to reform the presentation of forensic DNA evidence in criminal trials. *Brooklyn Law Rev* 74:325–374.

Kaye, D. H. 2000. Bioethics, bench, and bar: Selected arguments in Landry v. Attorney General. *Jurimetrics* 40:193–216.

Kaye, D. H. 2007. Please, let's bury the junk: The CODIS loci and the revelation of private information. *Northwest Univ Law Rev Colloq* 102:70–81. https://papers.ssrn.com/sol3/papers.cfm?abstract_id=1022938 (accessed February 22, 2020).

Kaye, D. H., and M. E. Smith. 2003. DNA identification databases: Legality, legitimacy, and the case for population-wide coverage. *Wis Law Rev* 2003:414–459.

Kelly, R. L. 2010. Bones of contention. New York Times, December 13. http://www.nytimes.com/2010/12/13/opinion/13kelly.html (accessed February 22, 2020).

Korn, D. 1999. Genetic privacy, medical information privacy, and the use of human tissue specimens in research. In: *Genetic Testing and the Use of Information*, ed. C. Long, 16–84. Washington, DC: AEI Press.

Krimsky, S., and T. Simoncelli. 2011. *Genetic Justice: DNA Databanks, Criminal Investigations, and Civil Liberties.* New York: Columbia University Press.

Lewis, M. H., A. Goldenberg, R. Anderson, E. Rothwell, and J. Botkin. 2011. State laws regarding the retention and use of residual newborn screening blood samples. *Pediatrics* 127(4): 703–712.

Lucas, D. 1989. The ethical responsibilities of the forensic scientist: Exploring the limits. *J Forensic Sci* 34(3):719–729.

Lucast, E. K. 2007. Informed consent and the misattributed paternity problem in genetic counseling. *Bioethics* 21(1):41–50.

Maclin, T. 2008. The good and bad news about consent searches. *McGeorge Law Rev* 39:27–82.

Maschke, K. J. 2008. DNA and law enforcement. In: *From Birth to Death and Bench to Clinic: The Hastings Center Bioethics Briefing Book for Journalists, Policymakers, and Campaigns*, ed. M. Crowley, 45–50. Garrison, NY: The Hastings Center.

Mund, C. 2005. Biobanks—Data sources without limits? *Jusletter* 3 (October).

Nakashima, E. 2008. From DNA of family, a tool to make arrests. Washington Post, April 21. https://ceadstorage.blob.core.windows.net/cead-images/NewsReportreDennisRader.PDF (accessed February 22, 2020).

Neuffer, E. 2007. Mass graves. In: *Crimes of War 2.0: What the Public Should Know*, eds. A. Dworkin, R. Gutman, and D. Rieff, 238–240. New York: W. W. Norton & Company.

NIST. 2012. Expert Working Group on human factors in latent print analysis. In: Latent Print Examination and Human Factors: Improving the Practice through a Systems Approach, ed. D. H. Kaye. Gaithersburg, MD: National Institute of Standards and Technology. https://www .nist.gov/publications/latent-print-examination-and-human-factors-improving-practice -through-systems-approach (accessed February 22, 2020).

Ossorio, P. N. 2006. About face: Forensic genetic testing for race and visible traits. *J Law Med Ethics* 32(2):277–292.

Pellegrino, E. 1993. The metamorphosis of medical ethics: A 30-year retrospective. *JAMA* 269(9):1158–1162.

Ramshaw, E. 2010. DNA deception. The Texas Tribune, February 22. http://www.texastribune.org /texas-state-agencies/department-of-state-health-services/dshs-turned-over-hundreds-of -dna-samples-to-feds/ (accessed February 22, 2020).

Ripley, A., T. Bates, M. Hequet, and R. Laney. 2005. The DNA dragnets. Time, January 16. http:// www.feldmeth.net/ethics/dna.html (accessed February 22, 2020).

Tozzo, P., L. Caenazzo, and M. J. Parker. 2013. Discovering misattributed paternity in genetic counselling: Different ethical perspectives in two countries. *J Med Ethics*. https://doi.org/10 .1136/medethics-2012-101062.

Wagner, J. K. 2009. Just the facts, ma'am: Removing the drama from DNA dragnets. *NC J Law Technol* 11:51–101.

Wambaugh, J. 1987. *The Blooding*. New York: Bantam.

Widdows, H., and S. Cordell. 2011. The ethics of biobanking: Key issues and controversies. *Health Care Anal* 19(3):207–219.

Williams, E. D., and J. D. Crews. 2003. From dust to dust: Ethical and practical issues involved in the location, exhumation, and identification of bodies from mass graves. *Croat Med J* 44(3):251–258.

Williams, R., and M. Wienroth. 2017. Social and ethical aspects of forensic genetics: A critical review. *Forensic Sci Rev* 29(2):145–169.

DNA in Immigration and Human Trafficking

22

SARA HUSTON KATSANIS

Contents

22.1 Introduction

Shortly after the first publication of DNA as a tool for identity (Jeffreys, Wilson, and Thein 1985), the first case of DNA testing for identity was published: it was an immigration case (Jeffreys, Brookfield, and Semeonoff 1985). Unlike other biometric identifiers (e.g., fingerprints, iris recognition, voice recognition), DNA analysis can confirm or refute claimed biological relationships. As such, DNA identity applications are especially relevant to border control and migration policies to investigate illicit intercountry adoptions, immigration fraud, and to identify victims of human trafficking.

Chapter 10 describes the application of genetic technologies to the identification of missing persons and victims of mass disasters. These applications are successful and valuable for identifying deceased persons in the wake of human rights atrocities and genocides (Holland et al. 1993; Wagner 2008; Pajnič, Pogorelc, and Balažic 2010), such as cases processed through the International Commission on Missing Persons (Davoren et al. 2007; Parsons et al. 2007; Huffine et al. 2001), but the value of DNA for identifying victims while still alive remains underutilized. As government programs using DNA to determine identity expand, the forensic community is developing programs to tap this technology for the

DOI: 10.4324/9780429019944-26

identification of human trafficking victims. At the same time, communities must grapple with the social implications of DNA collection of vulnerable populations and develop approaches that will protect these populations from secondary or unintentional harms.

In this chapter, we review the emerging and nascent applications of forensic DNA technologies to immigration procedures and human trafficking, and explore the ethical, legal, and social implications of collecting DNA from noncriminals.

22.2 Relationship Testing in Immigration

22.2.1 Relationship Testing

As described in Chapter 10, DNA is useful for establishing not just identity, but also familial relationships. Relationship testing typically involves analysis of autosomal short tandem repeats (STRs) and statistical analysis of likelihood ratios to predict relatedness (Debenham 2006). Sibship and more distant relationships also are detectable using autosomal STRs, but more readily established using single-nucleotide polymorphisms (SNPs) to establish haplotype inheritance and ancestry informative markers (AIMs). In addition, Y-STRs and mitochondrial DNA (mtDNA) analysis can be useful for kinship analyses.

Whereas exact comparisons between two DNA profiles can predict likelihood of two samples coming from the same biological source, comparison of nonidentical samples can predict likelihood of a biological relationship based on how many markers are shared (Daiger and Chakravarti 1983). Most relationship testing uses the likelihood ratio to estimate the probability of two genetic profiles if the two individuals are related versus unrelated, taking into account all of the typed markers. The fraction of shared alleles expected between two biological relatives depends on how they are related (see Table 22.1).

DNA plays a unique role in border security and immigration cases dependent on a claimed relationship of a petitioner to their relative. Having a unique and permanent biometric profile ensures objective and efficient border controls, the question of identity being removed from the individual assessment of border guards to a neutral automated procedure (Thomas 2006). Biometric identifiers present a number of advantages. Biometric identification can be used to detect document falsification for immigration fraud (Wasem 2007). They also might help facilitate return procedures of failed asylum seekers. Under international law, countries cannot deport individuals without knowing their country of origin, yet many asylum seekers will have lost or destroyed their travel documents upon arrival. Countries of origin then might then use this uncertainty about the background of failed asylum seekers to justify a refusal to accept the return of a citizen (Thomas 2006). Asylum seekers can provide credible, immutable evidence of their claim, and traffickers might be hindered in their attempts to use false identities (Thomas 2006).

Table 22.1 **Probabilities of Sharing Alleles at Multiple Markers**

Relationship	0 alleles	1 allele
Parent–child	0	1
Full siblings	1/4	1/2
Half siblings	1/2	1/2
Cousins	3/4	1/4
Uncle–nephew	1/2	1/2
Grandparent–grandchild	1/2	1/2

Proof of identity (POI) at the border includes three main "identity" attributes: biometric (e.g., fingerprints), attributed (e.g., your full name), and biographical (e.g., education or employment history). Traditionally, POI has relied on attributed and biographical data, but this approach is waning because of perpetrators' ability to easily compromise POI documents (Jamieson et al. 2008; Wenk 2011). Biometric attributes are becoming more important and are included on a number of different POI documentation.

In the United States, multiple programs apply biometric identification to border control efforts. The US-VISIT program, administered by the US Department of Homeland Security, conducts biometric identification and analysis, focusing on the collection and analysis of digital fingerprints and photographs from international visitors when they are issued US visas or at ports of entry. Data collected and analyzed by US-VISIT might be used to identify illegal migrants or individuals who might pose a security risk to the United States (National Science and Technology Council 2008). Movement toward inclusion of DNA data, as well as face recognition and other biometrics, began in 2011 (Lynch 2012).

The United States, Canada, United Kingdom, France, Australia, Finland, Sweden, Norway, the Netherlands, and Hong Kong Special Administrative Region of China, among other countries, now include DNA analysis and genetic relationship testing in border security measures. Typically, this involves confirmation of claimed biological relationships between immigrants and individuals who are already settled in the receiving country (Taitz, Weekers, and Mosca 2002a). Because family reunification is a primary motivator in many immigration petitions, the authenticity of an applicant's family relationship claim might determine admission or visa-issuance decisions. The verification of family relationship claims through DNA analysis involves relationship testing of putative family members by private or government laboratories. The relationship testing statistical results are used to judge the accuracy of the claimed relationship (Taitz, Weekers, and Mosca 2002a).

In 2000, the voluntary use of DNA in immigration cases was established in a memo from the US Immigration and Naturalization Service (INS)—which is now USCIS—to its field offices. It had previously been established that the INS could require blood parentage testing (through blood group antigen or human leukocyte antigen). However, the 2000 memo states that there is no "statutory or regulatory authority to require DNA testing." In 2006, the USCIS Ombudsman recommended granting local offices the authority to require DNA testing in specific circumstances and launching a pilot program to require DNA testing in the immigration context (Esbenshade 2010).

In some cases, biometric markers can identify criminals, terrorists, or known traffickers attempting to travel under false identities. Since January 2009, the United States has required DNA collection for inclusion in CODIS from non-U.S. citizens arrested or detained under the authority of the United States (73 Federal Register 74932 2008).

The implementation of DNA testing into refugee petitioning highlights some of the major challenges of systematically relying upon DNA for immigration decisions. The United Nations (UN) and individual countries are exploring methods to document refugees petitioning for immigration (United Nations High Commissioner for Refugees 2008; Taitz, Weekers, and Mosca 2002b). In the United States, refugee family reunification cases have been processed under the Priority 3 (P-3) Program since 2004. In 2008, DNA relationship testing was piloted as part of the P-3 Program in East Africa. The results of pilot DNA testing revealed high levels of fraud among East African P-3 refugees (Esbenshade 2010). As a result, applications under the P-3 Program were suspended (Bruno 2012). In 2010, the US Department of State passed new rules allowing DNA testing for international refugees seeking to reunify with their families in the United States (Holland 2011). The P-3 program resumed in 2012, requiring DNA confirmation of claimed relationships for the 20 countries admitted under the P-3 program (Bruno 2012).

A similar program was instituted by the Obama administration to verify parentage of unaccompanied minors petitioning to join their parents in the US (79 Federal Register 68343 2014); this Central American Migrant Minors program was ceased a couple years later by the Trump administration (82 Federal Register 38926 2017). Following the cancelation of the family separation policy at the US-Mexico border in 2018, the US Department of Health and Human Services used DNA testing to verify parentage claims (Farahany 2019). The parameters for testing nor the processes used were made public, but this action along with the Central American Migrant Minors DNA testing program, set a precedent to incorporate genetic testing into verification processes for claims of biological relationships.

22.2.2 Immigration Fraud and DNA

As the use of DNA relationship testing becomes common in immigration, it is necessary to understand the benefits and limitations to applying DNA technologies in international border security. Reported fraud in US refugee and immigration programs has motivated the use of DNA analysis and other biometric measures to assess claimed biological relationships (Esbenshade 2010). Immigrant petitioners might falsify genetic relationship test results by providing DNA samples of related persons in place of unrelated persons. For example, distant blood relatives might claim a parent–child relationship in an attempt to accelerate or increase their chances of admission into a receiving country. In nations where polygamy is permitted, a male petitioner might claim his that younger wives are his daughters (Wenk 2011). Fraudulent DNA relationship testing might involve inadvertent or intentional "genotype recycling," wherein an individual repeatedly provides a biological sample under different names (Wenk 2010).

There are several ways for laboratories performing relationship testing to detect potential fraud. For one, a laboratory might suspect sample substitution when two or more people demonstrate identical DNA profiles. Discrepancies in paperwork and genetic results might be obvious, such as differences between the claimed sex and the genetic test results. But in ordinary cases, fraud might not be obvious or easily detectable unless the laboratory is specifically looking for discrepancies, such as comparing new profiles to a database of prior profiled cases. This is simple for fraud perpetrators to circumnavigate by using multiple laboratories for testing. Ideally, an interlaboratory DNA profile database would be optimal for detecting genotype recycling and attempted fraud.

To deter DNA fraud in immigration programs, authorities are developing approaches to maintain control and integrity of DNA collection processes. In some cases, this means better control over collection of DNA, such as requiring DNA collection at embassies and analysis by accredited commercial laboratories, whereas in other cases this means processing relationship testing through government-controlled laboratories.

22.2.3 Rapid DNA Analysis

Standard forensic DNA analysis involves multiple steps and can take hours to days to process, depending on the laboratory and equipment. Rapid DNA analysis processes profiles with a more efficient turnaround time from minutes to hours, depending on the system.

Existing Rapid DNA instruments process five to eight samples in 45–90 minutes (Carney et al. 2019; Buscaino et al. 2018; Lounsbury et al. 2013; Estes et al. 2012). Turnaround time is likely to further reduce with improved technology. Most Rapid DNA systems use micro-fluidic technologies to isolate DNA, amplify loci, and detect STR alleles in a single inte-grated machine (Lounsbury et al. 2013; Estes et al. 2012; Foster and Laurin 2012; Laurin and Fregeau 2012; Hopwood et al. 2010). Instruments process reference samples from buc-cal swabs or bloodstain card punches.

This technology provides a range of opportunities for authorities to quickly evaluate a person's identity without detaining a person for an extended period. It might be used as on-site point-of-collection processing of relationship tests without the need for transporta-tion of specimens to an operational laboratory. It also permits comparisons of individuals for relationship testing without the need for databasing of DNA profiles, which might be problematic in some policies (such as DNA analysis of noncitizens). It is unclear whether DNA profiles resulting from Rapid DNA technologies will adhere to current evidentiary standards of DNA profiling and whether profiles will be of sufficient quality for databasing national databases such as CODIS. As the technology develops, limitations might become apparent, such as the limited throughput or incompatibility with databasing quality stan-dards. However, the ability to sample and DNA type an individual within minutes or hours has a host of applications, including anthropological and noncriminal applications.

Use of Rapid DNA instruments was piloted at the US-Mexico border to verify parent-age claims of family unit pairs (i.e., parent migrating with a minor) in 2019. The ability to document biological claims prompted a proposal to expand the program into routine use (Frazee 2019).

22.3 DNA Identification in Human Trade

22.3.1 Human Trade

Modern human trafficking includes involuntary domestic servitude, forced labor, child soldiers, forced marriages, and the sex trade. In the United States and much of Europe, adult and children trafficking victims are predominantly found in forced labor and in the sex trade, and child victims are often runaways, troubled, and homeless youth. Varying international definitions of "human trafficking" might include illicit intercountry adop-tion, the human organ trade, and child prostitution.

Under the United Nations Palermo protocols, trafficking in persons is defined as

> the recruitment, transportation, transfer, harboring or receipt of persons, by means of the threat or use of force or other forms of coercion, of abduction, of fraud, of deception, of the abuse of power or of a position of vulnerability or of the giving or receiving of payments or benefits to achieve the consent of a person having control over another person, for the pur-pose of exploitation.

(United Nations 2000)

The United States' Trafficking Victims Protection Act (TVPA) comprises both sex traf-ficking and labor trafficking (Public Law 106–386: Victims of Trafficking and Violence Protection Act of 2000). In addition, organ trade and illicit adoptions, for example, can involve the purchase of an individual without consent. Intercountry adoptions have increased in the past decades, which have led to crime and fraud in some countries. In most countries, a child might be adopted from another country if he or she has no parents, or if both parents relinquish their rights, obligations, and claims to the child. Criminals and corrupt officials have developed creative ways to circumvent the orphan status requirement, either through theft of children, or through bribery and coercion of relinquishing parents. Mass disaster and war can increase children's vulnerability to displacement from their biological families. Displaced children might be illicitly offered for adoption out of their home countries.

22.3.2 Relationship Testing Strategies to Detect or Investigate Human Trafficking

Trafficking in persons remains a critical challenge to the management of migration and cross-border crime. High incidence of human trafficking along with advances in human genetic identification technologies present opportunities for international law enforcement agencies to apply DNA collection to detect trafficking in persons (Birchard 1998; Katsanis 2010, 2017, 2019; Sherwell 2008; Kim and Katsanis 2013). In addition to traditional police work and training to identify trafficking victims, identification might be accomplished by the comparison of DNA profiles from potential victims identified through law enforcement, health, and social services to reference databases of profiles. Systematic collection of DNA from family members of missing persons will enable the establishment of such reference databases for comparison with potential trafficking victim profiles.

Sharing and analyzing DNA profiles internationally presents multiple technical and administrative challenges. Governments must cooperate to ensure adequate communication as well as safely handling and exchanging genetic information to protect individuals' privacy (McCartney and Williams 2011). Moreover, the broad types and clandestine nature of the slave trade requires diverse approaches to investigate identity. Table 22.2 lays out potential approaches and early programs for using DNA to identify trafficking and human trade victims.

Some early adopters of DNA to identify live victims of trafficking have had varying success. In China, a nationwide DNA database program established in 2009 has used DNA collection and comparison between parents of missing children and abducted or homeless children. By 2012, the program had facilitated the reunification of more than 2300 children with their families (Yan 2013). A program in the United Arab Emirates facilitates the collection of DNA from children possibly trafficked as camel jockeys (Truong and Angeles 2005). In Guatemala, the Alba–Kenneth Warning System, enacted by the national Congress, requires DNA databanking for missing children cases to identify stolen children (Decreto Número 28–2010 2010).

In intercountry adoption fraud cases, DNA identification technologies can efficiently and accurately assess the veracity of false relationship claims. The standard processes for adopting a child from another country might involve DNA testing, particularly in the case of adopting a sibling of an existing adopted child. In such cases, authorities might require DNA testing to confirm this relationship (US Citizenship and Immigration Services 2000). The number of international adoptions into the United States has accelerated since the 1990s and, in order to identify cases of adoption fraud, US agencies began to require DNA testing as part of adoption procedures from Guatemala (see Box 22.1) and Vietnam, and

Table 22.2   **Potential Applications of DNA in Human Trade**

Form of Human Trafficking or Trade	Potential DNA Application	Example Program
Illicit adoption	Confirm biological relationship of relinquishing parent to child Compare DNA of orphans to missing children database Collect DNA of intercountry adoptions for future identification purposes	USCIS required DNA for relinquishment cases in Guatemala (International Commission against Impunity in Guatemala 2010) DNA-PROKIDS (Eisenberg and Schade 2010)
Sex trafficking	Databank DNA for identification of sex workers at high risk of violent crime and trafficking Compare DNA of detained underage sex workers to missing children database	Dallas Prostitute Diversion Initiative (Felini et al. 2011; Katsanis 2017)
Child soldiers	Post-conflict family reunification of misplaced youth Collect DNA from regions at high risk of trafficking for future identification purposes	
Forced or bonded labor and migrant workers	Compare DNA of unidentified remains to missing persons database Compare DNA of alleged parents and children suspected to be trafficked Collect DNA from areas at high risk of trafficking for future identification purposes	United Arab Emirates program to identify trafficked camel jockeys (Truong and Angeles 2005)
Involuntary domestic servitude	Collect DNA of high-risk individuals issued diplomatic immigration visas	
Human organ trade	Databank DNA of donated human organs before transplantation for future identification purposes	

also recommended DNA testing in many other countries (US Citizenship and Immigration Services 2008). Despite required DNA testing, reports of adoption fraud from Guatemala and Vietnam continued, and adoptions from Guatemala have been closed since 2008 (Joint Council on International Children's Services and National Council for Adoption 2007).

BOX 22.1   DNA TESTING IN GUATEMALA ADOPTIONS

As international adoptions become increasingly prevalent, the risk of adoption fraud from the source countries increases, as was the case for Guatemalan adoptions, which rose over the course of a decade. From 2000 to 2008, nearly 30,000 Guatemalan children were placed for adoption with US families, until adoptions were halted after suspicion of fraud (Joint Council on International Children's Services and National

Council for Adoption 2007). Reports of fraud included birth mother coercion, fraudulent identities, and child theft (Siegal 2011; Birchard 1998; Katsanis 2010; Sherwell 2008).

DNA collection had been required for Guatemalan adoptions to demonstrate biological relatedness of a relinquishing mother to a child. Despite this, corrupt officials and bribery led to continued fraud, including substitutions of children at the DNA test and forced consent. In response, the US Embassy commenced a second DNA test before releasing the child to the adoptive parents. This brought to light several suspected cases of "genotype recycling" and possibly fraud at Guatemalan medical examiners' offices designated as DNA collection sites (International Commission against Impunity in Guatemala [CICIG] 2010; Corbett 2002; Groves 2009; Siegal 2011; Wenk 2011). Ultimately, several children were placed for adoptions despite the children being subject to investigations for kidnapping (Siegal 2011). One child was reunified with her family after being stolen more than a year prior and confirmed by DNA (Sherwell 2008; Siegal 2011).

To develop systematic approaches for identification of victims of crime and trafficked persons, academic centers are working alongside government authorities to collect and store DNA samples or profiles for identification of victims. Two such programs, DNA-PROKIDS (Program for Kids Identification with DNA Systems) and the Dallas Prostitute Diversion Initiative (PDI), facilitated prospective collection of noncriminals to identify trafficked children and postmortem high-risk victims of crime.

22.3.3 DNA-PROKIDS

DNA-PROKIDS, based in Granada, Spain, works internationally to promote DNA applications to combat human trafficking and adoption fraud. DNA-PROKIDS (www.dna-prokids .org) works with individual countries to establish DNA registries to facilitate the identification of trafficked children and reunification with their families, and to apply scientific methodology to develop law enforcement cases and police intelligence. DNA-PROKIDS was initiated at the University of Granada (UGR) Genetic Identification Laboratory, which continues to serve as the headquarters.

Countries might have samples collected through DNA-PROKIDS processed within their country or might send samples to UGR or the University of North Texas Center for Human Identification (UNTCHI) for DNA profile generation and databasing. DNA data are administered separately from forensic evidence and secured in registries divided by country. Laboratory case reports include information on statistical analysis and likelihood ratio or exclusion. A centralized database has not been developed for DNA-PROKIDS, so specimen and profiles are stored locally. Participation in the DNA-PROKIDS program is usually initiated through a memorandum of understanding (MOU) between DNA-PROKIDS and a country's law enforcement, forensic institute, or government authorities.

DNA-PROKIDS primarily distributes DNA sample collection kits, provides training and resources for law enforcement in participating countries, processes samples for countries requesting DNA analysis services, works with communities to encourage individuals to provide samples, and examines best practices for the establishment of international

DNA registries (Eisenberg and Schade 2010). DNA-PROKIDS aims ultimately to facilitate the use and exchange of DNA technologies to combat human trafficking both within and between countries. As of April 2012, more than 4200 samples had been processed, leading to hundreds of positive identifications and exclusions (Kim and Katsanis 2013).

The success of collaborative associations through DNA-PROKIDS is not always apparent. DNA-PROKIDS' experiences in Haiti (see Box 22.2) exposed a key program challenge: allowing collaborating institutions the autonomy to set their pace for sample collection while encouraging efficient resource allocation.

BOX 22.2   DNA-PROKIDS IN POST-EARTHQUAKE HAITI

Displacement of thousands during natural disaster, war, and other mass disasters often increases individuals' vulnerability to trafficking. After the January 2010 earthquake in Haiti, for example, traffickers offered displaced children food, water, and shelter before smuggling them into the Dominican Republic, where they became victims of slavery, sexual exploitation, illegal adoption, and organized crime (Katsanis 2010). In response, aid organizations spurred efforts to rapidly identify and protect homeless children. Identifying the need to proactively deter child theft or misplacement after the 2010 earthquake in Haiti, DNA-PROKIDS provided substantial resources to Haitian authorities, including thousands of DNA collection kits, laptop computers with software for sample collection and data storage, scanners to record information on sample collection cards, and equipment for instant photo documentation. Despite continuing reports of human trafficking, over the subsequent two years, DNA-PROKIDS received few if any samples for analysis from Haiti (Eisenberg and Schade 2010).

22.3.4 Dallas PDI

Sex workers are at an elevated risk for becoming victims of violent crime and trafficking. Because sex workers are rarely reported missing and typically do not carry identification, law enforcement often lack information necessary to pursue investigations and connect crimes involving sex workers (Quinet 2007). Sex work is considered a crime in most US jurisdictions, but, in some instances, sex workers might be identified as trafficked victims. For example, the US TVPA classifies any commercial sex act induced by force, fraud, or coercion or performed by minors as human trafficking (Public Law 106–386: Victims of Trafficking and Violence Protection Act of 2000). It is important for law enforcement to be able to distinguish victims of trafficking from consensual sex workers in order to avoid criminalization of trafficking victims.

The Dallas PDI (now an nonprofit program, New Life Opportunities, www.pdinew-life.org), was established by the Dallas Police Department in 2007, as a pre-booking program drawing effort from the Dallas County District Attorney's Office, City Attorney's Office, courts, Dallas County Sheriff's Office, Dallas County Health and Human Services, Parkland Hospital, and more than 30 social service and faith-based organizations. In order to facilitate sex workers' permanent exit from the sex industry, the PDI offered short-term residential treatment, including health care, counseling, and support for chemical dependency, as well as long-term support services to victims, such as offering alternative housing and career options.

The High-Risk Potential Victims DNA Database (HRDNA) was offered as part of the PDI's program as a voluntary DNA collection program, in collaboration with a local university (Katsanis 2017). This DNA collection program established a databank of voluntary samples from sex workers completing the PDI program. The databank was intended to assist postmortem identifications, and could be used by law enforcement as a "cold association" to open or pursue a criminal investigation. Buccal samples for the HRDNA were collected by law enforcement from consenting individuals and stored unprocessed unless or until an investigation was opened to warrant DNA profiling. By 2013, hundreds of HRDNA samples had been collected and stored for this purpose. The social and legal aspects of the PDI program continues, but the DNA sampling has been suspended pending sustainable funding.

22.4 Ethical, Legal, and Social Considerations with DNA Identification

As this text attests, DNA reliably and efficiently identifies individuals and assesses biological relationships. But as law enforcement and government uses of DNA applications expand, particularly to the identification of living noncriminal citizens, some have questioned whether routine DNA collection could create a slippery slope to privacy and human rights violations. Ensuring the socially responsible use of genetic information is key in developing model policies to implement DNA testing.

22.4.1 Defining "Family" in Immigration Procedures

Many national and international policies and recommendations for immigration and border security emphasize the importance of family reunification. The 1989 Convention of the Rights of the Child describes a family as "the fundamental group of society and the natural environment for the growth and well-being of all its members and particularly children," emphasizing that a family should be "afforded the necessary protection and assistance so that it can fully assume its responsibilities within the community" (United Nations Office of the High Commissioner for Human Rights 1989). After conflict and disaster, family reunification is a central focus of efforts and governments' migration policies often favor cohesion of a family unit (Taitz, Weekers, and Mosca 2002b). With the increased use of DNA relationship testing to support biological claims, immigration policies must take into consideration that families are social constructions, and not always biological entities. In some cultures, "family" might include a wide range of biological relatives as well as members who share religious, cultural, and social relationships (Parker, London, and Aronson 2013; Taitz, Weekers, and Mosca 2002b). What might appear to a DNA analyst as a misattributed relationship might actually be a miscommunication of terms to define relationships in a family (Parker, London, and Aronson 2013).

22.4.2 Privacy of Genetic Information

The collection and use of DNA from noncriminals can raise questions of privacy and human rights violations, such as Fourth Amendment infringements in the United States (Henning 2010). Despite widespread agreement that DNA applications could critically aid law enforcement or government authorities, improve public security, and promote justice, privacy concerns have been raised that the government might be intruding on individuals' right to privacy (Kim and Katsanis 2013; Ram 2018). One program to forcibly identify orphans of Argentina's "dirty war" raised questions about genetic privacy, citing concerns of valuing truth over privacy of the individuals concerned (Pertossi 2009). Although human rights activists intended to use DNA to identify hundreds of stolen children, some argued that the program was a privacy concern, particularly for children (now adults) unaware of their origins. Misattributed parentage can disrupt families and cause strife in communities peacefully living with the secrets of the past.

Developing policies ideally aim to balance individual privacy against public interest. Existing policies have not been resolved in most countries, with a range of court and legal opinions (Kaye 2013). The US Supreme Court heard one case on arrestee collection (*Maryland v. King*) in February 2013, deciding that collection from this population is no more an infringement on Fourth Amendment rights than dermatoglyphics.

22.4.3 Abuse of Power

In addition to privacy concerns are concerns for abuse of power by authorities maintaining databanks of DNA specimens and DNA databases. Although a typical DNA profile includes only markers that do not contain personal or health-related information (e.g., disease susceptibility, behavioral predispositions) (Katsanis and Wagner 2013), a stored sample harbors a slew of identifying, personal, and medical information. The public expects clear guidelines on how samples are to be used and assurance from misuse. In some cases, informed consent of participants might be sufficient, whereas in scenarios where vulnerable populations are sampled, additional protections might be necessary.

Although many countries and states have developed policies to protect identifiable biological specimen, some argue that the protections are insufficient. Indeed, the transportation of human specimens might itself be considered human trafficking if the purpose for acquisition requires coercion or deception and the specimens are exploited in any way. Those developing policies and laws regarding DNA collection often emphasize the importance of protecting the privacy of profiled individuals, carefully outlining the intended uses of specimens and the resulting profiles, and providing provisions for voluntary consent when appropriate. It is also important to establish whether and how specimens and profiles can be expunged and how individuals might request expungement of their personal information.

In the United States, public opinion regarding the use of genetic testing in medicine versus law enforcement varies, although most Americans support DNA testing in criminal contexts, even use of new approaches (Guerrini 2018). In 2007, a study revealed that although more than 90% of Americans over the age of 18 support the use of genetic information for medical research, 54% of respondents reported that they had little or no trust in law enforcement to protect their genetic privacy (Kaufman et al. 2009). Response from outside the United States also reveals concerns about genetic privacy. When the US government issued a notice of intent to collect DNA from immigrant detainees (28 CFR § 28.12), the Embassy of Canada formally commented that

in Canada, the creation of biometric templates for indefinite storage attracts a much higher reasonable expectation of privacy than the initial collection of live photographs and fingerprints given the much greater potential for linking disparate information, tracking and surveillance, especially given advances in computer processing technology that allow for quick matching or linking of data from numerous sources.

(Canadian Embassy 2008)

22.4.4 Incidental Findings

Any relationship testing has the potential of exposing hidden family relationships, such as nonpaternity or misattributed parentage. STR profiling alone has little chance of identifying medical traits that might be of import to the individual typed, but use of SNPs might lead to revelation of traits linked to particular markers. As well, use of AIMs in relationship testing might reveal family ancestries contrasting with family lore and oral history. To mitigate potential harms precipitated by such incidental findings, some laboratories have formal nondisclosure policies, whereas others have no formal policies but weigh out the situations on a case-by-case basis (Parker, London, and Aronson 2013) Improvements are needed in the consent processes for nonmedical applications in general (Katsanis 2018).

22.4.5 Managing International Interoperable DNA Databases

The development of global DNA databases could critically aid the detection and investigation of human trafficking. Global collaborations to share DNA data in human trafficking investigations present a slew of technical, legal, and administrative challenges. Possible interoperability of global DNA databases raises questions regarding international laws, policies, and guidelines on sample and DNA data storage as well as who maintains authority over data transferred between countries. Because an international anti-trafficking DNA database might present the prospect of conducting searches across databases from different countries, it might become necessary to establish a standard approach to approving and conducting international DNA database searches. The establishment of a central database would simplify the processes, but authority for such a database of international import might be difficult to establish and justify (Kim and Katsanis 2013). A network of databases among cooperating countries could suffice, but parameters for searching and access to such an infrastructure will need to be carefully established and controlled by authorities without conflicts of interest. The development and use of international DNA databases should be governed by clear guidelines to prevent misuse and malfunction, and to ensure security (Thomas 2006).

22.4.6 Cultural Perspectives on Genetic Information

As forensic genetics is applied to noncriminal purposes, and around the world, we must be mindful of the cultural dynamics of local communities. By and large, Western communities have embraced use of genomic information for crime solving and human rights investigations. This is more than evident in the US and European cultural media. However, cultural and political differences in other communities might hold different attitudes toward use of DNA. Some communities might fear the use of DNA by authorities or even the process of sample collection. Other communities might oppose the evaluation of DNA for clues into ancestry or evolution that undermine their cultural beliefs. Even in regions where use of DNA is embraced for law enforcement purposes, there persists distrust of how the genetic information might be used (Kaufman et al. 2009). For example, differences in social or political context might shape individuals' concerns regarding DNA sample submission and fear of retribution for reporting suspicions or evidence of ongoing crimes.

Collection of DNA from noncriminals can involve populations who might be more vulnerable to coercion, such as missing children and their families, or those whose legal standing is in question. Individuals approached by law enforcement or governmental authorities might believe they are required to provide samples as part of a criminal investigation. Moreover, even with informed consent, individuals might not fully understand the benefits, limitations, and potential risks associated with DNA collection and profiling, particularly when language or cultural barriers limit comprehension (Katsanis 2018). In fact, consent often is not in play for immigration or human trafficking contexts, rather acknowledgment of the participant's awareness of how their information is managed and what the options might be for removal of data in the future.

The predictive nature of genetic information, its relevance for family members, and its past use to support prejudice heighten existing concerns for privacy and intrusion of authorities. In regions of conflict where trafficking and human trade are prominent, populations are often distrusting of authorities as well as foreigners. No matter how well intended, persuasion of groups of individuals to participate in DNA collection programs might be unwelcome. Policies for use of genetic information for human rights purposes must be mindful of the varying cultural perspectives.

22.5 Summary

DNA technologies provide an efficient and accurate approach to individual identification and assessment of biological relationship approaches valuable to the identification of victims of human trafficking and for immigration decisions. In order to protect the privacy of individuals and citizens included in DNA collection programs, especially collection from noncriminals and vulnerable populations, it is necessary to develop policies to guide the administration and quality control of DNA applications. Programs for DNA collection of immigrants and to prevent human trafficking should consider first whether DNA collection is a necessary component for identification or whether other biometric measures work sufficiently. If DNA is a valuable approach for the population and circumstances, then the program should establish parameters for what authority is most appropriate for collection of DNA while maintaining the integrity of the collected specimens. A DNA program should define the retention policies including how long specimens will be retained, who will have access to the specimens, and how specimens will be destroyed. Similar but distinct policies are necessary for the DNA profile and access to any DNA profile databases.

Acknowledgments

The authors are grateful to Jose Lorente, Arthur Eisenberg, and Bruce Budowle for sharing insight into the DNA-PROKIDS program.

The authors are grateful to Martha Felini, Arthur Eisenberg, and Bruce Budowle for sharing insight into the Dallas PDI program.

References

73 Federal Register 74932. 2008. DNA-Sample Collection and Biological Evidence Preservation in the Federal Jurisdiction.

79 Federal Register 68343. 2014. Notice of Information Collection under OMB Emergency Review: Affidavit of Relationship (AOR) for Minors from Honduras, El Salvador and Guatemala.

82 Federal Register 38926. 2017. Termination of the Central American Minors Parole Program.

Birchard, K. 1998. Call for DNA testing for foreign adoptions. *Lancet* 352:9128.

Bruno, A. 2012. Refugee admissions and resettlement policies. Edited by Congressional Research Service. http://fpc.state.gov/documents/organization/187394.pdf (accessed March 25, 2013).

Buscaino, J., A. Barican, L. Farrales, et al. 2018. Evaluation of a rapid DNA process with the RapidHIT® ID system using a specialized cartridge for extracted and quantified human DNA. *Forensic Sci Int Genet* 34:116–127.

Canadian Embassy. 2008. Comment on FR Doc# E8-08339. *Might* 18, 2008. http://www.regulations.gov/#!documentDetail;D=DOJ-OAG-2008-0009-1201 (accessed March 25, 2013).

Carney, C., S. Whitney, J. Vaidyanathan, et al. 2019. Developmental validation of the ANDE™ rapid DNA system with FlexPlex™ assay for arrestee and reference buccal swab processing and database searching. *Forensic Sci Int Genet* 40:120–30.

Corbett, S. 2002. Where do babies come from? New York Times, July 16, 2002.

Daiger, S. P., and A. Chakravarti. 1983. Deletion mapping of polymorphic loci by apparent parental exclusion. *Am J Med Genet* 14(1):43–8.

Davoren, J., D. Vanek, R. Konjhodzic, J. Crews, E. Huffine, and T. J. Parsons. 2007. Highly effective DNA extraction method for nuclear short tandem repeat testing of skeletal remains from mass graves. *Croat Med J* 48(4):478–85.

Debenham, P. 2006. DNA fingerprinting, paternity testing and relationship (immigration) analysis. In: *Encyclopedia of Life Sciences*. (eds.) pp. 1–4. Chichester: John Wiley & Sons, Ltd.

Decreto Número. 2010. Ley del Sistema de Alerta Alba-Keneth. *El Congreso de la República de Guatemala*, September 13, 2010.

Eisenberg, A., and L. Schade. 2010. DNA-PROKIDS: Using DNA technology to help fight the trafficking of children. *Forensic Magazine*.

Esbenshade, J. 2010. *Special Report: An Assessment of DNA Testing for African Refugees*. Edited by Immigration Policy Center, American Immigration Council. http://www.immigrationpolicy.org/special-reports/assessment-dna-testing-african-refugees (accessed March 25, 2013).

Estes, M. D., J. Yang, B. Duane, et al. 2012. Optimization of multiplexed PCR on an integrated microfluidic forensic platform for rapid DNA analysis. *Analyst* 137(23):5510–9.

Farahany, N., S. Chodavadia, and S. H. Katsanis. 2019. Ethical guidelines for DNA testing in migrant family reunification. *Am J Bioeth* 19(2):4–7.

Felini, M., E. Ryan, J. Frick, S. Bangara, L. Felini, and R. Breazeal. 2011. *Prostitute Diversion Initiative Annual Report 2009–2010*. http://www.pdinewlife.org/wp-content/uploads/2012/01/PDI-Annual-Report-2011_Final-RS.pdf (accessed March 25, 2013).

Foster, A., and N. Laurin. 2012. Development of a fast PCR protocol enabling rapid generation of AmpFlSTR® Identifiler® profiles for genotyping of human DNA. *Investig Genet* 3:6.

Frazee, G. 2019. Watch: Acting homeland security chief McAleenan testifies on border security. PBS, June 11, 2019. https://www.pbs.org/newshour/politics/watch-live-acting-homeland-security-chief-mcaleenan-testifies-on-border-security (accessed July 10, 2019).

Groves, M. 2009. Trafficking reports raise heart-wrenching questions for adoptive parents. Los Angeles Times, November 11, 2009.

Guerrini C. J., J. O. Robinson, D. Petersen, and A. L. McGuire. 2018. Should police have access to genetic genealogy databases? Capturing the Golden State Killer and other criminals using a controversial new forensic technique. *PLoS Biol* 16(10):e2006906.

Henning, A. C. 2010. Compulsory DNA collection: A Fourth Amendment analysis. February 16, 2010. Edited by Congressional Research Council. http://www.fas.org/sgp/crs/misc/R40077.pdf (accessed March 25, 2013).

Holland, E. 2011. Moving the virtual border to the cellular level: Mandatory DNA testing and the U.S. refugee family reunification program. *California Law Rev* 99:1635–82.

Holland, M. M., D. L. Fisher, L. G. Mitchell, et al. 1993. Mitochondrial DNA sequence analysis of human skeletal remains: Identification of remains from the Vietnam War. *J Forensic Sci* 38(3):542–53.

Hopwood, A. J., C. Hurth, J. Yang, et al. 2010. Integrated microfluidic system for rapid forensic DNA analysis: Sample collection to DNA profile. *Anal Chem* 82(16):6991–9.

Huffine, E., J. Crews, B. Kennedy, K. Bomberger, and A. Zinbo, 2001. Mass identification of persons missing from the break-up of the former Yugoslavia: Structure, function, and role of the International Commission on Missing Persons. *Croat Med J* 42(3):271–5.

International Commission against Impunity in Guatemala (CICIG). 2010. Decree 77–2007: Report on players involved in the illegal adoption process in Guatemala since the entry into force of the adoption law. December 1, 2010. http://cicig.org/uploads/documents/informes/INFOR -TEMA_DOC05_20101201_EN.pdf (accessed March 25, 2013).

Jamieson, R. W. D., G. Stephens, and S. Smith. 2008. Developing a conceptual framework for identity fraud profiling. Paper read at 16th European Conference on Information Systems, at Galway, Ireland. http://is2.lse.ac.uk/asp/aspecis/20080120.pdf (accessed March 25, 2013).

Jeffreys, A. J., J. F. Brookfield, and R. Semeonoff. 1985. Positive identification of an immigration test-case using human DNA fingerprints. *Nature* 317(6040):818–819.

Jeffreys, A. J., V. Wilson, and S. L. Thein. 1985. Hypervariable 'minisatellite' regions in human DNA. *Nature* 314(6006):67–73.

Joint Council on International Children's Services and National Council for Adoption. 2007. U.S. adoptions from Guatemala to halt as of January 1, 2008. Adoption Social Work New York. http://www.adoptionsocialworkny.com/2007/09/ (accessed March 25, 2013).

Katsanis, S. 2010. Use DNA to stop child trafficking. *Toronto Globe and Mail*, February 23, 2010.

Katsanis, S. H., M. Felini, J. Kim, M. A. Minear, S. Chandrasekharan, and J. K. Wagner. 2017. Perspectives on DNA collection for a high-risk DNA database. *Int J Crim Justice Sci* 12(1):111–128.

Katsanis, S. H., E. Huang, A. Young, V. Grant, E. Warner, S. Larson, and J. K. Wagner. 2019. Caring for trafficked and unidentified patients in the EHR shadows: Shining a light by sharing the data. *PLoS One* 14(3): e0213766.

Katsanis, S. H., L. Snyder K. Arnholt A. Z. Mundorff. 2018. Consent process for family reference DNA samples. *Forensic Sci Int Genet* 32:71–79.

Katsanis, S. H., and J. K. Wagner. 2013. Characterization of the standard and recommended CODIS Markers. *J Forensic Sci* 58(Suppl 1):S169–72.

Kaufman, D. J., J. Murphy-Bollinger, J. Scott, and K. L. Hudson. 2009. Public opinion about the importance of privacy in biobank research. *Am J Hum Genet* 85(5):643–54.

Kaye, D. 2013. On the 'considered analysis' of collecting DNA before conviction. *Penn State Law 60 (Legal Studies Research Paper No. 27–2012)*.

Kim, J., and S. Katsanis. 2013. Brave new world of human rights DNA collection. *Trends Genet* 29(6):329–32.

Laurin, N., and C. Fregeau. 2012. Optimization and validation of a fast amplification protocol for AmpF/STR® Profiler Plus® for rapid forensic human identification. *Forensic Sci Int Genet* 6(1):47–57.

Lounsbury, J. A., A. Karlsson, D. Miranian, et al. 2013. From sample to PCR product in under 45 minutes: A polymeric integrated microdevice for clinical and forensic DNA analysis. *Lab Chip* 13(7):1384–93.

Lynch, J. 2012. *From Fingerprints to DNA: Biometric Data Collection in U.S. Immigrant Communities and Beyond.* Washington, DC, Immigration Policy Center, American Immigration Council.

McCartney, C. W., and R. Williams. 2011. Transantional exchange of forensic DNA: Viability, legitimacy, and acceptability. *Eur J Criminal Policy Res* 17:305–22.

National Science and Technology Council. 2008. Biometrics in government post-9/11: Advancing science, enhancing operations, October 2008. Edited by Office of Science and Technology Policy, Executive Office of the President. http://www.fas.org/irp/eprint/biometrics.pdf (accessed March 25, 2013).

Pajnič, I. Z., B. G. Pogorelc, and J. Balažic. 2010. Molecular genetic identification of skeletal remains from the Second World War Konfin I mass grave in Slovenia. *Int J Legal Med* 124(4):307–17.

Parker, L. S., A. J. London, and J. D. Aronson. 2013. Incidental findings in the use of DNA to identify human remains: An ethical assessment. *Forensic Sci Int Genet* 7(2):221–9.

Parsons, T. J., R. Huel, J. Davoren, et al. 2007. Application of novel "mini-amplicon" STR multiplexes to high volume casework on degraded skeletal remains. *Forensic Sci Int Genet* 1(2):175–9.

Pertossi, M. 2009. *Argentina Forces Dirty War Orphans to Provide DNA*. Associated Press, November 20, 2009.

Public Law 106–386: Victims of Trafficking and Violence Protection Act of 2000. October 28, 2000.

Quinet, K. 2007. The missing missing: Toward a quantification of serial murder victimization in the United States. *Homicide Stud* 11:319–339.

Ram, N., C. J. Guerrini, and A. L. McGuire. 2018. Genealogy databases and the future of criminal investigation. *Science* 360:1078–1079.

Sherwell, P. 2008. Guatemalan mother reunited with baby stolen and sold for adoption by U.S. couple. *The Sunday Telegraph*, July 26, 2008.

Siegal, E. 2011. *Finding Fernanda: Two Mothers, One Child, and a Cross-Border Search for Truth*. Portland, Oregon: Cathexis Press.

Taitz, J., J. E. Weekers, and D. T. Mosca. 2002a. DNA and immigration: The ethical ramifications. *Lancet* 359(9308):794.

Taitz, J., J. E. M. Weekers, and D. T. Mosca. 2002b. The last resort: Exploring the use of DNA testing for family reunitifcation. *Health Hum Rights* 6(1):20–32.

Thomas, R. 2006. Biometrics, international migrants and human rights. *Eur J Migration Law* 7(4):377–411.

Truong, T., and M. Angeles. 2005. Searching for best practices to counter human trafficking in Africa: A focus of women and children, March 2005. Edited by United Nations Educational, Scientific and Cultural Organization. http://unesdoc.unesco.org/images/0013/001384 /138447e.pdf (accessed March 25, 2013).

United Nations. 2000. Protocol to prevent, suppress and punish trafficking in persons, especially women and children, supplementing the United Nations Convention against transnational organized crime. http://www.uncjin.org/Documents/Conventions/dcatoc/final_documents _2/convention_%20traff_eng.pdf (accessed March 25, 2013).

United Nations High Commissioner for Refugees. 2008. UNHCR note on DNA testing to establish family relationships in the refugee context. June 2008. http://www.unhcr.org/refworld/pdfid /48620c2d2.pdf (accessed March 25, 2013).

United Nations Office of the High Commissioner for Human Rights. 1989. Convention on the rights of the child. http://www.ohchr.org/EN/ProfessionalInterest/Pages/CRC.aspx (accessed March 25, 2013).

US Citizenship and Immigration Services. 2000. Memorandum from Michael A. Pearson, Executive Associate Commissioner, Office of Field Operations to All Regional Directors, Director, International Affairs, FLETC/GLYNCO on Guidance on processing petitions for adopted alien children less than 18 years of age considered a "child" under the Immigration and Nationality Act through Public Law 1066–139. http://www.uscis.gov/files/pressrelease/ adoptchi.pdf (accessed March 25, 2013).

US Citizenship and Immigration Services. 2008. USCIS update: USCIS implements required DNA testing for Vietnamese adoptions. Might 29, 2008. http://www.uscis.gov/portal/site/uscis/ menuitem.5af9bb95919f35e66f614176543f6d1a/?vgnextoid=49f0bfd9dd43a110VgnVCM1 000004718190aRCRD&vgnextchannel=68439c7755cb9010VgnVCM10000045f3d6a1RCRD (accessed March 25, 2013).

Wagner, S. 2008. *To Know Where He Lies: DNA Technology and the Search for Srebrenica's Missing.* Berkeley: University of California Press.

Wasem, R. E. 2007. Immigration fraud: Policies, investigations, and issues. In *CRS Report for Congress*, edited by R. E. Wasem (accessed May 17, 2007).

Wenk, R. E. 2010. Sporadic genotype recycling fraud in relationship testing of immigrants. *Transfusion* 50(8):1852–3.

Wenk, R. E. 2011. Detection of genotype recycling fraud in U.S. immigrants. *J Forensic Sci* 56 (Suppl 1):S243–6.

Yan, Z. 2013. Database gives hope to abducted children. *China Daily USA*, January 22, 2013.

DNA Databases

23

ANDREA LEDIĆ, ADELA MAKAR,
AND IGOR OBLEŠČUK

Contents

23.1 Introduction

A rise in cross-border crime has shown an increasing need to develop forensic intelligence, which includes the collection, integration, and analysis of data obtained through forensic investigations, thus making the results of forensic investigations an additional source of information.

The application of forensic intelligence implies moving away from a reactive strategy aimed at solving individual criminal offenses and focusing on a proactive strategy of combining forensic data with a view to detecting criminal offenses and their perpetrators more efficiently, reducing the number of criminal offenses and preventing crime.

The phenomenon of using large quantities of data for a specific purpose, the so-called Big Data, within the context of criminal investigations, relies heavily on DNA databases. Thus, a DNA database was one of the first forensic intelligence databases. The development of DNA technology and the establishment of a corresponding DNA database on national and transnational levels is one of the most efficient ways to detect and prevent crime.

DOI: 10.4324/9780429019944-27

As pointed out by Hinmarsh and Prainsack (2010), "Forensic genetics has become a significant resource for criminal investigation and evidence-gathering activities for court proceedings in judicial systems around the world."

The legislation of each individual country regulates how genetic profiles are to be stored in national DNA databases (Asplen 2014; Santos et al. 2013).

In general, such databases contain reference DNA profiles of suspects, defendants, and victims, and DNA profiles of traces obtained from biological samples (blood, saliva, sperm, bones, soft tissue) found at the crime scene, whereas in some countries they also contain DNA profiles of other individuals who might be of interest during a criminal investigation (Machado and Granja 2020). The establishment of DNA databases and storing of genetic profiles is of major importance given the fact that such profiles, in accordance with law, remain available for search for many years, whereas in traditional criminal investigations DNA evidence could only be used until the suspect of a criminal offense is identified.

The process of creating forensic databases with genetic profiles began in mid-1990s (Machado and Granja 2020). The first country to start creating a DNA database was the UK (Amankwaa and McCartney 2019; Aronson 2005). At first, this involved DNA profiling of immigrants (1988), but it was soon realized that this technique could have much wider application in the criminal justice system. The first national criminal or forensic DNA database, the UK National DNA Database (NDNAD), was set up in England and Wales on April 10, 1995. As pointed out by Asplen (2014), England has, from the very beginning, become a world leader in discovering ways to use DNA to identify suspects, protect the innocent, and convict the guilty. By setting up the NDNAD, the English Forensic Science Service became a pioneer in the use of DNA profiling in forensic science. This database contains electronic records of deoxyribonucleic acid (DNA), known as DNA profiles, taken both from individuals and from crime scenes. In the period between April 2001 and March 2016, it produced 611,557 matches to unsolved crimes between individuals and crime scenes and between different crime scenes (NDNAD Strategy Board Annual Report 2015/2016).

At the same time as in the UK, the Federal Bureau of Investigation (FBI) developed the US's national (federal) DNA database, CODIS (Combined DNA Index System), for storing data in databases. In the US, CODIS databases exist on local, state, and national levels, and this multitier architecture allows forensic laboratories to control their own data, which means that each laboratory decides independently, in accordance with legislative frameworks, which profiles it will share with the rest of the country.

The National DNA Index System (NDIS) is part of CODIS set up on a national level, and it contains DNA profiles contributed by federal, state, and local forensic laboratories, namely, convicted offenders' profiles from all 50 US states, the District of Columbia, the federal government, US Army Criminal Investigation Laboratory, and Puerto Rico, but arrestee profiles from only 24 states as well as crime scene profiles (FBI, CODIS and NDIS Fact Sheet 2021; Asplen 2014).

The DNA Identification Act of 1994 (42 U.S.C. § 14132) authorized the establishment of the NDIS. However, it was amended several times to allow database services to be expanded (FBI, CODIS and NDIS Fact Sheet 2021).

The European Union Council Resolution of 9 June 1997 invited EU member states to consider the possibility of establishing DNA databases and later to exchange DNA analysis results. The Netherlands was the first country after the UK to set up a DNA database in 1997, followed by Austria, and in 1998 Germany and France. Since 2000, many European countries have established national DNA databases.

Somewhat less data is available for the so-called BRICS countries (Brazil, Russia, India, China, and South Africa), which are increasingly encouraging the creation of a new platform for development and cooperation. It is worth mentioning here that there remain vast developmental differences within those countries (Le Roux-Kemp 2018).

As further stated by Le Roux-Kemp (2018), South Africa has been conducting DNA profiling since 1991 and it established a national DNA database in 1997. However, it only recently promulgated legislation—the South African Criminal Law Forensic Procedures Amendment (Act 37 of 2013)—for the formal recognition and regulation of its DNA database. Russia adopted the National Genetic Registration, Russia Federal Act N242-FZ, in 2008, legally enabling the creation of a database containing DNA profiles of convicted offenders. However, this act only came into operation in 2010.

It was not until 2012 that Brazil promulgated legislation (Law No. 12654) for the establishment and regulation of a DNA database and DNA profiling, while India is still in the planning stages of its national DNA framework and legislation, even though it is already performing DNA profiling for other jurisdictions in Southeast Asia (Le Roux-Kemp 2018).

China was among the first in Asia and the South Pacific to start DNA profiling in criminal investigations in 1989, and its DNA database was established in the early 2000s. According to an Interpol survey, China's DNA database is the third biggest database in the world, immediately after the US and the UK.

23.2 Recommendations for Forensic DNA Databases

A DNA database is a collection of DNA profiles generated from biological samples (traces and individuals) that can be electronically stored for comparison with DNA profiles generated from samples found at a crime scene. When establishing a national DNA database, it is important to determine its purpose, that is, what one wishes to achieve by establishing it, and the types of DNA profiles that the database will contain, how the data contained in the database will be used, and by whom. Establishing a national DNA database is of major importance for solving and preventing crimes in a particular country and for identifying missing persons and unidentified bodies.

23.2.1 Legislation

All activities associated with the establishment and use of a DNA database should have solid foundation in national legislation. The collection and use of DNA profiles in a database, the retention time, and the security of information in a database should be governed by laws and other regulations. They should be defined by internal instructions. It is also possible to define within the law the possibility of monitoring the management of databases, who conducts that, and who is responsible for the data contained in databases and for the use of the database.

23.2.2 Contents of a DNA Database

A DNA database contains DNA profiles from individuals and DNA profiles from crime scenes. A database of DNA profiles from crime scenes should contain samples from all types of biological material. It is important to select DNA profiles from items that have probative value to maximize investigative outcomes and reduce investigative costs. Databases of DNA profiles mostly contain the following information:

- DNA profiles of criminal suspects
- DNA profiles of convicted offenders
- DNA profiles of traces from crime scenes

Forensic databases or specialized databases of missing persons also contain:

- DNA profiles of relatives or those searching for missing persons and/or items used by missing persons
- DNA profiles of unidentified human remains

DNA databases should also have an associated so-called elimination database containing DNA profiles of persons who come into the contact with material evidence, investigating officers, staff working in DNA laboratories, and those visiting laboratories, in order to compare their DNA profiles with DNA profiles used as material evidence in criminal proceedings.

To facilitate the search of DNA profiles and their hits in a database, the following conditions should be met:

- DNA profiles have to be linked to unique identifiers (by a reference number).
- DNA profiles should be of highest possible quality and include as many loci as possible (genetic markers).
- Standards for storing DNA profiles need to be established, such as the minimum number of loci, the maximum number of alleles per locus, and sample quality.
- Restrictions for matches and reporting requirements for matches need to be defined.

23.2.3 Data Protection

All information on individuals contained in a DNA database are personal data, and they should be in compliance with the national legislation governing personal data protection. As an example, we note the general European Union regulation governing data protection (the European Data Protection Directive).

23.2.4 Quality Management

Quality management is a key factor in building and maintaining an effective national DNA database and should be applied to the entire process, including sample collection, DNA analysis, and chain of custody. Laboratory accreditation is becoming the accepted benchmark of forensic DNA laboratories and is typically based on the International Standards Organization (ISO) Standard 17025 (ISO IEC 17025). At the database-management level, quality management should extend to data-entry processes to ensure accuracy of information, e.g., employing an automated system to upload DNA profiles and utilizing consistent nomenclature.

The Council of the European Union invited its member states to consider establishing DNA databases back in 1997, and a European Standard Set (ESS) of loci was established in

2001 (EU Council resolutions 2001) to enable the comparison of DNA profiles from differ-ent countries. The Interpol Standard Set of Loci (ISSOL) includes the same loci, plus the Amelogenin locus, which determines the sex.

Until late 2009, the ESS contained only seven loci, which was enough for occasional exchanges of DNA data between countries. However, when massive automated exchanges of DNA data began, the chance of adventitious matches became significant. Therefore, the European Network of Forensic Science Institutes (ENFSI) recommended that the ESS be expanded from seven loci by fivr additional loci (the Council of the European Union adopted this recommendation on November 30, 2009).

In the US, 13 loci were required to store a DNA profile of a person in a national DNA database (NDNAD, CODIS). In 2015, the CODIS core locus set was extended to 20 loci (Bodner et al. 2016).

23.2.5 Resources

One of the essential factors when establishing and maintaining a national DNA database are financial costs, which include recruitment and training of staff to work with databases, procurement and maintenance of IT infrastructure and of the system where the database is located, and finally the cost of DNA analysis so that DNA profiles could be stored in a database.

As pointed out by Amankwaa and McCartney (2021), "the DNA Expansion Programme (2000–2005) cost the government and the police about £300 million (Forensic Science and Pathology Unit 2005), and the annual cost of running the NDNAD is between £1.8 to 2 million" (FIND Strategy Board 2020).

23.3 Volume of Databases Worldwide

Ever since the first DNA databases were established in the second part of 1990s, most countries have been investing efforts in addressing main challenges—legislative, technical, personnel, and infrastructural ones—on their way to establishing databases that are of high quality and fit for purpose. Everybody aims to achieve balance between databases containing a large amount of data, on the one hand, which will enable the discovery of as many perpetrators of crime as possible, and the protection of personal data and ethical principles, on the other hand, in order to protect fundamental human rights (Machado and Granja 2020).

23.3.1 National Databases

National DNA databases mostly contain information about the analysis of autosomal STRs (short tandem repeats), which are mostly used for solving crimes and for human identification. They include information about individuals (convicts, defendants, suspects, victims, and other individuals of interest), as well as forensic DNA profiles and DNA profiles from crime scene evidence.

Every country determines in its national legislation the DNA data that will be stored in a database, their retention period, and the obligation to delete the DNA data from the database once this period expires (Machado and Granja 2020).

According to Interpol data from 2019, a survey was sent (Interpol Global DNA Profiling Survey Results 2019) to Interpol's 194 National Central Bureaus (NCBs) regarding DNA profiling and DNA databases for 2018. One hundred ten (57%) NCBs responded to the survey. The results of this survey were combined with the results of the Interpol Global DNA Profiling Survey 2016 to give a total of 130 NCBs, out of which 89 use DNA profiling in criminal investigations, 70 have a DNA database on a national level, 31 also have a specialized missing persons DNA database, 83 use Y-STR analysis, and 34 also use mitochondrial DNA analysis. At least 34 countries in the world are in the process of establishing a DNA database (Machado and Granja 2020) (Table 23.1).

The size of DNA databases varies by countries in accordance with their legislative frameworks. Data on the size of DNA databases in some of the countries around the world are shown in Table 23.2 and Figure 23.1.

According to the information available online at the end of 2020 (December 31, 2020), there are more than 6,600,000 DNA profiles of individuals and more than 660,000 profiles of traces from crime scenes in the UK NDNAD (NDNAD Statistics). NDNAD contains data from England, Wales, Northern Ireland, Scotland, and the Channel Islands.

According to FBI data, in January 2021 the US NDIS (CODIS-NDIS Statistics) contained 14,430,404 offender DNA profiles, 4,239,300 arrestee DNA profiles, and 1,082,436 forensic DNA profiles, and according to the information available it is the largest DNA database in the world.

The amount of DNA data in databases greatly varies among countries (Figure 23.2; Table 23.3). It would be simple to conclude that the size of a DNA database corresponds to the size of population of a given country. However, according to Filipe Santos and his colleagues (2013), who conducted a survey into the trends in European legislation, the size of a database depends on how restrictive the legislative framework is. If a specific law has few constraints on the insertion of DNA profiles into a DNA database (e.g., Austria, Denmark, England and Wales, Finland, Scotland) (Machado and Granja 2020), the country may be designated as having an expansionist tendency with respect to the development of a database. By contrast, in countries in which legislation restricts categories of individuals whose DNA profiles may be stored in the database (e.g., Belgium, France, Germany, Hungary, Netherlands, Portugal, Spain, Sweden) (Machado and Granja 2020), this is reflected in its size. The actual number of individuals whose DNA data are contained in databases is a result of a dynamic process of entry and deletion of DNA data, in accordance with the regulatory framework of each country. However, despite all this, even in countries with

Table 23.1 Countries That Reported Having a DNA Database

Africa	Algeria, Botswana, Egypt, Morocco, South Africa, Sudan, Tunisia
Americas	Bahamas, Bolivia, Brazil, Canada, Chile, Guatemala, Honduras, Panama, Saint Lucia, United States
Asia	Australia, Iraq, Japan, Jordan, Kazakhstan, Korea (Republic of), Lebanon, Malaysia, New Zealand, Qatar, Saudi Arabia, Singapore, Vietnam
Europe	Albania, Austria, Azerbaijan, Belarus, Belgium, Bulgaria, Croatia, Cyprus, Czech Republic, Denmark, Estonia, Finland, France, Georgia, Germany, Greece, Hungary, Iceland, Ireland, Israel, Italy, Latvia, Lithuania, Luxembourg, Malta, Montenegro, Netherlands, North Macedonia, Norway, Poland, Portugal, Romania, Russia, Serbia, Slovakia, Slovenia, Spain, Sweden, Switzerland, United Kingdom

Source: Interpol Global DNA Profiling Survey 2019 (combined results of 2016 and 2019 surveys).

Table 23.2 **Largest DNA Databases Worldwide by Number of Profiles According to Interpol (2019)**

Africa		Americas		Asia		Europe	
South Africa	1,240,168	Canada	535,236	Japan	1,213,928	France	4,247,382
Tunisia*	17,070	Chile	78,733	Saudi Arabia*	909,743	Germany	1,213,331
Egypt*	4,162	Brazil	18,064	New Zealand	237,269	Russia	734,373
Ghana*	1,193	Panama*	9,097	Republic of Korea	179,848	Spain	627,163
		Saint Lucia*	1,200	Bahrain	69,609	Austria	384,098
						Belarus	336,408
						Netherlands*	328,732

Based on the response to the Interpol Global DNA Profiling Survey 2016.

☐ Africa ☐ Americas ☐ Asia ☐ Europe

Figure 23.1 Largest DNA databases worldwide according to Interpol (2019).

restrictive legislation, such as France, Germany, Hungary, the Netherlands, and Sweden (Santos et al. 2013), the proportion of population included in a DNA database exceeds 1%.

There have been initiatives in some countries to establish DNA databases containing data of the entire population. However, it is not likely that such databases will be established because such laws would be contrary to the provisions that protect the rights of individuals to personal freedoms and privacy. For example, approximately 1.7 million DNA profiles were deleted from the DNA database in the United Kingdom as a result of European Court of Human Rights judgments (https://www.echr.coe.int/Pages/home.aspx?p=home) that indefinite retention of DNA data in the database is contrary to the provision of Article 8 of the European Convention on Human Rights (https://www.equalityhumanrights.com/en/human-rights-act/article-8-respect-your-private-and-family-life).

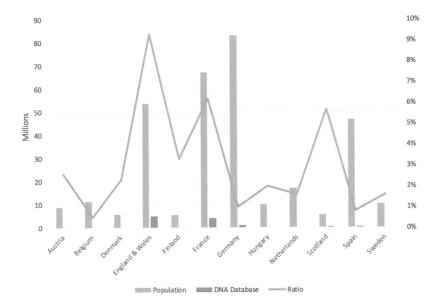

Figure 23.2 Size of DNA databases in some countries in Europe.

Table 23.3 **Size of DNA Databases in Some Countries in Europe (Size of the Population Compared to the Number of Individuals in the Database)**

Country	Population Number of Individuals	DNA Database Number of Individuals	Proportion Population/Individuals in Database
Austria	8,800,000	230,359	2.6%
Belgium	11,413,200	52,382	0.5%
Denmark	5,700,000	130,662	2.3%
England and Wales	53,700,000	5,018,728	9.3%
Finland	5,509,242	180,591	3.3%
France	66,992,699	4,167,132	6.2%
Germany	83,000,000	871,506	1.0%
Hungary	9,982,000	202,471	2.0%
Netherlands	17,000,000	267,407	1.6%
Scotland	5,500,000	311,107	5.7%
Spain	46,570,000	366,562	0.8%
Sweden	10,230,185	162,697	1.6%

Source: ENFSI DNA Working Group, 2019, (draft) *DNA Database Management Review and Recommendations.*

23.3.2 Interpol DNA Database

The International Criminal Police Organization (Interpol) (https://www.interpol.int/en) has been an indispensable partner since the very beginning of the international DNA data exchange. Interpol's DNA database was created in 2002 and it currently contains more than 247,000 profiles contributed by 84 countries worldwide. Those countries retain ownership of DNA profiles, and they control entry, access, and deletion of data in accordance with the provisions of their national legislation.

23.3.3 Database Efficiency

An important question that requires a reply is the efficiency of databases established in such a manner and the very indicators of this efficiency. According to Amankwaa and McCartney (2019), and previously published literature, there are seven indicators of DNA database efficiency. They are crime-solving capacity, incapacitation effect, deterrence effect, protection of privacy, legitimacy, implementation efficiency, and implementation cost. In this context, Amankwaa and McCartney state that

> the case resolution effectiveness of forensic DNA databases can be framed within two contexts ("effectiveness context"):
> a) Its contribution to the resolution of crimes where DNA from crime scene is loaded on the database ("DNA-related crime") and
> b) Its impact on the resolution of all recorded crime.

According to Amankwaa and McCartney (2021), "Enhancing the utility of forensic DNA analysis is imperative to justify its financial and ethical costs." Hence the need to monitor the efficiency of DNA databases to determine how much DNA databases contribute to crime resolution and what their overall impact is on crime resolution.

23.4 Sharing DNA Data across Borders

Global exchange of information and data has been recognized as an indispensable segment in criminal investigations. Cross-border comparison of data has opened a new dimension in combating crime. The information obtained through data comparison provides countries with new investigative approaches and plays a significant role in assisting law enforcement and judicial authorities.

The international dimension of criminal investigations also requires simplified cross-border data exchange, and it is essential to ensure that proper information is exchanged quickly and efficiently, and according to predetermined rules. Cross-border exchange of DNA profiles has great potential in improving the resolution of criminal offenses and international crime.

In order to achieve efficient international data exchange, it is vital to establish international standards for ensuring data accuracy and for controlling their quality. Instructions and guidelines related to international data exchange need to be developed in accordance with national legislation, namely, which type of DNA profiles are to be exchanged, how matches are to be reported, and which communication channels are to be used. Standards for the exchange and acceptance of DNA profiles should be defined in advance, as well as the minimum eligibility threshold for DNA profile matches.

Laboratories for biological material analysis whose results are stored in DNA databases in the European Union have to perform analyses in compliance with the EN ISO/IEC 17025 international standard. This provision provides for a more efficient DNA data exchange between countries by recognizing their equal reliability regardless of the country in which they were generated.

There are several ways to exchange data from national databases. Countries can make individual requests for verification through international legal assistance. Likewise, international police organizations such as Interpol and Europol (https://www.europol.europa

.eu) mediate the exchange of information and data between member states within the scope of their activities.

Automated exchange by providing direct insight into DNA databases significantly accelerates the existing procedures and enables countries to find out whether any other country has information that they require. Should the exchange generate a hit between DNA profiles, supply of further available personal data and other information are exchanged among countries upon request.

The biggest contribution to the automated DNA data exchange was achieved in the European Union through the so-called Prüm Decisions. Signing of the treaty in the town of Prüm in 2005 paved the way for the adoption of the EU Council Decision 2008/615/JHA of 23 June 2008 on the stepping up of cross-border cooperation, particularly in combating terrorism and cross-border crime. This initiated automated exchange of DNA and other data among European Union member states. Twenty-five European Union member states participated in automated DNA data exchange in late 2020, and the remaining two member states are in the process of establishing automated DNA data exchange. Other European countries (Iceland, Liechtenstein, Norway, Switzerland, UK) are also allowed to join the automated DNA data exchange.

Other international treaties also provide for automated DNA data exchange. Thus, the US arranges the exchange of biometric and biographic data and other information between law enforcement authorities through the so-called Prevention and Combating Serious Crime (PCSC) Agreements (https://www.google.com/search?q=pcsc+agreemnet+usa&ie =UTF-8&oe=UTF-8&hl=en-gb&client=safari). The Police Cooperation Convention for South East Europe (PCC SEE) also allows its signatories (Albania, Austria, Bosnia and Herzegovina, Bulgaria, Croatia, Hungary, Macedonia, Moldova, Montenegro, Romania, Serbia, and Slovenia) to exchange data from DNA databases.

23.5 Ethical Aspects of Forensic DNA Databases

According to Wallace (2014) and Cordner (2005), there are several ethical issues related to forensic DNA databases. These relate to the "collection and storage of DNA samples," "testing the samples/using the results," "access and retention of DNA data samples," and "misuse of genetic research in application of genetics to forensic sciences" (Yadav 2017).

Different countries have different criteria and regulatory frameworks to whose DNA samples and profiles, as a particularly sensitive type of personal data, may be obtained and stored in DNA databases, and for how long.

It is for this reason that DNA samples and profiles are protected by Article 8 of the Convention for the Protection of Human Rights and Fundamental Freedoms. Paragraph 2 of the article states that there should be no interference by a public authority with the exercise of this right, except in cases when this is necessary in a democratic society in the interests of national security, public peace and order, or the economic well-being of the country, and in cases when it is necessary to prevent disorder or crime, as well as in cases of protection of health or morals, and protection of rights and freedoms of others.

With the proper establishment of DNA databases and their maintenance as defined by law, ethical issues primarily related to potential threats to a number of civil rights, such as the violation of privacy of individuals whose profiles those databases contain, are minimized, as are the possibilities of other manipulations (hacking, etc.).

23.5.1 European Court of Human Rights Decision in UK Cases and North Macedonia Case

Over the last few years, the European Court of Human Rights (hereinafter: ECHR) based in Strasbourg, France, had to rule in several cases on potential violations of the European Convention for the Protection of Human Rights and Fundamental Freedoms related to the storage of DNA profiles and their retention period. According to the available data, *S and Marper v. United Kingdom* is the most cited ECHR case. This case involved two separate cases. The first case involved an 11-year-old boy who was arrested and charged with attempted robbery and acquitted five months later. Nonetheless, his DNA sample and fingerprints were taken, his profiles were entered into the database, and the samples retained for any further investigations. The second case involved a charge of harassment/abuse of a partner. Given the fact that in the meantime the partners reconciled, and the victim dropped the charges and the case was dismissed. In this case, DNA samples and fingerprints were also taken, and profiles were entered into databases. In both cases, requests were made to have their profiles removed from the databases and their biological samples destroyed. However, in both cases, those requests were denied by British authorities, namely, the police.

The ECHR determined that the legislation in force at the time in England, Wales, and Northern Ireland allowed indefinite retention of DNA profiles and fingerprints of any person suspected of any criminal offense, which could have harmful consequences for unconvicted persons, especially minors who are yet to integrate into the society.

According to Asplen (2014), the ECHR's opinion in this particular case was as follows:

> the retention of both cellular samples and DNA profiles amounted to an interference with the applicants' right to respect for their private lives, within the meaning of Article 8 § 1 of the Convention, concluding that fingerprints, DNA profiles, and cellular samples constituted personal data within the meaning of the Council of Europe Convention of 1981 for the protection of individuals with regard to automatic processing of personal data.

However, although this is the most widely cited decision, it is often cited completely incorrectly because the ECHR did not find any violation of human rights at arrest and in the collection of samples for molecular and genetic analysis and fingerprints. Rather, the court's decision was directed at the unlimited retention of those profiles once exonerated or deemed not prosecutable for any reason. As a conclusion, the ECHR recognized a legitimate purpose in collecting such data.

The ECHR judgment and the arguments raised have led to a change of attitude in the national judiciary practice. Thus, in its decision of May 2011, the UK Supreme Court ruled that DNA profiles entered the database after a person is arrested have to be destroyed if the person is later acquitted. Considering the guidelines issued by the Association of Chief Police Officers (ACPO), according to which DNA profiles should be destroyed only in exceptional circumstances, the police refused the supreme court's proposal. However, as pointed out by Derenčinović (2019), the supreme court held that this is not in accordance with the ECHR decision in the case of *S and Marper* nor with Article 8 of the convention, and accordingly it gave preference to the right of protection of private and family life.

UK authorities prepared a new Protection of Freedoms Act, and the Police and Criminal Evidence Act was amended. The most substantial amendment related to the destruction of

DNA samples as soon as a DNA profile has been derived or at the latest within six months of the taking of the DNA sample. The police are still authorized to retain for an indefinite time DNA profiles of individuals convicted of recordable offenses of more than five years and of those convicted of a qualifying offense (Derenčinović 2019).

In 2020, the ECHR delivered two judgments, *Gaughran v. the United Kingdom* (https:/hudoc.echr.coe.int) and *Trajkovski and Chipovski v. North Macedonia* (https://hudoc.echr.coe.int), in which the government practice in the collection, use, and retention of DNA profiles for criminal proceedings and crime prevention purposes was examined. As in the case of *S and Marper v. UK*, in those cases the ECHR also found a violation of the right to privacy in cases of applicants whose sample was taken when they were arrested for the purpose of molecular and genetic analysis to be retained for an indefinite period, and they were arrested on suspicion that they were perpetrators of criminal offenses and were later also found guilty. As pointed out by Derenčinović (2019):

> In the last two judgments, the ECHR concluded that the interference of the state in the private life of an individual by retaining DNA profiles indefinitely will not be allowed if it is not in compliance with the principle of proportionality which is *in concreto* derived from a number of guarantees aimed at preventing blanket and non-selective restriction of the right to privacy.

This would mean that the ECHR did not explicitly rule that indefinite retention of DNA profiles in cases of convicted persons is unlawful. However, in those cases states have to ensure that all guarantees are respected in order to prevent arbitrary interference of the state in citizens' privacy, especially when it comes to DNA profiles, data which, unlike some other biometric data, compromise both the right to privacy of the data subject and of a person biologically linked to the data subject.

23.5.2 United States Supreme Court in *Maryland v. King,* 569 U.S. 435 (2013)

So far, the most important decision of the US Supreme Court concerning DNA databases was reached in 2013 in the case of *Maryland v. King* (*Maryland v. Alonzo Jay King, Jr.*

As pointed out by Asplen (2014), although the constitutionality of DNA sampling from individuals convicted of crimes remains indisputable, the issue of the constitutionality of DNA sampling of arrestees has never been resolved, although 28 US states had laws on DNA sampling of arrestees and entry of profiles into NDIS already by 2013. Legal challenges emerged in California and Maryland stating that DNA sampling of individuals who have not been convicted violates the Fourth Amendment of the US Constitution.

In the case of *Maryland v. King*, the police arrested Alonzo Jay King Jr. for first-degree assault (raping an elderly person by threatening her with a gun) in 2009. In accordance with the Maryland law, King's sample was taken for DNA analysis and his DNA profile was made. King's DNA profile matched the traces found at the crime scene, and in accordance with this match Maryland accused and successfully convicted King of first-degree rape.

The Maryland Court of Appeals overturned the first-instance judgment, holding that the collection of King's samples for molecular and genetic analysis violated the Fourth Amendment. That decision also halted further collection of samples for molecular and genetic analyses from other suspects who were arrested for various other violent crimes.

Maryland applied for a stay of that judgment pending the Supreme Court's final decision. By a vote of 5 to 4, the Supreme Court ruled that the collection of samples for molecular and genetic analysis at the time of arrest for identification purposes did not violate the US Constitution.

In effect, the Supreme Court authorized the police to use the power of DNA technology earlier in the investigation and thus increase the efficiency of criminal investigations. Hence, the question of when the police can take samples to conduct DNA analysis was solved (Asplen 2014).

However, retention of data obtained from molecular and genetic analysis can be disputable when it comes to individuals who were acquitted or when criminal proceedings were suspended.

23.6 Quo Vadis Forensic DNA Database

The benefits of forensic DNA databases are enormous, particularly in improving the results of criminal investigations. However, with such success, there are new challenges emerging. There has been a rapid growth in the amount of data in databases worldwide, but this also requires additional infrastructure support.

As pointed out by Ge et al. (2014), to respond to such challenges, two different directions have been proposed for enhancing search capabilities, which include adding more autosomal short tandem repeat (STR) loci to the current core set of loci (FBI, EU countries), whereas China decided to establish a national Y-STR database.

Advanced forensic molecular and genetic analysis technology, in particular with regard to massively parallel sequencing (MPS) of forensic samples, generates an increasing amount of data in the field of STR, single-nucleotide polymorphism (SNP), and mitochondrial DNA (mtDNA) analysis. It is increasingly challenging to store in DNA databases large amounts of results generated in a short period of time, especially when it comes to defining the nomenclature under which new types of data will be stored. Close cooperation at the international level is therefore necessary, which will lead to solutions acceptable for future international exchange of that data.

23.6.1 Mitochondrial DNA

The basis of identification through mtDNA is the analysis of sequences of interest and their comparison (Amorim et al. 2019). When it cannot be excluded that an mtDNA of trace originates from a particular individual, the evaluation of the significance of evidence is expressed through the information on the frequency of mtDNA based on the number of a specific sequence within the population (Budowle et al. 1999).

To show statistical value, comparison is made with the already existing mtDNA haplotype databases.

The most important available global database used in forensics and human identification is the EDNAP Mitochondrial DNA Population Database (EMPOP). It was established in 2006 and it serves as a reference mtDNA population database. At the beginning of 2021, it contained more than 48,000 mtDNA sequences from around the world. One of the major features of this database is quality control of each sequence of mtDNA before its entry into the EMPOP. It also offers the possibility of quality control of population data and individual sequences obtained from individual samples (Parson et al. 2014).

Table 23.4   **Specialized Missing Persons DNA Databases Worldwide**

Africa	South Africa
Americas	Brazil, Canada, Chile, Guatemala, United States
Asia	Iraq, Japan, Kazakhstan, Lebanon
Europe	Austria, Azerbaijan, Belarus, Belgium, Cyprus, France, Germany, Greece, Ireland, Israel, Italy, Montenegro, Norway, Poland, Portugal, Romania, Slovakia, Slovenia, Spain, Switzerland, United Kingdom

Source: Interpol Global DNA Profiling Survey Results 2019.

23.6.2 Y-Chromosomal STR Markers

Y-chromosome analysis determines a haplotype, which due to lack of recombination and linear inheritance through a male line, needs to be compared with the Y-haplotype reference database (YHRD) in order to confirm affiliation.

YHRD is the largest global database of Y-chromosome haplotypes. It is characterized by online entries, quality control of haplotypes prior to their entry, and comparison of haplotypes. At the beginning of 2021, it contained more than 320,000 haplotype samples from around the world (Willuweit et al. 2015). It has been used as a global reference database since 2000, when it became publicly available online (Roewer et al. 2001).

23.6.3 Missing Persons

The purpose of missing persons DNA databases is to link a DNA profile of unidentified human remains with a DNA profile of a missing person or his/her relatives. They can also be used for disaster victim identification (DVI) and identification of human remains found in mass graves.

There are two strategies in establishing missing persons DNA databases, depending on the national legislation, regulatory framework for personal data protection, or specialized computer systems used for keeping records and processing DNA data on missing persons. In some states, DNA data on missing persons are kept together with other data in criminal databases, and in other states, separate missing persons and unidentified human remains databases are established. If two separate databases are used, it is recommended that DNA profiles of unidentified human remains be also compared with DNA profiles from the criminal database.

According to an Interpol report from 2019, 31 states have specialized missing persons DNA databases (Table 23.4).

References

Act No. 37 of 2013: Criminal Law (Forensic Procedures) Amendment Act. 2013. https://www.gov.za./sites/default/files/gcis_document/201409/37268act37of2013crimlawamend27jan2014.pdf

Amankwaa, A. O., and C. McCartney. 2019. The effectiveness of the UK national DNA database. *Forensic Sci Int: Synergy* 1: 45–55.

Amankwaa, A. O., and C. McCartney. 2021. The effectiveness of the current use of forensic DNA in criminal investigations in England and Wales. *WIREs Forensic Sci.* https://doi.org/10.1002/wfs2.1414.

Amorim, A., T. Fernandes, and N. Taveira. 2019. Mitochondrial DNA in human identification: A review. *PeerJ Life Environ* 7: e7314. https://doi.org/10.7717/peerj.7314.

Aronson, J. D. 2005. DNA fingerprinting on trial: The dramatic early history of a new forensic technique. *Endeavour* 29: 126–131. https://doi.org/10.1016/j.endeavour.2005.04.006.

Article 8 of the European Convention on Human Rights. https://www.equalityhumanrights.com/en/human-rights-act/article-8-respect-your-private-and-family-life.

Asplen, C. 2014. DNA databases. In *Forensic DNA Application: An Interdisciplinary Perspective*, edited by D. Primorac, and M. Schanfield, 557–568. CRC Press, Taylor & Francis Group.

Bodner, M., and W. Parson. 2020. The STRidER report on two years of quality control of autosomal STR population datasets. *Genes* 11(8): 901.

Bodner, M., I. Bastich, J. M. Butler, R. Fimmers, P. Gill, L. Gusmão, N. Morling, et al. 2016. Recommendations of the DNA Commission of the International Society for Forensic Genetics (ISFG) on quality control of autosomal short tandem repeat allele frequency databasing (STRidER). *Forensic Sci Int Genet* 24: 97–102.

Budowle, B., M. R. Wilson, J. A. DiZinno, C. Stauffer, M. A. Fasano, M. M. Holland, and K. L. Monson. 1999. Mitochondrial DNA regions HVI and HVII population data. *Forensic Sci Int* 103(1): 23–35.

CODIS and NDIS. https://www.fbi.gov/services/laboratory/biometric-analysis/codis/codis-and-ndis-fact-sheet.

CODIS-NDIS Statistics. https://www.fbi.gov/services/laboratory/biometric-analysis/codis/ndis-statistics#Tables (accessed March 2021).

Cordner, S. M. 2005. Ethics of forensic applications and databanks. In *Encyclopedia of Forensic and Legal Medicine Vol. 2*, edited by R. Byard and J. Payne-James, 178–184. Elsevier.

Derenčinović, D. 2019. *Projekt nedužnosti u Hrvatskoj, IP-2019-04-9893*, ORCID ID: orcid.org/0000-0002-4146-7905.

DNA Identification Act of 1994, codified at 42 U.S.C. § 14132.

EMPOP, EDNAP Mitochondrial DNA Population Database. https://empop.online (accessed March 2021).

ENFSI. https://enfsi.eu.

ENFSI DNA Working Group. 2017, April. *DNA DATABASE MANAGEMENT Review and Recommendations*.

ENFSI DNA Working Group. 2019, April. *Draft: DNA DATABASE MANAGEMENT Review and Recommendations*.

EU Council Decision 2008/615/JHA of 23 June 2008 on the stepping up of cross-border cooperation, particularly in combating terrorism and cross-border crime. *Official Journal of the European Union*.

EU Council Decision 2008/616/JHA of 23 June 2008 on the implementation of decision 2008/615/JHA on the stepping up of cross-border cooperation, particularly in combating terrorism and cross-border crime. *Official Journal of the European Union*.

EU Council Framework Decision 2009/905/JHA of 30 November 2009 on accreditation of forensic service providers carrying out laboratory activities. *Official Journal of the European Union*.

EU Council Resolution of 09 June 1997 on the exchange of DNA analysis results.

EU Council Resolution of 25 June 2001 on the exchange of DNA analysis results (2001/C 187/01). *Official Journal of the European Union*.

EU Council Resolution of 30 November 2009 on the exchange of DNA analysis results (2009/C 296/01). *Official Journal of the European Union*.

European Court on Human Right. https://www.echr.coe.int/Pages/home.aspx?p=home.

EUROPOL. https://www.europol.europa.eu.

FBI, CODIS and NDIS Fact Sheet. 2021.

FIND Strategy Board. 2020. *National DNA Database Strategy Board Biennial Report 2018–2020*. Forensic Information Database Strategy Board.

Forensic Science and Pathology Unit. 2005. *DNA Expansion Programme 2000–2005: Reporting Achievement.* Home Office.

Gaughran v. UK, application no. 45245/15, 13.02.2020.

Ge, J., H. Sun, H. Li, C. Liu, J. Yan, and B. Budowle. 2014. Future direction of forensic DNA databases. *Croat Med J* 55(2): 163–166.

General Data Protection Regulation (EU 2016/679).

Hares, D. R. 2015. Selection and implementation of expanded CODIS core loci in the United States. *Forensic Sci Inter Genet* 17: 33–34.

Hindmarsh, R. A., and B. Prainsack. 2010. *Genetic Suspects: Global Governance of Forensic DNA Profiling and Databasing.* Cambridge University Press.

INTERPOL. https://www.interpol.int/en.

INTERPOL. 2019. Global DNA Profiling Survey Result. https://www.interpol.int/How-we-work/Forensics/DNA.

INTERPOL (DNA Monitoring Expert Group). 2015. *Best Practice Principles: Recommendations on the Use of DNA for the Identification of Missing Persons and Unidentified Human Remains.*

ISO IEC 17025. https://www.iso.org/ISO-IEC-17025-testing-and-calibration-laboratories.html.

Le Rouxe-Kemp, A. 2018. Forensic DNA databases in Hong Kong and China: A BRICS comparative perspective. Indiana Int Comp Law Rev 28(2). https://doi.org/10.18060/7909.0061.

Machado, H., and R. Granja. 2020. DNA databases and big data. Chap. 5 in *Forensic Genetics in the Governance of Crime.* Singapore: Palgrave Pivot. https://doi.org/10.1007/978-981-15-2429-5_5

Maryland v. Alonzo Jay King, Jr.. 569 U.S. 435 (2013).

National DNA Database (NDNAD). https://www.gov.uk/government7groups7national-dna-database-strategy-board.

National DNA Database Strategy Bord Annual Report, 2015/2016. https://www.gov.uk/government/publications/national-dna-database-annual-report-2015-to-2016

NDNAD Statistics. https://www.gov.uk/government/statistics/national-dna-database-statistics (accessed March 2021).

Parson, W., and A. Dür. 2007. EMPOP-a forensic mtDNA database. *Forensic Sci Inter Genet* 1(2) :88–92.

Parson, W., et al. 2014. DNA Commission of the International Society for Forensic Genetics: Revised and extended guidelines for mitochondrial DNA typing. *Forensic Sci Inter Genet* 13:134–142.

PCSC Agreement USA. https://www.google.com/search?q=pcsc+agreemnet+usa&ie=UTF-8&oe=UTF-8&hl=en-gb&client=safari.

Regulation (EU) 2016/679 of the European Parliament and of the Council of 27 April 2016 on the protection of natural persons with regard to the processing of personal data and on the free movement of such data, and repealing. Directive 95/46/EC (General Data Protection Regulation).

Roewer, L., et al. 2001. Online reference database of European Y-chromosomal short tandem repeat (STR) haplotypes. *Forensic Sci Int* 118(2–3): 106–13.

S. and Marper v. UK, Nos. 30562/04 and 30566/04, § 41, 4 December 2008.

Santos, F., H. Machado, and S. Silva. 2013. Forensic DNA databases in European countries: is size linked to performance? *Life Sci Soc Policy* 9(12): 1–13.

Trajkovski and Chipovski v. North Macedonia, application no. 53205/13 and 63320/13, 13.2.2020.

Wallace, H. M., A. R. Jackson, J. Gruber, and A. D. Thibedeau. 2014. Forensic DNA databases— Ethical and legal standards: A global review. *Egypt J Forensic Sci* 4(3): 57–63.

Willuweit, S., and L. Roewer. 2015. The new Y Chromosome Haplotype Reference Database. *Forensic Sci Inter Genet* 15: 43–8.

Yadav, P. K. 2017. Ethical issues across different fields of forensic science. *Egypt J Forensic Sci* 7(1): 10.

YHRD, Y-Chromosome Haplotype Reference Database. https://yhrd.org (accessed March 2021).

Index